NEW SURVEY OF CLARE ISLAND

Volume 10: Land and Freshwater Fauna

Edited by T.K. McCarthy, Éamonn Lenihan
and John Breen

Acadamh Ríoga na hÉireann
Royal Irish Academy

First published in 2022
by the Royal Irish Academy,
19 Dawson St, Dublin 2.
ria.ie

© Royal Irish Academy 2022

All rights reserved. The material in this publication is protected by copyright law. Except as may be permitted by law, no part of the material may be reproduced (including by storage in a retrieval system) or transmitted in any form or by any means; adapted; rented or lent without the written permission of the copyright owners or a licence permitting restricted copying in Ireland issued by the Irish Copyright Licensing Agency CLG, 63 Patrick Street, Dún Laoghaire, Co. Dublin, A96 WF25.

ISBN 978-1-911479-87-1 (PB)
ISBN 978-1-911479-90-1 (PDF)
ISBN 978-1-911479-65-9 (epub)

Typesetting: Datapage International Ltd

Indexing: Julitta Clancy

Printed in Ireland by Walsh Colour Print

Cover photograph by John Breen

A NOTE FROM THE PUBLISHER

We want to try to offset the environmental impacts of carbon produced during the production of our books and journals. For the production of our books this year we will plant 45 trees with Easy Treesie.

The Easy Treesie – Crann Project organises children to plant trees. Crann – 'Trees for Ireland' is a membership-based, non-profit, registered charity (CHY13698) uniting people with a love of trees.
It was formed in 1986 by Jan Alexander, with the aim of 'Releafing Ireland'. Its mission is to enhance the environment of Ireland through planting, promoting, protecting and increasing awareness about trees and woodlands.

CONTENTS

In memoriam: T.K. McCarthy (1949 – 2019)   v
Committee members   vi
Preface and acknowledgements   vii
Notes on contributors   ix

Chapter 1: Introduction: The value of local surveys and their contribution to broader ecological research   1
Thomas Bolger

Chapter 2: The non-marine mollusca of Clare Island   7
Roy Anderson

Chapter 3: The Tardigrada of Clare Island   17
Erica DeMilio, Nigel J. Marley, Colin Lawton and Reinhardt Møbjerg Kristensen

Chapter 4: Spiders (Arachnida: Araneae) from Clare Island   41
Myles Nolan

Chapter 5: The Collembola and Acari (Oribatida and Mesostigmata) of Clare Island   69
Thomas Bolger

Chapter 6: Diplopoda, Chilopoda, Opiliones, Pseudoscorpiones and land Isopoda from Clare Island   79
Martin Cawley

Chapter 7: The aquatic Coleoptera of Clare Island   83
Stephen McCormack and T.K. McCarthy

Chapter 8: Aquatic beetle assemblages (Insecta: Coleoptera) in a Clare Island lough   93
Stephen McCormack and T.K. McCarthy

Chapter 9: The Carabidae (Coleoptera) of Clare Island   101
Roy Anderson and Stephen McCormack

Chapter 10: The Lepidoptera (butterflies and moths) of Clare Island   109
K.G.M. Bond

Chapter 11: Chironomidae (Insecta: Diptera) of Clare Island and the adjacent West Mayo mainland, Ireland   133
Declan A. Murray

Chapter 12: Snail-killing flies (Sciomyzidae) of Clare Island   185
Rory Mc Donnell, Mike Gormally and Stephen McCormack

Chapter 13: The Social Hymenoptera of Clare Island   189
John Breen, Robert Paxton and Audrey O'Grady

Chapter 14: Non-marine Ostracods of Clare Island 197
Gillian McCall

Chapter 15: Cladocera and Copepoda of the inland waters on Clare Island 209
Elvira de Eyto, Adriana Trojanowska-Olichwer, Jens Petter Nilssen and T.K. McCarthy

Chapter 16: Freshwater fish and their parasites in Clare Island 219
T.K. McCarthy, Karen Creed and Éamonn S. Lenihan

Chapter 17: The birds of Clare Island 231
Thomas C. Kelly

Chapter 18: The mammals of Clare Island 237
Colin Lawton and Kate McAney

Taxonomic Index 247

General Index 259

IN MEMORIAM

T.K. McCARTHY (1949 – 2019)

Michael Guiry

Thomas Kieran McCarthy, known all his life as 'Kieran', was born in Cork City. His family lived near The Lough, a parish of Cork named for a shallow spring-fed lake of about four hectares with a small island that has been a wildlife sanctuary since 1881. As a schoolboy, Kieran became passionately involved in the conservation of the wildlife there. Entering University College Cork in 1967, Kieran was a committed zoologist from the very first day, coming under the influence of Gerald A. Walton ('Doc Walton'), a gifted naturalist and inspiring teacher.

Graduating with a first in Zoology in 1971, Kieran signed up for a PhD on the leeches (Hirudinea) of Ireland, which was completed in 1974 and published in the *Proceedings of the Royal Irish Academy* in 1975. Kieran carried out post-doctoral research in Oxford, Helsinki and Dublin. In 1978, a lectureship in Zoology specialising in freshwater environments was advertised at University College Galway, now National University of Ireland, Galway (NUIG). Kieran was appointed and spent the remainder of his career at Galway. While Kieran's area of expertise was in the Hirudinea and the Corixidae (Water Boatmen), he was supremely knowledgeable on all aspects of the Irish freshwater fauna.

Whilst at NUIG he became closely involved in research on European eels in the Shannon Eel Fishery, funded by the Electricity Supply Board. Kieran supervised the masters and doctoral theses of a large number of students, many of whom now work in public and probate bodies concerned with freshwater environments. Following his retirement in 2011 Kieran continued to work with the Shannon Eel Fishery and represented Ireland at many international meetings, working particularly closely with Japanese eel researchers. He was a Visiting Professor at University of Łodz, Poland, and University of Tokyo, Japan. He also took the opportunity to travel the length and breadth of Ireland attending auctions and sales to purchase curiosities and books.

NEW SURVEY OF CLARE ISLAND

New Survey of Clare Island Executive Committee 2022

J. Breen	J.R. Graham	C. Manning (Deputy Secretary 1996–2004; Secretary, 2004–)
P. Coxon	M. Jebb	M.W. Steer (Managing Editor, 1994–; Chairman 2004–)
J. Feehan	T. Kelly	D. Synnott
Catherine Godson (Science Secretary, 2020–)	Mary Canning (President, 2020–)	

Former committee members: A. Clarke (President, 1990–3), D. Cabot (Chairman, 1989–97), G.J. Doyle (Secretary, 1991–2004), L. Drury (President, 2011–14; International Relations Secretary 2005–07, 2009–10; Council 2011–13) M.D.R. Guiry, Patrick Jerome Guiry (Science Secretary, 2017–20), J.S. Fairley, M. Herity (President, 1996–9), G.F. Imbusch, R.P. Kernan (Chairman, 1997–2004; Science Secretary, 1993–2000), C. Mac Cárthaigh, T.K. McCarthy, W.I. Montgomery, A.A. Myers, M.E.F. Ryan (President, 2002–05), J.O. Scanlan (President, 1993–96), J.A. Slevin (President, 2005–08), T.D. Spearman (President, 1999–2002), P.D. Sweetman, J. Waddell, K. Whelan.

Contributors to previous volumes in the New Survey of Clare Island series

Volume 1: History and Cultural Landscape: Timothy Collins, Críostóir Mac Cárthaigh (ed.), Nollaig Ó Muraíle, Kevin Whelan (ed.)

Volume 2: Geology: Peter Coxon, David Evans, John R. Graham (ed.), Kenneth T. Higgs, W.E.A. Phillips, C.J. Stillman, B.G.J. Upton, Michael Williams

Volume 3: Marine Intertidal Ecology: Michelle Cronin, Thomas Cross, Robert Cussen, Jane Delany Michael Guiry, Louise Harrington, Christine Maggs, David McGrath, Alan Myers (ed.), Julia Nunn, Sandy O'Driscoll, Ruth O'Riordan, Anne Marie Power, Fabio Rindi

Volume 4: The Abbey: Ann Buckley, Ian Cantwell, Fergus Gillespie, Paul Gosling (ed.), Conleth Manning (ed.), Karena Morton, Micheál Ó Comáin, Christoph Oldenbourg, Roger Stalley, John Waddell (ed.)

Volume 5: Archaeology: Markus Casey, Michelle Comber, Paul Gosling (ed.), Maureen McCorry, Conleth Manning (ed.), Karen Molloy, Sharon Nestor, Adrian Phillips, Mike Williams, Paula King, John Waddell (ed.)

Volume 6: The Freshwater and Terrestrial Algae: Jenny Bryant, Michael Guiry (ed.), David John (ed.), T.K. McCarthy (ed.), Fabio Rindi (ed.), Patricia Sims, Brian Whitton, David Williamson

Volume 7: Plants and Fungi: Donal Synnott (ed.), Pete Coxon, Ryan Corcoran, Paul Gibson, Stephen McCarron, Timothy Ryle, Matthew Jebb, M.R.D. Seaward, D.H.S. Richardson and David Mitchel

Volume 8: Soils and soil association: by Wies Vullings (ed.), James F. Collins (ed.), George Smillie (ed.)

Volume 9: Birds: R.H. Coombes, R.E. Cussen, R.E. Hutchinson, U. Keating, T.C. Kelly (ed.), E. McGreal, P. Smiddy, S. Warner, P. Winters

Clare Island: John Feehan

PREFACE AND ACKNOWLEDGEMENTS

The New Survey of Clare Island was undertaken by the Royal Irish Academy one hundred years after the original survey of Clare Island, with the overall aim being to assess the environmental changes that have taken place on the island during the intervening period. The New Survey has, so far, been published in nine volumes plus a synoptic volume. This volume (Volume 10), which examines the freshwater and land fauna of the island, is the final in the series. Some aspects of the fauna of Clare Island have already appeared in earlier volumes of the New Survey series, most notably in Marine Intertidal Ecology (Volume 3) and Birds (Volume 9). This volume compliments those volumes and presents studies on the freshwater and terrestrial fauna of the island. In addition, Tom Kelly has provided an overview of the birds of Clare Island for this volume (Chapter 17).

The investigations presented here began in 2001–02 and were co-ordinated, and subsequently edited, by the late T.K. (Kieran) McCarthy, who was joined by John Breen, and more recently by Éamonn Lenihan. In the early years of the project, Stephen McCormack was a fulltime research assistant, and he contributed sampling for some of the chapters. Several postgraduate theses were also conducted on the island's fauna and ecology. The studies reported in this volume continued intermittently until 2019.

The original survey of Clare Island sought to compare the fauna of the island with that of the neighbouring mainland. One of the studies presented here—Declan A. Murray's study of the Chironomidae (Chapter 11)—has also adopted this approach. However, the other studies presented in this volume confine their attention to Clare Island itself.

The studies initiated under Robert Lloyd Praeger over a century ago have acted as a catalyst for the present generation to complete this New Survey of Clare Island. Most of the taxa reported on here show an increase in numbers of species in the New Survey when compared to the original survey. This may be due to changes in taxonomy, more intensive sampling and the use of more modern techniques, such as DNA barcoding to separate cryptic species. We hope that the present volume will stimulate further studies in Irish taxonomy, especially in some of the more 'difficult' taxa.

The two current editors wish to acknowledge the major contribution of Dr T.K. (Kieran) McCarthy to this volume. Kieran was associated with the 'Zoology Volume' from the outset and was involved in the logistics of getting this part of the New Survey of Clare Island well established (McCarthy *et al.* 2002). He also supervised a number of postgraduate theses which contributed to this volume. He was actively involved in all aspects of the editorial process until his untimely death. We hope that this volume will be a fitting tribute to his memory.

The editors also wish to thank Professor Martin Steer, Managing Editor, for his diplomacy, discreet encouragement and sound advice, all of which helped get this volume to press. We also thank the staff of the Royal Irish Academy, especially, Róisín Jones in the early days and more recently Jonathan Dykes, who was very involved in the final preparation and publication of this volume.

The Editors also wish to thank the various authors whose contributions are included here. We thank them for their patience over a number of years and several revisions.

It is a pleasure to thank the many people from Clare Island who welcomed us to the island. We hope they will enjoy reading about this aspect of the life of their island.

REFERENCE

McCarthy, T.K., McCormack, S. and Cullen, P. 2002 *Terrrestrial and freshwater invertebrates of Clare Island*. Unpublished report to the Heritage Council. Dublin. Executive Committee, New Clare Island Survey.

Éamonn Lenihan
John Breen
April 2022

NOTES ON CONTRIBUTORS

Roy (Robert) Anderson studied for his primary degree in Pure Chemistry and PhD in Animal Nutrition at the Queen's University of Belfast. His research interests outside of a career in Agricultural Chemistry were primarily with invertebrate ecology and biogeography which he followed in his spare time and upon retirement. He has published widely on Irish (and British) non-marine Mollusca and Coleoptera with outputs also on Irish oniscidean Crustacea, Diplopoda and Chilopoda. He has a particular interest in Irish Carabidae (ground beetles) and terrestrial slug (Gastropoda) ecology and speciation.

Tom Bolger is a Science graduate and gained a Ph.D. in Agricultural Zoology both from University College Dublin. He later studied computer modelling at TCD. He was a member of staff at UCD (1981–2018) during which time he was head of the UCD School of Biology and Environmental Science. His primary research interests are in the effects of global change on terrestrial biodiversity and nutrient dynamics in soil systems. He was a Fulbright Scholar at the University of Georgia examining the relationship between biodiversity and ecosystem function. He is retired but continues as a Research Active Faculty Member at UCD.

Ken Bond Although he started his career in meteorology, Ken has long had an interest in insects, initially in the butterflies of a Dublin garden. Increasing interest in the distribution of insects finally prompted him to change course, becoming a mature student at Trinity College Dublin where he completed a degree in Natural Science (Zoology). Later at University College Cork he included insect pest identification and Museum curation in his activities. Since then, he has specialised in moths, especially Microlepidoptera, as well as having a special interest in the recording and conservation of the Marsh Fritillary butterfly.

John Breen obtained his BSc in zoology followed by a PhD in social insect ecology at University College Cork (National University of Ireland). He spent a year studying bumblebees and ants at the Zoological Museum, University of Bergen, Norway, followed by a three-year postdoctoral fellowship at Trinity College Dublin, before taking up a position at the University of Limerick. He is a retired Associate Professor. His main research interests are in the ecology of Irish social insects and, more recently, beekeeping.

Martin Cawley is a naturalist based in Sligo. He studied Zoology at the National University of Ireland, Galway and Optometry at the Dublin Institute of Technology. He has wide interests in natural history especially myriapods, arachnids, woodlice and beetles. He is currently working on bibliographical checklists of some invertebrate groups.

Karen Creed obtained her BSc and PhD from the National University of Ireland, Galway. She has carried out research on parasite assemblages of trout, eel and flounder in Lough Corrib, Lough Mask, Lough Carra, Lough Conn and on Clare Island. She has also carried out research on eel population dynamics in the south of Ireland. She is currently a Senior Manager in the Environmental Protection Agency.

Elvira de Eyto completed a PhD in 1999 at Trinity College Dublin, on the ecology of the family Chydoridae and its application to lake monitoring. Following postdoctoral research in the areas of lake ecology and salmon immunogenetics, she commenced work at the Burrishoole Research station in Co. Mayo, run by the Marine Institute. The focus of her current job is the biological monitoring of the Burrishoole catchment.

Erica DeMilio obtained a BSc (Honours) Zoology at the National University of Ireland, Galway. Her thesis investigated aspects of the ecology and methodologies for identification of tardigrades. She was awarded James Hardiman and Irish Research Council postgraduate research scholarships for further work focusing on the taxonomy and ecology of Irish tardigrades and macroecological questions relating to microscopic animals.

Mike Gormally is a Personal Professor in Environmental Science at the National University of Ireland, Galway and is Director of the BSc programme in Environmental Science. He obtained his primary degree from the National University of Ireland, Galway; his PhD from Trinity College Dublin and he is a fellow of the Royal Entomological Society. His research interests include habitat management for nature conservation, terrestrial invertebrate ecosystems services, biodiversity of agricultural landscapes, bioindicator species and biological control and he has worked on a range of Irish habitats including peatlands, woodlands, wet grasslands, flood meadows, turloughs and coastal habitats. He has as particular interest in the biology and ecology of marsh flies (Diptera: Sciomyzidae).

Tom Kelly Before, during and following his PhD studies at University College Cork (on the Ticks and viruses of seabirds In Ireland) T.C. Kelly was an active field ornithologist and, with others, especially the late Clive D. Hutchinson, and Richard T. Mills, added five or so new bird species to the Irish list. Published research includes the biology—especially the diets—of gulls (Laridae), behavioural responses of birds to moving aircraft, mathematical modelling of novel pathogen epidemics in seabird colonies (with Dr M.J.A. O'Callaghan and the late Professor Alexei Pokrovskii, Dept. of Applied Mathematics, UCC) and corvid shell dropping behaviour (with Professor J. Davenport).

Reinhardt Mjøberg Kristensen received his Ph.D. in Cell Biology and Anatomy from University of Copenhagen in 1979. He acted as scientific leader of Danish Arctic Station, Disko Island, Greenland from 1976-79 and director of Arctic Station, Greenland from 2004–14. He was a full professor at the Natural History Museum of Denmark from 1996–2019 and is an expert in phylogeny and ecology of tardigrades. The major body of research was the description, molecular phylogeny and ultrastructural investigations and discovery of three new phyla, the Loricifera (Kristensen, 1983), Cycliophora (Funch and Kristensen, 1995) and Micrognathozoa (Kristensen and Funch, 2000). He has led many expeditions to the Arctic, Australia/Solomon Islands and the Pacific Ocean. The largest part of Arctic Biology Research was done in co-operation with many PhD-students. It was mainly based on research concerning cryptobiosis in tardigrades including ESA space-programmes, where tardigrades were sent out in space; however, the main discovery has been: Loriciferans living in permanently anoxic condition in the deep sea.

Colin Lawton is a lecturer in Zoology at the National University of Ireland, Galway. His PhD was on the management of invasive grey squirrels in Ireland, and he has continued to research squirrel biology and conservation in Ireland and internationally. His other research interests include the conservation ecology of Irish mammals and other animals, the ecology of invasive species mammal monitoring techniques and their applications and mammal parasitology.

Éamonn Lenihan is a postdoctoral researcher in the School of Natural Sciences at the National University of Ireland, Galway (NUIG). He obtained his PhD in Zoology in 2020 from NUIG. His research focuses on the ecology and conservation of the critically endangered European eel (*Anguilla anguilla*) in hydropower impacted rivers, but he is also interested in the migration dynamics of other diadromous fish species.

Nigel Marley is based at the University of Plymouth, U.K. is a specialist in tardigrade taxonomy, nomenclature, and ecology. His past work has focused on diverse taxa including Milnesiidae and Isohypsibioidea. He is also interested in tardigrade biogeography and synecology. Dr Marley maintains a large personal collection of specimens from around the world.

Kate McAney obtained her PhD from University College Galway in 1987 for her research on the summer activity of the lesser horseshoe bat (*Rhinolophus hipposideros*) in County Clare and since 1991 has been working on this and other mammal species for Vincent Wildlife Trust, a specialist mammal conservation and research NGO. She wrote 'A conservation plan for Irish vesper bats: Irish Wildlife Manuals No. 20', edited a special publication on the lesser horseshoe bat that was published in 2013 by the *Irish Naturalists' Journal* and contributed to the *Atlas of Mammals in Ireland 2010–2015*.

NOTES ON CONTRIBUTORS

Gillian McCall graduated from Trinity College, Dublin and worked in the Biometrics Section of the Nature Conservancy (Natural Environment Research Council) in London for two years. She returned to Trinity to write a PhD thesis on the Entomostraca of Glenamoy, Co. Mayo, part of the International Biological Programme (Tundra Biome). She has also studied the Entomostraca of Co. Laois, Co. Carlow and North Tipperary. She has been an independent consultant, providing Environmental Impact Assessments of streams and ponds to the county councils of Co. Laois, Co. Kildare and Co. Dublin before the construction of motorway or dual carriageway.

T.K. (Kieran) McCarthy[†] (1949–2019) graduated from University College Cork, with a first-class honours' degree in zoology in 1971 and a PhD on the Irish freshwater Hirudinea (leeches) in 1974. Following post-doctoral research in Oxford, Helsinki, and Dublin, he was appointed to the Zoology Department, National University of Ireland, Galway. Throughout his career, and after retirement in 2011, he continued his research on a variety of aspects of freshwater ecology (limnology, entomology, fish parasitology and biogeography) and especially on European eels. He was a Visiting Professor at the University of Łodz, Poland, and University of Tokyo, Japan.

Stephen McCormack gained a BSc in Environmental Science at the National University of Ireland, Galway. His work on aquatic beetles on Clare Island was the basis of his MSc Zoology from the National University of Ireland, Galway. He studied University College Dublin for a PhD researching the parasitoid wasp communities of agricultural grasslands in Ireland. He co-authored *Insects of Ireland - A Field Guide*.

Rory Mc Donnell obtained his BSc and PhD in Environmental Science at the National University of Ireland, Galway. His PhD research was on the ecology and biological control potential of snail-killing flies (Sciomyzidae). He is currently an Assistant Professor at Oregon State University in the U.S. where his program is focused on 1) understanding the ecology of invasive slugs and snails in agriculture, horticulture, urban areas, the natural environment and at the interface of these systems, and 2) developing and implementing novel strategies for the management of these pests. He serves on the Governing Board of the International Branch of the Entomological Society of America which is the largest organization in the world serving the professional and scientific needs of entomologists. He has published over 50 scientific papers in leading journals in his field.

Declan Murray is Emeritus Associate Professor at the School of Biology and Environmental Science, University College Dublin. He completed his PhD with Carmel Humphries, D.Sc., M.R.I.A. on the distribution of Chironomidae (Diptera) of Ireland, at University College Dublin where he was appointed to the academic staff of the then Department of Zoology in 1968. Throughout his academic career he led research on the taxonomy, ecology and distribution of the Irish chironomid fauna and has also undertaken extensive studies abroad on Chironomidae from Nepal, Bathurst Island (Canada), Norway, Sweden, Germany, Portugal, the Atlantic islands of Iceland, Madeira, the Azores, the Canary Islands and Cape Verde and from Sulawesi, Australia, New Zealand and the Polynesian islands of Rarotonga and Tahiti in the southern hemisphere. As an authority on chironomid taxonomy and ecology, by invitation he co-authored sections of the diagnoses and identification keys to pupae and adult male Chironomidae of the Holarctic Region (1986, 1989) and the Chironomidae section of the Manual of Palaearctic Diptera (2000). He has published over 130 scientific papers, 70 specifically dealing with Irish Chironomidae, in national and international peer-reviewed scientific journals. He retired from academia in 2001 but has remained active in research. Most recently was lead author of the volume Chironomidae (Diptera) of Ireland – a review, checklist and their distribution in Europe (2018).

Jens Petter Nilssen is full-time research scientist at the Müller-Sars Society, a free scientific society where he works on taxonomy, biogeography and species-based ecology, history and theory of science, and humanities. He received his education from the University of Oslo and Polish Academy of Sciences in the 1960-70s. Since the mid-1970s, he has published, taught, supervised projects and students in the following fields: basic limnology, plankton ecology, acidification, eutrophication, palaeolimnology, species-based biology, diapause biology, and systematics, taxonomy and biogeography of Entomostraca (especially the genera Cyclops, Bosmina, Daphnia). He worked previously with the G. O. Sars' Collection at the University of Oslo and is now finishing the last Volume X of G. O. Sars' life-long series 'An account of the Crustacea of Norway'. He is also interested in 'science-on-science', especially theory, history, philosophy, sociology, and psychology of science.

Myles Nolan is a self-taught arachnologist who has added more than twenty spider species to the Irish list. He has a strong interest in the habitat preferences of Irish spiders, particularly in relation to conservation issues and has carried out extensive work on the spider collections held in the Natural History Museum, Dublin.

Audrey O'Grady is a Senior Lecturer in Biology and Science Education at the University of Limerick. Following a degree in Science Education, Audrey completed a PhD in Ecology (with a focus on Irish ants). Her main research interests at present are in Science Education and she is actively involved in a number of outreach programmes in STEM Education. Audrey is passionate about teaching and most recently was awarded the UL faculty award for excellence in teaching 2019.

Robert Paxton is a graduate of Sussex University, where he undertook a PhD on sex ratios in solitary wasps in 1985, Robert subsequently held postdoc positions at Cardiff University, researching bees and pollination, Uppsala University, working on bee population genetics, then University of Tübingen (Germany), researching the genetics of social evolution. In 2003 took up an academic post at Queen's University Belfast then, in 2010, he moved to the University of Halle (Germany), to take up a chair in zoology. His main research areas are social evolution, host-parasite relations, pollination and conservation genetics, with a taxonomic focus on bees.

Adriana Trojanowska-Olichwer is an Associated Professor in the University of Wrocław in Poland. She currently works in the field of biogeochemistry of water ecosystems and methane fermentation. She received her PhD from the university of Łodz in Poland. It was related to effect of enzyme - phosphatase on phytoplankton dynamics in lakes. Previously she also worked on ecology of zooplankton in dam reservoirs, which was the theme of her MSc.

CHAPTER 1

INTRODUCTION

THE VALUE OF LOCAL SURVEYS AND THEIR CONTRIBUTION TO BROADER ECOLOGICAL RESEARCH

Thomas Bolger

Studies of the fauna and flora of islands have a long and distinguished history in biology. Darwin's studies of the fauna and flora of the Galapagos Islands were seminal in the development of evolutionary theory (Darwin 1859, 1871) and research on the endemic species of those islands, most notably the finches, has continued to fascinate (e.g. Grant et al. 1976; Grant and Grant 2005, 2014). The publication of *The theory of island biogeograhy* (Mac Arthur and Wilson 1967) provided a hypothesis against which the assembly of island communities could be tested; later studies—including those of six small mangrove islands in the Florida Keys by students of E.O. Wilson, such as Dan Simberloff (Simberloff and Wilson 1969, 1970)—produced evidence supporting the theory. Continued studies of the bird assemblages of archipelagos in several parts of the world, such as those of Diamond (1975a), suggested that island bird communities were structured by assembly rules. These studies led to significant debates around the ecology and evolution of communities, particularly around the role of competition in their assembly with some arguing that the species composition and assembly structures could not be distinguished from assemblages generated randomly (c.f. Strong et al. 1984). This debate continues (e.g. Sanderson et al. 2008). The development of the unified neutral theory of biodiversity, which partially derived from these discussions, (Hubbell 2001) has added a new dimension to the debate. In addition, island biota have proved to be important natural experiments in the provision of ecosystem services, which are such a dominant focus in current biodiversity studies (e.g. Wardle et al. 1997; Vitousek 2004).

The Clare Island Survey, which was carried out during the early part of the twentieth century, was amongst the earliest and most extensive studies of an island fauna. It developed from a serendipitous study of Lambay Island, which produced interesting results and subsequently prompted Robert Lloyd Praeger and a group of colleagues to develop a more extensive study. Collins (1999) described the study as having taken place during a golden age of Irish natural history and how the rise of the amateur naturalist in the latter half of the nineteenth century paved the way for the multidisciplinary surveys of the island. The group that carried out the initial survey comprised talented amateurs as well as professional biologists. The team that conducted the New Survey of Clare Island one hundred years later had this same impressive mix of amateurs and professionals, although the reduced contribution from expert taxonomists from the museum sector is a sad reflection on the limited financial support for the Natural History Museum over recent decades (cf. Moriarty et al. 2005).

This volume, the final in the New Survey of Clare Island series, comprises an interesting collection of papers describing a diverse range of invertebrate and vertebrate taxa. The papers are variable in structure, ranging from straightforward reporting of the species found on the island to others that are interspersed with nuggets of ecological theory and descriptions of the life history of particular species. The value of this contribution of the enthusiastic talented natural historian to the development of science is delightfully illustrated in *Naturalist,* the autobiography E.O. Wilson (1994), described by Heneghan (1917) as 'an entomologist, taxonomist, conservationist, evolutionary biologist, social theorist, and countless other lofty things besides'. However, Wilson shows through his book how his detailed study of 'his pet taxon', ants, contributed significantly to the ecological and evolutionary theory mentioned above. This volume similarly illustrates the value of detailed faunistic studies in highlighting issues both in terms of science and policy deficiencies.

The original and new surveys of Clare Island not only provide a comprehensive list of the species occurring on an island off the Irish coast but also offer a great sense of just how difficult it is to inventory the fauna of any area and reflects the current sense of unease about just how many species there are globally. A review of our knowledge of Ireland's biodiversity carried out in 2010 (FitzPatrick *et al.* 2010) showed that our knowledge of many taxa is inadequate and, in some cases, almost non-existent. National checklists do not exist for many groups and local expertise is not available to identify many taxa. Indeed, not all the groups studied in the original survey were resurveyed in the New Survey because of a lack of expertise and the increasing specialisation of science. This is well illustrated in the case of one very diverse taxon, the Diptera, where only one family, the Chironomidae, is examined in detail in the New Survey and, within that family, the number of species has increased and the taxonomy has advanced considerably since the original survey.

This volume provides strong evidence of the need for increased investment in studies of Irish biodiversity, and its publication illustrates that we still do not have a comprehensive view of the species that occur in even an intensively studied area. Papers such as those by Murray on Chironomidae, Bond on Lepidoptera and Nolan on spiders list many species not previously recorded on the island. Some of this arises from changed taxonomy between the two surveys and shows that such studies are only possible where expert taxonomists, who are trained to deal with the intricacies of species identification, are available. Even in the case of relatively well-known taxa such as bees and ants, some of the species determinations required genetic analysis for definitive identification.

The extent to which this is a global problem is not clear (Costello *et al.* 2013) but it is generally believed to be significant. The fact that we do not know how many species exist globally has been a cause for concern for a considerably long time (*e.g.* May 1988) and our estimates do not appear to be improving (Caley *et al.* 2014). However, local surveys such as Clare Island are now contributing very significantly to revealing just how many species exist locally and globally. For example, 109 new species were described in the original survey (Herity 1999). In the case of mites, Halbert's (1915) paper is still considered significant because of the large number of species described based on the survey and its overall contribution to our knowledge of the Irish fauna (Evans and Browning 1953; Bolger *et al.* 2018). In the same way, such surveys are the basis for the meta-analyses currently being used to estimate global patterns of biodiversity of many taxa and these studies would not be possible if the local surveys did not exist; for example, surveys of Irish earthworms contribute to global estimates of their diversity and distribution (Phillips *et al.* 2019).

The overriding message from the papers in this volume is that the fauna of Clare Island is a subset of mainland fauna across a range of taxa from fish parasites to Chironomidae, birds and mammals. However, as shown by McCormack and McCarthy's study of the water beetles, the relative abundance of species with different traits differed on the island. These results are not surprising, but they illustrate very effectively the difficulty of using any one taxon as an indicator of overall animal biodiversity because the relative representation of the Irish mainland fauna on the island varied significantly. This is important because international initiatives such as the Convention on Biological Diversity (CBD) and Intergovernmental Platform for Biodiversity and Ecosystem Services (IPBES) (Larigauderie and Mooney 2010) highlight the urgent need for reliable information on trends in biodiversity as a basis for policy making. Given the difficulty of estimating the number of species on the planet (Costello *et al.* 2013) and the challenges of estimating even local species richness, it is clearly impossible to present a fully comprehensive assessment of biodiversity. Therefore, indicators are required; these tend to be biased towards groups such as plants, butterflies and birds that can be more easily monitored (Butchart *et al.* 2010). Taxa that are more difficult to survey, including those that contribute to major ecosystem service, are rarely used as biodiversity indicators (Feld *et al.* 2009). This bias has frequently been pointed out (Eglington *et al.* 2012; Larsen *et al.* 2012) and we see clear evidence of the potential for bias in the results of the zoological survey of Clare Island. The proportions of the different taxa occurring on the island and the changes in diversity and species composition over time vary significantly from group to group and therefore one cannot be used as an indicator of the other.

In 2000 Sala *et al.* argued that up to 50% of biotic diversity would be lost in the next century as a direct result of human activity and identified land use change as the main driver of the loss in many parts of the world. Our knowledge of the huge global losses of biodiversity has increased significantly since then with, for example, studies showing that extant vertebrate species monitored had an average decline of 25% in abundance (Dirzo *et al.* 2014) and that insect populations have declined by 75% over three decades (Sánchez-Bayo and Wyckhuys 2019). The IPBES (2019) recently reported that up to one million species face extinction (Tollefson 2019). All of this means that biodiversity loss is far outside humanity's safe operating space in terms of delivering ecosystem services (Rockström 2009). In this context, one of the potentially interesting aspects of the New Survey of Clare Island was the determination of the extent to which things changed over the century. However, taxon-specific issues prevent a meaningful comparison in this sense. As alluded to in Kelly's paper on birds, it is not always possible to ascertain whether there has been any species turnover between the two surveys. There has obviously been some turnover in bird species with the arrival of the Fulmar and the loss of the Arctic Tern since the first survey but the differences in species composition shown in Bolger's paper on soil mesofauna and Anderson and McCormack's paper on Carabidae are very likely to reflect, at least in part, differences in sampling techniques. These problems associated with determining whether differences between the fauna found in the two surveys might be due to changes in the environment or artefacts of the study indicate the need for systematic monitoring programmes whose value has been shown to be so vital in the identification of the current extinction crisis.

The papers in this volume add considerably to our knowledge of the fauna of Clare Island. Their importance cannot be overestimated in the context of their contributions to global studies, which are often perceived as more glamorous and high impact, but simply could not happen without data such as those collected in the Clare Island Survey. Furthermore, these papers illustrate the need for additional strategic policies in relation to biodiversity analysis in Ireland. While research in this area has been vigorously promoted in recent years, some aspects—most notably the cultivation of expertise and the infrastructures needed to record, analyse and curate the material—require greater support. As we move through the Anthropocene, with its many facets of global change, it is crucial that we understand what is happening to biodiversity in Ireland. There is essentially no long-term systematic monitoring of our biota (perhaps with the exception of birds and butterflies), which is a serious deficiency in our national responsibility to maintain a sustainable Earth.

REFERENCES

Bolger, T., Arroyo, J. and Pietrowska, K. 2018 A catalogue of the species of Mesostigmata (Arachnida, Acari, Parasitiformes) recorded from Ireland including information on their geographical distribution and habitats. *Zootaxa* **4519**(1), 1–220.

Butchart, S.H., Walpole, M., Collen, B., Van Strien, A., Scharlemann, J.P., Almond, R.E., Baillie, J.E., Bomhard, B., Brown, C., Bruno, J., Carpenter, K.E., Carr, G.M., Chanson, J., Chenery, A.M., Csirke, J., Davidson, N.C., Dentener, F., Foster, M., Galli, A., Galloway, J.N., Genovesi, P., Gregory, R.D., Hockings, M., Kapos, V., Lamarque, J.F., Leverington, F., Loh, J., McGeoch, M.A., McRae, L., Minasyan, A., Hernandez Morcillo, M., Oldfield, T.E., Pauly, D., Quader, S., Revenga, C., Sauer, J.R., Skolnik, B., Spear, D., Stanwell-Smith, D., Stuart, S.N., Symes, A., Tierney, M., Tyrrell, T.D., Vie, J.C. and Watson, R. 2010 Global biodiversity: indicators of recent declines. *Science* **328**, 1164–68.

Caley, M.J., Fisher, R. and Mengersen, K. 2014 Global species richness estimates have not converged. *Trends in Ecology and Evolution* **29**, 188–89.

Collins, T. 1999 The Clare Island Survey of 1909–1911: Participants, papers and progress. In C. MacCárthaigh and K. Whelan (eds) *New Survey of Clare Island. Volume 1: History and cultural landscape*, 1–40. Dublin. Royal Irish Academy.

Costello, M.J., May, R.M. and Stork, N.E. 2013 Can we name Earth's species before they go extinct? *Science* **339**, 413–16.

Darwin, C. 1859 *On the origin of species by means of natural selection, or the preservation of favoured races in the struggle for life*. London. John Murray.

Darwin, C. 1871 *Journal of researches into the natural history and geology of the countries visited during the voyage of the H.M.S. Beagle round the world*. New York. D. Appleton and Co.

Diamond, J.M. 1975 Assembly of species communities. In M.L. Cody and J.M. Diamond (eds) *Ecology and evolution of communities*, 342–444. Cambridge, Mass. Harvard University Press.

Dirzo, R., Young, H.S., Galetti, M., Ceballos, G., Isaac, N.J.B. and Collen, B. 2014 Defaunation in the Anthropocene. *Science* **345**, 401–06.

Eglington, S.M., Noble, D.G. and Fuller, R.J. 2012 A meta-analysis of spatial relationships in species richness across taxa: birds as indicators of wider biodiversity in temperate regions. *Journal for Nature Conservation* **20**, 301–09.

Evans, G.O. and Browning, E. 1953 Some inter-tidal mites of south-west England. *Bulletin of the British Museum of (Natural History) Zoology* **1**, 413–22.

Feld, C.K., Martins da Silva, P., Sousa, J.P., De Bello, F., Bugter, R., Grandin, U., Hering, D., Lavorel, S., Mountford, O., Pardo, I., Pärtel, M., Römbke, J., Sandin, L., Jones, K.B. and Harrison, P. 2009 Indicators of biodiversity and ecosystem services: a synthesis across ecosystems and spatial scales. *Oikos* **118**, 1862–71.

FitzPatrick, Ú., Regan, E. and Lysaght, L. 2010 *Ireland's biodiversity in 2010: state of knowledge*. Waterford. National Biodiversity Data Centre.

Grant, P.R. and Grant, B.R. 2005 Darwin's finches *Current Biology* **15**, R614-R615.

Grant, P.R. and Grant, B.R. 2014 *40 Years of evolution: Darwin's finches on Daphne Major Island*. Princeton N.J. Princeton University Press.

Grant, P.R., Grant, B.R., Smith, J.N., Abbott, I.J. and Abbott, L.K. 1976 Darwin's finches: population variation and natural selection. *Proceedings of the National Academy of Sciences* **73**, 257–61.

Halbert, J.N. 1915 Clare Island survey. 39. Acarinida ii. Terrestrial and marine Acarina. *Proceedings of the Royal Irish Academy* **31**(39), 45–136.

Heneghan, L. 2017 Preserving biodiversity, preventing climate disaster: childish dreams or audacious strategies? *Bulletin of the Atomic Scientists* **73**(4), 1–4.

Herity, M. 1999 Preface. In C. MacCárthaigh, and K. Whelan (eds) *New survey of Clare Island. Volume 1: History and cultural landscape*, v–vi. Dublin. Royal Irish Academy.

Hubbell, S.P. 2001 *The unified neutral theory of biodiversity and biogeography*. Princeton N.J. Princeton University Press.

IPBES 2019 Díaz, S. Settele, J., Brondízio, E.S., Ngo, H.T., Guèze, M., Agard, J., Arneth, A., Balvanera, P., Brauman, K.A., Butchart, S.H.M., Chan, K.M.A., Garibaldi, L.A., Ichii, K., Liu, J., Subramanian, S.M., Midgley, G.F., Miloslavich, P., Molnár, Z., Obura, D., Pfaff, A., Polasky, S., Purvis, A., Razzaque, J., Reyers, B., Roy Chowdhury, R., Shin, Y.J., Visseren-Hamakers, I.J., Willis, K.J. and Zayas, C.N. (eds). Summary for policymakers of the global assessment report on biodiversity and ecosystem services of the Intergovernmental Science-Policy Platform on Biodiversity and Ecosystem Services. Bonn, Germany. IPBES secretariat.

Larigauderie, A. and Mooney, H.A. 2010 The intergovernmental science-policy platform on biodiversity and ecosystem services: moving a step closer to an IPCC-like mechanism for biodiversity. *Current Opinion in Environmental Sustainability* **2**, 9–14.

Larsen, F.W., Bladt, J., Balmford, A. and Rahbek, C. 2012 Birds as biodiversity surrogates: will supplementing birds with other taxa improve effectiveness? *Journal of Applied Ecology* **49**, 349–56.

Mac Arthur, R.H. and Wilson, E.O. 1967 *The theory of island biogeography*. Princeton N.J. Princeton University Press.

May, R.M. 1988 How many species are there on earth? *Science* **247**, 1441–49.

Moriarty, C., Myers, A., Andrew, T., Bolger, T., Breen, J. and Parkes, M. 2005 *The Natural History Museum – present status and future needs*. Dublin. Royal Irish Academy.

Phillips, H.R.P., Guerra, C.A., Bartz, M.L.C., Briones, M.J.I., Brown, G., Crowther, T.W., Ferlian, O., Gongalsky, K.B., van den Hoogen, J., Krebs, J., Orgiazzi, A., Routh, D., Schwarz, B., Bach, E.M., Bennett, J., Brose, U., Decaëns, T., König-Ries, B., Loreau, M., Mathieu, J., Mulder, C., van der Putten, W.H., Ramirez, K.S., Rillig, M.C., Russell, D., Rutgers, M., Thakur, M.P., de Vries, F.T., Wall, D.H., Wardle, D.A., Arai, M., Ayuke, F.O., Baker, G.H., Beauséjour, R., Bedano, J.C., Birkhofer, K., Blanchart, E., Blossey, B., Bolger, T., Bradley, R.L., Callaham, M.A., Capowiez, Y., Caulfield, M.E., Choi, A., Crotty, F.V., Dávalos, A., Cosin, D.J.D., Dominguez, A., Duhour, A.E., van Eekeren, N., Emmerling, C., Falco, L.B., Fernández, R., Fonte, S.J., Fragoso, C., Franco, A.L.C., Fugère, M., Fusilero, A.T., Gholami, S., Gundale, M.J., López, M.G., Hackenberger, D.K., Hernández, L.M., Hishi, T., Holdsworth, A.R., Holmstrup, M., Hopfensperger, K.N., Lwanga, E.H., Huhta, V., Hurisso, T.T., Iannone, B.V., Iordache, M., Joschko, M., Kaneko, N., Kanianska, R., Keith, A.M., Kelly, C.A., Kernecker, M.L., Klaminder, J., Koné, A.W., Kooch, Y., Kukkonen, S.T., Lalthanzara, H., Lammel, D.R., Lebedev, I.M., Li, Y., Lidon, J.B.J., Lincoln, N.K., Loss, S.R., Marichal, R., Matula, R., Moos, J.H., Moreno, G., Morón-Ríos, A., Muys, B., Neirynck, J., Norgrove, L., Novo, M., Nuutinen, V., Nuzzo, V., Rahman P, M., Pansu, J., Paudel, S., Pérès, G., Pérez-Camacho, L., Piñeiro, R., Ponge, J.-F., Rashid, M.I., Rebollo, S., Rodeiro-Iglesias, J., Rodríguez, M.Á., Roth, A.M., Rousseau, G.X., Rozen, A., Sayad, E., van Schaik, L., Scharenbroch, B.C., Schirrmann, M., Schmidt, O., Schröder, B., Seeber, J., Shashkov, M.P., Singh, J., Smith, S.M., Steinwandter, M., Talavera, J.A., Trigo, D., Tsukamoto, J., de Valença, A.W., Vanek, S.J., Virto, I., Wackett, A.A., Warren, M.W., Wehr, N.H., Whalen, J.K., Wironen,

M.B., Wolters, V., Zenkova, I.V., Zhang, W., Cameron, E.K. and Eisenhauer, N. 2019 Global distribution of earthworm diversity. *Science* **366**, 480–85.

Rockström, J. 2009 A safe operating space for humanity. *Nature* **461**, 472–75.

Sánchez-Bayo, F. and Wyckhuys, K.A.G. 2019 Worldwide decline of the entomofauna: a review of its drivers. *Biological Conservation* **232,** 8–27.

Sala, O.E., Chapin, F.S., Armesto, J.J., Berlow, E., Bloomfield, J., Dirzo, R., Huber-Sanwald, E., Huenneke, L.F., Jackson, R.B., Kinzig, A., Leemans, R., Lodge, D.M., Mooney, H.A., Oesterheld, M., Poff, N.L., Sykes, M.T., Walker, B.H., Walker, M. and Wall, D.H. 2000 Biodiversity - global biodiversity scenarios for the year 2100. *Science* **287**, 1770–74.

Sanderson, J.G., Diamond, J.M. and Pimm, S.L. 2009 Pairwise co-existence of Bismarck and Solomon landbird species. *Evolutionary Ecology Research* **11**, 771–86.

Simberloff, D.S. and Wilson, E.O. 1969 Experimental zoogeography of islands: the colonization of empty islands. *Ecology* **50**, 278–96.

Simberloff, D.S. and Wilson, E.O. 1970 Experimental zoogeography of islands: a two-year record of colonization. *Ecology* **51**, 934–37.

Strong, D.R. Jr., Simberloff, D., Abele, L.G. and Thistle, A.B. 1984 *Ecological communities: conceptual issues and the evidence*. Princeton N.J. Princeton University Press.

Tollefson, J. 2019 One million species face extinction. *Nature* **569**, 171.

Vitousek, P.M. 2004 *Nutrient cycling and limitation: Hawai'i as a model system*. Princeton N.J. Princeton University Press.

Wardle, D.A., Zackrisson, O., Hornberg, G. and Gallet, C. 1997 The influence of land area on ecosystem properties. *Science* **277**, 1296–9.

Wilson, E.O. 1994 *Naturalist*. Washington D.C. Island Press.

CHAPTER 2

THE NON-MARINE MOLLUSCA OF CLARE ISLAND

Roy Anderson

ABSTRACT

The non-marine mollusca of Clare Island was first described by Stelfox (1911). An estimated 57 species were recorded including woodland molluscs associated with steep cliffs on the northern slopes of Knockmore. In recent surveying a total of 56 species was recorded but the Knockmore cliffs are now inaccessible due to rock falls and were not included. The relict woodland fauna of that area is therefore in need of assessment at some point in the future. A range of invasive species have arrived on the island within the past century, including at least three non-native slugs, a freshwater gastropod, two Mediterranean–Atlantic dune snails and one snail of disturbed ground. These non-native species reflect larger changes taking place in the molluscan fauna of Ireland during the twentieth century and the continuing increase of species associated with human activity and disturbance.

Introduction

The non-marine mollusca of Clare Island were characterised by Stelfox (1911). The present study aims to revisit this fauna with a view to characterising changes that may have taken place since 1911. A major problem with comparing faunas in this way is the inevitable advance of scientific study with multiple revisions of naming systems (taxonomy), better understanding of speciation and particularly of cryptic species in the fauna and better knowledge of the distribution patterns of species. There is also the issue, particularly relevant to molluscs, of changing faunas in relation to human influence and disturbance.

Methodology

Scope of survey

On 29–30 June 2002, a transect was followed across lower altitudes of the western, south-eastern, and eastern quadrants of the island. On 14–16 October 2002, a further transect was followed across higher ground in the centre and north-west (e.g. across the top of Knockmore to the lighthouse at Lachnacranny).

Methods

Transects were walked in clear weather and the following niches examined: undersides of stones, rubbish and wood in pasture or around dwellings; sandy shores (base of grasses examined; moss on tops of stone dykes (removed and sieved); grass or *Iris* tussocks (bases examined in situ); woodland leaf litter (sieved); dead tree branches (bark removed); freshwaters (sampled with pond net); upper tidal rocks (cracked open for interstitial fauna).

Literature

Terrestrial molluscs were identified using Cameron (2003), freshwater and eurysaline molluscs using Glöer (2002) and freshwater bivalves using Killeen *et al.* (2004). Terrestrial slugs were determined using a range of academic sources as summarised in Rowson *et al.* (2014).

Results

Species recorded and their site and status are summarised in Table 1. The nomenclature is that of Anderson and Rowson (2020).

Discussion

Name changes in non-marine mollusca

There has been much refinement of the species concept in European molluscs since the 1911 survey was published. Most of the changes are indicated as notes within Table 2. It is not always possible to align the names used in 1911 with their present-day equivalents. However, many of the newly recognised segregate species have habitat preferences which help decide the matter. For instance, *Balea heydeni* is much more likely to occur on the acid rocks of Clare Island than its segregate *Balea perversa*, from which it was separated within

Table 1
Non-marine molluscs recorded from Clare Island, post-1980

Species	Site	Status in Ireland (*common unless otherwise stated*)
Acanthinula aculeata (O. F. Müller, 1774)	Wood near The Mill	Near threatened (nt)
Aegopinella nitidula (Draparnaud, 1805)		
Aegopinella pura (Alder, 1830)		
Ampullaceana balthica (L., 1758)	General	
Ancylus fluviatilis O. F. Müller, 1774	Beetle Rocks; Lecarrow; The Mill coastline	
Anisus leucostoma (Millet, 1813)	Beetle Rocks	Local in temporary freshwater habitats and declining
Arion ater (L., 1758)	Cos Abhainn; Clare Harbour; Creggan L.; Granuaile's Tower; Knockmore; Portlea; The Mill fen; The Mill heathland	
Arion distinctus J. Mabille, 1868	General	
Arion hortensis A. Férussac, 1819	Clare Harbour	Very local, south-eastern
Arion intermedius Normand, 1852	Granuaile's Tower; Knock-more; wood near The Mill	
Arion owenii Davies, 1979	Stream at Ballytoohy	North-western
Arion rufus (L., 1758)	Beetle Rocks; Clare Harbour; Cos Abhainn; Granuaile's Tower; The Mill fen	
Arion subfuscus (Draparnaud, 1805)	Beetle Rocks; Granuaile's Tower; Knockmore; wood near The Mill; The Mill fen	
Balea heydeni Von Maltzan, 1881[1]	Granuaille's Tower	
Carychium tridentatum (Risso, 1826)	Ballytoohy	
Cepaea nemoralis (L., 1758)	Lighthouse	
Cernuella virgata (Da Costa, 1778)	Clare Harbour; The Mill fen	Alien; mainly coastal
Clausilia bidentata (Ström, 1765)	Lighthouse; stream at Ballytoohy; The Mill fen; The Mill coastine; The Mill heathland; wood near The Mill	
Cochlicella acuta (O. F. Müller, 1774)	Clare Harbour; The Mill fen	Mainly coastal
Cochlicopa cf. *lubrica* (O. F. Müller, 1774)	Clare Harbour; Granuaile's Tower; stream at Ballytoohy; The Mill fen	
Cochlicopa cf. *lubricella* (Rossmässler, 1834)	Clare Harbour; stream at Ballytoohy	
Cornu aspersum (O. F. Müller, 1774)	Clare Harbour; Granuaile's Tower; The Mill fen	Alien

continued

---

[1] Previously confused under the name *Balea perversa* (L., 1758). *Balea heydeni* has been recognised relatively recently but is much more widespread in Ireland than the nominal *Balea perversa* which is rare on walls and natural calcareous rocks. *Balea heydeni*, by contrast is almost uniquitous on tree trunks and under bark, only occasionally straying on to walls or natural stone and then mainly in non-calcareous areas.

Table 1. Cont.		
Species	**Site**	**Status in Ireland** (*common unless otherwise stated*)
Deroceras invadens Reise et al., 2011[2]	Granuaile's Tower	
Deroceras laeve (O. F. Müller, 1774)	Knockmore; Lighthouse	
Deroceras reticulatum (O. F. Müller, 1774)	Clare Harbour; Cos Abhainn; Craigmore; Granuaile's Tower; Lighthouse	
Discus rotundatus (O. F. Müller, 1774)	Cos Abhainn; Lighthouse; The Mill fen; The Mill coastline; Wood near The Mill	
Euglesa hibernica (Westerlund, 1894)	Knockmore	Near threatened (NT)
Euglesa lilljeborgii (Clessin, 1886)	Beetle Rocks	Vulnerable (VU)
Euglesa nitida (Jenyns, 1832)	Unknown	
Euglesa obtusalis (Lamarck, 1818)	Unknown	
Euglesa personata (Malm, 1855)	The Mill fen	
Euglesa subtruncata (Malm, 1855)	Unknown	
Galba truncatula (O. F. Müller, 1774)	Knockmore; Lighthouse	
Helicella itala (L., 1758)	The Mill fen	Mainly coastal; Vulnerable (VU)
Lauria cylindracea (Da Costa, 1778)	Cos Abhainn; Clare Harbour; stream at Ballytoohy; wood near The Mill	
Lehmannia marginata (O. F. Müller, 1774)	Beetle Rocks; Cos Abhainn; Creggan L.; Knockmore; The Mill heathland	
Leiostyla anglica (A. Férussac, 1821)	Portlea, Maum	Local, becoming less common; Vulnerable (VU)
Limacus maculatus (Kaleniczenko, 1851)	Granuaile's Tower	Alien
Limax maximus L., 1758	Clare Harbour; Granuaile's Tower; The Mill fen	
Milax gagates (Draparnaud, 1801)	Clare Harbour; Granuaile's Tower; The Mill fen	
Nesovitrea hammonis (Ström, 1765)	Craigmore; Lighthouse; The Mill fen	
Oxychilus alliarius (J.S. Miller, 1822)	Knockmore; Lighthouse; wood near The Mill; The Mill heathland	
Oxychilus cellarius (O. F. Müller, 1774)	Granuaile's Tower; The Mill fen	
Oxyloma elegans (Risso, 1826)	The Mill fen	
Potamopyrgus antipodarum (J. E. Gray, 1843)	Beetle Rocks; Knockmore	
Stagnicola. fuscus (C. Pfeiffer, 1821)	Beetle Rocks	
Tandonia budapestensis (Hazay, 1880)	Clare Harbour	Alien
Trochulus striolatus (C. Pfeiffer, 1828)	Granuaile's Tower; Clare Harbour; The Mill fen	Alien
Vallonia cf. *excentrica* Sterki, 1893	Lighthouse	
Vertigo antivertigo (Draparnaud, 1801)	The Mill fen	Becoming localised in wetlands; Vulnerable (VU)
Vertigo pygmaea (Draparnaud, 1801)	Stream at Ballytoohy; The Mill fen	Near threatened (nt)
Vertigo substriata (Jeffreys, 1833)	The Mill fen; stream at Ballytoohy	Near threatened (nt)
Vitrina pellucida (O. F. Müller, 1774)	Wood near The Mill	
Zonitoides excavatus (Alder, 1830)	The Mill fen; The Mill coastline	Western; acidic woods; Vulnerable (VU)
Zonitoides nitidus (O. F. Müller, 1774)	The Mill fen	

Sites visited: Beetle Rocks (L658843), 29.vi.2002, 14.x.2002; Clare Island Harbour (L714852), 29.vi.2002; Cos Abhainn Guest House (L663846), 29.vi.2002, 13.x.2002; Craigmore (L670844), 13.x.2002; Creggan lough (L689856), 13.x.2002; Granuaile's Tower (L714851), 29.vi.2002; Heathland near The Mill (L707863), 29.vi.2002; Knockmore (L663850, L664860, L667859, L668863), 13.x.2002; Lecarrow (L6986), 24.ix.2003 [G. Holyoak]; Lighthouse (L693880), 14.x.2002; Portlea, N of Maum (L705867), 24.ix.2003 [G. Holyoak]; Stream at Ballytoohy (L697867), 14.x.2002; The Mill coastline (L706865), 29.vi.2002; The Mill fen (L704864), 29.vi.2002, 24.ix.2003 [G. Holyoak]; wood near The Mill (L700869), 14.x.2002.

---

[2] Recent work on *Deroceras* in north-west Europe has shown that the former *Deroceras panormitanum* (Lessona and Pollonera, 1882) should now be referred to *D. invadens*.

Table 2
Non-marine molluscs on Clare Island, 2002, compared to Stelfox (1911)

Species	2002, lowland, S, E	2002, upland, N, W	Post-1980, all	1911
Acanthinula aculeata (O. F. Müller, 1774)		+	+	+
Acicula fusca (Montagu, 1803)				+[1]
Aegopinella nitidula (Draparnaud, 1805)				+
Aegopinella pura (Alder, 1830)				+
Ampullaceana baltica (L., 1758)		+	+	+
Ancylus fluviatilis O. F. Müller, 1774	+	+	+	+
Anisus leucostoma (Millet, 1813)		+	+	+
Arion ater (L., 1758)	+	+	+	[+]
Arion circumscriptus silvaticus Lohmander, 1937				[+][2]
Arion distinctus J. Mabille, 1868	+		+	[+][3]
Arion hortensis A. Férussac, 1819	+		+	
Arion intermedius Normand, 1852	+	+	+	+
Arion owenii Davies, 1979	+		+	
Arion rufus (L., 1758)	+		+	[+][4]
Arion subfuscus (Draparnaud, 1805)		+	+	+
Balea heydeni Von Maltzan, 1881	+	+	+	[+][5]
Candidula intersecta (Poiret, 1801)				+
Carychium minimum O. F. Müller, 1774				+
Carychium tridentatum (Risso, 1826)		+	+	+
Cepaea nemoralis (L., 1758)	+	+	+	+
Cernuella virgata (Da Costa, 1778)	+	+	+	
Clausilia bidentata (Ström, 1765)		+	+	+
Cochlicella acuta (O. F. Müller, 1774)	+	+	+	
Cochlicopa cf. *lubrica* (O. F. Müller, 1774)	+	+	+	[+][6]
Cochlicopa cf. *lubricella* (Rossmässler, 1834)	+		+	
Columella aspera Waldén, 1966				[+][7]
Cornu aspersum (O. F. Müller, 1774)	+	+	+	+
Deroceras invadens Reise et al., 2011	+		+	
Deroceras laeve (O. F. Müller, 1774)		+	+	+
Deroceras reticulatum (O. F. Müller, 1774)	+	+	+	+
Discus rotundatus (O. F. Müller, 1774)	+	+	+	+
Euconulus cf. *fulvus* (O. F. Müller, 1774)				[+][8]
Euglesa casertana (Poli, 1791)				+
Euglesa hibernica (Westerlund, 1894)		+	+	
Euglesa lilljeborgii (Clessin, 1886)		+	+	

continued

---

[1] Old woodland relict.
[2] It is not known which segregate of *Arion circumscriptus* agg. was recorded in the 1911 survey but is likely to have been *A. circumscriptus silvaticus*.
[3] It is not known which segregate of *Arion hortensis* agg. was recorded in the 1911 survey but is likely to have been *A. distinctus*.
[4] The records for 1911 do not discriminate members of *Arion ater* agg., now recognised as distinct i.e. *Arion ater* and *Arion rufus*.
[5] Recorded in the 1911 report as *Balea perversa*. *Balea heydeni* was recorded in 2002 and this is the most likely of the two segregates to occur. *Balea perversa* s.s. has however been recorded recently on Achill Island.
[6] Recorded in the 1911 survey as *Cochlicopa lubrica* but both segregates (*lubrica* and *lubricella*) are now known to occur.
[7] *Columella edentula* was recorded in 1911 but it is not known which of the segregates (*edentula* or *aspera*) was involved. However, it is very likely that *Columella aspera* was present as it is more typical of wet, exposed locations.
[8] *Euconulus fulvus* was recorded in the 1911 survey but it is not known which of the segregates (*E.* cf *fulvus* or *E.* cf *alderi*) was involved. *Euconulus* cf *alderi* seems the more likely.

Table 2. Cont.				
Species	2002, lowland, S, E	2002, upland, N, W	Post-1980, all	1911
Euglesa milium (Held, 1836)		+	+	
Euglesa nitida (Jenyns, 1832)		+	+	+
Euglesa obtusalis (Lamarck, 1818)		+	+	+
Euglesa personata (Malm, 1855)		+	+	+
Euglesa subtruncata (Malm, 1855)		+	+	+
Galba truncatula (O. F. Müller, 1774)	+	+	+	+
Gyraulus crista (L., 1758)				+
Helicella itala (L., 1758)		+	+	+
Lauria cylindracea (Da Costa, 1778)	+	+	+	+
Lehmannia marginata (O. F. Müller, 1774)	+	+	+	+
Leiostyla anglica (A. Férussac, 1821)		+	+	+[9]
Limacus maculatus (Kaleniczenko, 1851)	+		+	
Limax cinereoniger Wolf, 1803				+[10]
Limax maximus L., 1758	+		+	+
Milax gagates (Draparnaud, 1801)	+	+	+	+
Nesovitrea hammonis (Ström, 1765)		+	+	+
Oxychilus alliarius (J.S. Miller, 1822)	+	+	+	+
Oxychilus cellarius (O. F. Müller, 1774)		+	+	+
Oxyloma elegans (Risso, 1826)	+		+	+
Potamopyrgus antipodarum (J. E. Gray, 1843)	+	+	+	
Punctum pygmaeum (Draparnaud, 1801)				+
Spermodea lamellata (Jeffreys, 1830)				+[11]
Stagnicola fuscus (C. Pfeiffer, 1821)		+	+	[+][12]
Tandonia budapestensis (Hazay, 1880)		+	+	
Trochulus hispidus (L., 1758)				+
Trochulus striolatus (C. Pfeiffer, 1828)	+	+	+	
Vallonia cf. *excentrica* Sterki, 1893		+	+	+
Vallonia costata (O. F. Müller, 1774)				+
Valvata piscinalis (O. F. Müller, 1774)				+
Vertigo antivertigo (Draparnaud, 1801)		+	+	+
Vertigo pygmaea (Draparnaud, 1801)		+	+	+
Vertigo substriata (Jeffreys, 1833)		+	+	+
Vitrea crystallina (O. F. Müller, 1774)				[+][13]
Vitrina pellucida (O. F. Müller, 1774)		+	+	
Zenobiellina subrufescens (J. S. Miller, 1822)				+[14]
Zonitoides excavatus (Alder, 1830)	+		+	+
Zonitoides nitidus (O. F. Müller, 1774)	+		+	+
Totals	**29**	**42**	**56**	**57**

---

[9] Old woodland relict
[10] Old woodland relict
[11] Old woodland relict
[12] *Lymnaea palustris* was reported in 1911 but only its segregate *Stagnicola fuscus* is now known to be widespread in Ireland. The nominotypical segregate *Stagnicola palustris* has not yet been confirmed for Ireland.
[13] *Vitrea crystallina* was reported in 1911 but it is not known which of the segregates (*Vitrea crystallina* or *contracta*) was involved. *Vitrea crystallina* is the most likely of the two to occur on the Island.
[14] Old woodland relict

the last ten years. *Columella aspera* is more likely to occur in the prevailing ground conditions than its segregate *Columella edentula,* and so on. Notes on selected species at the end of this section will also help clarify the situation.

Gaps in coverage
There are some gaps in the present survey. Because of an absence of easy access to the cliffs on Knockmore, due to land slips within the last century, a small number of important woodland species were not recorded in 2002. There are five declining species within this category: *Acicula fusca; Zenobiellina subrufescens; Leiostyla anglica; Limax cinereoniger; Spermodea lamellata.* Only *Leiostyla anglica* has been recorded on the Island post-1980 (one specimen at Portlea, near Maum, September 2003). The removal of trees by human activity, acidification of ground layers, grazing and general disturbance in many areas have led to the virtual extinction of this fauna. Only where a substitute woodland habitat exists, usually on steep seaward cliffs in association with wood rush (*Luzula* sp.) is faunal survival assured in the long term. Investigation of this fauna was not possible in 2002. This will require a special effort to scale at least the upper part of the cliffs in order to reach significant areas of woodrush at some point in the future.

In addition, some aquatic species found in the original survey have not been seen recently (*Euglesa casertana; Valvata piscinalis;* Table 2). *Valvata piscinalis* is a freshwater prosobranch which can occur in soft, non-calcareous water but is typical of larger habitats, so its presence on Clare Island (if correct) is anomalous. *Euglesa casertana* is a very common pea mussel on the mainland and being unfussy i.e. eurytopic, its occurrence or otherwise is of little significance. Three additional species of *Euglesa* were found post-1980. J.G. Evans found *Euglesa milium* on a visit in 1983 to a site at L6986 which is near Lecarrow. In the present survey *E. lilljeborgi* was found in a small fenny pool near Beetle Rocks and *Euglesa hibernica* in peaty pools on the slopes of Knockmore. Unlike the other *Euglesa* spp. in Table 2 the latter two are biogeographically significant, having strongly western ranges in Europe and being commonest in Ireland and northern and western parts of the British Isles. *Euglesa lilljeborgii* is red listed in Ireland as 'vulnerable'.

In total 56 species were recorded post-1980, against an estimated 57 species in 1911 (NBDC and Conchological Society databases). In addition to the present author, two other conchologists have visited the island and are included in post-1980 records: J.G. Evans who collected *Pisidium* spp. at L6986 in 1983; and Geraldine Holyoak who collected at Lecarrow and the Mill Fen area in September 2003. One difference between 1911 and post-1980 lists relates to the absence of collecting on the northern cliffs, which became inaccessible after 1980. The relict woodland fauna recorded in this area pre-1911 is therefore in need of assessment at some point in the future. Another difference has been the arrival of a range of invasive species on the island, dealt with below.

Incursive or invasive species, added post-1980
Potamopyrgus antipodarum
Jenkin's spire shell is a small freshwater prosobranch originating from New Zealand and probably introduced to Irish freshwaters early in the nineteenth century. Stelfox and Welch (1980) report the finding of a shell in the Hyndman (now Ulster Museum) Collection, labelled 'Lough Neagh, --18—', which Stelfox thought referred to around 1837 (see Stelfox 1926). This is substantially earlier than the first British record of 1859, for Gravesend. It was first formally brought forward as Irish by Adams (1897), from the Lough Neagh outflow of the Lower Bann. It has since become one of the most widespread molluscs in Irish freshwaters. It was not seen by Stelfox on Clare Island in the 1911 survey but appears to have colonised the island subsequently. It was seen in a shallow, fenny pool at Beetle Rocks and in small streams along the south and east sides of the island as far as the lighthouse at Lacknacranny.

Arion hortensis seg.
This is the rarer and more restricted species of the two common segregates of *Arion hortensis* agg. It is highly likely that the *Arion 'hortensis'* reported in 1911 was in fact *Arion distinctus*, which is more tolerant of cold, moisture-saturated soils and whose European range extends farther north, to Scandinavia and the Faroe Islands (Holyoak and Seddon 1983). By contrast *Arion hortensis* has a strongly south-eastern range in Ireland and is less tolerant of wetness and exposure. Its discovery under waste wood by Granuaile's Tower in

June 2002 was unexpected. Probably a very recent introduction.

Tandonia budapestensis

The Budapest slug originally had a south-eastern range in Europe but has been spread north-westwards by human agency and was first recorded in the British Isles in 1921 (Philipps and Watson 1930). It has since spread to Ireland and become very common on richer, shaded soils and is a familiar garden species. Recorded only from the vicinity of Granuaile's Tower.

Limacus maculatus

The green cellar slug was first recognised as distinct from the introduced Spanish yellow cellar slug *Limacus flavus* in the 1970s. However, it is clear from examining plates in Scharff (1891) and references of the time to cellar slugs occurring in wild woodland that it was in Ireland in the late nineteenth century. However, it was not distinguished from the strongly anthropic *Limacus flavus*. *Limacus maculatus* is now known to have originated in eastern Europe (Wiktor and Norris 1982). It lives in parks and gardens but is also common in wild woodlands living under dead wood and feeding on fungi. The lack of natural woods on Clare Island does not seem to have prevented its establishment. The natural range of this species is imperfectly known but includes Romania, Bulgaria, Turkey and western Russia. Recorded only from the vicinity of Granuaile's Tower, but likely to be more widespread, especially around houses.

Deroceras invadens

The tramp slug is well named as it has become established through human trade and disturbance throughout west, central and northern Europe, North America and Australasia. Despite its abundance its country of origin is unknown and its true identity has only become established very recently. It was previously called *Agriolimax caruanae*, then *Deroceras caruanae* and finally *Deroceras panormitanum*, but recent work of Reise *et al.* (2011) in Germany suggests it may be of Italian origin and distinct from the true *Deroceras panormitanum*, which is a rare invasive species native to Sicily and Malta. It was first recorded in Ireland (Cork) in 1958 but has spread to become one of the commonest garden slugs there. Like most of the other slug species in this section it has been found only in the vicinity of Granuaile's Tower but is very likely to be found around occupied dwellings across the island.

Cernuella virgata

The striped snail is an old introduction to Britain and Ireland from the Mediterranean. It appears to have spread in Ireland during the historical period by colonising the ballast and dry slopes skirting railway lines. It has evidently expanded its range from sand dunes on mainland coasts to some of the western islands since the early part of the twentieth century. It was found around Granuaile's Tower, along the dunes bordering the harbour strand and also at The Mill fen.

Cochlicella acuta

The pointed snail, which is native to the Mediterranean, is a long-established introduction to Ireland, possibly in the late prehistoric period (Kerney 1999). It is common on dunes around the coast and has a similar distribution on Clare Island to the previous species.

Though not recorded in the 1911 survey it is now well established in the only suitable habitat on the island: in the dunes bordering the harbour strand. The Clare Island form is unusual in being relatively short-spired. Specimens had to be examined carefully as the Clare Island material showed some similarity to the closely related but smaller *Cochlicella barbara*, which is presently expanding its range in south-west Britain. The arrival of this species is interesting as it suggests that *Cochlicella* is still expanding its range in Ireland.

Trochulus striolatus

The strawberry snail is now widespread in Britain and Ireland but is probably native only in

Arion oweni. Photo: R. Anderson

south-west Britain. It has not been found either as a fossil or subfossil in Irish Postglacial deposits. It was evidently not as widespread in the nineteenth century as it is now and was not recorded from Clare Island in 1911. Found only at Granuaile's Tower in 2002, and along the harbour strand. Probably a fairly recent arrival.

Other post-1980 records of note

Euglesa (=Pisidium) hibernica, E. lilljeborgii
It is surprising that neither of these small bivalves was seen in the 1911 survey. Both are typical of western parts of Ireland, being partial to clean, wave-exposed lakeshores though on occasion smaller habitats. As already mentioned, *Euglesa lilljeborgii* was found in June 2002 in a small, rather rich and fenny pool near Beetle Rocks on the south-west of the Island. *Euglesa hibernica* was found in peaty runnels on the western flanks of Knockmore in October 2002.

Euglesa lilljeborgii is declining in Ireland according to Byrne *et al.* (2009) and has been red listed as 'vulnerable'. This has been ascribed to a decline in water quality in its preferred habitats.

Arion owenii
The tawny soil slug or Owen's slug was first recorded from Inishowen in Donegal (Buncrana) and described there, new to science (Davies 1979). It is widespread and locally common in old woodland, roadsides and gardens in Counties Donegal, Tyrone and Londonderry. Elsewhere it is extremely local and has not been found in large areas of east, central and south Ireland. The emerging range suggests a late introduction from Scotland possibly by natural means but much remains undetermined. It has never been reliably recorded outside Britain and Ireland and is therefore considered endemic to this geographical area.

The absence from early records for Clare Island is not surprising considering the above range, but also because it would previously have been either lumped with the *Arion hortensis* group of species or even dismissed as a form of *Arion subfuscus*. It was found in some numbers at the base of wild *Iris* by a stream at Ballytoohy (L697897) in relatively undisturbed surroundings. Unlike the other adventive slugs it is therefore established in semi-natural habitats and may be scattered in suitable places across the island.

Endangered species of the old wood fauna

Although the main area where the old wood fauna (*urwaldtiere*) was known pre-1911 could not be visited, it was thought to be worth underlining the worth of the rarer species in an Irish and wider context using the notes below.

At the same time there is a fauna of commoner woodland species that also seems to be confined to the *Luzula* cliff habitat and was not encountered elsewhere on the Island: *Aegopinella pura*; *Aegopinella nitidula*; *Arion circumscriptus silvaticus*; *Columella aspera*; *Euconulus* cf. *fulvus*; *Trochulus hispidus*; *Vitrea crystallina*. All of these species occur commonly throughout Ireland in a variety of woodland habitats. Their absence in most of the island, even from secondary holly/birch/hazel scrub underlines the long-term disruptive effect of intense cultivation and turf removal on woodland molluscs.

Acicula fusca - point snail
The point snail is listed as 'vulnerable' in Ireland. Usually found in calcareous soil in fragments of ancient woodland or scrub. It was found before 1911 in *Luzula* clumps on the northern slopes of Knockmore. Its status is unknown.

Zenobiellina subrufescens - brown snail
The brown snail is another vulnerable Irish woodland reflict, which is nowadays found mainly in old, landed estates or in steep with dense growths of *Luzula*. Found on the northern slopes of Knockmore pre-1911 but current status unknown.

Limax cinereoniger - ash-grey slug
The ash-black slug is also vulnerable in Ireland but may be locally common on western cliffs as well as

Cochlicella acuta. Photo: R. Anderson

semi-natural woodlands elsewhere. It appears to have undergone a substantial decline in southern Ireland but is stable in northern counties. Its status on the northern slopes of Knockmore is unknown.

Leiostyla anglica - English chrysalis snail

The English chrysalis snail is also classified as 'vulnerable'. A single specimen was found at Portlea north of Maum by Geraldine Holyoak in September 2003, so it may occur sparingly away from the *Luzula* habitat.

Spermodea lamellata - plaited snail

The plaited snail is endangered and confined, like most of the other species discussed here, to the coastal cliff habitat. It is very susceptible to drought and in Ireland has contracted in range dramatically in recent years, probably as a direct result of climate change.

All of these species have suffered disproportionately from man's influence on the landscape. Woodlands were a ready source of fuel for heating and cooking throughout Irish history and a large population, particularly before the Great Famine in the seventeenth and eighteenth centuries, saw most of the remaining wild woodlands disappear. This history is especially evident on Clare Island, which supported a large human population for several centuries and saw all the native woodland disappear at an early stage, resort being made to turf for fuel in the absence of wood.

As indicated above, the woodland fauna of Clare Island is now relict, surviving in patches of wood rush along steep slopes and cliffs on north-facing slopes. Dense patches of wood rush create a humid atmosphere at ground level, an essential survival factor for delicate species such as the plaited snail and the brown snail. The woodland relics are collectively confined to the Atlantic

Zonitoides excavatus. Photo: R. Anderson

fringes of Europe and are declining across their ranges owing to climate change, inappropriate forestry practices and more intensive land use. In the end it may be the relic woodland fauna of Clare Island living in an inaccessible habitat, which is woodland in name only, that will outlive the native forest fauna in most of the rest of Ireland.

Zonitoides excavatus - hollowed glass snail

This is the only obligate calcifuge ('lime-hating') land mollusc in Ireland and absent from the central plain and from eastern counties except for two very isolated colonies in Mourne Park and Rostrevor Forest, Co. Down and a few localities in the Wicklow Hills. It is widespread but uncommon in the west, favouring woodland within areas of acidic bedrock.

Lack of recent records from many areas has led to it being identified as at risk in Ireland with a red list category of 'vulnerable' (Byrne *et al.* 2009). A few specimens were found in and around a small fen and a marine foreshore near The Mill. Both habitats are atypical for the species which often occupies acidic oak or birch woodland elsewhere in Ireland.

REFERENCES

Adams, L.E. 1897 *Paludestrina jenkinsi* Smith in Ireland. *Journal of Conchology* **9**, 15.

Anderson, R. and Rowson, B. 2020 Annotated list of the non-marine Mollusca of Britain and Ireland. Conchological Society of Great Britain and Ireland. Available online at https://conchsoc.org/special_publications.

Byrne, A., Moorkens, E.A., Anderson, R., Killeen, I.J. and Regan, E.C. 2009 Ireland Red List No. 2 – Non-Marine Molluscs. National Parks and Wildlife Service, Department of Environment, Heritage and Local Government, Dublin, Ireland.

Cameron, R.A.D. 2003 Land snails in the British Isles. Aidgap Guide, Field Studies Council Occasional Publication 79, 82 pp. Preston Montford.

Davies, S.M. 1979 Segregates of the *Arion hortensis* complex (Pulmonata: Arionidae) with description of a new species, *Arion owenii. Journal of Conchology* **30**, 123–8.

Glöer, P. 2002 Die Süsswassergastropoden Nord- und Mitteleuropas, Bestimmungsschlüssel, Lebensweise, Verbreitung. Die Tierwelt Deutschlands Volume 73. 327 pp. Hackenheim, Germany. Goecke and Evers; Conch books.

Holyoak, D.T. and Seddon, M.B. 1983 Land Mollusca from Norway, Finland and Sweden. *Journal of Conchology* **31,** 190.

Kerney, M.P. 1999 *Atlas of the land and freshwater molluscs of Britain and Ireland,* 264 pp. Great Horkesley, Essex. Harley Books.

Killeen, I., Aldridge, D. and Oliver, G. 2004 Freshwater bivalves of Britain and Ireland. AIDGAP Guide, Field Studies Council Occasional Publication 82, 114 pp. Preston Montford.

Phillips, R.A. and Watson, H. 1930 *Milax gracilis* (Leydig) in the British Isles. *Journal of Conchology* **19,** 65–93.

Reise, H., Hutchinson, J.M.C., Schunack, S. and Schlitt, B. 2011 *Deroceras panormitanum* and congeners from Malta and Sicily, with a redescription of the widespread pest slug as *Deroceras invadens* n. sp. *Folia Malacologica* **19,** 201–33.

Rowson B., Turner J.A., Anderson, R. and Symondson W.O.C. 2014 Slugs of Britain and Ireland, identification, understanding and control, 136 pp. Telford. Field Studies Council Publications.

Scharff, R.F. 1891 The slugs of Ireland. *Transactions of the Royal Dublin Society* **4** (ser. II), 1–562.

Stelfox, A.W. 1911 A list of the Land and Freshwater Mollusks of Ireland. *Proceedings of the Royal Irish Academy* **29B,** 65–164.

Stelfox, A.W. 1926 *Paludestrina jenkinsi* in Lough Neagh and elsewhere. *Irish Naturalists' Journal* **1,** 174–5.

Stelfox, A.W. and Welch, R.J. 1980 A history of the land and freshwater Mollusca of Ulster. *Proceedings of the Royal Irish Academy* **80B,** 9–152.

Wiktor, A. and Norris, A. 1982 The synonymy of *Limax maculatus* (Kaleniczenko, 1851) with notes on its European distribution. *Journal of Conchology* **31,** 75–7.

CHAPTER 3

THE TARDIGRADA OF CLARE ISLAND

Erica DeMilio, Nigel J. Marley, Colin Lawton and Reinhardt Møbjerg Kristensen

ABSTRACT

The chapter on Tardigrada for the original Clare Island Survey of 1909–11 by James Murray accounts for approximately 70% of all records for the phylum in Ireland to the present day. The importance of Murray's contribution to the study of Irish tardigrades is acknowledged with a biography of the original survey author. The results of a new, small-scale survey of moss-inhabiting, limno-terrestrial Tardigrada on Clare Island are presented and compared to the findings of Murray. The majority of specimens collected during the new survey belonged to the *Macrobiotus hufelandi* C.A.S. Schultze, 1834 species group and displayed a spectrum of morphological variation. Ten other tardigrade species were recorded with two, *Hypsibius pallidoides* Pilato, Kiosya, Lisi, Inshina and Biserov, 2011 and *Isohypsibius sattleri* (Richters, 1902), being reported from Ireland for the first time. Based on the results of the new survey, further investigation into morphological variation in the *M. hufelandi* group present on Clare Island is recommended.

(The authors of this chapter state that references to literature and the taxonomic framework followed was accurate at the time of submission in November 2016.)

Introduction

Tardigrada is a phylum of aquatic, microscopic invertebrates commonly called 'water bears'. Up to the present, the status of Tardigrada has received little attention in Ireland. DeMilio *et al.* (2016) reviewed all known Irish records for the phylum coming from just fourteen references to tardigrades in the approximately one hundred years since their first recording in Ireland (Murray 1911a; Crisp and Hobart 1954; Le Gros 1959; Boaden 1966; Mitchell 1973; Morgan 1974, 1975, 1976, 1980; Morgan and King 1976; Baxter 1979; Kinchin 1990, 1992; Tumanov 2005; Guidetti *et al.* 2015). James Murray, as part of the original Clare Island Survey 1909–11, conducted the first of these studies and this work remains the greatest contribution to the knowledge of Irish tardigrades.

Murray's (1911a) chapter on tardigrades was entitled 'Arctiscoida'. Although the name Tardigrada had been widely used for this group of animals prior to the time of the Clare Island Survey (first by Spallanzani 1776), Murray considered the name to be invalid *sensu* Hay (1906) due to its previous applications to various mammal taxa. Shortly before his work on the Clare Island Survey, Murray (1910a) proposed that tardigrades be recognised within Arthropoda to the level of order under the name 'Arctiscoida', previously suggested by C.A.S. Schultze (1861) as a familial name. At that time all known tardigrade species belonged to a single family, Xenomorphidae Perty, 1835. Even the major division between the classes Heterotardigrada Marcus, 1927 and Eutardigrada Richters, 1926 had not yet been established. Limno-terrestrial species were assigned only to *Milnesium* Doyère 1840, *Diphascon* Plate 1888, *Echiniscus* C.A.S. Schultze, 1840 or *Macrobiotus* C.A.S. Schultze, 1834 until Thulin (1911, 1928) added several new genera. Tardigrada was assigned phylum status by Ramazzotti (1962a) and has since greatly increased in the number of taxa to approximately 1,200 described species and subspecies (Guidetti and Bertolani 2005; Degma and Guidetti 2007; Degma *et al.* 2016).

Murray (1911a) wrote, 'Irish Water-bears appear to have no history', referring to their complete absence in the literature to that point. The majority (approx. 70%) of the known Irish species to date are still those recorded by Murray. In this initial survey Murray (1911a) found thirty-five species of tardigrade from Clare Island and other nearby locations on mainland County Mayo. Four of these species: *Echiniscus columinis* Murray, 1911a, *Echiniscus militaris* Murray 1911a, *Paramacrobiotus richtersi* (Murray, 1911a) and *Murrayon hibernicus* (Murray, 1911a) were new to science. Thirteen sampling points are listed in his report: three on Clare Island itself ('moss', 'Croaghmore [Knockmore]', and 'salt marsh' at Kinacorra), four on Achill Island, one on Inishturk, and on the mainland, two in Castlebar ('moss' and 'lake') and one each at Louisburgh, Belclare and Westport. Murray (1911a) was not descriptive of the sampling locations, or of the sample material itself, some of which was collected by others on his behalf. He does not specify the quantity of sample material gathered at each point, and usually not the number of individuals per sample.

The present small-scale survey of Tardigrada includes sample sites only on Clare Island itself. The authors have collected additional samples from mainland County Mayo as part of a larger all-Ireland survey currently underway. As this study revisited Clare Island at a time approaching the 100th anniversary of Murray's untimely death in February 2014, and due to his place as the first to study tardigrades in Ireland, a short account of his extraordinary life follows.

The life of James Murray (1865–1914)
James Murray contributed two chapters to the original Clare Island Survey: Part 37. Arctiscoida [Tardigrada] and Part 52. Rotifera Bdelloida (Murray 1911a, 1911b). An accomplished, self-taught naturalist, Murray's dedication to the study of these then little-known phyla earned him enduring recognition as a pioneering tardigrade and rotifer expert. Murray was also a courageous explorer in some of the world's harshest environments at a time before the conveniences of modern field equipment.

The son of a grocer, James Murray was born in Glasgow on 21 July, 1865. From an early age he displayed a natural intelligence and aptitude for learning. Details of his early life are revealed in an obituary published following Murray's death in absentia (Macnair 1915). Murray enrolled in medical school, but withdrew in order to study sculpture at the Glasgow School of Art. Following this he entered something of a nomadic period, traveling around Italy and the United States taking work as an architectural sculptor. He started his own sculpting business in Hamilton upon his return to Scotland. Murray became a member of the Natural History Society of Glasgow in 1896. Deeply fascinated by the teeming microscopic world of life associated with bryophytes, he quickly earned a reputation of expertise on rotifers and tardigrades.

Not long after beginning the study of these phyla, Murray was invited to participate in the extensive Bathymetrical Survey of the Freshwater Lochs of Scotland (1902–07). Though he was to find the work physically taxing in its intensity, in this position he described numerous new tardigrade (Murray 1905a, 1907a) and rotifer (Murray 1906, 1908) species and made new ecological observations for both groups. His major contributions to Tardigrada and Rotifera biology during this period earned him the Neil Prize given by the Royal Society of Edinburgh.

In 1907, Murray was offered the position of Biologist on Earnest Shackleton's expedition to the Antarctic following a recommendation by the renowned oceanographer Sir John Murray, a fellow Scotsman of no relation, under whom he had worked on the Loch Survey. The *Nimrod* Expedition 1907–09 (Fig. 1), was the first of three Antarctic expeditions led by Shackleton. Its main aims were to find and reach the exact locations of the geographic and magnetic South Poles for the first time. Murray, a member of the shore party, was assigned command of the base camp at Ross Island, while other members of the expedition

Fig. 1 James Murray (seated) aboard *Nimrod* (1907–09). (Photo: National Library of Australia, nla.obj-140789574)

attempted to reach the Pole. He used this time to study the surrounding Antarctic life, making many interesting observations on lichens, tardigrades, rotifers and emperor penguins, among others. These observations and a general account of camp life can be read in Murray's summary biological report appendix in Shackleton's *The heart of the Antarctic* (1909), and in *Antarctic days*, a volume Murray subsequently co-wrote with fellow expedition member, artist George Marston (Murray and Marston 1913). Despite never reaching the South Pole, the expedition was regarded as successful for setting a new furthest south latitude record and for the collection of other new geographical and scientific data. Murray's contribution to this data was the first substantial study of the biology of Antarctic freshwater lakes and additional observations on physics, oceanography, optics and meteorology (Shackleton 1909 vol. 2).

Living and working for over a year in the Antarctic, Murray proved himself as a capable scientific explorer and this made him a desirable candidate for further expeditions. Murray was invited to join the Bolivia Boundary Commission 1911–13 as the team naturalist. Percy Fawcett, British soldier and explorer, led the expedition, which was to carry out a general survey of the area and to map the courses of rivers that were to form newly agreed borders between Bolivia and Brazil and Bolivia and Peru. The native people and local fauna were also to be studied as part of the survey, as much of the Amazonia region was yet unexplored.

The terrain was extremely difficult to traverse due to thick jungle cover, hot climate, prevalent tropical diseases and parasites, and potentially hostile contact with indigenous peoples. Despite the endurance he displayed while facing the challenging conditions of the Antarctic, the expedition in South America proved to be disastrous for Murray. He suffered from severe exhaustion and illness under the pace set by Fawcett. Murray developed a gangrenous infection in his leg following a superficial wound, which led him to become delirious and unfit to continue with the party.

He was left with a local man in the understanding that Murray was to be taken back to civilization on a mule. Murray's whereabouts however, remained unknown for several weeks during which time he was nursed back to health by a local family before he was well enough to be sent on to La Paz. This experience left Murray outraged, feeling that Fawcett had abandoned him to die, but was dissuaded from taking a lawsuit against him by friends in the Royal Geographical Society. Despite his tribulations with the Boundary Commission, Murray published on the Tardigrada, Rotifera, and other taxa, covering a large portion of the South American continent (Murray 1913a, 1913b, Murray and Wailes 1913). For a further account of the Bolivia Boundary Expedition based upon the diaries of Murray and other party members, see Grann (2009).

Seemingly undeterred from exploration by his near-fatal experience in the jungle, Murray joined another ambitious expedition in 1913, this time to explore the Arctic, where conditions were more similar to those of his successes with Shackleton. The Canadian Arctic Expedition (1913–16) was to be the first multidisciplinary research voyage into the region. Murray, aged 46, was a senior member of the large and distinguished, international scientific staff (Fig. 2).

Vilhjalmur Stefansson, who was later criticized for his questionable preparations and actions during the course of events that were to follow, commanded the expedition. The team was split between the Northern Party, which was to identify and explore new offshore islands, and the Southern Party, which was to conduct scientific studies along the northern Canadian

Fig. 2 *Karluk* leaders and scientific staff prior to sailing from Nome, Alaska, 1913. Front row: Alistair Forbes Mackay, first from left, Captain William Bartlett, second from left, Vilhjalmur Stefansson, third from left, James Murray, second from right. (Photo: Library and Archives Canada, PA-074063)

mainland. Murray should have sailed with the Southern Party, but along with four other scientists he was moved to the *Karluk*, the ship of the Northern party, as she had more room for passengers.

Niven (2000) provides an account of the fate of the Northern party. After departing amid great fanfare, on June 17 from Victoria, British Colombia the expedition was dominated by conflict between Stefansson, the *Karluk*'s Captain Bartlett, the scientific staff and the ship's crew. The scientists quickly became concerned about the behavior and leadership capabilities of Stefansson. During a stopover in Nome, Alaska in July several scientists, including Murray who acted as their spokesman, threatened to resign from the expedition, but were persuaded to continue.

Particularly severe ice conditions occurred that year off the coast of northern Alaska, with an earlier onset than was usual for the season. In August, the *Karluk* became lodged in pack ice and began to drift westerly. The ship was trapped for weeks amid severe weather conditions. Murray passed the time sampling for sea life and examining the unfamiliar fauna under his microscope. By September, Stefansson decided to secretly abandon the ship but told the others he was going only on a short hunting trip along with three other men. Days later a storm hit and caused the ship to drift further into the Arctic Ocean.

At this time, Murray, Alistair Forbes Mackay, who had also served under Shackleton in the Antarctic, along with a small number of others decided to begin preparations to leave the ship to search for land on their own. The *Karluk* remained trapped in ice throughout October and November into the Arctic winter. The trapped ship was dragged further west towards the coast of Siberia. Meanwhile, Stefansson and his companions had successfully reached land and met with the Southern Party, which was dismayed to learn of his abandonment of the others.

Aboard the *Karluk* in the last days of December 1913 the crew believed they could finally see land in the distance. The ship at this time had become very unstable and on 10 January, 1914 it finally cracked open on the ice and began to take on water. The party was forced to evacuate fully on 11 January. Some suffered from exhaustion, frostbite or other injuries. Relations between the Captain, Murray's group, and the other men deteriorated further.

On 31 January Murray and Dr Mackay announced to Captain Bartlett that they were finally going to attempt to reach land. They asked for 50 days' worth of supplies but declined to take their share of the sled dogs to help carry their equipment. Captain Bartlett requested that Murray and the three men who wished to accompany him produce a written and signed statement that they left on their own free will, absolving him of any further responsibility of their fate.

On 5 February Murray and his group set off on their attempt to reach land. That day they were seen by two of the other men who were on their way back to camp after a reconnaissance foray. Murray's group was seen to be making only slow progress and the men were struggling to carry their gear. On the group's tenth day away from camp they were seen again by other crewmembers out scouting. Murray and his companion's condition had worsened due to exhaustion and exposure. They had lost or discarded much of their supplies. The two other men traveling with Murray and Dr Mackay were on the verge of death, but the men refused to return to the main camp and insisted on pushing forward towards land. This was the last time Murray and his companions were seen and their remains were never found. It is believed that they never reached land. In total seventeen men died during the Canadian Arctic Expedition, eleven of these sailed on the *Karluk*.

Over the course of his career, Murray described one new genus (*Oreella* Murray, 1910b) and 60 new species and subspecies of tardigrade, 54 of which are considered valid (Degma *et al.* 2016). The genus *Murrayon* Bertolani and Pilato, 1988 and family Murrayidae Guidetti, Rebecchi and Bertolani, 2000 are named in recognition of his contributions to the study of the phylum.

Materials and methods
Field sampling
Sample collection was carried out on 20 September, 2013. One sample was taken at each of twelve sampling stations, which were selected to include various microhabitats occurring on Clare Island (Table 1, Fig. 3). Cryptogams were collected from

Table 1
Description of sampling points and summary of sample contents.
N= number of adult individuals, T= number of tardigrade taxa (including morphotypes) present.

Sample	Sampling point description	N	T
1	Western cliffs: Sheer cliff, west end of island, approx. 60m high, acrocarpous moss on exposed rock face.	1	1
2	Western bog: West end of island close to the signal tower, lowland bog with stagnant pools at the western base of Knockmore, *Sphagnum* sp. on soil, 20cm from a bog pool.	2	1
3	Knockmore slope: West-facing lower slope of Knockmore, pleurocarpous moss on exposed rock face.	31	7
4	Knockmore cascade 1: Lotic cascade approx. 3m high, located at base of Knockmore, aquatic moss on rock 2.5m from foot of cascade, fully submerged at time of sampling.	0	0
5	Knockmore cascade 2: Lotic cascade approx. 3m high, located at base of Knockmore, aquatic moss on rock at base of cascade, fully submerged at time of sampling.	0	0
6	Ballytoohy Beg: Wet grassland at riverbank, area of high grass among sheep grazing pastureland, pleurocarpus moss on soil.	47	7
7	Dorree (Mill) River: 1.5m into river from bank, aquatic moss on rock, permanently submerged.	0	0
8	Knocknaveen slope: Southeast facing lower slope of Knocknaveen, acrocarpous moss on exposed rock.	9	2
9	Apple Tree: Apple tree sheltered by two other apple trees and high shrubs, pleurocarpous moss on south-facing trunk.	32	6
10	Harbour: Rocky outcrop opposite harbour, sample material pleurocarpous moss on rock approx. 1m from tidal maximum with groundwater drainage flowing over sample, partially submerged.	0	0
11	Kinnacorra 1: East island, saltmarsh, pleurocarpous moss on soil among high grass.	4	2
12	Kinnacorra 2: East island, saltmarsh, pleurocarpous moss on soil among high grass, approx. 7m from Kinnacorra 1.	5	3

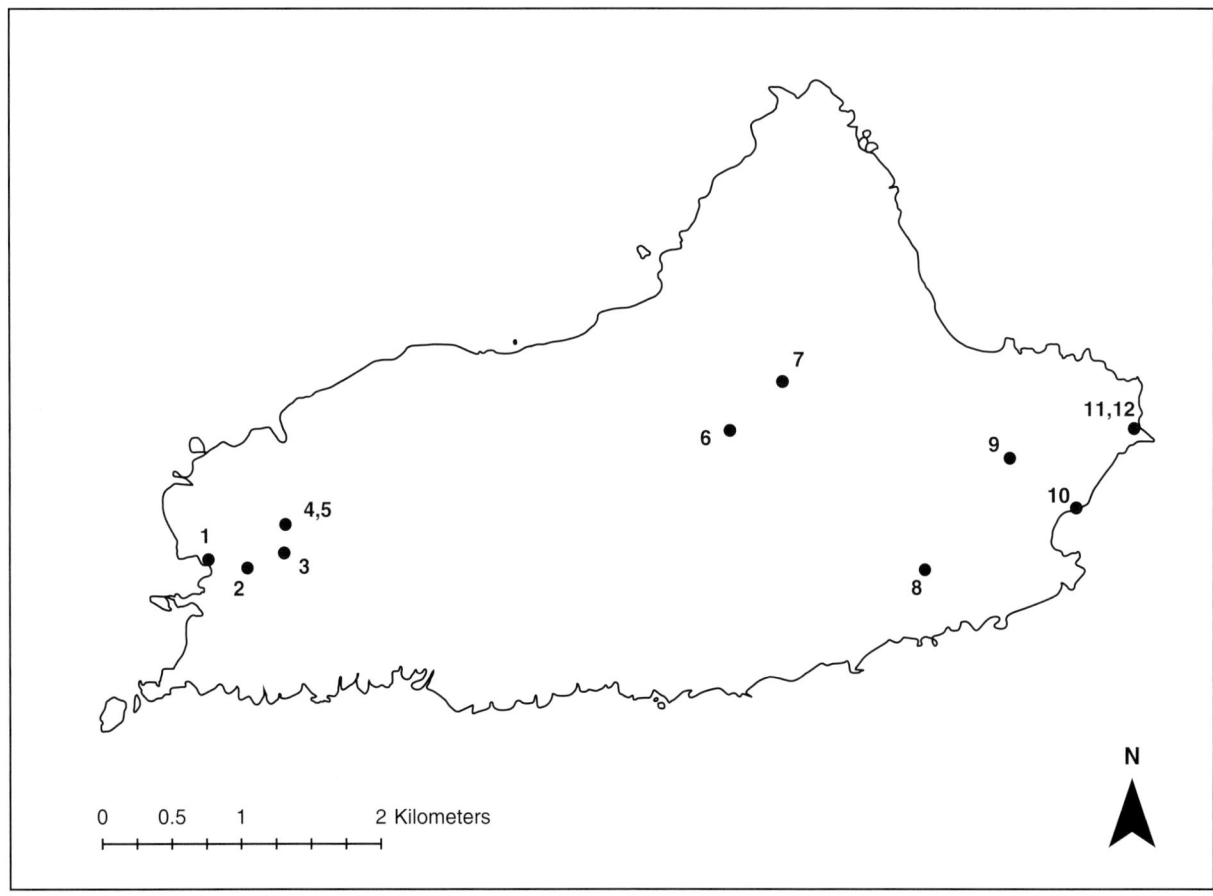

Fig. 3 Clare Island sampling sites, current survey

freshwater courses, exposed rock surfaces, grass covered ground, peatland and tree trunks. Two stations, the southern slope of Knockmore and the salt marsh area at Kinacorra were selected in attempt to replicate as closely as possible the locations sampled by Murray. Samples of host moss or lichen were taken down to the substratum surface within a 4 x 4cm^2 quadrat using forceps. Samples were stored in brown paper envelopes and allowed to air dry.

Laboratory analysis

Samples were soaked in spring water for 24 hours prior to analyses. Anoxybiosis was induced to allow specimens to fall free from the sample matrix. Sample material was vigorously shaken for three minutes, squeezed, sieved and collected on 36µm mesh. Eggs and exuvia were retained when present to aid in species identification. Individuals, eggs and exuvia were slide mounted in polyvinyl alcohol. These were observed and identified using DIC imaging on an Olympus BX51 with oil immersion up to 1000× with a 2× magnification changer or on an Olympus BX53 with 1000× oil immersion. Measurements were taken using Olympus cellSens imaging software (Standard Version 1 CS-ST-V1). Body length measurements for individual specimens were taken excluding the fourth pair of legs. Species identifications were made by comparisons with the original taxonomic descriptions. Other relevant literature used in the identification of species is given below.

Results

Tardigrades were recovered from eight out of twelve samples (zero freshwater, eight terrestrial). In total, 170 individuals and nine eggs were extracted from all sample material. Of these 170 individuals, 39 were not analysed further due to lack of adult characteristics or accidental damage during processing, resulting in a total of 131 identified adult individuals. The sample sites and a summary of contents of samples are given in Table 1.

Notes on species

Heterotardigrada Marcus, 1927

Echiniscoidea Richters, 1926

Echiniscidae Thulin, 1928

Echiniscus quadrispinosus quadrispinosus **Richters, 1902**

Seven adult specimens. Body length 169–264µm. Gonopore structure visible on three individuals only, all female. Orange colour. Red eyes present. Double granulation ('*quadrispinosus* type') structure clearly visible, with characteristic cross-shaped zones of clearing on scapular and terminal plates. Scapular plate with small accessory plates. Internal claws with basal spurs. Dentate collar on leg pair IV with either eight or nine teeth. All individuals possess lateral filaments A, B, C, D and E, expect for one individual (body length 169µm) that lacked B on both sides, and C and E from one side. This specimen also possessed supplementary plates so is not treated as the subspecies *cribrosus* Murray, 1907a, which lacks both filament B and supplementary plates. All specimens with spines at Cd and Dd, with Cd longer that Dd.

Three larvae, body lengths 141–155µm, with two claws per leg were found in the same sample. Each morphologically similar, possessing lateral filaments A and E only and Cd and Dd with Cd longer than Dd. Dentate collar with eight teeth present on leg pair IV. Larvae excluded from count data.

Clare Island distribution: Sample 9

Pseudechiniscus suillus **(Ehrenberg, 1853)**

Three specimens. Body length 192–200µm. All female. Red-orange colour. Black eyes present. Mean length of internal cirrus, 7.4µm and external cirrus, 12.7µm. Cephalic papilla dome shaped, length 3.8µm and conical clava, 6.5µm (both suitable for measurement on one specimen only). Cirrus A mean length 29.2µm, without bifurcated tip. No other lateral or dorsal appendages present. Plate margins indistinct. All plates without lobes or projections. Terminal plate not facetted, without notches. Fine granulation uniformly distributed on and between dorsal plates. Ventral cuticle with bands of granulation arranged in an irregular network similar to that reported by Dastych (1984). Basal portion of all legs with patch of fine granulation. Leg pair IV papilla small. Dentate collar absent. Claws small, 6.2–7.0µm long. Internal claws of all legs with minute basal spur.

As a result of the lack of detail in the original description, varying morphologies have been reported for *P. suillus* (e.g. Thulin 1911;

Franceschi 1952; Morgan and King 1976; Ramazzotti and Maucci 1983; Dastych 1984; Dastych 1988; Miller *et al.* 1994). Additionally, there exists a complex of cryptic species known as the *suillus* group. These species are often difficult to separate and require revision (Fontoura and Morais 2011). The Clare Island specimens do not match well with the descriptions of any of the other similar species in the *suillus* group and fit within the broad limits of *Pseudechiniscus suillus (senso lato)*. A modern redescription of *P. suillus* is necessary. Following this, previous records including this from Clare Island, should be re-evaluated.

Clare Island distribution: Sample 3

Eutardigrada Richters, 1926

Parachela Schuster, Nelson, Grigarick and Christenberry, 1980

Hypsibioidea Pilato, 1969 in Sands, McInnes, Marley, Goodall-Copestake, Convey and Linse, 2008

Hypsibiidae Pilato, 1969

Diphasconinae Dastych, 1992

Diphascon chilenense **Plate, 1888**
Three specimens. Body length 213–319µm. Eyes absent. Cuticle smooth. Granular macroplacoids almost equal in size but with a slight increase in length from first to third. Miniscule, round microplacoid present. Evident septulum and pharyngeal apophyses. Other features of the buccal-pharyngeal apparatus and claw structure of the Clare Island examples fit well with the morphometric data given for *D. chilenense* by Pilato (1987) and with the corrected diagnosis for the species by Pilato and Binda (1998).

DeMilio *et al.* (2016) recommended that Murray's (1911a) record for *D. chilenense* from Clare Island be confirmed because of the brevity of the original description of *D. chilenense* and the possibility of historical confusion with other similar species. *D. chilenense* shares some features to members of the *Diphascon pingue* group defined by Fontoura and Pilato (2007) as *Diphascon* species possessing a smooth cuticle, three rod-shaped macroplacoids, a microplacoid, septulum and lacking lunules, cuticular thickenings on the legs, and indentation of the claw bases. The Clare Island specimens could be distinguished from other species of the *pingue* group by the short, rounded, granular placoids in a shorter placoid row. Among other *Diphascon* species that can be considered to have granular macroplacoids, the Irish *D. chilenense* specimens could be distinguished most apparently from: *D. langhovdense* (Sudzuki, 1964) by its lack of septulum; *D. puniceum* (Jennings 1976) by its granulation of the caudal cuticle and lack of septulum; *D. sanae* Dastych, Ryan and Watkins, 1990 by its lack of the microplacoid, *D. stappersi* Richters, 1911 by the presence of eyes and its lack of septulum, and *D. tenue* Thulin, 1928 by its lack of pharyngeal apophyses, microplacoid and septulum.

Despite the great geographic distance separating the populations no morphological difference could be found between the description of South American *D. chilenense* material of Pilato and Binda (1998) and the Clare Island specimens. Genetic analyses could be useful in confirming the cosmopolitan distribution of this species.

Clare Island distribution: Sample 6

Hypsibiinae Pilato, 1969
Hypsibius dujardini **(Doyère, 1840)**
One specimen. Body length 192µm. Eyes present. Two rod-shaped macroplacoids, the first with a median constriction, the second more posteriorly constricted. Microplacoid absent. No cuticular bars on legs pairs I–III. Cuticular bar present between the anterior and posterior claws of leg pair IV. This feature, the placoid morphology, the presence of eyes and a large septulum indicate *dujardini* (*senso lato*) itself rather other similar species in the *dujardini* group (Miller *et al.* 2005). Clare Island distribution: Sample 12

Hypsibius pallidoides **Pilato, Kiosya, Lisi, Inshina and Biserov, 2011**
Two specimens. Body length 212 and 225µm. Eyes present. Cuticle without granulation. Buccal tube narrow (1.5µm). Pharyngeal bulb round with apophyses, two macroplacoids, the first a short rod with a median constriction, the second more granular in shape, length 2:3 to the first, and a very small, scarcely visible septulum. Primary branches of all external and internal claws with thin accessory points. Cuticular bars absent on all legs.

H. pallidoides was described by Pilato *et al.* (2011) as having external claws similar to those of *H. pallidus* Thulin, 1911 (redescribed along with the close species *H. microps* Thulin, 1911 by Kaczmarek and Michalczyk (2009) from the type material). In order to aid in the identification of the Clare Island specimens these were compared to the type material of *H. pallidus* at the Zoological Museum, Natural History Museum of Denmark, Copenhagen. The external claw morphology of the Clare Island examples was found to be of the *H. pallidus* type. The combination of this type of external claw morphology and the small dot-like septulum of *H. pallidoides* particularly distinguish it from other similar species of *Hypsibius* as discussed by Pilato *et al.* (2011) and Pilato *et al.* (2012). Clare Island distribution: Sample 6

Itaquasconinae Bartoš in Rudescu, 1964
Adropion scoticum scoticum **(Murray, 1905b)**
Five specimens. Body length 209–399μm. Eyes absent. All specimens typical of the species. Ratio of pharyngeal bulb length to width is approx. 2:1. Cuticular bars are present on leg pairs I–III between the internal and external claw and behind the internal claws, also on leg pair IV between the anterior and posterior claws. Clare Island distribution: Sample 11, Sample 12

Isohypsibioidea Sands, McInnes, Marley, Goodall-Copestake, Convey and Linse, 2008

Isohypsibiidae Sands, McInnes, Marley, Goodall-Copestake, Convey and Linse, 2008

Isohypsibius sattleri **(Richters, 1902)**
Three specimens. Body length 117–224μm. Eyes not observed. Two macroplacoids, the first being so constricted at the median point as to appear as three small, granular macroplacoids, particularly in the smallest individual. No microplacoid. Fits well with redescription by Dastych (1991). All claws with lunules, similar in size on each leg. No tubercles on the legs. Laterally on the body at each pair of legs is a pointed projection bearing inconspicuous, fine hairs. Clare Island distribution: Sample 11, Sample 12

Macrobiotoidea Thulin, 1928 in Sands, McInnes, Marley, Goodall-Copestake, Convey and Linse, 2008

Macrobiotidae Thulin, 1928
Macrobiotus crenulatus **Richters, 1904**
Two specimens. Body length 187 and 380μm. Eyes not observed, otherwise the morphology of these specimens is in full accordance with the redescription of the species and the distinctive combination of characters discussed by Binda (1988).

Clare Island distribution: Sample 2

Macrobiotus hufelandi **C.A.S. Schultze, 1834 group species**
The majority of specimens (89/131) were identified as *Macrobiotus* species falling within the *hufelandi* group. Among these specimens subtle morphological variation was observed in the following features: buccal tube length, buccal tube width, stylet support insertion point, shape of the pharyngeal bulb, presence and pattern of granulation on leg pair IV, respective size of lunules on pair IV, and the shape, size and arrangement of cuticular pores. It could not be determined in all cases if this variation was intraspecific in nature, therefore these features were used to group the specimens into morphotypes referred to '*Macrobiotus* morphotypes 1–9' (Figs. 4–8). This was done in order to account for the variation observed among the specimens belonging to the *Macrobiotus hufelandi* group, as the morphotypes could not be confidently attributed to any described *Macrobiotus* species following comparisons with the literature.

A full morphometric study of the collected *Macrobiotus* morphotypes was not within the scope of this faunistic survey. Instead, a basic comparative approach was taken. Three representative specimens of each morphotype were selected at random for morphometric measurement of selected characters in μm (Table 2): body length (excluding leg pair IV), buccal tube length (the anterior margin of the stylet sheath to tube end excluding apophysis), stylet support insertion point (anterior point of stylet insertion along the buccal tube), buccal tube width (diameter from the external walls of the buccal tube at the point of insertion of the stylet supports), and length of the posterior claws of leg pair IV (primary branch length to claw base excluding lunules). *Macrobiotus* morphotype 6 was an exception as the only two specimens of this type were collected. For purposes of comparisons between *Macrobiotus* morphotypes and comparisons of morphotypes to described *Macrobiotus* species *pt* values (i.e., the ratio of a length of a structure to the length of the buccal tube given as a percentage, after Pilato 1981) were calculated for the selected characters.

Morphotype 1   Morphotype 2

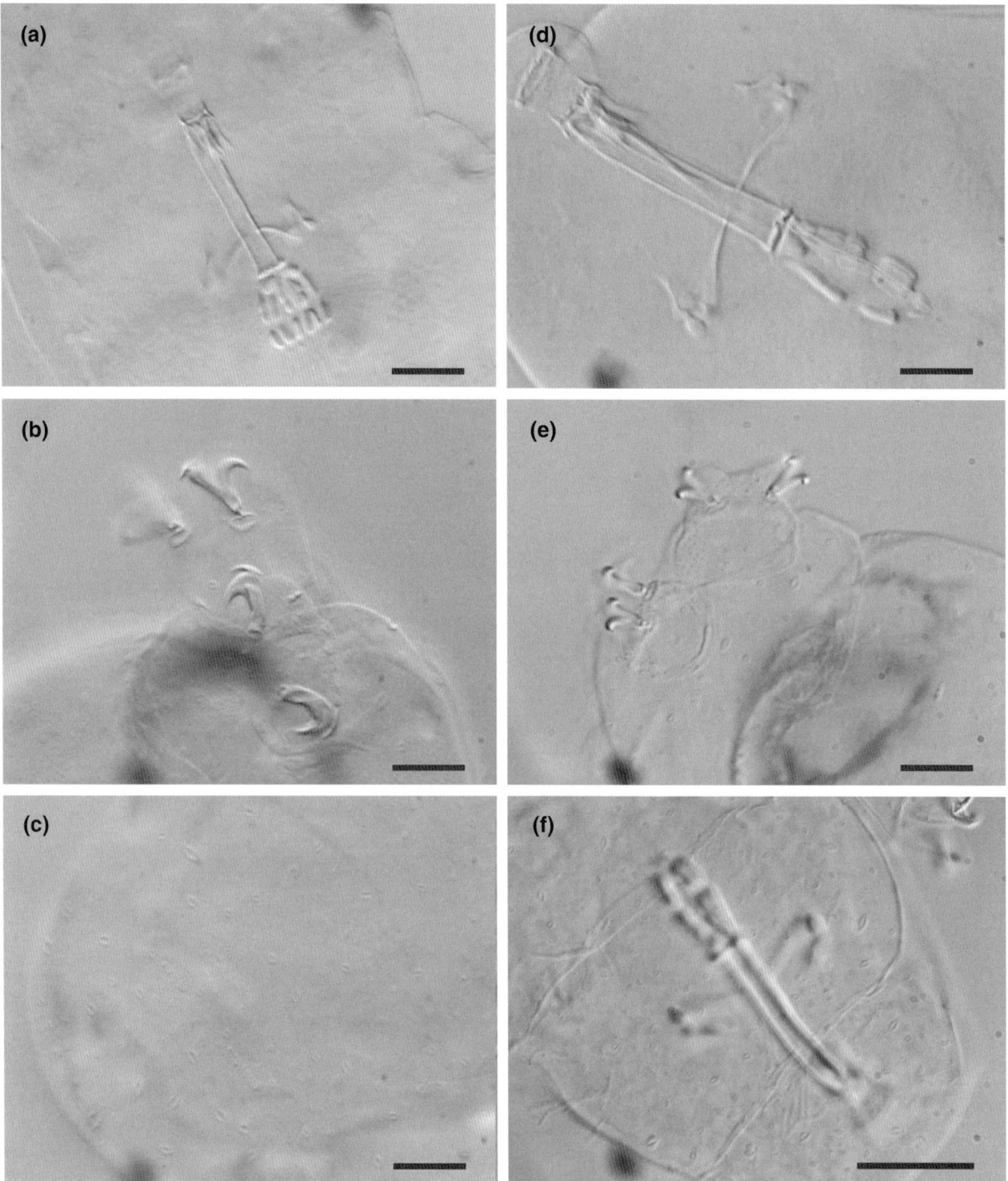

Fig. 4 *Macrobiotus* morphotype 1 (a–c) and morphotype 2 (d–f). a, d buccal-pharyngeal apparatus; b, e claws of leg pair IV; c, f cuticular pores. Scale bar = 10μm (Photos: the authors)

Macrobiotus morphotype 1 (Figs. 4a–c)

Ten specimens. Body length 171–342μm. Eyes not observed. Buccal tube narrow. Buccal armature weak with no bands of teeth visible. Dorsal and ventral transverse ridge systems thin and indistinct. Pharyngeal bulb round. Macroplacoid row passes midline of bulb, microplacoid indistinct. Cuticle with large, abundant, round and elliptical pores up to 3μm diameter. Lunules of leg pair IV crenate, small, and narrow. Granulation on leg pair IV coarse but sparse behind the claws, between the legs, and external surface of legs. Clare Island distribution: Sample 3, Sample 9

Morphotype 3   Morphotype 4

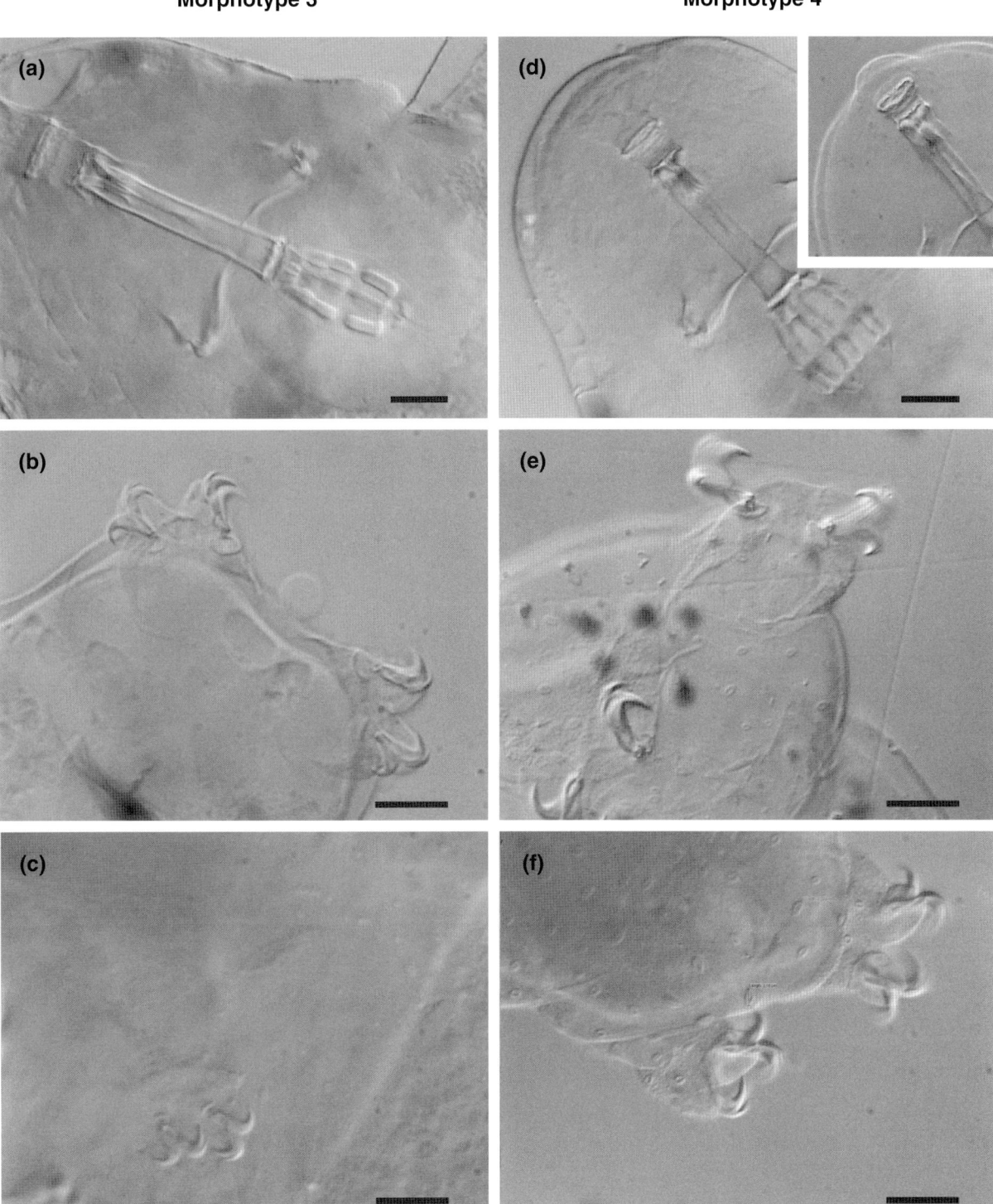

Fig. 5 *Macrobiotus* morphotype 3 (a–c) and morphotype 4 (d–f). a, d buccal-pharyngeal apparatus; b, e claws of leg pair IV; c, f cuticular pores. Scale bar = 10µm (Photos: the authors)

Macrobiotus morphotype 2 (Figs. 4d–f)

Eleven specimens. Body length 209–251µm. Eyes present. Anterior and posterior bands of teeth not visible. Three distinct dorsal and three ventral ridges, medio-ventral ridge rounded. Pharyngeal bulb round. Macroplacoid row short, microplacoid distinct. Cuticle with abundant round and elliptical pores up to 3µm diameter but often less. Lunules of leg pair IV small, narrow and weakly dentate. Granulation on leg pair IV coarse behind the claws, between the legs, and external surface of legs. Clare Island distribution: Sample 3, Sample 9

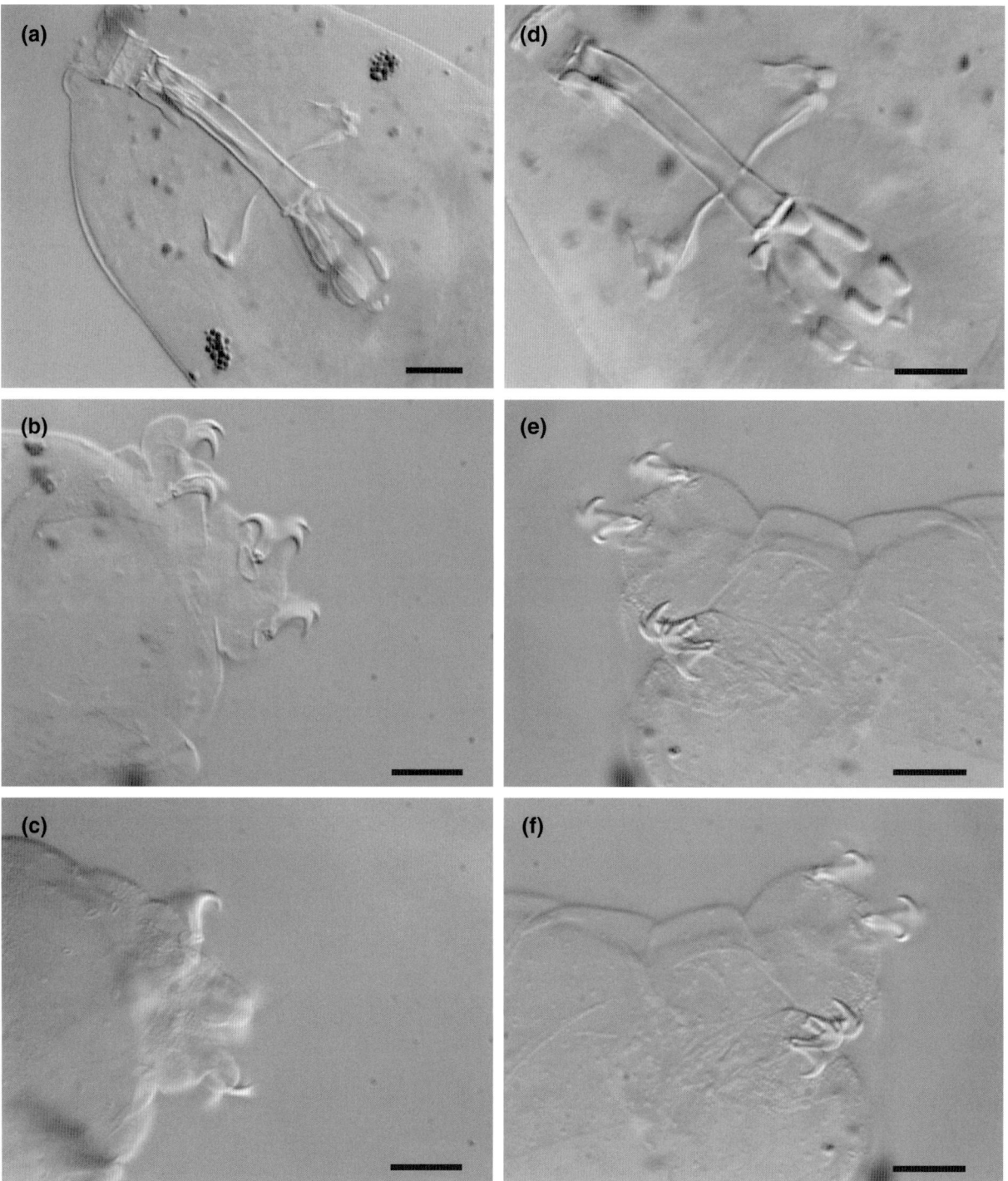

Fig. 6 *Macrobiotus* morphotype 5 (a–c) and morphotype 6 (d–f). a, d buccal-pharyngeal apparatus; b, e claws of leg pair IV; c, f cuticular pores. Scale bar = 10µm (Photos: the authors)

Macrobiotus morphotype 3 (Figs. 5a–c)

Twenty-two specimens. Body length 212–393µm. Eyes present. Anterior band of teeth not visible, posterior band faintly visible, or clearly visible in larger specimens. Three dorsal and three ventral ridges, medio-ventral ridge shaped as a flattened oval. Pharyngeal bulb oval. Macroplacoid row passes midline of bulb, microplacoid small but distinct. Cuticular pores scarce, round and small ≤2µm. Claws of leg pair IV short. Lunules of leg pair IV large, wide and with smooth margins. Granulation on leg pair IV fine and diffuse behind

Morphotype 7 | Morphotype 8

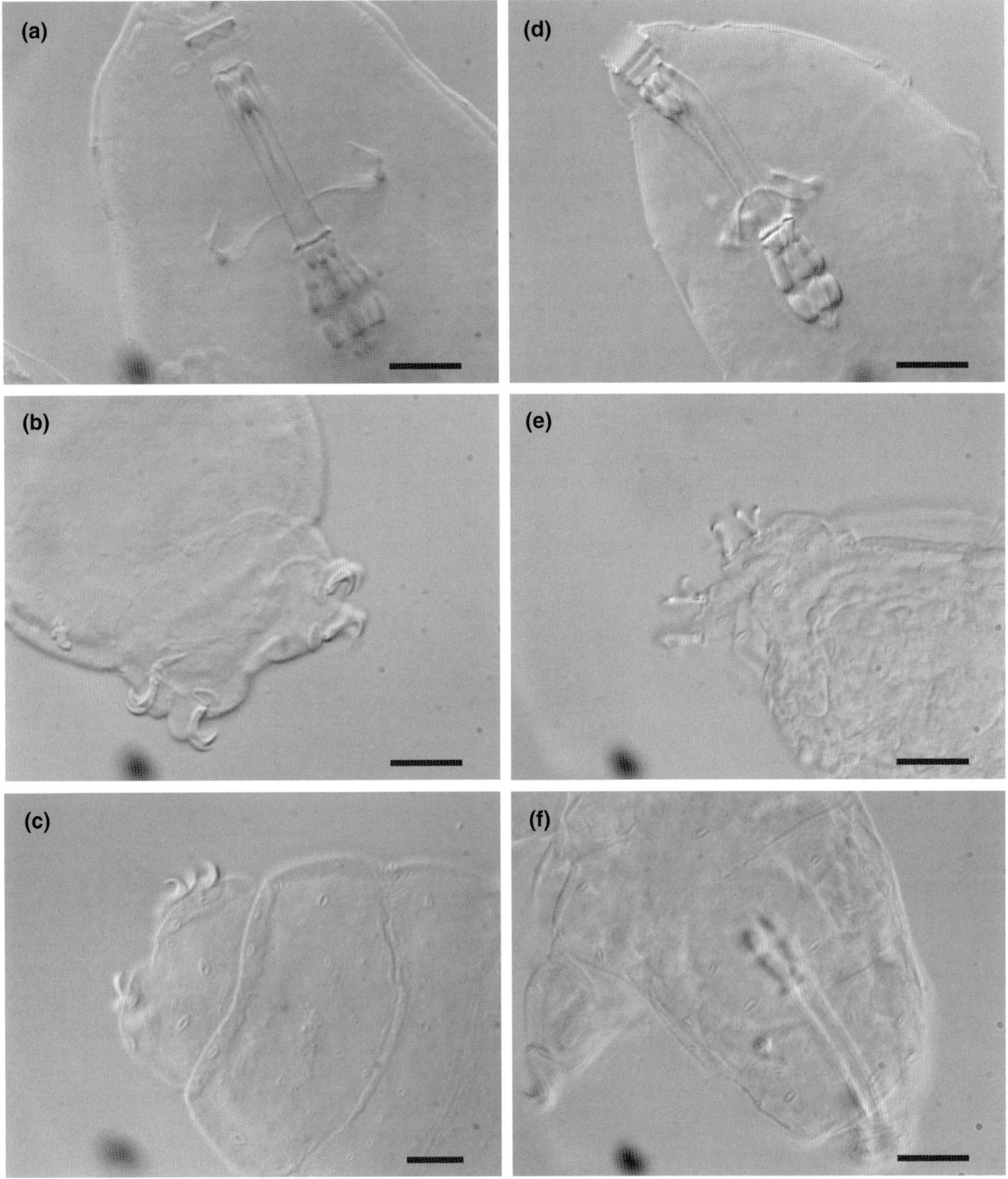

Fig. 7 Macrobiotus morphotype 7 (a–c) and Type morphotype 8 (d–f). a, d buccal-pharyngeal apparatus; b, e claws of leg pair IV; c, f cuticular pores. Scale bar = 10μm (Photos: the authors)

claws only, not present between leg pair IV. Clare Island distribution: Sample 3, Sample 6

***Macrobiotus* morphotype 4 (Figs. 4d–f)**
Thirteen specimens. Body length 200–400μm. Eyes present. Anterior band of teeth not visible, posterior band of teeth is faintly or clearly visible in larger specimens. Three dorsal and three ventral ridges, medio-ventral ridge rounded. Pharyngeal bulb round. Macroplacoid row passes midline of bulb, distinct microplacoid. Cuticle with distinct, abundant, round and elliptical pores of medium

Morphotype 9

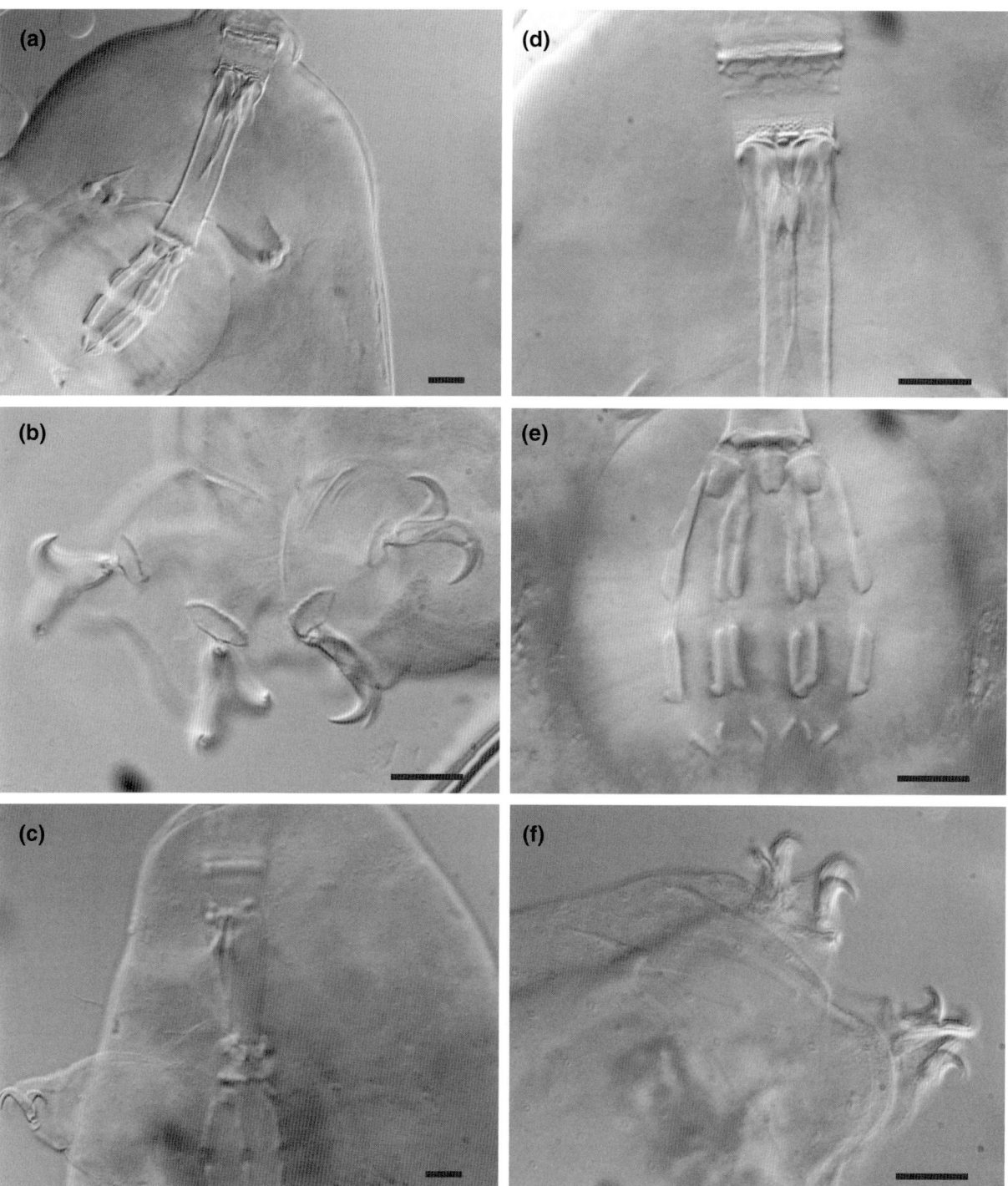

Fig. 8 *Macrobiotus* morphotype 9. a, d, e buccal-pharyngeal apparatus; b claws of leg pair IV; c, f cuticular pores. Scale bar = 10µm (Photos: the authors)

size, <2.5µm. All specimens with a ring of small round pores approx. 1.3µm diameter surrounding the buccal aperture. Claws of leg pair IV long. Lunules medium size and width, dentate. Granulation on leg pair IV coarse but diffuse over both the external surfaces of the legs, between the legs and sparse behind and in front of claws. Clare Island distribution: Sample 3, Sample 9

Macrobiotus morphotype 5 (Figs. 6a–c)
Four specimens. Body length 255–334µm. Eyes present. Both anterior and posterior bands of teeth

Table 2
Summary of selected measurement data (μm) for *Macrobiotus* morphotype characters. BL= body length, BTL= buccal tube length, BTW= buccal tube width, SSIP= stylet support insertion point, Claw IV= length of posterior claw on leg IV.

Morphotype	1	2	3	4	5	6	7	8	9
BL									
Range	171–342	209–327	212–393	200–400	255–334	314–346	176–394	168–360	319–572
Mean/SD	283/50	278/61	304/45	342/51	290/33	330/23	260/78	236/61	481/99
BTL									
Mean/SD	35.7/10.2	38.3/8.9	43.3/12.9	35.0/4.4	33.3/4.7	36.5/2.1	36.3/2.1	30.7/3.1	39.7/15.3
SSIP									
Mean/SD	27.3/9.1	27.3/7.1	34.3/9.3	26.3/2.9	26.7/4.2	29.5/2.1	30.0/2.0	19.7/2.1	32.7/12.7
pt	75.8	71.6	79.6	75.3	79.9	80.8	82.6	64.7	82.26
BTW									
Mean/SD	4.0/0.0	3.67/6	6.1/0.1	5.2/0.3	5.3/0.6	5.5/0.7	4.7/0.6	3.0/0.5	8.2/2.8
pt	12.0	10.12	14.6	14.9	16.7	15.0	12.9	9.8	20.98
Claw IV									
Mean/SD	7.0/1.8	8.3/2.7	5.5/0.6	11.3/0.6	7.0/1.1	7.2/0.1	6.1/0.1	5.5/0.8	12.9/1.4
pt	19.8	20.9	13.3	31.8	21.2	19.21	16.55	16.41	38.25

faintly visible in the largest specimens. Three dorsal and three ventral ridges, medio-ventral ridge elongated. Pharyngeal bulb slightly elongated to oval. Macroplacoid row passes midline of bulb, small microplacoid. Cuticle with large and elliptical pores up to 3μm with largest concentrated in the dorso-caudal region, ventral side with smaller, less distinct round pores. Claws of leg pair IV long. Large, wide and smooth lunules. No granulation behind claws or between legs of leg pair IV. Coarse granulation covers external surface of leg pair IV extending onto the dorso-caudal region. Coarse granulation on dorsum in nine horizontal bands with brown pigment separated by areas free from granulation. Clare Island distribution: Sample 6

Macrobiotus **morphotype 6 (Figs. 6d–f)**
Two specimens. Body length 314 and 346μm. Eyes present. Anterior band of teeth not visible, teeth only faintly visible in posterior band. Three dorsal and three ventral ridges, medio-ventral ridge elongated. Pharyngeal bulb is round. Macroplacoid row narrowly passes midline of bulb, distinct microplacoid. All cuticular pores are small, <2.5μm diameter, faint, and are abundant caudally, with only few elliptical pores present. Claws of leg pair IV medium sized. Lunules leg pair IV long, narrow, and distinctly dentate with teeth of irregular length. Coarse granulation on leg pair IV present behind claws and more diffuse between the legs, on external surface of leg and faintly in front of claws. Clare Island distribution: Sample 3

Macrobiotus **morphotype 7 (Figs. 7a–c)**
Fifteen specimens. Body length 176–394μm. Eyes present. Buccal armature weak, contains no visible bands of teeth or only a few teeth faintly visible in the posterior band position. Three dorsal and three ventral ridges, medio-ventral ridge rounded. Pharyngeal bulb round. Macroplacoid row passes midline of the bulb, microplacoid distinct. Cuticle with large round and elliptical pores typically >2.5μm, more plentiful and distinct on dorsum, few present ventrally. Claws of leg pair IV short. Lunules of leg pair IV are large, wide, and with smooth margins. No granulation observed on leg pair IV. Clare Island distribution: Sample 6

Macrobiotus **morphotype 8 (Figs. 7d–f)**
Seven specimens. Body length 168–360μm. Eyes not observed. Buccal armature weak with no visible bands of teeth. Dorsal and ventral transverse ridge systems indistinct. Pharyngeal bulb slightly elongated. Macroplacoid row passes midline of bulb, microplacoid small. Cuticle with abundant, large elliptical pores up to 3μm diameter, smaller round pores, less distinct, more rarely present. Claws of leg pair IV short. Lunules small, narrow and weakly dentate. Leg pair IV with sparse but coarse granulation behind claws only. Clare Island distribution: Sample 3, Sample 8, Sample 9

Macrobiotus **morphotype 9 (Figs. 8a–f)**
Five specimens. Body length 319–572μm. Eyes present. Buccal tube short and wide. Stylet support insertion point low on the buccal tube.

Buccal armature robust with both anterior and posterior bands of teeth clearly visible. Three dorsal and three ventral transverse ridges, medioventral ridge of flattened oval shape. Pharyngeal bulb oval. Macroplacoid row passes midline of bulb. First macroplacoid with slight anterior constriction and stronger median constriction. Second macroplacoid subterminally constricted. Microplacoid distinct but slender. Cuticle with abundant, small round pores <1.5μm. Claws of leg pair IV long with large, wide lunules with frayed margins. Fine diffuse granulation only on the external surfaces of leg pair IV. Clare Island distribution: Sample 9

Macrobiotus hufelandi group eggs (Fig. 9)

Nine *hufelandi* group eggs were collected during the current survey from two of the four samples that contained *Macrobiotus* morphotypes (Samples 3 and 9), but it was not possible to match them with certainty to any particular adult morphotype. The nine eggs collected varied in egg diameter, process shape and height, and diameter of basal and distal discs. Three types of egg could be distinguished. Many *hufelandi* group species share similar egg morphologies to those collected from Clare Island. For each egg type we have discussed below similarities to described *hufelandi* group species.

Egg Type 1 (Fig 9a,b)

Two examples, Sample 9. Diameter not including processes 72 and 80μm. Egg surface reticulated. Reticulation tight with mesh only slightly larger in the peribasal ring than in areas between processes. Processes smooth and shaped as short, inverted goblets. Mean process height 3.2 and 4.2μm. Diameter of process basal discs slightly larger than distal disc diameter. Mean basal disc diameter 4.7μm for both examples. Mean distal disc diameter 2.5 and 3.5μm. Distal disc margins dentate.

Egg Type 1 is somewhat similar in general appearance to descriptions of the eggs of: *M. almadi* Fontoura, Pilato and Lisi, 2008 in diameter and the egg surface reticulation having larger mesh around the processes but the teeth of distal disc margins of the eggs of *M. almadi* are small and indistinct and the process lengths and basal disc diameter are lower for Egg Type 1 and *M. martini* Bartels, Pilato, Lisi and Nelson, 2009 in process shape and the appearance of the reticulation of the egg surface but the processes of Egg Type 1 are never convex as in *M. martini*. There is also a superficial resemblance to the eggs of *M. terminalis* Bertolani and Rebecchi, 1993 in the process shape and the egg surface reticulation having larger mesh around the processes but the mesh network is thinner in *M. terminalis* and the margins of the distal discs are jagged, not dentate in *M. terminalis*. However, the most similar to eggs of this type are those of *M. hufelandi hufelandi* in the egg diameter, the shape of the processes, the basal disc being larger in diameter than the distal disc margin, the dentate margins of the distal discs and the appearance of the egg surface reticulation.

Egg Type 2 (Fig 9c,d)

Three examples, Sample 3. Diameter not including processes 56–75μm. Egg surface reticulated with larger mesh than Egg Type 1 with largest areas of mesh present in areas between the egg processes. Processes smooth and inverted goblet shaped. Diameter of process bases equal or nearly equal to distal dish diameter. Distal disc margins appear smooth from certain angles but are ragged, never dentate.

Egg Type 2 is similar in general appearance to descriptions of the eggs of *M. iharosi* Pilato, Binda and Catanzaro, 1991 and *M sapiens* Binda and Pilato, 1984 in the shape of the processes and that the basal and distal disc diameters are nearly equal but both of species have eggs with distal disc margins dentate unlike the ragged margins of the discs of Egg Type 2. Also, the processes of Egg Type 2 are not spaced as closely as those figured for *M. iharosi*. The distal dis margins are of similar shape to those of the eggs of *M. sandrae* Bertolani and Rebecchi, 1993 and *M. vladimiri* Bertolani, Biserov, Rebecchi and Cesari, 2011 but the reticulation surface of eggs of Type 2 is not uniform or with the mesh size surrounding the processes of larger size than the intermediate area as described for *M. sandrae* type eggs and *M. vladimiri* respectively. Clare Island Egg Type 2 also differs from the eggs of these species in having shorter processes 3.0–4.5μm compared to 4.3–7.6μm in *M. sandrae* and 6.5–8.0μm for *M. vladimiri*. No described *hufelandi* group species egg was determined to match well to Clare Island Egg Type 2.

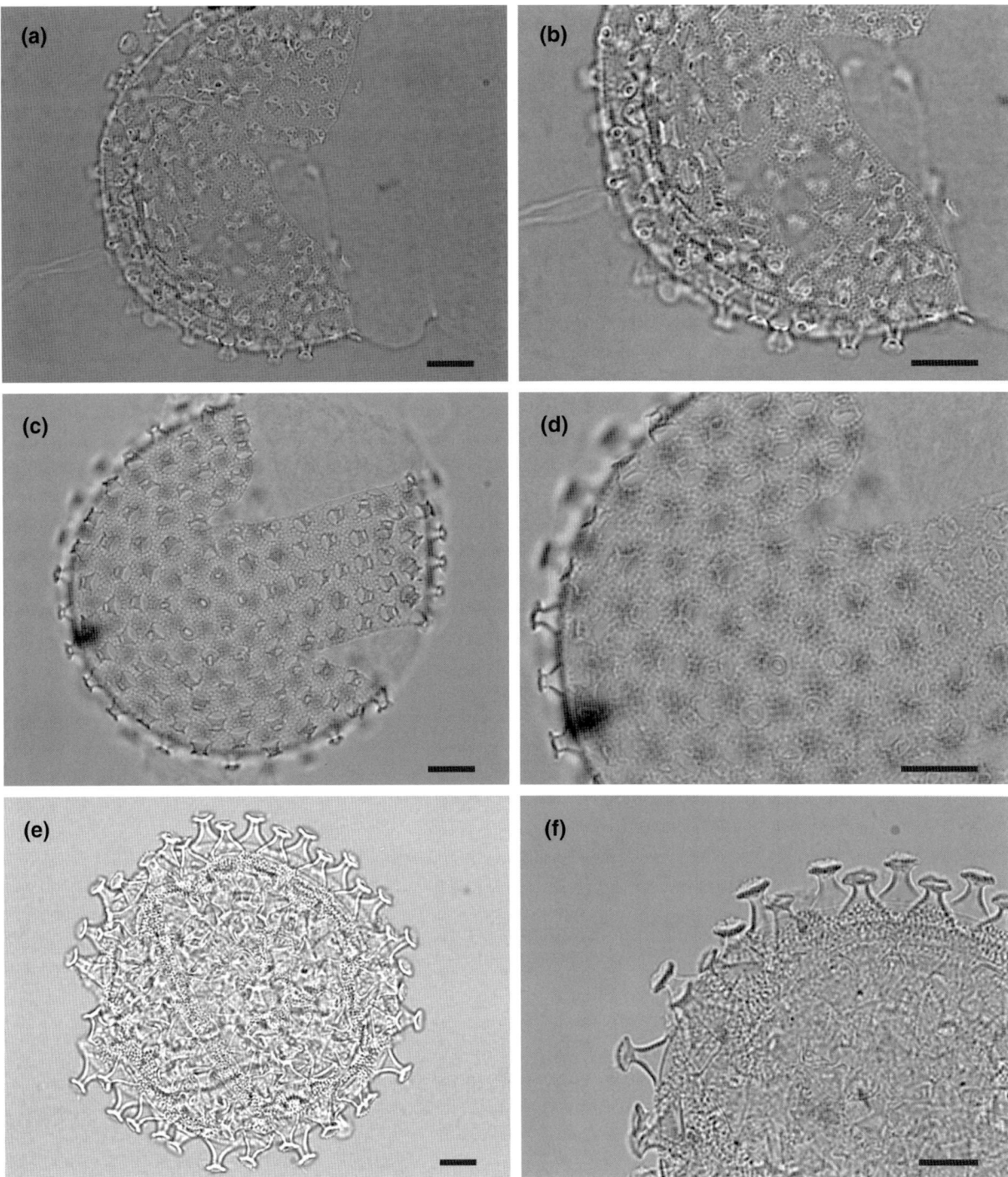

Fig. 9 *Macrobiotus hufelandi* group egg types at two different focal planes. a, b Egg Type 1; c, d Egg Type 2; e, f Egg Type 3. Scale bar = 10μm (Photos: the authors)

Egg Type 3 (Fig 9e,f)
Four examples, Sample 9. Diameter not including processes 62 and 79μm. Egg surface dotted. Processes flask shaped with narrowed necks. Diameter of process bases clearly larger than distal dish diameter. Distal disc margins ragged.

The eggs of *M. serratus* Bertolani, Guidi and Rebecchi, 1996 and *M. trunovaue* Biserov, Pilato and Lisi, 2011 are described as having dotted surfaces but they do not resemble Type 3 eggs in any other character. Type 3 eggs are most similar to the description of the eggs of *M. joannae* Pilato and Binda, 1983 in the dotted appearance of the egg surface and general shape of the processes. However the diameter of the eggs of *M. joannae* is much larger (86–111μm). In an egg of *M. joannae* with a diameter of 103µm the process base diameter reaches 6.8µm while in a smaller diameter egg of Type 3 (79µm) the basal discs have a mean diameter of 8.7µm. The distal dish margins of the eggs of *M. joannae* are described as being strongly dentate with 12–41 clear teeth while the distal dish margins of Type 3 egg are ragged, or appearing torn in an irregular manner but never toothed. As a result of these differences, eggs of *M. joannae*, the most similar to Clare Island Type 3, could not be further considered as a possible species identification for Egg Type 3.

Mesobiotus cf. *harmsworthi* (Murray, 1907b)

Three specimens. Body lengths 265–346µm. While the specimens superficially resemble the nominal species of the *harmsworthi* group in having eyes, a smooth unpigmented cuticle without pores or granulation, three nearly equal macroplacoids in the shape of short rods arranged in an arced row, microplacoid, and lightly indented lunules, a more precise identification was not possible due to the absence of eggs, which is necessary in the differentiation of species of the *harmsworthi* group (Kaczmarek *et al.* 2011).

Vecchi *et al.* (2016) moved the *harmsworthi* group and other species from the genus *Macrobiotus* to the new genus *Mesobiotus* Vecchi, Cesari, Bertolani, Jönsson, Rebecchi and Guidetti, 2016 following morphological and molecular analyses. The establishment of *Mesobiotus* raises the number of tardigrade genera known from Ireland to 25. Clare Island distribution: Sample 6

Minibiotus intermedius (Plate, 1888)

Thirteen specimens. Body length 145–299µm. Eyes present. All specimens conform to the redescription of the species by Claxton (1998).

Minibiotus intermedius has a smooth cuticle without dots, pores, spines or other projections, three granular macroplacoids, a small microplacoid, claws with high accessory points and smooth lunules. According to Claxton (1998) *Mi. intermedius* completely lacks cuticular granulation or it may be present sparsely on leg pair IV or at the bases of the claws of pair IV only. The Clare Island specimens all have featureless cuticles devoid of granulation unlike similar species that are described as having cuticular granulation or dots on other areas of the body or other leg pairs: *Mi. aquatilis* Claxton, 1998; *Mi. continuus* Pilato and Lisi, 2006; *Mi. decrescens* Binda and Pilato, 1995; *Mi. floriparus* Claxton, 1998; *Mi. hispidus* Claxton, 1998; *Mi. maculartus* Pilato and Claxton, 1988; *Mi. milleri* Claxton, 1998; *Mi. stuckenbergi* (Dastych, Ryan and Watkins 1990); and *Mi. taiti* Claxton, 1998. While not noted in the original descriptions, the diagnosis for *Mi. weinerorum* (Dastych, 1984) was emended by Claxton (1998) to include the presence of both cuticular pores and granulation surrounding all claws. Additionally, *Mi. subintermedius* (Ramazzotti 1962b) was determined to have cuticular pores upon re-examination of the type material by Guidetti *et al.* 2007. Each of the aforementioned species is distinct from *Mi. intermedius* by other various characters.

Most similar to *Mi. intermedius* in terms of cuticle morphology might be *Mi. allani* (Murray, 1913c) and *Mi. crassidens* (Murray, 1907c). These species can be separated from each other and from *Mi. intermedius* on the basis of egg morphology but as these descriptions occurred before the establishment of modern taxonomic criteria, they both lack details on the adult form. However, *Mi. intermedius* can be distinguished from the adults of *Mi. crassidens* as it lacks eyes, has broader macroplacoids with the third being larger than the first two, a larger microplacoid, and more slender claws than *Mi. intermedius*. *Mi. allani* was described as lacking eyes and being generally indistinguishable from *Mi. crassidens* except for in the egg morphology. Clare Island distribution: Sample 1, Sample 3, Sample 8

General discussion

At 131 individuals, the number of tardigrades collected during the present survey was not high. However, all samples collected from terrestrial situations contained at least one tardigrade. This is in contrast to the sample material that was either fully submerged (Samples 4, 5, and 7) or partially submerged (Sample 10) in freshwater, from which no tardigrades were recovered. As well as the sediment of freshwater bodies, aquatic mosses are known microhabitats of tardigrades (reviewed by Nelson and Marley 2000; McFatter *et al.* 2007; and Garey *et al.* 2008). Although Murray (1911a) typically did not provide great detail on the sampling sites of the original Clare Island Survey, two of these seem to be associated with freshwater: 'Tarn on Slievemore, Achill Island' and 'Castlebar, Lake'. The 'Tarn on Slievemore' material consisted of moss from the margin of a tarn on Slievemore at an altitude of approximately 1,000m; Murray recorded eight taxa present there. At least some of the material from the 'Castlebar, Lake' material was aquatic moss on a stone at the margin of the lake. It contained six taxa. The degree of submersion of the material at these two locations is not clear. Murray (1911a) did not include any freshwater samples from Clare Island. As none of the freshwater material, but all of the terrestrial material from the present survey was positive for tardigrades, there is a possibility that the collection or processing methodology employed was not suitable for freshwater habitats. Tardigrades from limnic habitats on Clare Island remain unknown.

It is difficult to make in depth comparisons between the species collected during the original and present surveys due to the difference in the number of samples (Murray 1911a listed only three sites on Clare Island itself) and because the quantity of sample material collected by Murray at each site is unknown. Table 3(A–B) lists the taxa recorded by Murray from either Clare Island or other locations as part of the original Clare Island Survey 1909–11 and the present survey. Murray recorded 35 taxa during the original Clare Island Survey. The precise location from which *Fractonotus caelatus* (Marcus, 1928) was collected is not known. Thus, either 12 or 13 taxa were recorded from Clare Island itself, with the other records from other locations included in the original survey by Murray. He recorded *Calohypsibius verrucosus* (Richters, 1900), *Diphascon chilenense*, and *I. papillifer bulbosus* (Marcus, 1928) from Clare Island only. The present survey found ten species and nine Macrobiotus morphotypes of the *hufelandi* group. *Echiniscus quadrispinosus quadrispinosus*, *Hypsibius pallidoides* and *Isohypsibius sattleri* were not recorded by Murray (1911a). The latter two have not been previously reported from Ireland.

Although the new survey included two samples from Kinacorra, the Clare Island type location of *Paramacrobiotus richtersi*, no new specimens were collected. However, Guidetti *et al.* (2015) did obtain *P. richtersi* from the type location and include it an integrative analysis of the *richtersi* complex. This was the only species described from the original Clare Island survey with Clare Island as the type location. *Echiniscus columinis* and *Murrayon hibernicus* were described from Achill Island and *Echiniscus militaris* from Castlebar, Co. Mayo. None of these were recorded in the new survey. Taking into account only those from Clare Island, the original and present survey found in common none of the same heterotardigrade species (which were only represented by one species

Table 3
A Heterotardigrada Marcus, 1927 and B Eutardigrada Richters, 1926 species reported from the Original and New Clare Island Surveys. * indicates type population, cf. indicates uncertainty of species identity as reviewed by DeMilio *et al.* (2016) not the original recording author, sl. indicates that the specimens fit within the broad limits of the current species description.

3A. Heterotardigrada Marcus, 1927

Taxon	Murray 1911a, Clare Island	Murray 1911a, Other locations	Present survey
Echiniscoides sp.	-	X	-
Bryodelphax parvulus Thulin, 1928	-	X	-
*Echiniscus columinis** Murray, 1911a	-	X	-
Echiniscus granulatus (Doyère, 1840)	-	X	-
*Echiniscus militaris** Murray, 1911a	-	X	-
Echiniscus quadrispinosus quadrispinosus Richters, 1902	-	-	X
Echiniscus testudo (Doyère, 1840)	-	X	-
Echiniscus trisetosus Cuénot, 1932	-	X	-
Pseudechiniscus suillus (Ehrenberg, 1853)	-	cf.	sl.
Hypechiniscus gladiator gladiator (Murray, 1905a)	-	X	-
Hypechiniscus exarmatus (Murray, 1907a)	X	X	-
Total	1	10	2

continued

Table 3. Cont.

3B. Eutardigrada Richters, 1926

Taxon	Murray 1911a, Clare Island	Murray 1911a, Other locations	Present survey
Milnesium sp.	-	X	-
Calohypsibius ornatus (Richters, 1900)	X	X	-
Calohypsibius verrucosus (Richters, 1900)	X	-	-
Diphascon chilenense Plate, 1888	cf.	-	X
Hypsibius arcticus (Murray, 1907b)	X	X	-
Hypsibius dujardini (Doyère, 1840)	-	cf.	sl.
Hypsibius pallidoides Pilato, Kiyosa, Lisi, Inshina and Biserov, 2011	-	-	X
Adropion scoticum scoticum (Murray, 1905b)	X	X	X
Platicrista angustata (Murray, 1905a)	-	X	-
Fractonotus caelatus (Marcus, 1928)	?	?	-
Isohypsibius annulatus annulatus (Murray, 1905a)	X	X	-
Isohypsibius papillifer bulbosus (Marcus, 1928)	X	-	-
Isohypsibius schaudinni (Richters, 1909)	-	X	-
Isohypsibius sattleri (Richters, 1902)	-	-	X
Isohypsibius tuberculatus (Plate, 1888)	-	X	-
Thulinius augusti (Murray, 1907a)	-	X	-
Macrobiotus crenulatus (Richters, 1904)	X	X	X
Macrobiotus echinogenitus Richters, 1903	-	X	-
Macrobiotus harmsworthi harmsworthi Murray 1907b	cf.	cf.	cf.
Macrobiotus hufelandi group species	X	X	X
Macrobiotus occidentalis occidentalis Murray, 1910b	-	X	-
Macrobiotus virgatus Murray, 1910b	-	X	-
Minibiotus intermedius (Plate, 1888)	cf.	cf.	X
Paramacrobiotus areolatus (Murray, 1907b)	-	X	-
*Paramacrobiotus richtersi** (Murray, 1911a)	X	-	-
Murrayon hastatus (Murray, 1907a)	-	X	-
*Murrayon hibernicus** (Murray, 1911a)	-	X	-
Total	**12–13**	**20–21**	**9**

in the original survey and two in the present). The two surveys collected six of the same eutardigrade species from Clare Island: *Diphascon chilenense*, *Adropion scoticum scoticum*, *Macrobiotus crenulatus*, *M. harmsworthi*, *M. hufelandi*, and *Minibiotus intermedius*. Differences in species composition found in the original and the present survey are likely due to patchy distributions of species (e.g. Kathman and Cross 1991; Meyer 2006) and the low number of samples collected.

The presence of *Pseudechiniscus suillus*, *Diphascon chilenense*, *Hypsibius dujardini* and *Minibiotus intermedius* in Ireland was again suggested by the present study. These species were reported previously (Murray 1911; Morgan 1975; Baxter 1979; Kinchin 1990) from Ireland but there has been uncertainty about these records (DeMilio et al. 2016) as they occurred before additional characters useful in separating these from morphologically similar taxa were established (see e.g. Claxton 1998; Pilato et al. 2000; Miller et al. 2005; Kaczmarek and Michalczyk 2009; Kaczmarek et al. 2011). However, the present records for *Pseudechiniscus suillus* and *H. dujardini* should be re-examined upon published redescriptions of these species. The presence of *Milnesium tardigradum tardigradum* remains unconfirmed. No *Milnesium* species were collected during the present survey, or from Clare Island during the original survey. This is notable as '*Milnesium tardigradum*' along with '*Minibiotus intermedius*' and '*Macrobiotus hufelandi*' have been the most often recorded in Ireland (see DeMilio *et al.* 2016).

Both Morgan (1975) and Baxter (1979) commented on the absence of *Ramazzottius oberhauseri* (Doyère, 1840) from Ireland. This species has a global distribution and is commonly reported from Britain and mainland Europe (McInnes 1994) yet no representative of this genus has been found in Ireland. However, it is still uncertain if this absence is only due to the limited sampling to have occurred in Ireland.

The results of the present survey suggest that the Clare Island tardigrade fauna is dominated by *Macrobiotus hufelandi* group species. It is possible that specimens exhibiting widely varying morphologies as observed among the *Macrobiotus* morphotypes collected by this survey could have been attributed by historical authors such as Murray (1911a) only to '*Macrobiotus hufelandi*' as the extent of interspecific diversity within the *hufelandi* group had not yet been realised. Murray recorded *M. hufelandi* as present in ten of thirteen sample sites, including two of three on Clare Island, making it the most often recorded species among the samples of the original survey.

Given the morphological similarity of the adults of the *hufelandi* group species, the egg is usually required to confirm the precise species identity. Yet, in the present study, in which three types of eggs were collected the observed morphotypes were still not identifiable beyond belonging to the *hufelandi* group. Bertolani and Rebecchi (1993) revised the *hufelandi* group and redefined *Macrobiotus hufelandi sensu stricto*. Egg Type 1 was determined to be most similar to *M. hufelandi hufelandi* as described by Bertolani and Rebecchi (1993) in the reticulation of the egg surface, egg diameter, the process shape and the dentate distal disc margin. However, the process length was much shorter in the Clare Island Type 1 eggs (mean 3.2–4.2µm) than the range given for *M. hufelandi s. s.* (5.5–7.4µm) (Bertolani and Rebecchi 1993). It is not clear if this difference in process length could be accounted for as variation between populations. Despite the similarities between Egg Type 1 and the described eggs of *Macrobiotus hufelandi* (Bertolani and Rebecchi 1993) none of the adult morphotypes that co-occurred in the same sample as Egg Type 1 (i.e., Morphotypes 1, 2 and 4) were similar to *Macrobiotus hufelandi*.

Morphotypes 1, 2, 4 (as well as Morphotypes 5, 7, 8) have conspicuously large cuticular pores (up to 3µm in diameter) in contrast to *M. hufelandi s. s.* which has small cuticular pores with a maximum diameter of 1.5µm. The weak buccal armature and transverse crest morphology of Morphotype 1 are clearly different to *M. hufelandi s. s.* All specimens of Morphotype 4 possess a ring of pores surrounding the buccal aperture, which is not a character described for *M. hufelandi s. s.* The anterior and posterior bands of teeth were not visible in specimens of Morphotype 2 but are considered to be well defined in *M. hufelandi s. s.* When compared to the morphometric values given for *M. hufelandi s. s.* by Bertolani and Rebecchi (1993) Morphotype 2 is also dissimilar in that it has the stylet supports inserted more rostrally on the buccal tube (mean *pt* value 71.6 in Morphotype 2, 80.9 in *M. hufelandi*) and a shorter claw length of leg pair IV (mean *pt* value 20.9 in Morphotype 2, 28.5 in *M. hufelandi*). Morphotypes 1, 2, and 4 all differ to *M. hufelandi* in the distribution of cuticular granulation. The described pattern of granulation for *M. hufelandi s. s.* is granulation present on leg pair III and localised on leg pair IV to behind the claws while the three morphotypes all have cuticular granulation on leg pair IV only but located on additional areas of the leg.

Taking into account adult specimens from other samples where Type 1 eggs were not collected, Morphotype 3 fits most closely with *M. hufelandi s. s.* in terms of cuticular pore diameter, lunule morphology, and granulation pattern of leg pair IV. Additionally, the mean *pt* value given by Bertolani and Rebecchi (1993) for *M. hufelandi* stylet insertion point (80.9) is similar to that of *Macrobiotus* morphotype 3 (79.2). However, *Macrobiotus* morphotype 3 specimens do not have a well-defined anterior band of teeth and the *pt* value for the length of claws of leg pair IV of *Macrobiotus hufelandi* (28.5) are more than twice that of the same *pt* value for *Macrobiotus* morphotype 3 (13.3). Therefore, despite the presence of a similar egg, we cannot confirm the presence of *M. hufelandi hufelandi* on Clare Island. This further confirms the need to reconsider historical records and distributions of cosmopolitan species, such as *M. hufelandi* as some records may actually represent cryptic species. While it is possible that some of the variation observed among the *Macrobiotus* morphotypes may be due to within species ontogenic variation, or other intraspecific variation, at least some of the *Macrobiotus* morphotypes observed during the current survey represent independent species, though these were not identifiable to known species. As Egg Types 2 and 3 could not be confidently matched to any described *hufelandi* group eggs, there is potential that they originate from unknown species present on Clare Island.

Dastych (2002) recognised the important progress made by previous authors toward taxonomic clarification of the *hufelandi* group species but acknowledged the continually problematic nature of the group. At the time Dastych estimated the group to contain 24 species, though the number of described *hufelandi* group species has continued to increase (e.g. Biserov *et al.* 2011; Pilato *et al.* 2012; Guidetti *et al.* 2013; Meyer *et al.* 2014). Although the boundaries between intraspecific and interspecific morphological variability in adult specimens are often not well determined for this group, the validity of the *Macrobiotus hufelandi* group as a large and widely distributed species complex of morphologically similar adults with intraspecific variation mostly in the egg has been confirmed through a combination of morphological and molecular data (Bertolani *et al.* 2011; Cesari *et al.* 2011; Cesari *et al.* 2013; Guidetti *et al.* 2013).

In order to help resolve the taxonomic meaning of the variation observed between specimens of this group on Clare Island, a further integrative study involving full morphometrics and DNA barcoding is recommended with increased sampling. Currently, the tardigrade species richness of Clare Island, and more generally, of Ireland cannot be well estimated due to the limited scope of past sampling. However, a larger scale All-Ireland survey of terrestrial tardigrades is currently in progress (unpublished data) with an aim of providing more data on Irish tardigrade biodiversity.

Acknowledgements

The authors remember Dr T. Kieran McCarthy's kindness in overseeing this contribution prior to his passing. Mr Dariuz Nowak and Mr Joe O'Malley assisted in field sampling. Dr Jesper G. Hansen, Dr Margaret Flaherty, Katherine Burton (National Library of Australia), and Mr Kelvin Gillmor are acknowledged for their assistance and advice on figures. Prof Roberto Bertolani and Dr Roberto Guidetti contributed valuable comments. We also thank Dr Eamonn Lenihan and the Royal Irish Academy staff that worked to bring this volume to fruition.

REFERENCES

Baxter, W.H. 1979 Some notes on the Tardigrada of North Down, including one new addition to the Irish fauna. *Irish Naturalists' Journal* **19**(11), 389–91.

Bertolani, R. and Rebecchi, L. 1993 A revision of the *Macrobiotus hufelandi* group (Tardigrada, Macrobiotidae), with some observations on the taxonomic characters of eutardigrades. *Zoologica Scripta* **22**(2), 127–52.

Bertolani, R. Rebecchi, L., Giovannini, I. and Cesari, M. 2011 DNA barcoding and integrative taxonomy of *Macrobiotus hufelandi* C.A.S. Schultze, 1834, the first tardigrade species to be described, and some related species. *Zootaxa* **2997**, 79–36.

Biserov, V., Pilato, G., and Lisi, O. 2011 *Macrobiotus trunovae* sp. n., a new species of tardigrade from Russia. *Invertebrate Zoology* **8**(1), 57–62.

Boaden, P.J.S. 1966 Interstitial Fauna from Northern Ireland. *Veröffentlichungen des Instituts für Meeresforschung in Bremerhaven* **II**, 125–30.

Cesari, M., Giovannini, I., Bertolani, R. and Rebecchi, L. 2011 An example of problems associated with DNA barcoding in tardigrades: a novel method for obtaining voucher specimens. *Zootaxa* **3104**, 42–51.

Cesari, M., Guidetti, R., Rebecchi, L., Giovannini, I. and Bertolani, R. 2013 A DNA barcoding approach to the study of tardigrades. *Journal of Limnology* **72**(s1), 182–98.

Claxton, S. 1998 A revision of the genus *Minibiotus* (Tardigrada: Macrobiotidae) with descriptions of eleven new species from Australia. *Records of the Australian Museum* **50**(2), 125–60.

Crisp, D.J. and Hobart, J. 1954 A note on the habitat of the marine tardigrade *Echiniscoides sigismundi* (Schultze). *The Annals and Magazine of Natural History* **7**, 554–60.

Dastych, H. 1991 *Isohypsibius sattleri* (Richters 1902), a valid species (Tardigrada), *Senckenbergiana Biologica* **71**, 181–89.

Dastych, H. 2002 A new species of the genus *Macrobiotus* Schultze, 1834 from Îles Kerguélen, the sub-Antarctic (Tardigrada). *Mitteilungen Hamburgisches Zoologisches Museum und Institut* **99**, 11–27.

Degma, P. and Guidetti, R. 2007 Notes to the current checklist of Tardigrada. *Zootaxa* **1579**, 41–53.

Degma, P., Bertolani, R. and Guidetti, R. 2015 Actual checklist of Tardigrada species. http://www.tardigrada.modena.unimo.it/miscellanea/Actual%20checklist%20of%20Tardigrada.pdf, 41 pp. Accessed 01/02/16.

DeMilio, E., Lawton, C. and Marley, N.J. 2016 Tardigrada of Ireland: a review of records and an updated checklist of species including a new addition to the Irish fauna. *ZooKeys* **616**, 77–101.

Fontoura, P. and Pilato, G. 2007 *Diphascon (Diphascon) faialense* sp. nov. a new species of Tardigrada (Eutardigrada, Hypsibiidae) from the Azores and a key to the species of the *D. pingue* group. *Zootaxa* **1589**(1), 47–55.

Fontoura, P. and Morais, P. 2011 Assessment of traditional and geometric morphometrics for distinguishing cryptic species for the *Pseudechiniscus suillus* complex (Tardigrada, Echiniscidae). *Journal of Zoological Systematics and Evolutionary Research* **49**(s1), 26–33.

Garey, J.R., McInnes, S.J., and Nichols, P.B. 2008 Global diversity of tardigrades (Tardigrada) in freshwater. *Hydrobiologia* **595**, 101–06.

Guidetti, R. and Bertolani, R. 2005 Tardigrade taxonomy: an updated checklist of the taxa and a list of characters for their identification. *Zootaxa* **845**, 1–46.

Guidetti, R., Peluffo, J.R., Rocha, A.M., Cesari, M. and de Peluffo, M.C.M. 2013 The morphological and molecular analyses of a new South American urban tardigrade offer new insights on the biological meaning of the *Macrobiotus hufelandi* group of species (Tardigrada: Macrobiotidae). *Journal of Natural History* **47**(37–38), 2409–26.

Guidetti, R., Cesari, M., Rebecchi, L., Altiero, T. and Bertolani, R. 2015 Integrative taxonomy approach to the *Paramacrobiotus richtersi* group (Eutardigrada, Macrobiotidae). *Thirteenth International Symposium on Tardigrada 2015* 23–26 June, Modena, Italy.

Grann, D. 2009 *The lost city of Z: a tale of deadly obsession in the Amazon*. New York. Doubleday.

Hay, W.P. 1906 A bear animalcule renamed. *Proceedings of the Biological Society of Washington* **19**, 41–8.

Kaczmarek, Ł. and Michalczyk Ł. 2009 Redescription of *Hypsibius microps* Thulin, 1928 and *H. pallidus* Thulin 1911 (Eutardigrada: Hypsibiidae) based on the type material from the Thulin collection. *Zootaxa* **2275**, 60–8.

Kaczmarek, Ł., Gołdyn, B., Prokop, ZM. and Michalczyk, Ł. 2011 New records of Tardigrada from Bulgaria with the description of *Macrobiotus binieki* sp. nov (Eutardigrada: Macrobiotidae) and a key to the species of the *harmsworthi* group. *Zootaxa* **2781**, 29–31.

Kathman, R.D. and Cross, SF. 1991 Ecological distribution of moss dwelling tardigrades on Vancouver Island, British Colombia, Canada. *Canadian Journal of Zoology* **69**, 122–29.

Kinchin, I.M. 1990 The cosmopolitan tardigrade *Milnesium tardigradum* Doyère: An observation from Northern Ireland. *Microscopy* **36**, 412–14.

Kinchin, I.M. 1992 An introduction to the invertebrate microfauna associated with mosses and lichens with observations from maritime lichens on the west coast of the British Isles. *Microscopy* **36**, 721–31.

Le Gros, A.E. 1959 *Pseudechiniscus cornutus* (Richters) and *Hypsibius spitbergensis* (Richters), tardigrades new to Ireland. *Irish Naturalists' Journal* **13**, 44.

Macnair, P. 1915 James Murray, naturalist and explorer. *The Glasgow Herald*, 16 Sep. p. 9.

McFatter, M.M., Meyer, HA. and Hinton, JG. 2007 Nearctic freshwater tardigrades: a review. *Journal of Limnology* **66**(s1), 84–9.

Meyer. H.A. 2006 Small-scale spatial distribution variability in terrestrial tardigrade populations. *Hydrobiologia* **558**, 133–39.

Meyer, H.A., Domingue, MN. and Hinton, JG. 2014 Tardigrada of the West Gulf Coastal Plain with descriptions of two new species from Louisiana. *Southeastern Naturalist* **13**(5), 117–30.

McInnes, S.J. 1994 Zoogeographic distribution of terrestrial/freshwater tardigrades from current literature. *Journal of Natural History* **28**, 257–352.

Miller, W.R., McInnes, S.J. and Bergstrom, D.M. 2005 Tardigrades of the Australian Antarctic: *Hypsibius heardensis* (Eutardigrada: Hypsibiidae: *dujardini* group) a new species from sub-Antarctic Heard Island. *Zootaxa* **1022**, 57–64.

Mitchell, D. 1973 *Hypsibius (Diphascon) pinguis* Marcus, a tardigrade new to the British Isles. *Irish Naturalists' Journal* **17**, 395.

Murray, J. 1905 Tardigrada of the Scottish Lochs. *Transactions of the Royal Society of Edinburgh* **41**, 677–98.

Murray J. 1906 Rotifera of the Scottish Lochs. *Transactions of the Royal Society of Edinburgh* **45**, 151–91.

Murray, J. 1907a Scottish Tardigrada, collected by the Lake Survey. *Transactions of the Royal Society of Edinburgh* **45**(3), 641–88.

Murray, J. 1907b Arctic Tardigrada, collected by Wm.S. Bruce. *Transactions of the Royal Society of Edinburgh* **45**(3), 669–681.

Murray, J. 1907c Some South African Tardigrada. *Journal of the Royal Microscopical Society* **27**(5), 515–524.

Murray, J. 1908 Scottish Rotifers, collected by the Lake Survey (Supplement). *Transactions of the Royal Society of Edinburgh* **45**, 151–91.

Murray, J. 1910 Water bears, or Tardigrada, (suppl. notes). *The Journal of the Quekett Microscopical Club* **11**, 180–81.

Murray, J. 1911a Part 37. Arctiscoida. *Proceedings of the Royal Irish Academy* **31**(37), 1–16.

Murray, J. 1911b Part 52. Rotifera Bdelloida. *Proceedings of the Royal Irish Academy* **31**(52), 1–52.

Murray, J. 1913a South American Rotifera-Part I. *Journal of the Royal Microscopical Society* **33**, 229–46.

Murray, J. 1913b South American Rotifera-Part II. *Journal of the Royal Microscopical Society* **33**, 341–362.

Murray, J. and Marston, G. 1913 *Antarctic days*. London. Andrew Melrose.

Murray, J. and Wailes, GH. 1913 *Notes on the natural history of Bolivia and Peru*. Edinburgh. The Scottish Oceanographical Laboratory.

Morgan, C.I. 1974 Studies on the biology of tardigrades. Unpublished PhD thesis, University of Swansea.

Morgan, C.I. 1975 Some notes on the Tardigrada of the Mullet Peninsula, including four additions to the Irish fauna, and a key to the Irish species. *Irish Naturalists' Journal* **18**(6), 165–77.

Morgan. C.I. 1976 Studies on the British tardigrade fauna, some zoogeographic and ecological notes. *Journal of Natural History* **10**, 607–632.

Morgan, C.I. 1980 Notes on the Distribution and Abundance of the Irish Marine Tardigrada, including two new additions to the Irish fauna. *Irish Naturalists' Journal* **20**(4), 148–152.

Morgan, C.I. and King, P.E. 1976 *British Tardigrades*, Synopses of the British Fauna *(New Series) vol. 9*. London. Academic Press.

Nelson, D.R. and Marley, N.J. 2000 The biology and ecology of lotic Tardigrada. *Freshwater Biology* **44**, 93–108.

Niven, J. 2000 *The ice master*. London. Macmillan.

Pilato, G. 1981 Analisi di nuovi caratteri nello studio degli Eutardigradi. *Animalia* **8**, 51–7.

Pilato, G. 1987 Revision of the genus *Diphascon* Plate, 1889, with remarks on the subfamily Itaquasconinae (Eutardigrada, Hypsibiidae). In R. Bertolani (ed.) *Biology of Tardigrades, Proceedings of the fourth International Symposium on Tardigrada, Modena, September 3–5, 1985, Selected Symposia and Monographs*, 337–57. Unione Zoologica Italiana, Mucchi.

Pilato, G. and Binda, MG. 1998 A comparison of *Diphascon (D.) alpinum* Murray, 1906, *D. (D.) chilenense* Plate, 1889 and *D. (D.) pingue* (Marcus, 1936). *Zoologischer Anzeiger* **236**, 181–185.

Pilato, G., Binda, M.G., Napolitano, A. and Moncada, E. 2000 The specific value of *Macrobiotus coronatus* De Barros 1942, and description of two new species of the *harmsworthi* group (Eutardigrada). *Bollettino delle sedute della Accademia Gioenia di Scienze Naturali* **33**, 103–20.

Pilato, G., Kiyosa, Y., Lisi, O. and Sabella, G. 2012 New records of Eutardigrada from Belarus with the description of three new species. *Zootaxa* **3179**, 39–60.

Ramazzotti, G. 1962a Il Phylum Tardigrada. *Memorie dell'Istituto Italiano di Idrobiologia* **14**, 1–595.

Ramazzotti, G. 1962b Tardigradi del Cile, con descrizione di quattro nuove specie e di una nuova varietà. *Atti della Società Italiana di Scienze Naturali e del Museo Civico di Storia Naturale in Milano* **101**, 275–87.

Ramazzotti, G. and Maucci, W. 1983 The Phylum Tardigrada— 3rd Edition, English translation by CW Beasley. *Memorie dell'Istituto Italiano di Idrobiologia* **41**, 1–1014.

Schultze, C.A.S. 1861 *Echiniscus creplini*, animalculum e familia arctiscoidum. Gryphiae. F.G. Kunike.

Shackleton, E.H. 1909 *The heart of the Antarctic*. London. William Heinemann.

Spallanzani, G. 1776 Opuscoli di Fisica Animale e Vegetabile. *Società Tipografica*, 203–85.

Thulin, G. 1911 Beiträge zur Kenntnis der Tardigraden fauna Schwedens. *Arkiv för Zoologi* **7**, 1–60.

Thulin, G. 1928 Über die Phylogenie und das system der Tardigraden. *Hereditas* **11**, 207–66.

Tumanov, D.V. 2005 *Isohypsibius panovi*, a new species of Tardigrada from Ireland (Eutardigrada, Hypsibiidae). *Zootaxa* **812**, 1–4.

CHAPTER 4

SPIDERS (ARACHNIDA: ARANEAE) FROM CLARE ISLAND

Myles Nolan

ABSTRACT

The spiders of Clare Island were surveyed by the author with additional records by other fieldworkers; 141 species (and two *forma*) belonging to 23 families were recorded. One hundred and fifty-nine species in total have been collected on the island to date. All species recorded are known from mainland Ireland. Comparison is made of the most abundant species collected from woodland, wetland and grazed habitats on the island and shows the fauna associated with each to be quite different from the others. The fauna associated with the short-sward grazed land seems to bear similarities to upland grazed-lands and ancient sheep pasture in Ireland. Some rarities are noted as are some introduced synanthropic species.

Introduction

Spiders were surveyed by Dennis Robert Pack-Beresford (DRPB) on the original Clare Island Survey in July 1909 and in a proximate area of the Mayo mainland in September 1911 (Pack-Beresford, 1911b). He listed 108 taxa in total in his paper (1911b), 73 of which were collected on Clare Island and 35 on the Mayo mainland (in his paper two species collected on Clare Island are omitted from the checklist but appear in the comments following). His collections were supplemented by contributions from other individuals.

On invitation from the Royal Irish Academy a number of collecting trips were made to Clare Island by the author (MN) during the summer of 2002 and the catch from these trips was supplemented by specimens from independently set traps and from small collections made by another individual. This note offers a brief account of the spiders recorded, examines their habitat preferences and compares these with their general habitat preferences in Ireland. Notable species are discussed as are the similarities and differences between the faunas recorded on both surveys.

Methods

A trapping methodology had been initiated in 2001 by individuals working through the National University of Ireland, Galway. Traps were emptied, sorted and maintained by Stephen McCormack (SMcC) who operated as a field assistant on the project. This involved the placement of pitfall traps, Malaise traps and water traps. These methods were supplemented by hand collections. **Pitfall traps** (PT): traps were set, emptied and sorted by SMcC who then forwarded the catch to MN. Traps consisting of single cups were set at 48 locations scattered around the island. Locations, trap numbers and habitat information (as supplied by SMcC) are tabulated in Appendix 1. **Malaise tray traps** (MTT): several water-traps were set by SMcC at two sites in association with Malaise traps. These consisted of foil catering trays set at the base of one axis of the Malaise trap. They were set within the small area of mixed deciduous wet woodland at Portlea (MTT woodland) and amongst the extensive *Phragmites* beds dominating the swampy in-filled Lough Avullin (MTT *Phragmites*). **Hand collecting:** this included beating vegetation, sweep-netting, grubbing through vegetation, sifting material and looking under stones and debris (natural and man-made). Collections were made by MN during three trips to Clare Island on the following dates in 2002: 20–24 May, 23–27 July and 22–26 September. Specimens were also collected by SMcC in late 2001 and through 2002. Martin Cawley (MC) collected and identified some spiders during 2002 while targeting other groups, adding an additional seven species to the list, and he very kindly allowed me include his data in this paper. Specimens were collected by MN and SMcC by visiting specific areas and by sampling opportunistically while following various roads and trackways crossing the island, including human habitations and anthropogenic structures or debris. An effort was made to sample as many habitats and microsites as possible in order to obtain the widest range of species. Many specimens that were identifiable by eye (MN) were ignored if a specimen had already been collected. Specimens were generally preserved immediately in 70% Industrial methylated spirits but in a few cases, immatures were retained alive until they moulted.

Spiders were identified using Roberts (1993) and Schikora (1993, 1995). County records and distributional information were derived from Helsdingen (1996), Cawley (2009), Nolan (2016) and a small number of subsequent publications. Nomenclature follows Lavery (2019), which followed the World Spider Catalog (World Spider Catalog 2019).

A complete catalogue of all species and specimens recorded is given in Appendix 2. Species that are new (i.e. unpublished) county records for Co. Mayo are indicated with a dagger †. New records from Clare Island are indicated with an asterisk *. Species recorded by DRPB from mainland Mayo but not Clare Island and which were found on the resurvey are indicated with a degree sign °.

Results

Altogether 143 taxa from 23 families are noted from 2,562 specimens identified to species (two taxa are generally considered *forma*: *Dicymbium nigrum* f. *brevisetosum* Locket, 1962 and *Oedothorax gibbosus* f. *tuberosus* (Blackwall, 1841)). Eighty-six species are new records for Clare Island, and 21 are new records for Co. Mayo. Thirty-four species were recorded by DRPB on the Mayo coast but not from Clare Island, and nineteen of these were collected during the resurvey. The most abundantly collected species are shown in Table 1. This tabulates species occurring >3%, offering a representative proportion of species whilst avoiding presenting much longer lists of substantially less abundant taxa. Ten species represented just over 50% (n=1,300) of the total catch.

Analysis

Passive traps were located in differing habitat types: MTT in wet, deciduous woodland and *Phragmites* swamp; pitfalls in a range of habitats but primarily open and short sward grazed lands (hand collections were more random and often targeted). This allowed the spider fauna to be assessed in respect of three major landscape types: woodland, wetland and open/grazed land. The numbers of specimens/species collected from each habitat type and by each method are: MTT woodland (n=456/34), MTT Avullin (n=322/33), pitfall trap (n=1,081/74), hand (n=703/114). The most abundant species (>3%) occurring at each location or collected by each method are shown in Table 2. Hand collecting (including use of a sweep-net) allowed a range of microsites that could not be targeted by passive trapping methods to be effectively sampled; vertical surfaces such as rock faces, cryptic situations such as rock crevices, epiphytic mosses and stones, and synanthropic situations such as buildings and anthropogenic debris. Sweeping and beating of vegetation was used to target taller vegetation and situations where aerial webs should be available.

Table 1 Species occurring at >3% total catch (n=2,562)		
Species	number	% total
Hypomma bituberculatum	277	10.82
Trochosa terricola	149	5.81
Tenuiphantes zimmermanni	134	5.23
Pachygnatha degeeri	133	5.19
Pardosa pullata	121	4.72
Pachygnatha clercki	113	4.41
Tenuiphantes alacris	103	4.02
Tiso vagans	99	3.86
Pirata piraticus	90	3.51
Oedothorax fuscus	81	3.16
Totals	**1,300**	**50.73**

Table 2
Species occurring at ≥3% in habitat types or collected by different methods. Columns detail numbers of specimens and % (rounded down in parentheses) collected at each habitat type. Totals collected by all methods at all locations and their percentages.

Species	MTT Woodland	MTT *Phragmites*	Pitfalls All	Pitfalls Short sward	Hand collections	Total specimens	% of total specimens
Tenuiphantes alacris	89 (27)					103	4.02
Tenuiphantes zimmermanni	66 (20)				50 (7)	134	5.23
Dicymbium tibiale	42 (13)					52	2.02
Monocephalus fuscipes	28 (8)					38	1.48
Neriene montana	19 (5)					22	0.85
Hypomma bituberculatum		238 (52)			37 (5)	277	10.82
Pachygnatha clercki		75 (16)				113	4.41
Gnathonarium dentatum		33 (7)				45	1.75
Piratula hygrophila		25 (5)				30	1.17
Trochosa terricola			132 (12)	48 (9)		149	5.81
Pachygnatha degeeri			121 (11)	85 (17)		133	5.19
Pardosa pullata			96 (8)	36 (7)	25 (3)	121	4.72
Tiso vagans			96 (8)	69 (13)		99	3.86
Pirata piraticus			70 (6)	25 (6)		90	3.51
Oedothorax fuscus			64 (5)	37 (7)		81	3.16
Oedothorax retusus			57 (5)	25 (6)		61	2.38
Pardosa amentata			43 (3)			51	1.99
Walckenaeria vigilax			38 (3)	21 (5)		40	1.56
Metellina mengei					29 (4)	30	1.17
Pardosa nigriceps				12 (3)		24	0.93
Erigone dentipalpis				17 (4)		24	0.93
Totals	244 (75.75%)	371 (81.34%)	825 (66.3%)	275 (55.63%)	141 (20.04%)	1,669	65.09

Most abundant species

Ten species occurred at abundances >3% of the total catch and these are detailed in Table 1. The most abundant three spiders, *Hypomma bituberculatum* (Wider, 1834), *Pirata piraticus* (Clerck, 1757) and *Pachygnatha clercki* Sundevall, 1823 broadly prefer wet/damp habitats. *Tenuiphantes alacris* (Blackwall, 1853) is a woodland species and can be very abundant in this habitat. Species such as *Tenuiphantes zimmermanni* Bertkau, 1890, *Trochosa terricola* Thorell, 1856 and *Oedothorax fuscus* (Blackwall, 1834) can be relatively common in a wide range of habitats (*T. zimmermanni* strongly preferring shaded situations) while *Pachygnatha degeeri* Sundevall, 1830, *Pardosa pullata* (Clerck, 1757) and *Tiso vagans* (Blackwall, 1834) are associated with low sward open habitats and grasslands. All these species are common and abundant in Ireland, however their abundances across Clare Island were very uneven, and some of them were restricted to a particular habitat type where they occurred in relatively very large numbers. This can be clarified by comparing occurrences in relation to habitat type and collection method (Table 2).

Trochosa terricola. Photo: Aleksandrs Balodis. This item is distributed under the terms of the CC BY 3.0 Unported licence

Portlea woodland (MTT)

In all 39 species were collected at the woodland site, 34 by MTT with an additional five collected by hand and pitfall trap. The dominant species here differ quite strongly from those dominating the total catch. Five species accounted for over 75% of the specimens recorded though they constituted less than 14 % of the total catch of spiders. Of these species *T. alacris*, noted above as a woodland spider, was the most abundant and it is notable that the fourteen additional specimens collected by other methods were all taken within or immediately adjacent to the woodland area. The same applies to *Neriene montana* (Clerck, 1757) where the three additional specimens were collected from the woodland, and to a lesser extent to *Dicymbium tibiale* (Blackwall, 1836) where 50% of the additional specimens (n=10) also came Portlea woodland. Unlike the other two species, *N. montana* is somewhat anthropophilic, occurring in gardens or parks where it will utilise ivy *Hedera* or similar dense, shading shrubbery. *Monocephalus fuscipes* (Blackwall, 1836) to some extent favours woodland habitat but is a common species occurring in a range of other habitats. *T. zimmermanni* is one of the more abundant linyphiiid spiders found in Ireland and again is found in a wide variety of habitats. However, it is also a shade loving species and it is telling that very few specimens (n=11) were collected in pitfall traps while an additional 50 were collected by hand. Many of the latter were collected by delving deep into vegetation or grubbing into well shaded microsites where webs were set.

Other typical woodland species collected include *Bathyphantes nigrinus* (Westring, 1851), *Gongylidium rufipes* (Linnaeus, 1758), *Tenuiphantes tenebricola* (Wider, 1834), *Monocephalus castaneipes* (Simon, 1884), *Neriene peltata* (Wider, 1834) and *Parapelecopsis nemoralis* (Blackwall, 1841) all from the Linyphiidae and *Clubiona comta* C.L. Koch, 1839 from the Clubionidae. Many of these are strongly arboreal and were found only or predominantly in the woodland on Clare Island. *B. nigrinus* and *G. rufipes* prefer shaded woodland habitat with the former typically occurring in webs on grasses while *G. rufipes* is often found on taller shrubs and bushes or the lower branches of trees. *T. tenebricola* sets webs amongst low vegetation. *N. peltata* usually sets its webs amongst the twiggy lower branches of trees. *C. comta* was also recorded only in the woodland and, like *N. montana*, is adept at making use of well-developed ivy in gardens and other built-up areas and can be expected in such anthropogenic situations.

One species of potential conservation interest was associated with tree-trunk mosses. *M. castaneipes* prefers epiphytic mosses on tree-trunks especially but is also occasionally found amongst epilithic mosses on boulders or under stones in more upland situations. This species has been flagged as possibly having undergone a significant decline in Britain (SHRS 2022) and the relevant microsites need to searched to detect the spider.

Two of the three specimens of *P. nemoralis* collected were also gathered from tree-trunk mosses. This spider is also essentially found in woodlands but it has a rather local distribution and is not known from wide areas of Britain where it is more common in northern England and Scotland. It seems to be rarely found in continental Europe generally (Nentwig *et al.* 2020) and does not occur in much of southern Europe.

Taranucnus setosus (O.P.-Cambridge, 1863) occurred only in the woodland but can be found in more open habitats such as bogs where it preferentially sets its web in very heavily shaded and damp, spacious cavities. The theridiid spider *Rugathodes instabilis* (O.P.-Cambridge, 1871) also prefers rather damp or humid situations, occurring regularly on moist grasslands within woodlands; it was collected on Clare Island only within the woodland.

Lough Avullin *Phragmites* swamp (MTT)

The fauna collected in the MTT traps set amongst *Phragmites* at Lough Avullin differs radically from the woodland catch. Four hundred and fifty-six specimens comprising 33 species were collected and four species accounted for over 80% of the catch (Table 2), all of which prefer very damp or wetland habitats. *H. bituberculatum* accounted for over 50% of the catch; it occurs widely but in a restricted range of habitats in such relatively large numbers. *P. clercki* occurs in abundance in association with wet grasslands, marshes and swampy or flooding woodlands. *Gnathonarium dentatum* (Wider, 1834) similarly occurs in various wet habitats and can be locally abundant. *Piratula hygrophila* Thorell, 1872 prefers wetlands with good shade and occurs abundantly in *Phragmites* beds and damp/wet woodlands. None of these species are rare but all are somewhat restricted in the range of habitats they occupy.

It will be noted from Table 2 that 39 specimens of *H. bituberculatum* were collected in addition to those from the MTT; 24 of these came from Lough Avullin and a further ten specimens from saltmarsh on the eastern edge of the island. Thus, while this species was the most abundantly collected *in toto* it is restricted in distribution, occurring uncommonly across much of the island. Most of the other specimens of *G. dentatum* and *P. hygrophilous* were also collected at Lough Avullin while most of the additional *P. clercki* came from pitfall traps in wet situations such as *Schoenus* dominated flush. Other relatively common wetland spiders found at Lough Avullin included *Antistea elegans* (Blackwall, 1841), *Clubiona phragmitis* C.L. Koch, 1893, *C. stagnatilis* Kulczynski, 1897 and the common species *P. piraticus*. Four species were collected which have been shown in relatively recent years to be far more widespread in Ireland than earlier records indicated: *Aphileta misera* (O.P.-Cambridge, 1882), *Bathyphantes approximatus* (O.P.-Cambridge, 1871), *Ceratinella brevis* (Wider, 1834) and *Walckenaeria unicornis* O.P.-Cambridge, 1861 (Cawley 2009) all of which can be sifted from wet or damp mosses and litter. *C. brevis* is not uncommon but has a rather local distribution in Ireland and Britain and is more usually associated with woodland habitat. *A. misera* is perhaps the most interesting of these being largely restricted to acidic wetlands across its European range. It can occur in abundance on bogs that have undergone significant drainage. It has a rather limited and local distribution in Britain (SHRS 2022).

One notable wetland species that was not recorded at Lough Avullin is *Silometopus elegans* (O.P.-Cambridge, 1872) which was collected amongst *Molinia* on blanket bog and from rushes at a lake margin.

Pitfall traps
One thousand and eighty-one specimens comprising 74 species were collected in total from pitfall traps. Nine species occurring at >3% abundance of the total pitfall catch constituted 66.3% of the total pitfall catch (Table 2). The wolf spider (Lycosidae) *T. terricola* was the most abundant at 12.21% closely followed by *P. degeeri*. *Pardosa pullata* and *T. vagans* each comprised over 8% of the catch. Twenty-seven species occurred in the pitfall traps as singletons or doubletons and eleven species were recorded only by this method of collecting. It will be noted that the first six most abundant species were amongst the most abundantly collected species overall. All nine species are common spiders and *T. terricola*, *P. degeeri*, *P. pullata*, *T. vagans*. *P. piraticus* and *O. fuscus* have already been addressed. *Oedothorax retusus* (Westring, 1851) is typically found on grasslands, both natural and anthropogenic. Both *Pardosa amentata* (Clerck, 1757) and *Walckenaeria vigilax* (Blackwall, 1853) prefer relatively open humid to damp situations. Taken together these nine species are broadly reflective of the open grazed habitats that dominate Clare Island. None of them would be found abundantly within either woodland or inundated wetland habitat with the exception of *P. piraticus*.

Amongst the other species of interest collected in pitfall traps were *Agyneta mossica* (Schikora, 1993) which was first recorded quite recently from Ireland (Nolan and McCormack 2004) and of which there are still relatively few records (Nolan 2016b). It is noted in Nolan (2016b) that the specimen of *A. saxatilis* collected by DRPB on Clare Island in fact refers to *A. mossica*, so the specimen of *A. saxatilis* collected on the resurvey is counted a new island record while *A. mossica* is not. A single specimen of *Allomengea vidua* (L. Koch, 1879) was trapped on a river bank and this wetland species seems to have undergone a significant decline in recent years in Britain (SHRS 2022). There are a handful of recent records from Ireland (Cawley 2009). There are additional records of *Agyneta cauta* (O.P.-Cambridge, 1902) from Ireland which are not reflected in the published county lists (Helsdingen 1996; Cawley 2009) and three specimens were trapped on Clare Island on blanket bog and grasslands.

Pitfalls – short sward
Pitfall traps, it will be seen from Appendix I, were set in a wide variety of habitat types on the island and this is reflected in the wide variety of species collected. A proportion of traps were set on short sward grasslands, a vegetative structural type that covered a large proportion of the island. It was of interest to see how the fauna from this subset of traps compared with that from the

woodland and wetland habitats discussed above. Consequently, the catch trapped in pitfalls set in locations described as dry grassland (##4, 5, 6, 41), heavily grazed grassland (##7, 47, 48), short sward grassland (##17, 18, 29, 32) and *Plantain* sward (##19, 43, 44, 45, 46) were segregated from the pitfall dataset and the most abundant species (>3%) were tabulated (Table 2). There are obviously strong similarities with the total pitfall catch however the relative position of the most abundant species has shifted with a substantially higher proportion of *P. degeeri* and *T. vagans* appearing in the short sward traps while the proportion of *T. terricola* has dropped. Two species are introduced through the isolation of this subset of traps, *Erigone dentipalpis* (Wider, 1834) and *Pardosa nigriceps* (Thorell, 1856). All but one specimen of *E. dentipalpis* caught in pitfalls (n=18) were collected in this subset of traps as were 50% of *P. nigriceps* specimens. Neither species was collected on Clare Island in large numbers. Both are very common species in Ireland, *E. dentipalpis* occurring sometimes in massive abundance on managed grasslands and heavily disturbed ploughed/pastoral soils and *P. nigriceps* also occurring on pastoral grasslands but generally proximate to taller vegetation, which it also utilises.

Hand collections

One hundred and fourteen species were identified from 699 specimens collected by hand. Only four species were collected at abundances >3% (n=143) and these accounted for 20.04% of the catch. The species involved are *T. zimmermanni*, *H. bituberculatum*, *Metellina mengei* (Blackwall, 1869) and *P. pullata*. Seven species collected at abundances of 2–3% (n=82) represented a further 13.71% of the catch. Forty-seven species were collected as singletons or doubletons. These collections gave a very high rate of a new species for every 6.13 specimens collected and 45 species were collected only by this method.

The high numbers in which *T. zimmermanni* and *H. bituberculatum* were collected by hand have already been noted. *P. pullata* has been noted as a very abundant species of open habitats. *M. mengei* is an orb-weaver from the family Tetragnathidae and was swept in abundance from tall grasses, rushes and bracken. Many more specimens could have been collected.

Sweep-net collections, combined with looking for webs, resulted in the recording of a number of orb-web builders such as *Araneus diadematus* Clerck, 1757, *Araniella opistographa* (Kulczynski, 1905), *Larinioides cornutus* (Clerck, 1757), *Zygiella atrica* (C.L. Koch, 1845) and *Zygiella x-notata* (C.L. Koch, 1845) from the Araneidae and *M. mengei, Metellina merianae* (Scopoli, 1763), *M. segmentata* (Clerck, 1757), *Tetragnatha extensa* (Linnaeus, 1758) and *T. montana* Simon, 1874 from the Tetragnathidae. While some of these species e.g. *A. diadematus* are possibly not common on the island due to the absence of favourable conditions (lack of hedgerows, heavily grazed gorse that offered poor structure for web placement), and it would have been possible to collect specimens of *T. extensa* and *M. merianae* in great abundance. The former of these was especially common amongst stands of rushes and the latter abundant in association with overhanging *Calluna* and exposed rock faces.

Two species, *Phylloneta sisyphia* (Clerck, 1757) and *Linyphia triangularis* (Clerck, 1757) can often be collected in great numbers in Ireland; the former from twiggy shrubs and both from *Ulex*, but neither was particularly abundant on Clare Island.

Sampling of field layer vegetation including *Pteridium, Calluna* and *Ulex* produced nearly all records of *Bolyphantes luteolus* (Blackwall, 1833), *Peponocranium ludicrum* (O.P.-Cambridge, 1861), *Clubiona trivialis* C.L. Koch, 1843, *Cheiracanthium erraticum* (Walckenaer, 1802), *Enoplognatha ovata* (Clerck,1757) and *Philodromus cespitum* (Walckenaer, 1802). Other than *B. luteolus* and *C. erraticum* these are essentially common species. *C. trivialis* is found most frequently on *Calluna* heaths and only occasionally on other open habitats.

Searching by hand and eye at ground level produced specimens of a range of active ground hunters, cryptic species and web-dwelling spiders. These included *Amaurobius ferox* (Walckenaer, 1830), *Amaurobius similis* (Blackwall, 1861), *Centromerus prudens* (O.P.-Cambridge, 1873), *Entelecara erythropus* (Westring, 1851), *Harpactea hombergi* (Scopoli, 1763), *Heliophanus cupreus* (Walckenaer, 1802), *Oonops pulcher* Templeton, 1835, *Phrurolithus festivus* (C.L. Koch, 1835), *Robertus arundineti* (O.P.-Cambridge, 1871), *R. instabilis* and *Xysticus erraticus* (Blackwall, 1834). Specimens of *H. cupreus* were taken from coastal *Armeria* vegetation which was notably difficult to find in accessible areas and the small area sampled seemed to be inaccessible to sheep.

Argyroneta aquatica. Photo: Norbert Schuller Baupi. This item is distributed under the terms of the CC BY 3.0 Unported licence

Another notable species collected by hand is the water spider *Argyroneta aquatica* (Clerck, 1757) of which few specimens were collected but many seen in some of the larger pools on the island, where it seemed to be very abundant.

Rarer spiders

Hand-collecting produced arguably the most interesting species with respect to general rarity in Ireland and elsewhere. These include *Liocranoeca striata* (Kulczynski, 1882) (Liocranidae), *Argenna subnigra* (O.P.-Cambridge, 1861) (Dictynidae), *Euophrys petrensis* C.L. Koch, 1837 (Salticidae), *Halorates reprobus* (O.P.-Cambridge, 1879) and *Porrhomma montanum* Jackson, 1913 (both Linyphiidae). Three specimens of *L. striata* were collected from reed and litter debris at the edge of a saltmarsh while a single specimen of *A. subnigra* was winkled out of a coastal rock crevice within the splash zone. A female of *E. petrensis* with an egg-sac was found under a stone on coastal grassland at the western end of the island. The status of these three species, all uncommon in Ireland, has been discussed previously (Nolan 2016b). *A. striata* is now known from four Irish counties. It is rather uncommon in Britain though has been found in a wider range of habitats than in Ireland. *A. subnigra* is also known from only four Irish counties; it prefers well-draining insolated soils and seems to be most abundant on dune systems on Ireland's east coast. *E. petrensis* is known from only three Irish counties but has been found at two locations in Co. Galway (Pack-Beresford 1911a; Cawley and Nolan 2007) suggesting Ireland's west coast is its primary base in Ireland. In England it is associated primarily with southern heathlands it occurs in montane situations in northern Britain. MC collected a specimen of *H. reprobus* from fissures in coastal rock near the high-water mark. This species' global distribution is restricted to the coastlines of north-western Europe so Ireland and Britain almost certainly maintain a significant proportion of the spider's global populations. It is now known from twelve Irish counties (Helsdingen 1996; Cawley 2009). Three specimens of *P. montanum* were found under deeply embedded stones on Clare Island. This spider was first noted from Ireland quite recently (Nolan 2002) and it too is now known from four counties (Nolan 2016b).

Synanthropic species

Several synanthropic species were collected in or around built structures. The commonest of these was the sector spider *Z. x-notata*, which frequents the outside of houses and is uncommonly found away from built environments. It builds a characteristic orb-web with an open sector, resembling a cake with a single slice removed, and these were seen frequently on the outside of windows. *Pholcus phalangioides* (Fuesslin, 1775), the daddy long-legs spider is now known from all over Ireland, including Tory Island off the Donegal coast (Cawley 2007). This species has spread across the northern half of Ireland since the 1970s, at which time it was largely confined to the southern half of Ireland (Locket *et al*. 1974). *Steatoda grossa* (C.L. Koch, 1838) in 1996 was known from only three Irish counties (Helsdingen, 1996) but is now known to be very common in southern Ireland and has been found in some northern counties also (Anderson 2009). Both of these species are susceptible to transportation by humans and this is almost certainly how they got to Clare Island. Specimens of the smallest of the common house spiders, *Tegenaria domestica* (Clerck, 1757), were found in the Abbey on the southern side of the island and in a cowshed. It is very likely that some of its larger relations (now in the genus *Eratigena*) have also been transported onto Clare Island. Much less familiar is the small linyphiiid *Microctenonyx subitaneus* (O.P-Cambridge, 1875), which MC collected from hayshed debris having recently added it to the Irish list (Cawley 2001). Because this is the species' preferred habitat it seems reasonable to consider that it too may have been introduced by humans. Found in a similar situation was *Thyreosthenius parasiticus* (Westring, 1851) which MC collected from debris in a chicken coop, a somewhat unusual situation for this semi-troglodytic spider. Both these last two species have few Irish records.

Differences from the original survey

The following fourteen species were collected on Clare Island by Pack-Beresford but were not recorded in 2002. *Dysdera crocata* C.L. Koch, 1838, *Drassodes lapidosus* (Walckenaer, 1802), *Haplodrassus signifer* (C.L. Koch, 1839), *Micaria pulicaria* (Sundevall, 1831), *Hahnia montana* (Blackwall, 1841), *Araeoncus humilis* (Blackwall, 1841), *Erigone arctica* (White, 1852), *Hilaira excisa* (O.P.-Cambridge, 1871), *Lepthyphantes leprosus* (Ohlert, 1865), *Maso sundevalli* (Westring, 1851), *Tapinopa longidens* (Wider, 1834), *Meta menardi* (Latreille, 1804), *Trochosa ruricola* (Degeer, 1778), and *Euophrys frontalis* (Walckenaer, 1802). Given that the collecting carried out in 2002 was limited and rather sporadic it is not surprising that species previously found will have been missed and with one exception there is little reason to suggest that these species are no longer to be found on Clare Island. The exception is *M. menardi*, a cave-dwelling species but which also makes regular use of similarly dark anthropogenic structures. Pack-Beresford found only a single egg-sac under a large boulder, and it might be the case that a breeding population never established itself on the island. Of the remaining species, all but *H. excisa*, a northern spider found in wetland habitats, are relatively common species. It is possible that *D. lapidosus* may have been a misidentified *D. cupreus* (which was recorded in 2002), with which it can be easily confused. *L. leprosus* is a relatively common synanthropic species and it may be the case that houses and other built structures were insufficiently examined in 2002. *D. crocata*, the woodlouse spider occurs both synanthropically and on coastal dunes. *E. arctica* frequents beaches and other coastal habitats where it can often be found under debris, and the jumping spider *E. frontalis* also often occurs in coastal situations, on grasslands or amongst herbaceous vegetation.

Thirty-four species were found by DRPB on the Mayo coast that he did not collect on Clare Island. Of these nineteen were collected during the resurvey, though there is little reason to consider that any of these may have colonised the island over

Drassodes cupreus. Photo: Gail Hampshire. This item is distributed under the terms of the CC BY 2.0 licence

the last century. Fifteen remain undetected on the island. Of interest amongst these are *Arctosa perita* (Latreille, 1799), *Hypsosinga pygmaea* (Sundevall, 1831) and *Neon reticulatus* (Blackwall, 1853). It is possibly the case that the area of *Ammophila* dune on Clare Island is too small to provide a toehold for *A. perita*, which strongly prefers open sands in areas of mobile and fixed dune (though it will colonise bare peats on cutaway bogs). *H. pygmaea* has a rather restricted distribution but can be collected (sometimes in abundance) from *Calluna* on peatlands (Nolan 2013); it might occur on areas of blanket bog on the eastern side of the island, as might the small salticid *N. reticulatus*, which can be found amongst *Cladonia* lichens and dryer mosses on bog and other habitats.

Conclusions

The resurvey recorded 141 species (and two *forma*) and added 86 species to the list for Clare Island, bringing the total recorded from the Island to 159. All species recorded on the island occur on the Irish mainland. Both surveys were relatively brief and there are almost certainly additional species to be found. In 2002, while pitfall traps collected substantial numbers of specimens and species, the traps were set singly and were quite widely dispersed; concentrating larger numbers of traps in a wider range of habitats might have added to the species list. With respect to those species collected by DRPB but not on the resurvey, there is little substantial reason to conclude they are no longer resident on Clare Island and a longer survey period in 2002 might have recorded them.

It is difficult to make very substantial comparisons between the two surveys. Dennis Robert Pack-Beresford presented his results as an annotated species list and offered practically no quantitative data. One substantial difference lies in the number of species from the family Linyphiidae recorded: 82 species on the resurvey compared to Pack-Beresford's 32, accounting for over 60% of the new records for the island. Undoubtedly the use of pitfall traps over an extended period assisted in the collection of additional species. It is notable however that some of the dominant woodland species such as *T. alacris* and *N. montana* do not appear on DRPB's list, suggesting the woodland, which was extant on the island then, had not been sampled. It is not possible to state with certainty that some of the spiders newly recorded there have colonised since the previous survey, other than some of the strongly synanthropic species noted above.

The use of passive traps on the resurvey allowed the collection of an active spider fauna over time, making it possible to show that the faunas associated with different habitat types are quite distinct. Several relatively uncommon species such as *A. misera*, *A. vidua*, *S. elegans* and *W. unicornis* show that a wetland fauna of some interest is present on Clare Island. Though the area of woodland is quite small it has a typical albeit truncated fauna. None of the species found in it can be considered particularly rare, but as noted above, a species such as *M. castaneipes* might be of conservation interest given that it is possibly in decline in Britain (SHRS 2022). Two uncommon spiders, *L. striata* and *A. mossica*, and one rare species *E. petrensis* were recorded from more open habitats.

The low numbers in which some common pastoral/agricultural species were recorded is noteworthy. Species such as the wolf-spider *Pardosa palustris* (Linnaeus, 1758) (n=3) and the linyphiids *E. dentipalpis* (n=26) and *Erigone atra* Blackwall, 1833 (n=18) can be superabundant in areas where soils are severely disturbed (e.g. by ploughing or heavy grazing). The low numbers in which these species were recorded might suggest that the heavy sheep-grazing practiced across much of the island does not have the effects typical of dairy and tillage agriculture. Only nine specimens of *E. atra* were collected in passive traps, the rest by hand.

The most abundant linyphiid in the short sward traps, *T. vagans*, was shown in a study of Northern Irish Natural Heritage Areas to be strongly associated with species-rich grassland and heather moorland (Cameron *et al.* 2004). Whereas a single specimen of the spider was recorded in a survey of agricultural grasslands in Ireland's south and east (Anderson *et al.* 2008), a short study of the Curragh showed *T. vagans* to be the most abundant spider from open grassland on this ancient sheep pasture (Nolan 2016a), and the Curragh grassland has been characterised as most closely resembling low-nutrient upland grasslands in Ireland and Britain (Feehan 2007). Thus, the abundance of *T. vagans* on Clare Island might suggest that the grazing effects of the abundant sheep on the island are not so deleterious to the landscape as other grazing regimes.

Acknowledgements

Thanks are due primarily to Stephen McCormack for his assistance and company during fieldwork in 2002 and for the supply of numerous specimens; to Mary and John Moran whose hospitality I enjoyed while on the island; to Jim O'Connor and Nigel Monaghan, past Keepers of the Natural History Museum, Dublin, for access to the Museum collections, library and laboratory facilites; to Michel Dugon who read this paper in an earlier draft. Thanks are due also to the late Dr. Kieran McCarthy, to Dr. John Breen and to the Royal Irish Academy for the kind invitation to execute the survey.

REFERENCES

Anderson, A., Helden, A., Carnus, T., Gleeson, R., Sheridan, H., McMahon, B., Melling, J., Lovic, Y. and Purvis, G. 2008 Arthropod diversity of agricultural grassland in south and east Ireland: introduction, sampling sites and Araneae. *Bulletin of the Irish biogeographical Society* **32**, 142–59.

Anderson, R. 2009 *Steatoda grossa* (C.L. Koch) (Araneae: Theridiidae), a false black widow, in Belfast. *The Irish Naturalists' Journal* **30**, 144.

Cameron, A., Johnston, R.J. and McAdam, J. 2004 Classification and evaluation of spider (Araneae) assemblages on environmentally sensitive areas in Northern Ireland. *Agriculture, Ecosystems and Environment* **102**, 29–40.

Cawley, M. 2001 Distribution records for uncommon spiders (Araneae) including five species new to Ireland. *Bulletin of the Irish biogeographical Society* **25**, 135–43.

Cawley, M. 2007 The spider (Arachnida: Araneae) fauna of Tory Island, Co. Donegal, Ireland. *Bulletin of the Irish biogeographical Society* **31**, 20–43.

Cawley, M. 2009 A summary of new Irish county records for spiders (Araneae). *Bulletin of the Irish biogeographical Society* **33**, 184–221.

Cawley, M. and Nolan, M. 2007 *Clubiona frutetorum* L. Koch from Ireland. *Newsletter of the British arachnological Society* **110**, 2.

Feehan, J. 2007 *Cuirrech Lifé: The Curragh of Kildare, Ireland*. School of Biology and Environmental Science at UCD in association with the Department of Defence.

Helsdingen, P.J. Van 1996 The county distribution of Irish spiders, incorporating a revised catalogue of the species. *Irish Naturalists' Journal* Special Zoological Supplement.

Lavery, A. 2019 A revised checklist of the spiders of Great Britain and Ireland. *Arachnology* **18**, 196–212.

Locket, G.H., Millidge, A. F and Merrett, P. 1974 *British Spiders, volume III*. London. Ray Society.

Nentwig, W., Blick, T., Gloor, D., Hänggi, A., Kropf, C. 2020 Spiders of Europe. (Version of May 2020). Online at https://araneae.nmbe.ch.

Nolan, M. 2002 Spiders (Araneae) of montane blanket bog in Co. Wicklow, Ireland. *Bulletin of the Irish biogeographical Society* **26**, 39–59.

Nolan, M. 2013 Spiders (Araneae) of Irish raised bogs: Clara Bog, Co. Offaly and Carrowbehy Bog, Co. Roscommon. *Bulletin of the Irish biogeographical Society* **37**, 172–203.

Nolan, M. 2016a Spiders of the Curragh. A Report for the Heritage Section of Kildare County Council.

Nolan, M. 2016b Records of some rare and uncommon Irish spiders (Arachnida: Araneae). *Bulletin of the Irish Biogeographical Society* **40**, 142–63.

Nolan, M. and McCormack, S. 2004 First Irish record of *Meioneta mossica* Schikora, 1993 (Araneae, Linyphiidae). *Bulletin of the Irish biogeographical Society* **28**, 204–06.

Pack-Beresford, D.R. 1911a Some new Irish spiders. *Irish Naturalist* **20**, 173–7.

Pack-Beresford, D.R. 1911b Clare Island Survey: Araneida. *Proceedings of the Royal Irish Academy* **31**(35), 1–8.

Roberts, M.J. 1993 *The Spiders of Great Britain and Ireland*. Compact edition. I-III (2 vols). Colchester. Harley Books.

Schikora, H.-B. 1993 *Meioneta mossica* sp.n., a new spider close to *M. saxatilis* (Blackwall) from northern and central Europe. *Bulletin of the British Arachnological Society* **9**: 157–63.

Schikora, H.-B. 1995 Intraspecific variation in taxonomic characters, and notes on distribution and habitats of *Meioneta mossica* Schikora and *M. saxatilis* (Blackwall), two closely related spiders from northern and central Europe (Araneae, Linyphiidae). *Bulletin of the British Arachnological Society* **10**, 65–74.

SHRS 2022 Spiders and Harvestman Recording Scheme http://SHRS.britishspiders.org.uk (Version of June 2022).

World Spider Catalog 2019 *World Spider Catalog Version 20.0.*, Natural History Museum, Bern, online at http://wsc.nmbe.ch

APPENDIX 1

Trap number, location (Irish Grid) and habitat information of sampling sites on Clare Island

Trap number	Grid ref	Habitat characterisation
1	L709864	Short vegetation, *Nardus* and mosses near *Sphagnum*-filled pool
2	L706857	Bracken near Poirtin Fuinch L.
3	L705856	In rushes on wet ground with cultivation ridges
4	L705856	In bracken on dryish grassland
5	L695854	Dry grassland on SE slope Knocknaveen
6	L695854	In bracken on dryish grassland SE slope Knocknaveen
7	L689852	Tightly grazed grassland
8	L688854	Short sward of mosses and lichen near margin of infilled lake
9	L689857	In rushes at margin of Creggan L.
10	L689857	Grassy slope above Creggan L.
11	L689856	Amongst rushes
12	L688861	Tightly grazed bog vegetation on area of deep peat deposits
13	L687863	Amongst rushes near Dooree stream
14	L687871	Ungrazed sward outside fence at top of cliff
15	L692874	Tall heather at bottom of slope in Ballytoohy
16	L693877	Wet flush with *Schoenus nigricans*
17	L693880	Coastal grassland-type vegetation short sward of *Festuca* sp
18	L693881	Coastal grassland-type vegetation short sward of *Festuca* sp
19	L695883	Plantain sward near light house
21	L700878	Blanket bog vegetation, *Molinia* dominated
22	L701876	Blanket bog vegetation, *Molinia* dominated
23	L704870	Heathery slope north facing with *Erica erigena*
24	L702870	Woodland, good ground cover *Ranunculus ficaria*, *Oxalis*, ferns plus moss
27	L702870	Woodland
28	L704868	Rushy field behind boulder beach at Portlea
29	L719864	Short grass sward on low exposed promontory
30	L721859	Margin of saltmarsh
31	L721859	Margin of saltmarsh
32	L714853	Short sward on sandy soil in campsite
33	L713853	Meadows at Fawnglass
34	L713853	Meadows at Fawnglass
35	L703864	Ungrazed Festuca sward near castle
36	L703864	Bank of Dooree Rr. Near L. Avullin
37	L703865	Bank of Dooree Rr. Near L. Avullin
38	L671863	Ungrazed *Luzula* outside fence near top of Knockmore
41	L662856	350m up S side Knockmore, dry grassland mostly
43	L654852	Plantain sward at west end of island
45	L665843	Plantain sward at west end of island near peaty pools
46	L665843	Plantain sward at west end of island near sea
47	L712860	Tightly grazed sward amongst Gorse bushes
48	L712859	Tightly grazed sward amongst Gorse bushes

APPENDIX 2

SPECIES CATALOGUE
Spiders are listed alphabetically by family, genus, species. Nomenclature follows Lavery (2019) which followed the World Spider Catalog (WSC 2019) (with a small number of exceptions noted in the text). Also noted are the number and gender of specimens identified, collection dates, trap-methods, collector/s, grid references and location/habitat-type/microsite. Records are detailed in chronological order other than occasionally for the sake of brevity. An asterisk * indicates a species new to Clare Island; a dagger † indicates a new county record for Co. Mayo; a degree sign ° indicates a species recorded on Clare Island on the resurvey but by DRPB only on the Mayo coast. Numbers of specimens identified are shown in parentheses after the species name.

Abbreviations: H – hand, MC – Martin Cawley, MN – Myles Nolan, MT – malaise trap, MTT – Malaise tray-trap, PT – Pitfall trap, SMcC – Stephen McCormack, SP – Sweep-net.

AGELENIDAE

Tegenaria domestica (Clerck, 1757) (2)
1♀ 23 May, H, MN, L706864, cowshed; 1♂ 23 May, H, MN, L688844, under a slate in the church. Both specimens moulted to adulthood in June.

Textrix denticulata (Olivier, 1789) (9)
7 May, H, MC L649841 exposed coastal headland, 1 immature 7 May, H, MC L652853 Signal Tower, heavily grazed heathy grassland; 1♂3♀ 20 May, H, MN, L714854 in webs under stones at back of beach; 1♂ 7–28 June, PT, L693877, wet flush with *Schoenus nigricans*; 1♂ 7–28 June, PT, L712860 grazed sward amongst *Ulex*; 1♂ 31 July-23 August, PT, L693880 coastal short *Festuca* sward.

AMAUROBIIDAE

†**Amaurobius ferox* (Walckenaer, 1830) (1)
1♀ 20 September, H, MC, L714850, waste ground at castle ruins.

Amaurobius similis (Blackwall, 1861) (2)
1♀ 23 May, H, MN, L706864 from inside a cowshed; 1♂ 22 September, H, MN, L713852, *Ammophila*, caught while entering the web of a female (presumably of the same species), the latter rapidly retreated sensing the struggle.

ARANEIDAE

Araneus diadematus Clerck, 1757 (2)
1 immature 23 September, SP, MN, L665846, roadside rushes; 1 immature 24 September, SP, MN, L711858, *Fuchsia*.

**Araniella opistographa* (Kulczynski, 1905) (1)
1♂ 21 May, SP, SmcC, L702870, on *Pteridium* near woodland.

*°*Larinioides cornutus* (Clerck, 1757) (3)
1♀, 7 May, H, MC, L713852, sandy foreshore; 1♀ 23 May, SP, MN, L696876, from web on *Calluna*; 1♀ 27 July, H, MN, L703867, immature from retreat on sapling at woodland edge – specimen matured in captivity.

*°*Zygiella atrica* (C.L.Koch, 1845) (1)
1♂ 17 September, H, MC, L7286, in a depression on a disturbed rocky area at high water mark.

Zygiella x-notata (C.L.Koch, 1845) (2)
1♀ 17 February, H, SMcC, L705864, house; 1♀ 22 September, H, MN, L715855, exterior wall of hotel.

CHEIRACANTHIIDAE

**Cheiracanthium erraticum* (Walckenaer, 1802) (2)
2♀ 18 September, SP, MC, L702870 Portlea, swept from rank vegetation bordering deciduous woodland.

CLUBIONIDAE

**Clubiona comta* C.L.Koch, 1839 (4)
4♂ 18 April-21 May, MTT, L702870, woodland.

Clubiona neglecta O.P.-Cambridge, 1862 (3)
1♂ 7 May, H, MC, L676841 Craigmore, under a piece of wood on a low sea-cliff; 1♂ 10 May, H, MC, L676863, damp grassland at 250m; 1♂ 7–28 June, PT, L705856, rushes on wet ground.

Clubiona phragmitis C.L.Koch, 1893 (23)

From MTT, L703863, Lough Avullin, *Phragmites* swamp: 1♂ 25 March-5 April, 6♀ (3 specimens in MT bottle) 4–18 April, 4♀ ?-26 July; 1♀ 21 May, H, MN, L721859, saltmarsh under man-made debris; 1♀ 24 May, H, MN, L705867, under stones in brood cocoon; 1♀ 27 July, H, MN, L705867, under boulder; 3♂6♀ 22 September, H, MN, L716856, under and amongst slates on cliff bedrock – numerous silk cells.

Clubiona reclusa O.P.-Cambridge, 1863 (4)

3♀ 9–10 July, H, SMcC, L713853 ditch vegetation and meadow; 1♂ 26 July, H, MN, L713852, *Ammophila* dune.

†*Clubiona stagnatilis* Kulczynski, 1897 (2)

1♂ 22 May, SP, MN, 1♂ 25 March-5 April, MTT, L703863, *Phragmites* beds at Lough Avullin, *Phragmites* swamp.

Clubiona trivialis C.L.Koch, 1843 (6)

All 23 May, H: 1♂1♀ MN, L690874, shaken from *Calluna*; 2♂ MN, L699877, *Molinia* on blanket bog; 1♀ SMcC, L696876 shaken from *Calluna*, **1♂** 16 September, H, MC, L705864 Maum, under stones on a roadside bank.

CYBAEIDAE

*°*Cryphoeca silvicola* (C.L.Koch, 1834) (8)

2♂ 27 March-10 April, PT, tall *Calluna*; 2♀ 23 May, H, MN, L690874, shaken from *Calluna*; 1♂ 7–27 June, PT, L665843, *Plantago* sward; 1♂2♀ 16 September, H, MC, L705864 Maum, roadside bank.

DICTYNIDAE

Argenna subnigra (O.P.-Cambridge, 1861) (1)

1♀ 24 July, H, MN, L700844, in a web set in a crack on *Verrucaria* encrusted coastal rocks (thus within splash zone).

Argyroneta aquatica (Clerck, 1757) (3)

1♀ 8 February, SP, SMcC, L691855, Lough Leinapollbruty, submerged lakeshore vegetation; 2♂ 24 and 25 July, SP, MN and SMcC, L651842 submerged lakeshore vegetation.

DYSDERIDAE

Harpactea hombergi (Scopoli, 1763) (3)

1♀ 20 May, H, MN, L714854 on shed near beach; 1♀ 24 July, H, MN, L700844 from egg-cocoon in a nook in cliff wall; 1♀ 27 July, H, MN, L711856 from a stone wall.

GNAPHOSIDAE

Drassodes cupreus (Blackwall, 1834) (5)

1♂ 10 May, H, MC, L712859, from a rocky scrape in a field; 1♂ 7–28 June, PT, L712859, grazed sward amongst *Ulex*; 1♂ 7–28 June, PT, L700878, *Molinia* dominated blanket bog; 1♀ 27 July, H, MN, L705867 female in brood cocoon with hatched spiderlings, amongst boulders at back of beach; 1♂ no details.

HAHNIIDAE

*°*Antistea elegans* (Blackwall, 1841) (6)

1♀ 25 March-5 April, 1♀ 4–18 April, MTT, Lough Avullin, *Phragmites* swamp; 1♀ 7–28 June, PT, L687863 streamside rushes; 2♂ 30 July-23 August, PT, L705856, rushes on wet ground; 1♂ 31 July-23 August, PT, L704870, heathery area with *Erica*.

LIOCRANIDAE

Liocranoeca striata (Kulczynski, 1882) (3)

1♂ 19 May, 2♀ 21 May, H, MN and SMcC, L721859, sieved from reed litter and from under tidal detritus on stones at edge saltmarsh.

*°*Scotina gracilipes* (Blackwall, 1859) (1)

1♂ 31 July-23 August, PT, L700878, *Molinia* dominated blanket bog.

LINYPHIIDAE

Agyneta cauta (O.P.-Cambridge, 1902) (3)

1♂ 7–28 June, PT, L700808, *Molinia* dominated blanket bog; 1♂ 7–26 June, PT, L693880, short coastal *Festuca* sward; 1♂ 7–28 June, PT, L712859, grazed sward amongst *Ulex*.

Agyneta conigera (O.P.-Cambridge, 1863) (8)

3♂2♀ 23 May, SP and H, MN, L699877, *Molinia* on blanket bog and L696876, *Calluna* on blanket bog; 2♂ 7–28 June, PT, L712859, 1♂ 7–28 June, PT, L712860, heavily grazed sward amongst *Ulex*.

Agyneta decora (O.P.-Cambridge, 1871) (13)

1♂ 27 March-10 April, PT, L703864, bank of the Dooree river near Lough Avullin; 1♂ 4 April-18 April, MTT, L702870, woodland; 1♂ 18 April-24 May, MTT, L702870, woodland; 1♂ 4 April-18 April, MTT, L703864, Lough Avullin, *Phragmites* swamp; 1♀ 24 May, H, MN, L706866, under stone; 1♂ 7 June-28 June, PT, L689856, rushes; 1♂ 7 June-28 June, PT, L692874, tall *Calluna*; 1♂ 24 July, H, MN, L700844 coastal vegetation; 1♂ 30 July-23 August, PT, L709864 *Nardus* and mosses near *Sphagnum*; 1♂1♀ 30 July-23 August, PT, L665843 coastal *Plantago* sward; 2♂ PT, L692874, tall *Calluna* (no dates).

Agyneta mossica (Schikora, 1993) (1)

1♂ 7–28 June, PT, L700878, *Molinia* dominated blanket bog;

†**Agyneta saxatilis* (Blackwall, 1844) (1)

1♂ ?-26 July, MTT, L703863, Lough Avullin, *Phragmites* swamp.

Agyneta subtilis (O.P.-Cambridge, 1863) (10)

All 7–27/28 June, PT, 1♂ L692874, tall *Calluna*, 1♂ L705856, rushes on wet ground, 1♂ L665843, *Plantago* sward, 7♂ L712859, L712860, heavily grazed sward amongst *Ulex*.

†*Allomengea vidua (L.Koch, 1879) (1)

1♂ 31 July-23 August, PT, L703864, bank of the Dooree river near Lough Avullin.

†*Aphileta misera (O.P.-Cambridge, 1882) (1)

1♂ ?-26 July, MTT, L703863, Lough Avullin, *Phragmites* swamp.

**Baryphyma trifrons* (O.P.-Cambridge, 1863) (1)

1♀ 22 May, SP, MN, L711856 from *Ulex*.

†**Bathyphantes approximatus* (O.P.-Cambridge, 1871) (1)

1♂ 4–18 April, MTT, L703864, Lough Avullin, *Phragmites* swamp.

*°*Bathyphantes gracilis* (Blackwall, 1841) (36)

1♀ 14 November 2001, H, SMcC, L6987, under stones; 3♀ 19 February, H, SMcC, L703863, Lough Avullin, *Phragmites* litter. From MTT, L703863, Lough Avullin, *Phragmites* swamp: 2♂1♀ 25 March-5 April; 4♂3♀ 4–18 April; 1♂1♀ 22 May, SP, MN; 1♀ 26 July, H, MN. 1♀ 27 March-10 April, PT, L704868, marshy field. From MTT, L702870 mixed deciduous woodland: 1♂2♀ 4–18 April; 2♂ 18 April-21 May; 1♂1♀ 28 June-19 July. 1♀ 20 May, H, MN, L714854 overhanging vegetation on beach; 1♀ 22 May, H, MN, L713852, *Ammophila*; 1♀ 24 May-7 June, PT, L689856, rushes; 1♀ 7–28 June, PT, L695854, dryish grassland; 1♀ 7–27 June, PT, L703865, bank of Dooree river; 1♀ 7–27 June, PT, L713853 meadow; 1♀ 30 July-23 August, PT, L705856, rushes on wet ground; 1♀ 30 July23 August, PT, L712859, grazed sward amongst *Ulex*; 2♀ 31 July-23 August, PT, L703864, bank of Dooree river; 1♀ 22 September, H, MN, L713852, *Ammophila*.

*°*Bathyphantes nigrinus* (Westring, 1851) (11)

All specimens from woodland L702870: 3♂ 24 March-4 April, 3♂ 4 April-18 April, 3♂ 18 April-21 May, all MTT; 1♀ 19 May, H, SMcC, moss and leaf litter, 1♂ 27 July SP, MN.

**Bathyphantes parvulus* (Westring, 1851) (21)

2♂ 24 March-4 April, MTT, L702870, woodland; 1♀ 7–27 June, PT, L703865, bank of Dooree river; 4♂1♀ 7–27 June, PT, L704870, heathy slope with *Erica*; 1♂1♀ 7–27 June, PT, L713853 meadow; 4♂ 7–27 June, PT, L714853 short sandy sward;

1♂ 7–28 June, PT, L689856, rushes; 3♂1♀ 7–28 June, PT, L704868, marshy field;

1♂ 26 July, H, MN, L711856 *Ulex* along Maum road; 1♀ 31 July-23 August, PT, L703864, bank of Dooree river.

*°*Bolyphantes luteolus* (Blackwall, 1833) (5)

1♂2♀ 16 September, H, MC, L705864 Maum, roadside bank; 1♂ 18 September, H, MC, L7286 Ooghduff, coastal headland; 1♀ 23 September, H, MN, L694864 from *Calluna* overhanging a streamlet.

*°*Centromerita bicolor* (Blackwall, 1833) (8)

1♂1♀ 14 November 2001, H, SMcC, L694882 coastal grassland near cliff; 1♂ 1 December 2001, H, SMcC, L698856 *Juncus*; 1♀ 14 February, H, SMcC, L721859, saltmarsh. 27 March-10 April: 1♀, PT, L654852 *Plantago* sward; 3♀, PT, L721859 edge of saltmarsh.

Centromerita concinna (Thorell, 1875) (20)

1♂1♀ 14 November 2001, H, SMcC, L694882 *Plantago* near lighthouse; 1♀ 1 December 2001, H, SMcC, L675850 *Juncus* tussock beside turf clamp @ 264 m; 1♂ 1 December 2001, H, SMcC, L705867, *Molinia* at rear of boulder beach. Many remaining specimens from 27 March-10 April, PT: 1♀, L662856, dryish grassland, 1♀, L688861, grazed bog vegetation on deep peat; 1♀, L689852, heavily grazed grassland; 1♀, L689857, rushes at margin Lough Creggan; 1♀, L689857, grassy slope; 1♀, L703864, bank of Dooree river. Remaining specimens on 23 September, H, MN: 1♂ immature L667858, Knockmore, under stone at cairn; 2♀, L668860 Knockmore, amongst dense cushion moss; 1♀, L668860, Knockmore, under stone in area of cut peat; 1♀, L668860, Knockmore, under stone; 1♂, L669861, Knockmore, on web deep under stone; 2♂, L694864, Dooree streamlet, shaken from overhanging *Calluna*; 1♀ 24 September, H, MN, L655845, gravel plain.

Centromerus dilutus (O.P.-Cambridge, 1875) (1)

1♀ 25 March-5 April, MTT, L703863, Lough Avullin, *Phragmites* swamp.

Centromerus prudens (O.P.-Cambridge, 1873) (12)

1♂ 18 September, H, MC, L672857 Knockmore, under stones on gravelly grazed hillside; 1♀ 18 September, H, MC, L668861, summit of Knockmore under stones; 2♂3♀ 23 September, H, MN, L668860, under stones at summit Knockmore, one from a sheet web; 2♂3♀ 24 September, H, MN, L653848 from scree and loose soil on cliffs.

Ceratinella brevipes (Westring, 1851) (10)

1♂ 28 November 2001, H, SMcC, L675850 amongst *Juncus*; 1♂ 15 February, H, SMcC, L704874 heathy cliff; 1♀ 24 March-4 April, 1♂ 4–18 April, MTT, L702870, woodland; 2♂ 25 March-5 April, 1♂ 4–18 April, MTT, L703863, Lough Avullin, *Phragmites* swamp; 1♂ 27 March-10 April, PT, L703864, bank of river Dooree; 2♀ 7–28 June, L712859, L712860 grazed sward amongst *Ulex*.

Ceratinella brevis (Wider, 1834) (3)

1♂ 25 March-5 April, MTT, L703863, Lough Avullin, *Phragmites* swamp; 1♀ ?-26 July, MTT, L703863, Lough Avullin, *Phragmites* swamp; 1♀ 31 July-23 August, PT, L688854, lacustrine short sward of mosses and lichens.

†*Cnephalocotes obscurus* (Blackwall, 1834) (1)

1♂ 7 June-28 June, PT, L700878, *Molinia* dominated blanket bog.

Dicymbium nigrum (Blackwall, 1834) (31)

Two records of *Dicymbium brevisetosum* Locket, 1962 are included here. The species is considered a form of *D. nigrum* by Roberts (1993) and this view is accepted here. Only males can be assigned to the taxon.

1♀ 28 November 2001, H, SMcC, L675850 amongst *Juncus*; 1♂ 27 March-10 April, PT, L688854, lakeside moss and lichen sward; 2♀ 27 March-10 April, PT, L689856, rushes; 1♂ 27 March-10 April, PT, L693877, wet flush with *Schoenus*; 1♂ 25 March-5 April, MTT L703863, Lough Avullin; 1♂ f. *brevisetosum* 4–18 April, MTT, L703863, Lough Avullin; 2♂ 4–18 April, MTT, L702870, woodland; 1♀ 24 May-7 June, PT, L689852, heavily grazed grassland; 2♀ 7–28 June, PT, L689856, rushes; 2♀ 23 May, H, L696876, *Calluna*; all specimens 23/24 May-7 June, PT, 1♀ L695854, dryish grassland, 1♂ L688861, grazed bog vegetation on deep peat, 1♂1♀ L705856, rushes on wet ground; 2♀ 7–28 June, PTx2, L705856, rushes on wet ground; 1♀ 7–27 June, PT L703864, bank of the Dooree river near Lough Avullin; 1♀ 24 July, H, L695843, Portnakilly, under straw debris on cliff top; 1♂1♀ 31 July-23 August, PT, L703864, bank of the Dooree river near Lough Avullin; 3♀ 31 July-23 August, PT, L705856, rushes on wet ground; 1♀ 31 July-2 September, PT, L695854 *Pteridium* grassland; 1♂1♀ 30 July-22 August, PT, L702870, woodland; 1♂ f. *brevisetosum* 20 September, H, MC, L715849, sheltered coastal headland.

Dicymbium tibiale (Blackwall, 1836) (52)

The vast majority of specimens from MTT in woodland L702870, 4♂3♀ 24 March-4 April; 3♂4♀ 4–18 April; 1♂ 18 April-21 May; 11♂3♀ 28 June-19 July; 10♂3♀ 19 July-15 August; also from this location 1♀ 21 May, H, MN, sieved from moss on rocks and 2♂2♀ 7–27 June, PT; 1♀ 27 March-10 April, PT, L705856 *Pteridium* on grassland; 1♂1♀ 27 March-10 April, PT, L662856, grassland on Knockmore; 1♀ 24 May-7 June, PT, L698957, rushes at Lough Creggan; 1♀ 7–28 June, PT, L688854, lakeside moss and lichen sward.

†*Diplocephalus cristatus* (Blackwall, 1833) (1)

1♂ 15 February, H, SMcC, L703871, heath at Ballytuohy cliffs.

Diplocephalus permixtus (O.P.-Cambridge, 1871) (15)

1♀ 25 March-5 April, MTT, L703863, Lough Avullin, *Phragmites* swamp. The following all from L721859, saltmarsh: 2♀ 27 March-10 April, PT; 2♀ 21 May, H, MN, under stone embedded on very wet mud; 1♀ 7–28 June, PT, 1♀ 30 July-23 August, PT. 1♂ 27 March-10 April, PT L654852 *Plantago* sward; 1♀ 10 April-21 May, 2♀ 7–27 June, 3♀ (dates uncertain) PT, L703864, bank of Dooree river; 1♂ 24 September, H, MN, L655844, shaken from marginal lake mosses.

Diplostyla concolor (Wider, 1834) (12)

1♂1♀ 7 May, H, MC, Strake L667845, under a piece of wood on road verge; 1♀ 7 May, H, MC, L713852, sandy foreshore; 2♂2♀ 20 May, H, MN and SMcC, L714854, under stones at back of beach; 1♂1♀ 25 July, H, MN, L655851 Loughanaphuca, under stones - the female moulted to adult on 26 July. 1♂1♀ 18 September, H, MC, L673848 south-east side of Knockmore at base of a stone wall; 1♂ 19 September, H, MC, L673848 Fawnglass, base of stone wall.

°Dismodicus bifrons (Blackwall, 1841) (12)

4♂2♀, 7 May, H, MC, L713852, sandy foreshore; 2♀ 19 and 20 May, H, MN and SMcC, L714854, under plastic and at shed base; 1♀ 22 May, H, MN, L713852 amongst *Ammophila*; 1♂ 7–28 June, PT, L704868, marshy field; 1♀ 28 June-19 July, MTT, L702870, woodland; 1♀ 23 September, H, MN, L667864 northern-facing slope amongst *Luzula*.

Drepanotylus uncatus (O.P.-Cambridge, 1873) (7)

1♂1♀ 25 March-5 April, MTT, L703863, Lough Avullin, *Phragmites* swamp; 3♀ 27 March-10 April, PT, L719864, short exposed sward; 1♂ 27 March-10 April, PT, L712859, saltmarsh margin; 1♂ 4–18 April, MTT, L703863, Lough Avullin, *Phragmites* swamp.

†*Entelecara erythropus* (Westring, 1851) (7)

2♂2♀ 20 May, H, MN, L714854, under logs and concrete pipe close to a shed; 2♀ 21 May, H, MN, L705864 in a shed; 1♀ 24 May, L706866 Portlea, under a stone.

Erigone atra Blackwall, 1833 (18)

1♂1♀ 15 November 2001, H, SMcC, L695843 cliff base grassland; From L721859, saltmarsh: 1♀ 14 February, H, SMcC, 4♀ 27 March-10 April, PT, 2♂ 30 July-23 August, PT; 1♂ 25 March-5 April, MTT, L703863, Lough Avullin, *Phragmites* swamp; 2♀ 7 May, H, MC, L713852, sandy foreshore; 1♂ 22 May, H, SMcC, L714854 on car at Community centre; 1♂ 31 July-2 September, PT, L695854 *Pteridium* on dryish grassland; 4♂ 22 September, H, MN, L713852, *Ammophila* at beach.

Erigone dentipalpis (Wider, 1834) (24)

1♂ 28 November 2001, H, SMcC, L675850 *Juncus*. All PT, L689852, heavily grazed grassland, 1♂ 27 March-10 April, 1♂ 24 May-7 June, 3♂ 7–28 June, 1♀ 31 July-23 Julyi; 1♀ 7–28 June, PT, L693877, wet flush with *Schoenus*; All PT, short coastal *Festuca* sward, 1♂ 7–28 June, L693881, 1♂1♀ 31 July-23 August, L693880, 1♂ 31 July-23 August, L693881; 1♂1♀ 31 July-23 August, PT, SMcC, L665843, *Plantago* sward; 4♂1♀ 31 July-2 September, PT, SMcC, L695854, dryish grassland; 1♂ 22 September, H, MN (no other data available); 1♀ 17 September, H, MC, L7286, in a depression on a disturbed rocky area at high water mark; 1♀ 19 September, H, MC, L673848 Fawnglass, base of stone wall; 1♂ 19 September, H, MC, L699870, overgrown garden; 1♂ 24 September, H, MN, L653848 Loughanaphuca from scree cliffs.

Erigone promiscua (O.P.-Cambridge, 1872) (12)

2♂ 15 February, H, SMcC (no location); 2♂1♀ 18 January-5 February, 1♂ 7–28 June, PT, L709864 *Nardus* and mosses near *Sphagnum*; 1♂ 7–28 June, PT, SMcC, L693877, wet flush with *Schoenus*; 4♀ 25 July, H, MN, L655845, marshy area from amongst stones and wet mosses; 1♂ 23 September, H, MN, L666859, Knockmore from slope below cairn level.

Erigonella hiemalis (Blackwall, 1841) (4)

2♂ 4–18 April, MTT, L702870, woodland; 1♀ 23 May-7 June, PT, L695854, 1♀ 24 May-7 June,

PT, L705856, both traps amongst *Pteridium* on grassland.

†*_Gnathonarium dentatum_ (Wider, 1834) (45)

All specimens from L703863, *Phragmites* beds at Lough Avullin, *Phragmites* swamp: 14♂3♀ 25 March-5 April, MTT; 15♂1♀ 4 April-18 April, MTT; 1♂10♀ 22 May, SP, MN; 1♀ 26 July, H, MN, from litter.

*°*Gonatium rubens* (Blackwall, 1833) (7)

1♀ 15 February, H, SMcC, L695843, *Armeria* tussock; 1♀ 27 March-10 April, PT, L692874, tall *Calluna*; 1♀ 24 May-7 June, PT, L692874, tall *Calluna*; 1♀ 7–28 June, PT, L700878, *Molinia* dominated blanket bog; 1♀ 7–28 June, PT, L7112859, grazed sward amongst *Ulex*; 1♀ 26 July, SP, MN, L703863 from *Ulex*; 1♀ 16 September, H, MC, L694879, south of Lecknacurra on a hillside at base of *Calluna*.

Gongylidiellum vivum (O.P.-Cambridge, 1875) (9)

1♀ 28 November 2001, H, SMcC, L675850 *Juncus*; 1♂ 4–18 April, MTT, L703863, Lough Avullin, *Phragmites* swamp; 1♂ 10 April-21 May, PT, L703864, bank of Dooree river; 1♂ 7 June-28 June, PT, L695883, *Plantago* sward; 1♂ 7–28 June, PT, L693877, wet flush with *Schoenus*; ♀1, 7 June, 28 June, PT, SMcC, L695854, dryish grassland; 1♂ 30 July-23 August, PT, L721859, saltmarsh margin; 1♀ 23 September, H, MN, L857684, under stone. Another specimen was collected in PT L693881 on short coastal *Festuca* sward – no collection dates available.

†*_Gongylidium rufipes_ (Linnaeus, 1758) (1)

1♂ 18 April-21 May, MTT, L702870, woodland.

*°*Halorates reprobus* (O.P.-Cambridge, 1879) (1)

1♀ 9 May, H, MC, L701845 Portnakilly, in low fissures at high water mark on coast. 1♂ collected 7 December – location uncertain.

Hypomma bituberculatum (Wider, 1834) (277)

The vast majority of specimens came from the *Phragmites* swamp at Lough Avullin L703863: 10♂3♀ 25 March-5 April, MTT; 168♂47♀ 4–18 April, MTT; 3♂6♀ 4–18 April, MT; 3♂21♀ 22 May, SP, MN and SMcC, 1♀ ?-26 July, MTT.

1♂ 9 May, H, MC, L695843, Portnakilty, damp coastal grassland; 2♀ 19 May, 2♂6♀ 21 May, H, MN, L721859, saltmarsh, from under debris and discarded objects or stones; 1♀ 21 May, SP, SMcC, L704863 from thistle near *Phragmites* beds; 1♀ 7–27 June, PT, L713853 meadow; 1♀ 24 July, H, SMcC, L712859 marshy area; 1♀ 31 July-23 August, PT, L703864, bank of Dooree river.

Leptorhoptrum robustum (Westring, 1851) (12)

1♂ 11 May, H, MC, L712851, under wood in abandoned garden; 1♂2♀ 7–27 June, PT, L714853 short sward on sandy soil in campsite; 1♀ 7–27 June, PT, L713853 meadow; 1♂ 7–27 June, PT, L703865, bank of Dooree river; 2♀ 24 July, H, MN, L691841, one under a concrete pipe with two egg-sacs, the other under a stone with one egg-sac; 4♂ 31 July-23 August, PT, L703864, bank of Dooree river.

Linyphia triangularis (Clerck, 1757) (7)

1♂ 26 July, H, M, L703863, *Ulex*; 3♂ 27 July, H, MN, L702870, woodland from a tree, grasses and a sheet-web; 1♂ 22 August, light-trap, SMcC, L702870 near woodland; 1♂1♀, H, MN, L711856 from sheet-webs.

Lophomma punctatum (Blackwall, 1841) (18)

2♂1♀ 24 March-4 April, MTT, L702870, woodland; 4♂2♀ 25 March-5 April, MTT, L703863, Lough Avullin, *Phragmites* swamp; 1♂ 27 March-10 April, PT, L702870, woodland; 1♀ 27 March-10 April, PT, L654852 *Plantago* sward; 2♂1♀ 4–18 April, MTT, L703863, Lough Avullin, *Phragmites* swamp; ♀1 21 May, H, MN, L721859, under stone on very wet mud in saltmarsh; ♀1 23 May, H, MN, L692878 summit; 1♀ 7–27 June, PT, L703865, bank of Dooree river; 1♀ 30 July-23 August, PT, L712859, grazed sward amongst *Ulex*.

Micrargus herbigradus (Blackwall, 1854) (6)

1♂ 24 March-4 April, MTT, L702870, woodland; 2♂ 25 March-5 April, MTT, L703863, Lough Avullin, *Phragmites* swamp; 1♀ 4 April-18 April, MTT, L702870, woodland; 1♂ 7–28 June, PT, L705856, rushes on wet ground;1♀ 28 June-19 July, MTT, L702870, woodland.

Micrargus subaequalis (Westring, 1851) (4)

2♂ 7–28 June, PT, L712859, lightly grazed sward amongst *Ulex*; 1♀ 30 July-23 August, PT, L709864, short *Nardus* and moss sward; 1♂ 24 July, H, MN, L691841, under stone on grass.

**Microctenonyx subitaneus* (O.P-Cambridge, 1875) (2)

2♀ 20 September, H, MC, L714859 Capnagower, sieved from hayshed debris.

*°*Monocephalus castaneipes* (Simon, 1884) (5)

All specimens were from the patch of mixed deciduous woodland at Portlea, L702870: 1♀ 18 April-24 May, MTT; 3♀ 25 May, H, MN, sieved from tree-trunk mosses; 1♀ 19 September, H, MC.

Monocephalus fuscipes (Blackwall, 1836) (38)

Most specimens were collected from L702870 Portlea, mixed deciduous woodland: 1♀ 16 February, H, SMcC; 12♂ 24 March-4 April, MTT; 1♂ 27 March-10 April, PT; 14♂1♀ 4–18 April, MTT; 2♀ 21 May, H, MN, from mosses on rocks; 1♀ 28 June-19 July, MTT. Also 1♂ 14 February, H, SMcC, L705867, *Pteridium* litter; 1♀ 22 May, H, MN, L712859, under roadside stone; 1♀ 7–28 June, PT, L712860 grazed sward amongst *Ulex*; 1♀ 19 September, H, MC, L702870 Portlea, mixed deciduous woodland; 1♂ 23 September, H, MN, L667858 summit of Knockmore under stone. 1♀ PT, L671863, ungrazed *Luzula* outside fence near top of Knockmore (no dates).

Neriene clathrata (Sundevall, 1830) (3)

1♂ 4–18 April, 1♂ 18 April-21 May, MTT, L702870, woodland; 1♂ 22 September, SP, MN, L716856 cliff vegetation.

†*Neriene montana (Clerck, 1757) (22)

All specimens were collected in MTT in woodland L702870.

7♂1♀ 4–28 April; 1♂ 18 April-21 May; 7♂ 18 April-21 May, MTT; 1♀ 21 May, H, MN; 1♀ 23 May, H, MN; 1♀ 27 July, H, MN, who produced an egg-sac after capture.

†*Neriene peltata (Wider, 1834) (1)

1♂ 21 May, H, SMcC, L702870, *Betula* in woodland.

* *Obscuriphantes obscurus* (Blackwall, 1841) (11)

1♂ 24 March-4 April, 1♂ 4–18 April, MTT, L702870, woodland; 1♂ 22 May, SP, MN, L711856, roadside *Ulex*; ♂3♀3 23 May, H, MN and SMcC, L694877, shaken from *Calluna*, 1♂ L696876 from *Calluna*, 1♀ L692878 summit.

*°*Oedothorax fuscus* (Blackwall, 1834) (59)

1♀ 28 November 2001, H, SMcC, L675850 *Juncus*; 1♀ 1 December 2001, H, SMcC, L698856 Knocknaveen, *Juncus*. From PT in *Plantago* sward: 3♀ 27 March-10 April, L654852; 4♀ 7–28 June, L695883; 2♀ 31 July-23 August, L665843; 4♂3♀ 31 July-23 August, L695883; From PT in short coastal sward L719864: 1♀ 27 March-10 April; 2♀ 24 May-6 June; 2♂1♀ 6 June-27 June. 8♀ 27 March-10 April, PT, L721859, saltmarsh margin; 1♀ 21 May, H, MN, L704863 on a stone near ditch in field; 1♀ 22 May, H, MN, L713852, *Ammophila*; 1♀ 24 May-7 June, PT, SMcC, L714853 short sandy sward; 3♀ 7–26 June, L693880, short coastal *Festuca* sward; 5♀ 7–27 June, PT, L703864, bank of Dooree river; 2♀ 7–27 June, PT, L703865, bank of Dooree river; All 7–28 June, PT: 1♀ L689857, rushes at margin Lough Creggan; 1♀ L687863, rushes near stream; 1♂ L689852, heavily grazed grassland; 1♂1♀ L695854, dryish grassland. 1♂3♀ 7–28 June, PT, L693877, wet flush with *Schoenus*; 2♀ 7–28 June, PT, SMcC, L721859, saltmarsh margin; 1♂1♀ 24 July, H, MN, L691841, under stone and debris on grass; 1♂ 25 July, H, MN, L655851 Loughanaphuca, under stone; 1♀ 30 July-23 August, PT, L712859, grazed sward amongst *Ulex*; 1♀ 30 July-23 August, PT, L721859, saltmarsh margin; 1♀ 31 July-2 September, PT, L695854 *Pteridium* on dryish grassland; 1♂1♀ 31 July-23 August, PT, L689852, heavily grazed grassland; 1♂ 1 August, SP, SMcC, L713853 meadow; 1♂1♀ 16 September, H, MC, L694879, south of Lecknacurra on a hillside at base of *Calluna*; 1♂ 17 September, H, MC, L7286, in a depression on a disturbed rocky area at high water mark; 1♀ 19 September, H, MC, L673848 Fawnglass, base of stone wall; 4♀ 22 September, H, MN, L713852, *Ammophila*; 1♀ 23 September, H, MN, L694864, shaken from *Calluna* overhanging a streamlet. No dates for the following: 1♀ PT, L693881 short coastal *Festuca* sward; 3♀ PT, L695883, *Plantago* sward; 3♀ PT, L703864, bank of Dooree river.

Oedothorax gibbosus (Blackwall, 1841) (6)

Two specimens of *Oedothorax gibbosus* forma *tuberosus* (Blackwall, 1841) are included here. Only males can be assigned to the taxon.

1♂ 10 May, H, MC, L676863 damp grassland at 250m; 1♂ f. *tuberosus* 7–28 June, PT, L689856, rushes; 1♂ f. *tuberosus* ?-26 July, MTT, L703863, Lough Avullin, *Phragmites* swamp; 2♀ 30 July-23 August, PT, L705856 *Pteridium* on dryish grassland; 1♀ 31 July-23 August, PT, L703864, bank of Dooree river.

Oedothorax retusus (Westring, 1851) (61)

1♀ 21 May, H, MN, L704863, under stone in field; 1♀ 23 May, SP, SMcC, L694877 Ballytuohy, from *Calluna*. 2♂ 4–18 April, MTT, Lough Avullin, *Phragmites* swamp. All the following from PT amongst *Pteridium* on dryish grassland: 1♀ 23 May-7 June, L705856; 1♂1♀ 24 May-7 June, L695854; 4♀ 7 June-28 June, L695854; 1♂1♀ 30 July-23 August, L705856; 4♀ 31 July-2 September, L695854. 1♂2♀ 7–27 June, PT, L703864, bank of Dooree river; 3♂4♀ 7–27 June, PT, L702870, woodland; 1♀ 7–27 June, PT, L703865, bank of Dooree river; All the following 7–28 June, PT: 1♂ L695883, *Plantago* sward; 5♂ L689852, heavily grazed grassland; 3♂1♀ L705856, rushes on wet ground; 1♂2♀ L695854, dryish grassland. 1♀ 30 July-23 August, PT, L706857, *Pteridium*; 2♀ 30 July-23 August, PT, L688861, grazed bog vegetation on deep peat. All 31 July-23 August, PT: 1♂ L689852, heavily grazed grassland; 1♀ L687863, rushes near stream; 8♀ L703864, bank of Dooree river. 1♀ 31 July-2 September, PT, L695854, dryish grassland; 3♂4♀ L703864, bank of Dooree river (no dates).

Palliduphantes ericaeus (Blackwall, 1853) (19)

2♂ 4–18 April, MTT, L702870, woodland; 1♂, 7 May, H, MC, L713852, sandy foreshore; 1♀ 21 May, H, MN, L701868 near woodland on *Calluna*; 2♂1♀ 21 May, H, MN, L696876, under stones, on *Calluna* and at wall base; 1♀ 23 May, H, MN, L696876, *Calluna*; 1♂ 7–28 June, PT, L712859, grazed sward amongst *Ulex*; 1♂1♀ 26 July, H, MN, L710854 amongst stones; 1♀ 30 July-23 August, PT, L705856 *Pteridium* on dryish grassland; 1♀ 30 July-23 August, PT, L721859, saltmarsh margin; 1♂ 31 July, SP, SMcC, L711856 *Pteridium*; 1♂ 31 July-23 August, PT, L704868, marshy field; 1♂1♀ 31 July-23 August, PT, L689852, heavily grazed grassland; 1♀ 23 September, SP, MN, L667864 Knockmore amongst *Luzula* on cliffs. 1♂ PT, L712859, grazed sward amongst *Ulex* (no dates).

†*Parapelecopsis nemoralis* (Blackwall, 1841) (3)

1♀ 21 May, H, MN, L696876, from vegetation at base of wall; 2♀ 25 May, H, MN, L702870, woodland, sieved from epiphytic mosses.

*°*Peponocranium ludicrum* (O.P.-Cambridge, 1861) (17)

1♀ 18 April, H, SMcC, L704866, *Ulex*; 2♀ 21 May, H, MN, L696876 from *Calluna* amongst rocks; 3♀ 22 May, SP, MN, L711856, *Ulex* on Maum road; 6♀ 23 May, H, SP, MN (4), SMcC (2), L696876, roadside *Calluna*; 4♀ 26 July, SP, MN, L703863, *Ulex* at Lough Avullin, *Phragmites* swamp; 1♀ 26 July, SP, MN, L711856, *Ulex* on Maum road.

Pocadicnemis pumila (Blackwall, 1841) (6)

1♀ 15 February, H, SMcC, L701866, vegetation base of wall; 1♂ 23 May, SP, MN, L696876, amongst *Calluna*; 1♂ 24 May, H, MN, L705867, under a stone; 1♀ 7–27 June, PT, L704870, heathery slope with *Erica*; 1♂ 7–28 June, PT, L687863, amongst rushes near stream; 1♀ 7–28 June, PT, L712860, grazed sward amongst *Ulex*; 1♀ 7–28 June, PT, L712859, grazed sward amongst *Ulex*.

Poeciloneta variegata (Blackwall, 1841) (15)

1♀ 15 February, H, SMcC, L701866, stone wall; 1♂ 7 May, H, MC L652853, heavily grazed heathy grassland at Signal Tower; 10 May, H, MC, L698854 Knocknaveen, under stones on summit; 1♀ 10 May, H, MC, L676863, damp grassland at 250m; 1♀ 21 May, H, MN, L704864, under a stone roadside; 1♀ 25 July, 1♂ submature 25 July, H, MN, L655851 Loughanaphuca, from under stones (female had three egg sacs in her web); 1♀ 30 July-23 August, PT, L665843, *Plantago* sward; 1♀ 18 September, H, MC, L672857 Knockmore, under stones on gravelly grazed hillside; 2♂3♀ 23 September, H, MN, L857684, under stones alongside trackway; 1♂ 24 September, H, MN, L652853, under stone.

Porrhomma montanum Jackson, 1913 (3)

2♀, 25 July, H, MN, L655851 Loughanaphuca, under deeply embedded stones; 1♀ 18 September, H, MC, L668861, summit of Knockmore under stones.

Saaristoa abnormis (Blackwall, 1841) (7)

1♀ 23 May, H, MN, L692878, under a plank at summit; 2♂ 7–28 June, PT, L712860 grazed sward amongst *Ulex*; 1♂ 28 June-19 July, MTT, L702870, woodland; 1♂ 19 July-15 August, MTT, L702870, woodland; 1♀ 25 July, H, MN, L655851, under a stone near Loughanaphuca; 1♀ 27 July, H, MN, L702870, woodland leaf litter.

Silometopus elegans (O.P.-Cambridge, 1872) (5)

2♂1♀ 23 May, H, MN, L699877, *Molinia* on blanket bog; 1♀ 7–28 June, PT, L689857, rushes at margin of Lough Creggan; 1♀ 7–28 June, PT, *Molinia* dominated blanket bog.

Taranucnus setosus (O.P.-Cambridge, 1863) (1)

1♂ 28 June-19 July, MTT, L702870, woodland.

* *Tenuiphantes alacris* (Blackwall, 1853) (103)

All specimens from L702870, woodland (or immediately adjacent).

1♂4♀ 15 November 2001, H, SMcC, leaf litter; 1♀ 6 February, H, SMcC, L704868 just outside woodland from *Calluna*; 23♂7♀ 24 March-4 April, MTT; 1♀ 27 March-10 April, PT; 31♂7♀ 4–18 April, MTT; 10♂ 18 April-21 May, MTT; 1♀ 21 May, H, MN; 5♂5♀ 28 June 19 July, MTT; 1♂ 19 July-15 August, MTT; 1♀ 30 July-23 August, PT; 2♀ 27 July, SP, MN, from grasses; 1♂2♀ 19 September, H, MC, L702870 Portlea, mixed deciduous woodland.

* *Tenuiphantes flavipes* (Blackwall, 1854) (1)

1♀ 20 May, H, MN, L714854, under stone on grassy area of beach.

*° *Tenuiphantes mengei* Kulczynski, 1887 (11)

1♂ 14 November 2001, H, SMcC, L6987 on grass under stones. From MTT L703863, Lough Avullin, *Phragmites* swamp: 1♂ 25 March-5 April, 1♀ 4–18 April, 1♂ ?-26 July. 2♂1♀ 21 May, SP, H, MN, L696876, *Calluna*, base of a wall and under a stone; 1♂1♀ 23 May, SP, MN, L699877 *Calluna* leading into blanket bog (♂) and *Molinia* on blanket bog (♀); 1♀ PT, L701876 *Molinia* dominated blanket bog (no dates); 1♀ 7–28 June, PT, L705856, rushes on wet ground.

* *Tenuiphantes tenebricola* (Wider, 1834) (1)

1♂ 4 April-18 April, MTT, L702870, woodland.

Tenuiphantes tenuis (Blackwall, 1852) (36)

1♀ 14 November 2001, H, SMcC, L694882 coastal grassland; 1♀ 28 November 2001, H, SMcC, L675850 from *Juncus*; 1♂ 1 December 2001, H, SMcC, L698856, Knocknaveen slopes from *Juncus*; 2♂ 24 March-4 April, 2♂ 4 April-18 April, MTT, L702870, woodland; 2♂2♀ 25 March-5 April, 1♂1♀ 4 April, 1♂ ?-26 July, MTT, L703863, Lough Avullin, *Phragmites* swamp; 1♀, 7 May, H, MC, L713852, sandy foreshore; 1♂1♀ 20 May, H, MN, L714854, under stones on beach; 1♀ 20 May, H, MN, L714854 from vegetation at base of shed; 1♀ 21 May, H, MN, L701868 on *Calluna* near woodland; 1♀ 21 May, H, SMcC, L706864 near a bridge; 1♀ 25 July, 1♂2♀ 22 September, H, MN, L713852, *Ammophila* dune; 1♀ 27 July, SP, MN, L702870 from a tree in woodland; 1♀ 27 July, H, MN, L704864 small 'bridge' over drainage pipe; From 30 July to 23 August, PT: 2♀, L705856, rushes on wet ground; 1♀, L706857, *Pteridium*; 1♀, L721859, saltmarsh margin; 1♀ L703864, bank of Dooree river. 1♂ 31 July, SP, SMcC, L711856 *Pteridium*; 1♀ 22 September, SP, MN, L716856 cliffs; 1♂ 23 September, SP, MN, L665846, roadside *Digitalis*; 1♂ 24 September, H, MN, L654849 Loughanaphuca under stone; 1♀ L700878, *Molinia* dominated blanket bog (no dates).

Tenuiphantes zimmermanni Bertkau, 1890 (134)

A shade loving species, the majority of specimens were caught in woodland L702870: 4♂4♀ 24 March-4 April, MTT; 16♂3♀ 4–18 April, MTT; 7♂ 18 April-21 May, MTT; 1♀ 19 May, H, SMcC, from moss and leaf litter; ♂3♀3 21 May, H, MN, sieved from rock moss, base of birch tree in web; 19♂4♀ 28 June-19 July, MTT; 1♀ 27 July, SP, MN grasses; 7♂2♀ 19 July-15 August, MTT; 1♂ 30 July-23 August, PT; 3♂1♀ 31 July-22 August, PT; 1♂ PT (no dates). Other specimens: ♀1 14 November 2001, H, SMcC, L694882 cliff coastal grassland; 5♂ 4 April-18 April, MTT, SMcC,

L703863, Lough Avullin, *Phragmites* swamp; 2♂2♀ 21 May, H, MN, L705864, Gill's shed; ♀1 21 May, H, MN, L708860, commonage, under stone amongst *Calluna*; 1♂ 23 May-7 June, PT, SMcC, L705856, rushes on wet ground; 2♂ 23 May, H, SMcC, L694877, from dry vegetative debris in a crack amongst *Calluna* rocks; ♀1 23 May, SP, MN, L694877, *Calluna*; 4♂2♀ 24 May, H, MN, L705867, under stones; 1♂4♀ 24 May, H, MN, L706866, under stones; 1♀ 24 July, H, MN, L695843, amongst hay debris; 3♂2♀ 25 July, SP, MN, L657846, Loughanaphuca, shaken from *Erica*; 1♂ 25 July, H, MN, L713852, from dune; 1♂ 7–28 June, PT, L712859, grazed sward amongst *Ulex*. All 26 July, H, SP, MN: 1♀ L689857, Lough Creggan, under tree trunk remnants; 2♂ L698856, Knocknaveen amongst stones and moss; 1♀ L703863, *Ulex*; 3♂ L710854, under stones on greenway; 3♀ 27 July, H, MN, L706866, Portlea; 2♂ ?-26 July, MTT, L703863, Lough Avullin, *Phragmites* swamp; 1♂ 31 July-23 August, PT, SMcC, L704870, heathy slope with *Erica*; 2♂1♀ 19 September, H, MC, L702870, Portlea, mixed deciduous woodland; 1♀ 23 September, H, MN, L694864, shaken from *Calluna* overhanging a streamlet; ♂1 2 October, H, SMcC, L652853, heavily grazed heathy grassland at Signal Tower; 1♂1♀ L692874, tall *Calluna* (no dates).

Thyreosthenius parasiticus (Westring, 1851) (1)

1♂ 19 September, H, MC, L688847, sieved from chicken coop debris.

Tiso vagans (Blackwall, 1834) (99)

All but the following three specimens from PT: 1♂ 15 February, H, SMcC, L709864, *Nardus* and mosses near *Sphagnum* pool; 1♂ 26 July, H, MN, L698856, Knocknaveen under stone. All 27 March-10 April: 2♂ L654852, *Plantago* sward; 2♂ L688861, grazed bog vegetation on deep peat; 4♂ L703864, bank of Dooree river. 2♀ 10 April-21 May, L703864, bank of Dooree river; 2♂3♀ 23 May-7 June, L705856 *Pteridium* on dryish grassland; 1♂ 23 May-7 June, L709864 *Nardus* and mosses near *Sphagnum*. All 24 May-7 June: 1♂ L688861, grazed bog vegetation on deep peat; 1♂ L689857, rushes at margin Lough Creggan; 1♀ L689857, grassy slope; 3♂ L695854, dryish grassland. 6♂4♀ 6–27 June, L719864, short sward; 1♂3♀ 7 June-26 June, L693880, short coastal *Festuca* sward; 1♂2♀ 7–27 June, L703864, bank of Dooree river. All 7–28 June: 1♂ L688854, lacustrine mosses and lichens; 2♀ L689856, rushes; 1♂ L693877, wet flush with *Schoenus*; 1♂2♀ L695854 *Pteridium* on dryish grassland; 3♂1♀ L695854, dryish grassland; 2♂ L695883, *Plantago* sward; 1♂ L705856, rushes on wet ground; 5♂6♀ L712859, grazed sward amongst *Ulex*; 1♂ L712860, grazed sward amongst *Ulex*. 1♂2♀ 30 July-23 August, L665843, *Plantago* sward; 1♀ 30 July-23 August, L712859, grazed sward amongst *Ulex*; 1♀ 31 July-2 September, L695854 *Pteridium* on dryish grassland; 1♂ 31 July-2 September, L695854, dryish grassland. All 31 July-23 August, PT: 1♀ L665843, *Plantago* sward; 2♂4♀ L693880, short coastal *Festuca* sward; 2♂1♀ 31 July-23 August, L695883, *Plantago* sward; 1♂ L703864, bank of Dooree river; 1♀ 18 September, H, MC, L668861 on summit of Knockmore amongst mosses. The following have no dates: 2♂1♀ PT L693880, short coastal *Festuca* sward; 1♂ PT L693881 short coastal *Festuca* sward; 2♂4♀ PT L703864, bank of Dooree river near Lough Avullin; 4♂ PT L712859, grazed sward amongst *Ulex*.

* *Trichopternoides thorelli* (Westring, 1861) (1)

1♀ 22 May, H, MN, L713852, from *Ammophila* on the small dune system.

†**Troxochrus scabriculus* (Westring, 1851) (1)

1♀ 27 March-10 April, PT, L703864, bank of Dooree river near Lough Avullin.

Walckenaeria acuminata Blackwall, 1833 (8)

1♂1♀ 1 December 2001, H, SMcC, L698856, amongst *Juncus*; 1♀ 24 March-4 April, MTT, L702870, woodland; 1♀ 24 May-7 June, PT, L695854, *Pteridium* on grassland; 1♀ 21 May, H, MN, L705864, under rotten log on roadside; 1♀ 23 May, H, MN, L692878, amongst rocks; 1♂ 18 September, H, MC, L672857, Knockmore, under stones on gravelly grazed hillside; 1♀ 19 September, H, MC, L699870, abandoned garden.

Walckenaeria antica (Wider, 1834) (3)

1♀ 24 May-7 June, PT, L689857, rushes at margin of Lough Creggan; 1♀ 7–28 June, PT, L695854, dryish grassland; 1♀ 7–28 June, PT, L712860, grazed sward amongst *Ulex*.

Walckenaeria clavicornis (Emerton, 1882) (4)

1♂ 27 March-10 April, PT, L662856, dryish grassland at 350m; 1♂ 27 March-10 April, PT, L689856, amongst rushes; 1♂ 26 July, H, MN, L698856, amongst stones and mosses near summit of Knocknaveen; 1♂ 18 September, H, MC, L668861, on summit of Knockmore amongst mosses.

Walckenaeria cuspidata Blackwall, 1833 (9)

1♂ 25 March-5 April, MTT, L703863, Lough Avullin, *Phragmites* swamp; all others MTT, L702870, woodland; 4♂ 24 March-4 April, 3♂ 4–18 April, 1♂ 18 April-21 May.

Walckenaeria nudipalpis (Westring, 1851) (2)

1♀ 27 March-10 April, PT, L689857, grassy area near Lough Creggan; 1♂ 4–18 April, MTT, L703863, Lough Avullin, *Phragmites* swamp.

Walckenaeria unicornis O.P.-Cambridge, 1861 (3)

2♂ 25 March-5 April, 1♂ 4–18 April, MTT, L703863, Lough Avullin, *Phragmites* swamp.

Walckenaeria vigilax (Blackwall, 1853) (40)

Most specimens caught in PT. 1♂ 24 May-7 June, 1♂ 7–28 June, L688861, grazed bog vegetation on deep peat; 7♂ 7–28 June, 1♂1♀ 31 July-2 September, L695854 on open, and amongst *Pteridium* on, dryish grassland; 2♂ 7–28 June, L695883, *Plantago* sward; 13♂ 7–27 June, L703864, bank of Dooree river; 1♀ 30 July-23 August, L706857, amongst *Pteridium*; 5♂1♀ 7–28 June, L712859, grazed sward amongst *Ulex*; 4♂ 6–27 June, L719864, short sward; 1♂ PT, L703864, bank of Dooree river (no dates). Hand-caught specimens: 1♀ 19 February, H, SMcC, L703863 amongst *Phragmites* debris at Lough Avullin, *Phragmites* swamp; 1♀ 24 September, H, MN, L655844, shaken from moss at edge of large lake.

LYCOSIDAE

Alopecosa pulverulenta (Clerck, 1757) (20)

1♂ 23 May, H, MN, L692876 coastal sward; 1♀ 23 May, H, MN, L692878 summit; 1♂ 23 May, H, MN, L693877, wet flush with *Schoenus*; 1♂ 23 May, H, SMcC, L694877 *Calluna*; 2♂1♀ 24 May-7 June, PT, L689857, rushes at margin Lough Creggan; 1♂ 24 May-7 June, PT, L689857, grassy slope; 1♂ 24 May, H, MN, L703867, woodland edge; 1♂1♀, 7–28 June, PT, L687863, rushes near stream; 1♂1♀, 7–28 June, PT, L689857, grassy slope; 1♀ 7–28 June, PT, L693877, wet flush with *Schoenus*; 1♂ 7–26 June, PT, L693880, short coastal *Festuca* sward. 3♂1♀ PT, L693877, wet flush with *Schoenus* near *Calluna* slope, 1♂ L700878, *Molinia* dominated blanket bog (no dates).

†*Arctosa leopardus* (Sundevall, 1833) (6)

24 May-6/7 June, PT: 1♂, L688861, grazed bog vegetation on deep peat; 1♂, L689875 margin Lough Creggan; 1♂, L719864, short sward on exposed promontory. 7–28 June, PT: 1♂ L705856, rushes on wet ground; 1♂ L700878, *Molinia* dominated blanket bog; 1♀ L688854 moss and lichen sward lake edge.

Pardosa amentata (Clerck, 1757) (51)

♂2 18 April, H, SMcC, L712854, dry bank; 1♂ 11 May, H, MC, L712851, under wood in abandoned garden; 1♂1♀ 21 May, H, MN, L704863 on a plank in a field; ♂3♀2 24 May-6 June, PT, L719864, short sward; 2♂2♀ 24 May-7 June, ♀1 7–27 June, PT, L714853 short sandy sward; 1♂ 4 April-18 April, MTT, L703863, Lough Avullin, *Phragmites* swamp; 7♂4♀ 7–27 June, PT, L703864, bank of Dooree river; ♀2 7–27 June, PT, L713853 meadow; ♀1 ?-26 July, MTT, L703863, Lough Avullin, *Phragmites* swamp; 1♀ 27 July, H, MN, L703863, Lough Avullin, *Phragmites* swamp (carrying an egg-sac); 1♂ 31 July-23 August, PT, SMcC, L703864, bank of Dooree river; 11♂6♀ L703864, bank of Dooree river (no dates); 2♂ L704870, heathy slope with *Erica* (no dates).

Pardosa nigriceps (Thorell, 1856) (24)

1♀ 23 May, H, MN, L692878 summit; 1♀ 23 May, H, MN, L697877 *Calluna* close to blanket bog; 1♀ 23 May, H, MN, L696876, *Calluna*; 1♂ 23 May, H, SMcC, L694877 *Calluna*; 4♂ 7–28 June, 1♂2♀ 30 July-23 August, PT, L712859, grazed sward amongst *Ulex*; 1♂ 7–28 June, PT, L700878, *Molinia* dominated blanket bog; 5♂ 7–28 June, PT, SMcC, L712860 grazed sward amongst *Ulex*; 4♀ 26 July, H, MN, L703863 on and near *Ulex* near Lough Avullin, three specimens with egg-sacs; 1♀ 27 July, H, MN, L711856, roadside *Ulex* with spiderlings on her back; 1♀ 16 September, H, MC, L694879, south of Lecknacurra on a hillside at base of *Calluna*. 1♂ L701876 *Molinia* dominated blanket bog (no dates).

Pardosa palustris (Linnaeus, 1758) (3)

1♂ 24 May-6 June, PT, L719864, short grass sward close to sea; 1♀ 7–28 June, PT, L695854, dryish grassland; 1♀ 24 July, H, MN, L691841, under a stone on grassland, carrying an egg-sac.

Pardosa pullata (Clerck, 1757) (110)

1♀ 7 May, H, MC, Strake L667845, road verge; 1♀ 9 May, H, MC, L695843, Portnakilty, damp coastal grassland; 1♀ 21 May, H, MN, L696876, *Calluna*. All 23 May, H, SP, MN and SMcC: 2♀ 23 May, H, MN, L690874 *Calluna* steep slope; 1♂ L692876 coastal sward; 2♂1♀ L692878 summit, one under plank; 1♀ L693877, wet flush with *Schoenus* near *Calluna* slope; 1♀ L694877 *Calluna*; 1♂1♀ L696876, *Calluna*. All 24 May, H, MN and SMcC: 2♂1♀ L705867 on vegetation; 1♂ L700878, on bog amongst rubbish. 4♂4♀ 23 May-7 June, PT, SMcC, L705856, *Pteridium* on dryish grassland; 4♂1♀ 23 May-7 June, PT, SMcC, L705856, rushes on wet ground. All 24 May-7 June, PT: 2♂ L688861, grazed bog vegetation on deep peat; 1♂1♀ L689856, rushes; 5♂ L689857, rushes at margin Lough Creggan; 2♂ L689857, grasses at margin Lough Creggan; 3♂7♀ L695854, *Pteridium* on dryish grassland 5♂1♀ L714853 short sandy sward. 1♂2♀ 7-26 June, PT, SMcC, L693880, short coastal *Festuca* sward; 1♂ 7–27 June, PT, SMcC, L704870, heathy slope with *Erica*. All 7–28 June, PT: 2♂1♀ L687863, rushes near stream; 1♀ L688854, lacustrine mosses and lichens; 1♀ L688861, grazed bog vegetation on deep peat; 2♂ L689852, heavily grazed grassland; 1♀ L689856, rushes; 2♂1♀ L689857, grassy slope; 1♀ L693877, wet flush with *Schoenus*; 1♀ L695854 *Pteridium* on dryish grassland; 1♂ L700878, *Molinia* dominated blanket bog; 4♂ L705856, rushes on wet ground; 3♂ L712859, grazed sward amongst *Ulex*; 1♂ L712860 grazed sward amongst *Ulex*; 1♂ L721859, saltmarsh margin. All 26 July, H, MN: 2♀ L698856 Knocknaveen, amongst stones and moss with two egg-sacs; 1♀ L703863, Lough Avullin, *Phragmites* swamp, on wet mosses; 1♂ L711856 amongst *Ulex*. All 27 July, H, MN, amongst stones at Portlea: 2♀ L705867 one with egg sac; 1♂ L706866. All 30 July-23 August, PT: 2♀ L688861, grazed bog vegetation on deep peat; 1♀ L705856, rushes on wet ground; 1♂2♀ L706857, *Pteridium*; 1♀ L712859, grazed sward amongst *Ulex*; 1♀ L721859, saltmarsh margin. All 31 July-23 August, PT: 2♀ L688854, lacustrine mosses and lichens; 1♀ L689856, rushes; 1♀ L689857, grassy slope; 1♂3♀ L693877, wet flush with *Schoenus*; 1♂2♀ L693880, short coastal *Festuca* sward; 1♀ L700878, *Molinia* dominated blanket bog; 1♀ L703864, bank of Dooree river. 1♀ 31 July-2 September, PT, L695854 *Pteridium* on dryish grassland; 1♀ 16 September, H, MC, L694879, south of Lecknacurra on a hillside at base of *Calluna*. 5♂1♀ PT, L693877, wet flush with *Schoenus nigricans* (no dates); 1♀ PT, L700878, *Molinia* dominated blanket bog (no dates).

Piratula hygrophila (Thorell, 1872) (30) (previously in *Pirata*)

1♂ 21 May, H, MN, L703863 wet vegetation close to River Dooree; 1♀ 7–28 June, PT, L688881, grazed bog vegetation on deep peat; 1♂ 7–28 June, PT, L693877, wet flush with *Schoenus nigricans*; 12♂13♀ ?-26 July, MTT, L703863, Lough Avullin, *Phragmites* swamp. 2♀ 31 July-23 August, PT, L703864, bank of Dooree river near Lough Avullin.

Pirata piraticus (Clerck, 1757) (90)

1♂2♀ 22 May, H, SMcC, L703863, Lough Avullin, amongst *Phragmites* debris; 1♀ 23 May, SP, MN, L699877, *Molinia* on blanket bog; 23♂ 24 May-6 June, PT, L719864, short sward; All 7–27 June, PT: 6♂ L713853 meadow; 2♂ L687863, rushes near stream; 1♂ L688854, lacustrine mosses and lichens; 1♀ L689857, rushes at margin Lough Creggan; 1♂ L695854, dryish grassland; 3♂ L700878, *Molinia* dominated blanket bog; 6♂1♀ L705856, rushes on wet ground; 9♂1♀ L721859, saltmarsh margin; 3♂5♀ 25 July, H, MN (one SMcC), L655845, marshy area, three females with egg-sacs. 4♀ ?-26 July, MTT, L703863, Lough Avullin, *Phragmites* swamp; 2♀ 26 July, H, MN, L703863, Lough Avullin, *Phragmites* swamp, both with egg-sacs; 2♀ 27 July, H, MN, L721859, saltmarsh, both under a hubcap, with egg sacs. All 30 July-23 August, PT: 1♂ L665843, *Plantago* sward; 1♂ L705856, rushes on wet ground; 9♂3♀ L721859, saltmarsh margin. 1♂1♀ 31 July-23 August, PT, L688854, lacustrine mosses and lichens. No dates for the following: 1♂ L692874, tall *Calluna*; 2♂ L704870, heathery area with *Erica*; 1♂ L703864, bank of the Dooree river.

Trochosa terricola Thorell, 1856 (149)

1♂ 15 November 2001, H, SMcC, L701877, base of a stone wall; 1♀ 15 November 2001, H, SMcC, L705867, amongst grasses and bracken; 1♀, 01 December 2001, H, SMcC, L698856, *Juncus*; 1♀ 20 May, H, MN, L714854, under a stone at back of beach; 1♀ 25 July, H, MN, L655845, stony area on grassland; 1♀ 25 July, H, MN, L655851 Loughanaphuca; 1♂ 4–18 April, 2♂ 18 April-21 May MTT, L702870, woodland. All other specimens PT: 3♂ 27 March-10 April, 1♀ 7–28 June, 2♀ 30 July-23 August, L706857, amongst *Pteridium*; 1♂ 23 May-7 June, L705856, rushes on wet ground; 3♂ 27 March-10 April, 1♀ 30 July-23 August, L705856, *Pteridium* on grassland; 1♀ 7–28 June, 1♀ 31 July-2 September, L695854, dryish grassland; 1♀ 24 May-7 June, 1♀ 31 July-2 September, 1♀ 7–28 June, L695854 *Pteridium* on dryish grassland: 2♂2♀ 27 March-10 April, L689852, heavily grazed grassland: 3♂ 27 March-10 April, L688854, mosses and lichens near lakeshore; 2♂3♀ 27 March-10 April, 1♀ 24 May-7 June, 1♀ 7–28 June, L689857, rushes at margin of Lough Creggan; 3♂1♀ 27 March-10 April, 2♀ 24 May-7 June, 3♀ 7–28 June, 4♀ 31 July-23 August, L689857, grassland: 3♂ 27 March-10 April, 1♀ 7–28 June, 4♀ 31 July-23 August, L689856, rushes: 2♂ 27 March-10 April, L688861, grazed bog vegetation on deep peat: 7♂3♀ 27 March-10 April, 3♀ 7–28 June, L687863, rushes near Dooree stream: 4♂ 27 March-10 April, L687871, ungrazed sward: 1♂ 27 March-10 April, 1♂ 24 May-7 June, 1♀ 31 July-23 August, L692874, tall *Calluna*: ♂4 27 March-10 April, ♀5 31 July-23 August, L693877, wet flush with *Schoenus nigricans*: 2♀ 31 July-23 August, L693880, short coastal sward of *Festuca*: 2♀ 31 July-23 August, L693881 short coastal sward of *Festuca*: 1♀ 7–28 June, L701876, 1♀ 31 July-23 August, L700878, *Molinia* dominated blanket bog: 5♂ 27 March-10 April, L704870, heathery area with *Erica*: 1♀ 30 July-23 August, L721859, saltmarsh margin: 1♀ 24 May-7 June, L714853 short sward on sandy soil: 2♂ 27 March-10 April, 1♀ 31 July-23 August, L703864, bank of Dooree river: 1♀ ?-26 July, MTT, L703863, Lough Avullin, *Phragmites* swamp; 14♂2♀ 27 March-10 April, L662856 dryish grassland at 350m: 7♂ 27 March-10 April, L654852 *Plantago* sward: 1♀ 30 July-23 August, L665843, *Plantago* sward: 5♂ 27 March-10 April, 1♀ 7–28 June, L712859, grazed sward amongst *Ulex*; 1♂ 16 September, H, MC, L694879, south of Lecknacurra on a hillside at base of *Calluna*. No dates for the following: 1♂ PT, L692874, tall *Calluna*; 2♀ PT, L693877, wet flush with *Schoenus nigricans*; 1♀ PT, L703864, bank of Dooree river near Lough Avullin.

NESTICIDAE
*°*Nesticus cellulanus* (Clerck, 1757) (1)

1♀ 14 February, SP, SMcC, Portlea L705867, amongst *Pteridium* litter.

OONOPIDAE
**Oonops pulcher* Templeton, 1835 (6)

7 May, H, MC, L652853, heavily grazed heathy grassland at Signal Tower; 1♀ 19 May, H, SMcC, L714854, in a shed; 1♂1♀ 23 May, H, MN and SMcC, L688844, in grounds of Church, male on underside of wood, female under slates; 1♀ 23 May, H, SMcC, L694877, from dried decaying vegetation in a crack in *Calluna* overgrown rocks; 1♂ 24 September, H, MN, under a stone on grassland.

PHILODROMIDAE
†**Philodromus cespitum* (Walckenaer, 1802) (5)

2♂1♀ 22 May, H, MN, L711856, beaten from *Ulex*; 1♂ 7–28 June, PT, L712859, grazed sward amongst *Ulex*; 1♀ 25 July, H, MN, L714854, on car at community centre.

PHOLCIDAE
**Pholcus phalangioides* (Fuesslin, 1775) (2)

Exuvium of immature 23 May, H, MN, L714854 Community Centre; 1♀ 16 September, H, MC, L715854, toilet of Clare Island Hotel.

PHRUROLITHIDAE
**Phrurolithus festivus* (C.L.Koch, 1835) (3)

1♂ 9 May, H, MC, L689844, under stones in a graveyard at Church; 1♀ 24 July, H, MN, L700844, amongst coastal vegetation; 1♀ 25 July, H, MN, L652842, on beach amongst stones.

SALTICIDAE
**Euophrys petrensis* (C.L.Koch, 1837) (1)

1♀ 25 July, H, MN, L652843, with egg-sac under stone on grassy sward close to small beach.

Heliophanus cupreus (Walckenaer, 1802) (6)

1♂ 9 May, H, MC, L714854, low eroding sea-cliff; 1♂4♀ 24 July, H, MN, L700844, all the females (one dead) were in brood-cocoons in a crevice in coastal rock overgrown by *Armeria* dominated vegetation.

SEGESTRIIDAE

Segestria senoculata (Linnaeus, 1758) (5)

7 May, H, MC, L649841, exposed coastal headland, L652853, heavily grazed heathy grassland at signal tower; 10 May, H, MC, L698854, Knocknaveen, under stones on summit; 1♀ 19 May, H, SMcC, L721859, saltmarsh under stones; 1♂ 21 May, H, MN, L708860, near a house; 1♀ 28 June-19 July, MTT, L702870, woodland.

TETRAGNATHIDAE

Metellina mengei (Blackwall, 1869) (30)

1♀ 5 February, H, SMcC, L702870, heath near woodland; 1♂1♀ 6 February, SP, SMcC, L702870, woodland; 2♂ 14 February, H, SMcC, L705867, *Pteridium* litter; All 15 February, H/SP, SMcC: 3♂ L703871, heathy cliffs above shore; 2♀ L704869 heathy cliffs above shore; 1♂4♀ L704874 heathy cliffs above shore. 4♀ 17 February, SP, SMcC L703863, *Ulex*; 1♂1♀ 4 April, SP, SMcC, L703868, rushy field near woodland; 1♂ 18 April-21 May, MTT, L702870, woodland; 1♀ 19 May, H, SMcC, L702870, woodland moss and leaf litter; 1♂4♀ 22 May, SP, MN, L711856, roadside *Ulex*; 1♂ 23 May, SP, MN, L696876, *Calluna*; 1♂ 22 September, SP, MN, L716856 cliffs.

Metellina merianae (Scopoli, 1763) (8)

1♀ 15 November 2001, H, SMcC, L702870, woodland leaf litter; 1♀ 20 May, H, MN, L714854, under boulder on beach; 2♂1♀ 21 May, H, MN, L705864 inside a shed – one male *var. celata*; 1♀ 24 May, H, MN, L709865 very wet cliff face; 1♀ 28 June-19 July, MTT, L702870, woodland; 1♀ 16 September, H, MC, L694879, south of Lecknacurra on a hillside at base of *Calluna*.

Metellina segmentata (Clerck, 1757) (9)

1♂2♀ 15 November 2001, H, SMcC, L702870, woodland from leaf litter; 1♀ 6 February, SP, SMcC, L703868 *Ulex* near woodland; 1♂1♀ 22 September, SP, MN, L717857, roadside vegetation; 1♂ 23 September, H, MN, L711858 from *Fuchsia* growing near roadside, specimen was at female web; 1♂1♀ 24 September, SP, MN, L665846 from rushes at roadside.

Pachygnatha clercki Sundevall, 1823 (113)

Most specimens were from MTT L703863, Lough Avullin, *Phragmites* swamp: 3♂30♀ 25 March-5 April; 11♂29♀ 4 April-18 April; 2♀ 4 April-18 April. A smaller number were from MTT, L702870, woodland: 1♀ 24 March-4 April; 1♂2♀ 18 April-21 May; 1♀ 28 June-19 July. A substantial number of specimens from the period 27 March-10 April, PT: 1♂ L654852 *Plantago* sward; 1♂9♀ L693877, wet flush with *Schoenus*; 3♀ L703864, bank of Dooree river; 1♂1♀ L704868, marshy field; 1♂ L704870, heathy slope with *Erica*; 1♀ L719864, short sward; 1♂ L721859, saltmarsh margin. 1♂3♀ 7–26 June, PT, L665843, *Plantago* sward; 1♀ 7–27 June, PT, L713853 meadow; 1♂ 7–28 June, PT, L704868, marshy field; 1♀ 7–28 June, PT, L705856, rushes on wet ground. 1♀ 15 November 2001, H, SMcC, L695843, grassland at cliff base; 1♀ 28 November 2001, H, SMcC, L675850 *Juncus*; 1♂1♀ 14 February, H, SMcC, L721859, saltmarsh; 1♀ 22 May, SP, MN, MTT L703863, Lough Avullin, *Phragmites* swamp 1♀ 22 September, H, MN, L713852, *Ammophila*; 1♀ 23 September, SP, MN, L705864, amongst *Monbretia*.

Pachygnatha degeeri Sundevall, 1830 (133)

1♀ 27 March-10 April, PT, L704870, heathy slope with *Erica*; 1♀ 27 March-10 April, PT, L721859, saltmarsh margin; 1♂ 9 May, H, MC, L695843, Portnakilty, damp coastal grassland; 1♀ 20 May, H, MN, L714854, stony soil on beach; 1♀ 22 May, H, MN, L711859, woodpile; 1♂1♀ 23 May-7 June, PT, L705856, rushes on wet ground; 2♂2♀ 23 May-7 June, PT, L705856 *Pteridium* on dryish grassland; 2♂3♀ 23 May, H, MN, L692878, summit amongst rocks; 2♀ 23 May, SP, MN, L696876, *Calluna*. From 24 May-7 June, PT: 2♂2♀ L688861, grazed bog vegetation on deep peat; 1♂ L689857, rushes at margin Lough Creggan; 1♂1♀ L692874, tall *Calluna*; 1♂ L695854, dryish grassland; 2♀ L695854, *Pteridium* on dryish grassland; 1♂2♀ L714853 short sandy sward. From 7–26 June, PT: 7♂16♀ L693880, short coastal *Festuca* sward; 2♂1♀ L704870, heathy slope with *Erica*; 1♀ L713853 meadow. From 7–28 June,

PT: 1♀ L689852, heavily grazed grassland; 1♂ L689857, grassy slope; 1♀ L692874, tall *Calluna*; 1♀ L693877, wet flush with *Schoenus*; 1♂3♀ L693881 short coastal *Festuca* sward; 1♀ L695854 *Pteridium* on dryish grassland; 1♀ L695854, dryish grassland; 2♂1♀ L695883, *Plantago* sward; 3♀ L700878, *Molinia* dominated blanket bog; 2♂ L705856, *Pteridium* on dryish grassland; 2♂3♀ L705856, rushes on wet ground; 6♂9♀ L712859, grazed sward amongst *Ulex*; 2♀ L712860 grazed sward amongst *Ulex*; 1♂ L721859, saltmarsh margin. 3♀ 30 July-23 August, PT, L712859, grazed sward amongst *Ulex*; 1♂ 31 July-23 August, PT, L693877, wet flush with *Schoenus*; 1♂1♀ 31 July-23 August, PT, L693880, short coastal *Festuca* sward; 1♀ 31 July-23 August, PT, L700878, *Molinia* dominated blanket bog; 1♂ 22 September, SP, MN, L716856 cliffs; 1♀ 23 September, H, MN, L712859 from grassy ground. Other specimens with no dates: 2♂ L693877, wet flush with *Schoenus*; 2♂5♀ L693880, short coastal *Festuca* sward; 2♀ L693881 short coastal *Festuca* sward; 4♀ L700878, *Molinia* dominated blanket bog; 1♀ L701876 *Molinia* dominated blanket bog; 3♂6♀ L712859, grazed sward amongst *Ulex*.

Tetragnatha extensa (Linnaeus, 1758) (16)

1♀ 21 May, SP, SMcC, L702870 from *Pteridium* near woodland. All 22 May, MN: 1♀, H, L713852, *Ammophila*; 1♂, SP, L703863, *Phragmites* at Lough Avullin, *Phragmites* swamp; 1♀, SP, L704863, rushes in wet field. 1♂ 23 May, SP, SMcC, L693877, wet flush with *Schoenus*; 1♂ 7 July, H, SMcC, location uncertain, marshy area near lakes; 1♂ 23 May -7 June, PT, L693877, wet flush with *Schoenus*; 1♂ 9 July, H, SMcC, L705868; 1♀ 23 July, SP, L6986 field; 1♂5♀ 24 July, SP, MN, L714851 rushes near pier; 1♀ 26 July, H, MN, L699857, steep rushy slope on Knocknaveen.

Tetragnatha montana Simon, 1874 (1)

1♂ 27 July, SP, MN, L702870, from grasses in woodland.

THERIDIIDAE

Enoplognatha ovata (Clerck, 1757) (15)

2♀ 10 July, SP, SMcC, L713853, ragwort and meadow; 1♂ 19 July-15 August, MTT, L702870, woodland; 1♀ 23 July, Light-trap, SMcC, L714852 meadow; 1♂ 26 July, H, MN, L711856, roadside *Ulex*; 1♀ 26 July, SP, MN, L714852, *Ulex* near *Phragmites* beds; 2♀ 27 July, H, MN, L703868, trackside vegetation near woodland; 2♀ 27 July, H, MN, L705867, *Salix* behind boulder beach; 1♀ 27 July, H, MN, L711856, roadside *Ulex*; 1♀ 22 September, H, MN, L713852, amongst *Ammophila* – post-partum female in retreat; 3♀ 24 September, SP, MN, L665846, roadside rushes and fern.

Enoplognatha thoracica (Hahn, 1833) (1)

1♀ 25 July, H, MN, L655845, under stone at Beetle head.

Pholcomma gibbum (Westring, 1851) (4)

1♀ 21 May, H, MN, L696876, from web under dense *Calluna*; 1♀ 22 May, SP, MN, L711856 from *Ulex*; 1♀ 23 May, SP, MN, L696876, *Calluna*; 1♂ 7–28 June, PT, L712859, grazed sward amongst *Ulex*.

°Robertus arundineti (O.P.-Cambridge, 1871) (10)

1♀ 6 May, H, MC, Capnagower L715865, damp overgrazed heathy grassland; 2♀ 25 July, 1♂6♀, 24 September, H, MN, L655844, very wet mosses at edge of a small lake.

Robertus lividus (Blackwall, 1836) (39)

1♂ 15 February, H, SMcC, L701877, base of stone wall; 1♂ 24 March-4 April, MTT, L702870, woodland; 2♀ 9 May, H, MC, L714854, low eroding sea-cliff; 1♀ 10 May, H, MC, L676863, damp grassland at 250m; 1♂ 23 May-7 June, PT, L705856, rushes on wet ground; 1♀ 23 May, H, MN, L692878, under fallen wooden post at summit; 1♂ 23 May, H, MN, L694877, under damp stone; 4♂ 24 May-28 June, PT, L687871, ungrazed sward; 1♂ 24 May-7 June, PT, L688861, grazed bog vegetation on deep peat; 1♂ 24 May-7 June, PT, L692874, tall *Calluna*; 4♂ 7–27 June, 1♂ 31 July-23 August, PT, L665843, *Plantago* sward; 1♀ 7–27 June, 1♂ 31 July-23 August, PT, L704870, heathy slope with *Erica*; 1♂ 7–28 June, PT, L687863, rushes near stream; 2♂ 7–28 June, PT, L689856, rushes; 2♂ 7–28 June, PT, L705856, rushes on wet ground; 2♂ 28 June-19 July, 1♂ 19 July-15 August, MTT, L702870, woodland; 1♂ 26 July, H, MN, L698856 Knocknaveen, under deeply embedded stone in peat; 1♂ ?-26 July, MTT, L703863, Lough Avullin, *Phragmites* swamp; 1♀ 16 September, H, MC, L694879, south of Lecknacurra on a hillside at base of *Calluna*; 1♀ 18 September, H, MC, L672857, Knockmore, under stones on gravelly grazed hillside; 1♂1♀ 18 September, H, MC, L668861, summit of Knockmore under stones; 1♀ 23 September, H, MN, L667858, Knockmore, under stone at Cairn. 3♂ PT, L692874, tall *Calluna*, 1♀ PT, L703865, bank of Dooree river (no dates).

†*Rugathodes instabilis* (O.P.-Cambridge, 1871) (1)
1♂ 19 May, H, SMcC, L702870, Woodland, from moss and leaf litter.

†*Steatoda grossa* (C.L.Koch, 1838) (1)
1♀ 10 February H, SMcC, L705864, from the bathroom in a house.

*° *Phylloneta sisyphia* (Clerck, 1757) (8)
2♂3♀ 22 May, 3♀ 26 July, H, MN, L711856, *Ulex*.

THOMISIDAE
Ozyptila atomaria (Panzer, 1801) (2)
1♂1♀ 31 July-23 August, PT, L700878, *Molinia* dominated blanket bog.

†*Ozyptila brevipes* (Hahn, 1826) (1)
1♂ 7 June-28 June, PT, L700878, *Molinia* dominated blanket bog.

Ozyptila trux (Blackwall, 1846) (4)
1♂ 7 June-28 June, PT, L688861, grazed bog vegetation on deep peat; 1♂ 7 June-27 June, PT, L704870, north-facing *Erica* dominated slope; 2♀ 16 September, H, MC, L694879, south of Lecknacurra on a hillside at base of *Calluna*.

Xysticus cristatus (Clerck, 1757) (26)
1♀ 14 February, H, SMcC, L705867, *Pteridium* litter; 2♀ 14 February, H, SMcC, L717857 cliff between beach and saltmarsh; 1♂ 18 April, H, SMcC, L711854, at stone wall; 1♀ 7 May, H, MC, Strake L667845, road verge; 1♀ 9 May, H, MC, L689844, under stones in graveyard at Church; 1♀ 23 May, SP, MN, L694877 *Calluna*; 1♂ 23 May, SP, MN, L696876, *Calluna*. All other specimens from PT: 1♂ 24 May-7 June, L714853 short sandy sward; 2♂2♀ 6–27 June, L719864, short sward; 2♀ 7–26 June, SMcC, L693880, short coastal *Festuca* sward; 1♂2♀ 7–28 June, 1♂ 31 July-23 August, L689857, grassy slope; 1♂2♀ 7–28 June, L693877, wet flush with *Schoenus*; 1♀ 7–28 June, L706857, *Pteridium*; 1♀ 7–28 June, L721859, saltmarsh margin; 1♂ 18 September, H, MC, L6586 Ooghduff, coastal headland.

†*Xysticus erraticus* (Blackwall, 1834) (2)
2♀ 1 December 2001, H, SMcC, L705867, *Molinia* behind boulder beach.

CHAPTER 5

THE COLLEMBOLA AND ACARI (ORIBATIDA AND MESOSTIGMATA) OF CLARE ISLAND

Thomas Bolger

ABSTRACT

Soil fauna are one of the most species-rich components of terrestrial ecosystems and play a critical role in the functioning of one of the two major integrating systems, the decomposer system. Therefore, understanding the biodiversity of this component of the biota is essential in terms of sustainability. While soil-dwelling mites and springtails have received quite a lot of attention in Ireland, studies have been largely confined to agricultural and forest systems. Recent studies of other habitats suggest that much of the fauna have yet to be inventoried and new species are added to the Irish species list on a regular basis. This indicates a serious gap in our knowledge since, in common with terrestrial habitats generally, by far the greatest amount of biodiversity occurs underground. The studies of the Clare Island mites and Collembola support this contention. While the original and present surveys found approximately the same numbers of species on the island, the current survey found 34 species of mites and 20 species of Collembola not found in the original survey. This is unlikely to be due to species turnover but simply to the difficulty of gathering complete inventories of the species present. While most of the species of Collembola found are common and all of the species added during this survey are widespread and common in Ireland, several of them live in the soil and could therefore have been overlooked in the original survey because of the sampling method used. In the case of the mites most of the species are similarly common but four of the species recently recorded have been confirmed as new to the Irish species list.

Introduction

Soil fauna are one of the most species-rich components of terrestrial ecosystems and play a critical role in the functioning of one of the two major integrating systems, namely the decomposer system (Coleman and Hendrix 2000). Therefore understanding the biodiversity of this component of the biota is essential in terms of sustainability.

While soil-dwelling mites and Collembola have received quite a lot of attention in Ireland, studies have largely been confined to agricultural and forest systems (e.g. Curry 1969; Heneghan and Bolger 1996; O'Connell and Bolger 1997; Arroyo *et al.* 2013b; Bolger *et al.* 2013). Recent studies suggest that much of the fauna have yet to be inventoried and new species are added to the Irish species list on a regular basis (Sterzyńska and Bolger 2004; Arroyo *et al.* 2008; Arroyo and Bolger 2008; 2010; Moraza *et al.* 2009). This indicates a serious gap in our knowledge as—in common with terrestrial habitats generally—the greatest amount of biodiversity occurs underground

(Giller *et al.* 1997; Hågvar 1998) and the complex interactions between soil microbes, animals and plant roots are essential for the functioning and maintenance of biological communities in nutrient-poor soils (Brussaard *et al.* 1997; Wardle 1999; Loreau *et al.* 2001).

Collembola are, together with mites, the most diverse group of soil mesofauna. They are important functionally as decomposers of soil organic matter: they exert a great influence on the mineralisation/immobilisation of soil nutrients (Verhoef and De Goede 1985, and as selective grazers of fungal hyphae and spores, they play a key role in dispersing fungi and promote fungal succession (McLean *et al.* 1996).

Soil mites are the most abundant microarthropods in many types of soil. In rich forest soil, a 100g sample may contain as many as 500 mites representing almost 100 genera (Coleman and Crossley 1996), and a handful of forest humus contains many mite species (Walter and Proctor 1999). Although 40,000 to 50,000 species have been described, their real species richness is estimated to be more than a million (Walter and Proctor 1999).

Both Collembola and mites were studied in the original Clare Island survey. In the original survey the work was led by J.N. Halbert (mites) and G.H. Carpenter (Collembola) (Carpenter 1913; Halbert 1915). Both worked at the Natural History Museum in Dublin and were experts in their respective fields, having written multiple papers in the areas and both, in particular Halbert, described new species (Beirne 1985; Bolger 1996; Luxton 1998; Collins 1999). Carpenter collected on the island for a single week during July 1911 but was also supplied with material by other fieldworkers who collected both on the island and the adjacent mainland. It would appear, from his descriptions, that the specimens he examined were collected by simply turning stones or hand searching on surfaces. These forms of collecting would mean that while many species of Collembola would be captured many other species that are either fast moving or live within the substrate (e.g. soil or litter) would be missed. Halbert generally collected surface materials but did not appear to take structured soil samples. Therefore in this study we have used both soil coring and suction sampling as a method of collecting and concentrated on the soil and soil surface fauna. In addition, our work was confined to Clare Island and did not extend to the adjacent mainland.

Methods

Several visits were made to the island during the period 1994 to 2001. Extensive soil sampling was carried out by collecting intact soil cores 5cm in diameter and 5cm in depth. The animals from these cores were extracted using a Kempson High Gradient Extractor. Suction samples were collected at 17 locations in 2002 using a Vortis described in Arnold (1994); however, although this method recovered a large amount of material, many of the specimens were damaged to such an extent that they could not be identified and therefore the description of these data will be limited.

Results and discussion

Collembola

We collected 24 species of Collembola from the island, whereas Carpenter had recorded 18 species (Table 1). However, surprisingly only four species, *Anurida maritima*, *Isotomurus palustris*, *Tomocerus minor* and *Sminthurides aquaticus*, were found in both surveys. All four of these species are common and occur in a wide range of habitats. Thirty-eight species have now been recorded from Clare Island.

The species found most frequently in the current study were *Folsomina quadrioculata*, *Supraphorura furcifera*, *Friesea mirabilis* and *Parisotoma notabilis*, all of which are very abundant, widespread species but were not found in the original survey. This is quite unexpected but may simply reflect the sampling methods used which, as mentioned above, appeared to depend largely on hand sorting. It is also perhaps surprising that *S. furcifera* should be the most frequently found member of the Onychuridae because, although widespread, many other species have been found more frequently in soil in Ireland.

All of the species found are common in Ireland and many of them have been found in almost every habitat studied in Ireland. For example, *P. notabilis*, *Isotoma viridis*, *I. palustris*, *F. quadrioculata*, *Folsomia candida* and *T. minor* are already known from habitats as diverse as grassland, woodland, limestone pavement, blanket bog and sand dunes. On Clare Island they were found in a broad range of habitats including brackish sandy grassland, outside the fence at the top of the cliffs on Croaghmore and from plantain grassland.

Table 1
Species of Collembola recorded by G.H. Carpenter for the original survey (Carpenter 1913) and from the current survey. Where synonyms are indicated, they are those used in the original survey.

	Carpenter 1913	Current survey
Order Poduromorpha		
Family Hypogastruridae		
Ceratophysella armata (Nicolet) *Achorutes armatus* (Niciolet)	+	
Ceratophysella denticulata (Bagnall)		+
Ceratophysella longispina (Tullberg) *Achorutes longispina* Tullberg	+	
Hypogastrura viatica (Tullberg) *Achorutes viaticus* (Linn.) Tullberg	+	
Xenylla humicola (O. Fabricius)		+
Xenylla maritima Tullberg	+	
Family Neanuridae		
Anurida granaria (Nicolet)		+
Anurida maritima (Guerin)	+	+
Anuridella marina Willem		+
Friesea mirabilis (Tullberg)		+
Neanura muscorum (Templeton)		+
Family Onychiuridae		
Supraphorura furcifera (Börner)		+
Family Tullbergiidae		
Mesaphorura krausbaueri Börner		+
Stenaphorura denisi (Bagnall)		
Order Entomobryomorpha		
Family Entomobryidae		
Entomobrya nicoletii (Lubbock)	+	
Lepidocyrtus cyaneus Tullberg	+	
Orchesella cincta (Linn.)	+	
Orchesella villosa (Geoffroy)	+	
Family Isotomidae		
Ballistrua schoetti (Della Torre)		+
Folsomia candida Willem		+
Folsomia fimetaria (Linnaeus)		+
Folsomia quadrioculata (Tullberg)	+	
Isotoma anglicana (Lubbock)		+
Parisotoma notabilis Schäffer		+
Isotoma olivacea Tullberg, var *grisescens* Schäffer	+	
Isotoma viridis Bourlet	+	
Isotomiella minor (Schäffer)		+
Isotomurus palustris (Müller)	+	+
Proisotoma minuta (Tullberg)		+
Pseudoisotoma sensibilis (Tullberg) *Isotoma sensibilis* (Tullberg)	+	

continued

Table 1. Cont.

Tetracanthella brachyuran Bagnall		+
Family Tomoceridae		
Tomocerus longicornis (Müller)	+	
Tomocerus minor (Lubbock)	+	+
Order Neelipleona		
Family Neelidae		
Megalothorax minimus Willem		
Order Symphypleona		
Family Dicyrtomidae		
Dicyrtomina minuta (O. Fabr.)	+	
Dicyrtoma fusca (Lubbock)		+
Family Katiannidae		
Sminthurinus elegans (Fitch)		+
Family Sminthurididae		
Sminthurides aquaticus (Bourlet)	+	+
Sphaeridia pumilis (Krausbauer)		+

Table 2
Species of Mesostigmata recorded by J.N. Halbert for the original survey (Halbert 1915) and from the current survey. Where synonyms are indicated, they are those used in the original survey.

	Halbert 1915	Current study
Order Mesostigmata		
Suborder Gamasina		
Family Ascidae		
Arctoseius cetratus (Sellnick)		+
Arctoseius minutus (Halbert)	+	+
Seiulus minutus sp. nov.		
Cheiroseius (Posttrematus) serratus (Halbert)	+	+
Paraseius serratus sp. nov.		
Zerconopsis remiger (Kramer)		+
Zercoseius spathuliger (Leonardi)		+
Family Epicriidae		
Epicrius mollis (Kramer)		+
Family Laelapidae		
Cosmolaelaps claviger (Berlese)		+
Laelaps agilis C.L. Koch	+	
Eulaelaps stabularis (C.L. Koch)	+	
Laelaps (Eulaelaps) stabularis (C.L. Koch)		
Ololaelaps venetus (Berlese)	+	
Laelaps (Ololaelaps) tumidulus C.L. Koch		
Pseudoparasitus meridionalis G. and R. Can.	+	
Laelaps (Pseudoparasitus) meridionalis G. and R. Can.		
Family Macrochelidae		
Geholaspis aeneus Krauss		+
Geholaspis longispinosus (Kramer)	+	+
Holostaspis longispinosus (Kramer)		
Longicheles mandibularis (Berlese)	+	+
Holostaspis longulus Berlese		

continued

Table 2. Cont.		
Macrocheles glaber (Müller)		+
Macrocheles opacus (Koch)		+
Macrocheles submotus Falconer	+	+
Macrocheles tridentinus (G. and R. Canestrini)		
Family Pachylaelapidae		
Pachyseius humeralis Berlese		+
Pachylaelaps (Onchodellus) sp.		+
Family Parasitidae		
Gamasodes spiniger Trägårdh	+	
Gamasoides spinipes (C.L. Koch)		
Gamasodes fimbriatus Karg	+	
Gamasoides spinipes (C.L. Koch)		
Holoparasitus inornatus Berlese	+	+
Gamasus (Ologamasus) inornatus Berl.		
Holoparasitus stramenti Karg	+	
Gamasus (Ologamasus) pollicipatus Berl.		
Lysigamasus armatus (Halbert)		+
Lysigamasus celticus Bhattacharyya		+
Lysigamasus lapponicus Trägårdh	+	+
Gamasus (Pergamasus) lapponicus Trägårdh		
Lysigamasus runcatellus (Berlese)	+	
Gamasus runcatellus Berl.		
Pergamasus robustus (Oudemans)	+	+
Gamasus (Pergamasus) robustus Oudms.		
Pergamasus crassipes Berlese	+	+
Gamasus (Pergamasus) crassipes L.		
Pergamasus longicornis Berlese		+
Pergamasus septentrionalis (Oudemans)		+
Family Rhodacaridae		
Rhodacarus roseus (Oudemans)		+
Family Veigaiidae		
Veigaia agilis (Berlese)		+
Veigaia kochi Trägårdh	+	+
Cyrtolaelaps kocki Trägårdh		
Veigaia nemorensis (Koch)	+	+
Cyrtolaelaps nemorensis (C.L. Koch)		
Veigaia transisalae (Oudemans)	+	+
Cyrtolaelaps transisalae Oudms.		
Family Zerconidae		
Zercon triangularis Koch	+	+
Family Trachyuropodidae		
Trachyuropoda formicaria (Lubbock)	+	
Urotrachytes formicarius (Lubbock)		
Suborder Uropodina		
Family Uropodidae		
Uropoda (Cilliba) cassida (Herm.)	+	
Uropoda (Uropoda) halberti Hirschmann,	+	
1993 *Haluropoda minor* sp. nov.		
Uropoda (Uropoda) minima Kramer, 1882	+	
Discopoma integra Berl.		

While the species definitions of most of the Collembola recovered have remained relatively stable, there are some exceptions which may have reduced the amount of overlap between the two surveys. For example, although it was described in 1862, *Isotoma anglicana* was probably frequently recorded as *I. viridis* before the publication of Fjellberg's (1980) key and this may have been the case in Carpenter's study. These species are distinguished based on the structure of the apical edge of the manubrium, which is illustrated in Plate 1.

Acari - Oribatida and Mesostigmata
Mites from the acarine orders Oribatida and Mesostigmata were studied in the current survey and 64 species were recovered (Tables 2, 3). This is similar to the number of species (60) from these groups found by Halbert but the studies included only 30 species in common. Between the two studies 94 species have now been recorded from the island. The fact that such a small proportion of the species were found in both surveys suggests yet again that different sampling techniques recovered different species but also suggests that neither survey is providing an accurate estimate of the number of species present on the island. In this survey we concentrated on soil mites and used the Chao-1 index to estimate the total number of species present in the community. This analysis suggests that about six additional species would be expected to occur in the soil on the island. Thus we appear to have recovered more than 90% of the soil dwelling species from these taxa occurring on the island.

The most frequently recovered species, *Lysigamasus celticus*, *L. armatus*, *L. lapponicus*, *Pergamasus crassipes*, *Longicheles mandibularis* and *Macrocheles submotus*, are all predatory species which are common in Ireland, and four of these six species were found by Halbert.

The majority of the other species are also common in Ireland. However, three species were recorded for the first time in Ireland. These are *Melanozetes meridianus* Sellnick, which is a widespread Holarctic species occurring in moss and forest litter; *Nothrus anauniensis* Can. et. Fanz., which has an essentially worldwide distribution, particularly in moist and deciduous forests; and *Liebstadia humerata* Sellnick, which is a Holarctic species known to occur particularly in lichens on tree bark woodland (Weigmann 2006).

The discovery of three new national records amongst a collection of 64 species is likely to be an indication of the generally understudied national fauna rather than suggesting that the fauna of the island is in any way unusual. Virtually all recent studies of the Irish acarine fauna have added species to the national species list (e.g. Arroyo and Bolger 2008, 2010; Arroyo *et al.* 2008, 2013a) and indeed a number of species new to science have recently been described by Moraza *et al.* (2009).

In summary, this survey has added considerably to the number of species recorded from Clare Island and indeed from Ireland. However, the differences between the fauna recovered would

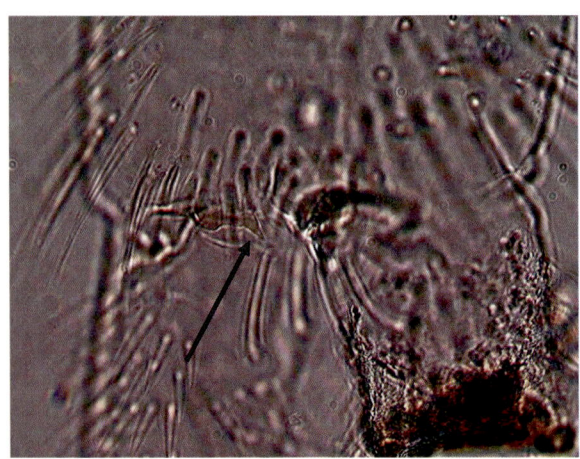

Plate 1a: Manubrial edges of *Isotoma viridis* (1 point). Photo: T. Bolger

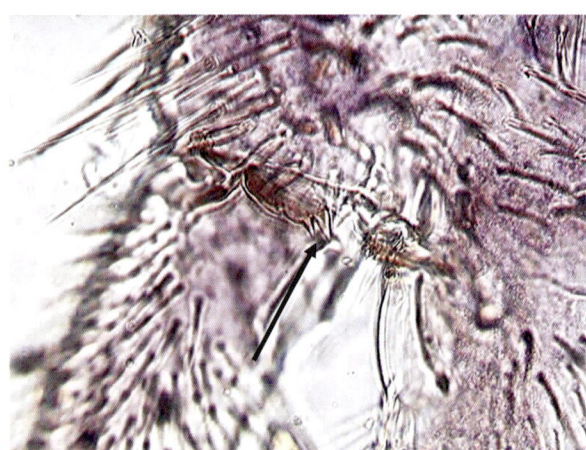

Plate 1b: *Isotoma anglicana* (2 points). Photo: T. Bolger

Table 3
Species of Oribatida recorded by J.N. Halbert for the original survey (Halbert 1915) and from the current survey. Where synonyms are indicated, they are those used in the original survey.

	Halbert 1915	Current survey
Order Oribatida		
Cohort Enarthronota		
Family Brachychthoniidae		
Brachychthonius berlesei Willmann		+
Family Hypochthoniidae		
Hypochthonius rufulus Koch	+	+
Cohort Mixonomata		
Family Euphthiracaridae		
Rhysotrichia ardua (C.L. Koch)	+	
Phthiacarus arduus (C.L. Koch)		
Cohort Desmonomata		
Family Camisiidae		
Camisia segnis (Hermann)	+	
Nothrus segnis (Herm.)		
Camisia spinifer (C.L. Koch)	+	
Nothrus spinifer C.L. Koch		
Platynothrus peltifer (Koch)	+	+
Hermannia bistriata (Nicolet)		
Family Hermanniidae		
Hermannia convexa (C.L. Koch)	+	
Hermannia gibba (Koch)		
Hermannia reticulata (Thorell)	+	+
Hermannia scabra (L. Koch)	+	
Family Malaconothridae		
Malaconothrus monodactylus (Michael)	+	
Trimalalconothrus tardus (Michael)	+	
Family Nanhermanniidae		
Nanhermannia nana (Nicolet)	+	+
Hermannia nanus (Nicolet)		
Family Nothridae		
Nothrus anauniensis Can. Et. Fanz.		+
Nothrus palustris Koch	+	+
Nothrus silvestris Nicolet	+	+
Family Perlohamiidae		
Epilohmannia sp.		+
Family Phthiracaridae		
Atropacarus wandae (Niedbała)		+
Phthiacarus affinis (Hull)		+
Steganacarus magnus (Nicolet)	+	+
Hoploderma magnum (Nicolet)		
Cohort Brachypylina		
Family Carabodidae		
Carabodes marginatus (Michael)		+
Carabodes willmanni Bernini		+
Odontocepheus elongatus (Michael)	+	+
Carabodes elongatus (Michael)		

continued

Table 3. Cont.

Family Ceratoppiidae		
Ceratoppia bipilis (Hermann) *Notaspis bipilis* Michael	+	+
Ceratoppia quadridentata (Haller)		+
Family Damaeidae		
Porobelba spinosa Sellnick		+
Damaeus (Paradamaeus) clavipes (Hermann)	+	+
Family Liacaridae		
Adoristes ovatus (C.L. Koch) *Liacarus ovatus* (C.L. Koch)	+	
Liacarus coracinus (C.L. Koch)	+	
Xenillus tegeocranus (Hermann) *Cepheus tegeocranus* (Herm.)	+	
Family Oppiidae		
Dissorhina ornata (Oudemanns)		+
Medoppia obsoleta (Paoli)		+
Family Thyrisomidae		
Banksinoma lanceolata (Michael)		+
Cohort Poronota		
Family Achipteriidae		
Parachipteria italicus (Nicolet)	+	+
Parachipteria ovalis (C.L. Koch)[1] *Oribata ovalis* C.L. Koch	+	
Family Ceratozetidae		
Ceratozetes gracilis (Michael) *Oribata gracilis* Michael (var. major Berlese)	+	+
Ceratozetes mediocris Berlese		+
Edwardzetes edwardsi (Nicolet) *Oribata edwardsi* Nicolet	+	
Melanozetes meridianus Sellnick		+
Melanozetes mollicomus (Koch) *Oribata mollicoma* C.L. Koch	+	+
Melanozetes stagnatilis (Hull) *Oribata alpina* sp. nov.	+	
Family Chamobatidae		
Chamobates cuspidatus (Michael) *Oribata cuspidata* Michael	+	
Family Euzetidae		
Euzetes globulus (Nicolet) *Oribata globula* Nicolet	+	+
Family Galumnidae		
Acrogalumna longipluma (Berlese)		+
Family Mycobatidae		
Minonthozetes pseudofusiger (Schweizer)		+
Minunthozetes semirufus (C.L. Koch) *Oribata fusigera* Michael	+	
Mycobates sp.		+
Family Oribatulidae		
Liebstadia humerata Sellnick		+
Liebstadia similis (Michael) *Notaspis similis* Michael	+	+

continued

---

[1] This is a best estimate of the synonymy based on the literature. However, Subías (2004) believes this to be 'sp. Inq.'

Table 3. Cont.		
Phauloppoa lucorum (C.L. Koch) *Notaspis lucorum* (C.L. Koch)	+	
Zygoribatul exilis (Nicolet) *Notaspis exilis* Michael	+	
Family Phenopelopidae		
Eupelops acromios (Herm.) *Pelops acromias* (Herm.)	+	
Eupelops plicatus (Koch) *Pelops fuliginosus* C.L. Koch	+	+
Family Scutoverticidae		
Scutovertex sculptus Michael	+	+

appear to reflect differences in the sampling methods and in the components of the ecosystems sampled. They also indicate that the Irish acarine fauna is understudied and requires significant further work in order to get a realistic impression of the species composition of the fauna and the extent of the biodiversity which occurs nationally.

Acknowledgements

I would like to acknowledge the assistance with sample processing and fieldwork provided by Liz Raeburn, Aoife Brennan, Stephen McCormack and T.K. McCarthy.

REFERENCES

Arnold, A.J. 1994 Insect suction sampling without nets, bags or filters. *Crop Protection* **13**, 72–5.

Arroyo, J. and Bolger, T. 2008 *Serratoppica serrata* and *Eupelops major* (Arachnida: Acari: Oribatida), two new records for Ireland with some comments on their biogeography. *Revista Ibérica de Aracnología* **16**, 119–22.

Arroyo, J. and Bolger, T. 2010 The Mite (Arachnida: Acari) Fauna inhabiting Irish Machair: a European Union priority coastal habitat. *Journal of Coastal Conservation* **15**, 181–94

Arroyo, J., Baars J.-R., O′Driscoll, L., Blackmore, M. and Bolger, T. 2013a First records of *Heminothrus (Capillonothrus) thori* (Berlese, 1904) and *Perlohmannia (Perlohmannia) dissimilis* (Hewitt, 1908) (Arachnida: Acari: Oribatida) in Ireland. *Bulletin of the Irish Biogeographical Society* **37**, 12–16.

Arroyo, J., Kenny, J. and Bolger, T. 2013b Variation between mite communities in Irish forest types – Importance of bark and moss cover in canopy. *Pedobiologia* **56**, 241–50.

Arroyo, J., Moraza, M.L. and Bolger, T. 2008 First records of *Leptogamasus (Leptogamasus) obesus* (Holzmannm, 1969) and *Leptogamasus (Valigamasus) pannonicus* (Willmanm, 1951) (Acari: Gamasida: Parasitidae) in Ireland. *Bulletin of the Irish Biogeographical Society* **32**, 213–15.

Beirne, B.P. 1985 Irish entomology: The first hundred years. *Irish Naturalists' Journal – Special Entomological Supplement*, 1–40.

Bolger, T. 1997 The life and work of G.H. Carpenter. *Pedobiologia* **14**, 1–2.

Bolger, T., Kenny, J. and Arroyo, J. 2013 The Collembola fauna of Irish forests - a comparison between forest type and microhabitat within the forests. *Soil Organisms* **85**, 61–7.

Brussaard, L., Behan-Pelletier, V.M., Bignell, D.E., Brown, V.K., Didden, W., Folgarait, P., Fragoso, C., Freckman, D.W., Gupta, V.V.S.R., Hattori, T., Hawksworth, D.L., Klopatek, C., Lavelle, P., Malloch, D.W., Rusek, J., Soderstrom, B., Tiedje, J.M. and Virginia, R.A. 1997 Biodiversity and ecosystem functioning in soil. *Ambio* **26**, 563–70.

Carpenter, G.H. 1913 Aptera. Clare Island Survey. *Proceedings of the Royal Irish Academy* **31**B, 1–13.

Coleman, D.C. and Crossley, J.D.A. 1996 *Fundamentals of Soil Ecology*. Academic Press.

Coleman D.C. and Hendrix P.F. (eds) 2000 *Invertebrates as webmasters in ecosystems*. Wallingford, Oxon. CABI Publishing.

Collins, T. 1999 The Clare Island Survey of 1909–11: participants, papers and progress. In C. Mac Cárthaig and K. Whelan (eds) *New Survey of Clare Island. Volume 1: History and cultural landscape*, 1–40. Dublin. Royal Irish Academy.

Curry, J.P. 1969 The qualitative and quantitative composition of the fauna of an old grassland site at Celbridge, Co. Kildare. *Soil Biology and Biochemistry* **1**, 219–27.

Fjellberg, A. 1980 *Identification keys to Norwegian Collembola*. Norsk Entomologisk Forening. Ås.

Giller, K.E., Beare, M.H., Lavelle, P., Izac, A.-M.N., Swift, M.J. 1997. Agricultural intensification, soil biodiversity and agroecosystem function. *Applied Soil Ecology* **6**, 3–16.

Hågvar, S. 1998 The relevance of the Rio-convention on biodiversity to conserving the biodiversity of soils. *Applied Soil Ecology* **9**, 1–7.

Halbert, J.N. 1915 Clare Island Surrey, 39. Acarinida. Section ii: Terrestrial and marine Acarina. *Proceedings of the Royal Irish Academy* **31**(39), 45–136.

Heneghan, L. and Bolger, T. 1996 Effects of acid rain components on soil microarthropods: a field manipulation. *Pedobiologia* **40**, 413–38.

Loreau, M., Naeem, S., Inchausti, P., Bengtsson, J., Grime, J.P., Hector, A., Hooper, D.U., Huston, M.A., Raffaelli, D., Schmid, B., Tilman, D. and Wardle, D.A. 2001 Biodiversity and ecosystem functioning: current knowledge and future challenges. *Science* **294**, 804–08.

Luxton, M. 1998 The oribatid and parasitifor mites of Ireland, with particular reference to the work of J.N. Halbert (1872–1948). *Bulletin of the Irish Biogeographical Society* **22**, 2–71.

McLean, M.A., Kaneko, N. and Parkinson, D. 1996 Does selective grazing by mites and Collembola affect litter fungal community structure? *Pedobiologia* **40**, 97–105.

Moraza, M.L., Arroyo, J. and Bolger, T. 2009 Three new species of mites (Acari: Zerconidae) from canopy habitats of Irish forests. *Zootaxa* 2019, 29–39.

O'Connell, T. and Bolger, T. 1997 Stability, ephemerality and dispersal ability: Soil microarthropod assemblages on fungal sporophores. *Biological Journal of the Linnean Society* **62**, 111–31.

Sterzyńska, M. and Bolger, T. 2004 Collembola of North Bull Island – new records for the Irish coast. *Fragmenta Faunistica* **47**, 47–50.

Subías, L.S. 2004 Listado sistemático, sinonímico y biogeográfico de los ácaros oribátidos (Acariformes, Oribatida) del mundo (1758–2002). *Graellsia* **60**, 3–305.

Verhoef, H.A. and De Goede, R.G.M. 1985 Effects of collembolan grazing on nitrogen dynamics in a coniferous forest. In A.H. Fitter, D. Atkinson, D.J. Read and M.B. Usher (eds), *Ecological interactions in the soil*, 367–76. Oxford. Blackwell Scientific.

Walter, D.E. and Proctor, H.C. 1999 *Mites: ecology, evolution and behaviour*. Wallingford, UK. CABI Publishing International.

Wardle, D.A. 1999 How food webs make plants grow? *Trends in Ecology and Evolution* **14**, 418–29.

Weigmann, G. 2006 *Hornmilben (Oribatida). Die Tierwelt Deutschlands* **76**. Goecke and Evers, Keltern.

CHAPTER 6

DIPLOPODA, CHILOPODA, OPILIONES, PSEUDOSCORPIONES AND LAND ISOPODA FROM CLARE ISLAND

Martin Cawley

ABSTRACT

This paper gives details of surveys undertaken by the author in 2002 of five groups of generally cryptozoic, terrestrial arthropods: Diplopoda (millipedes), Chilopoda (centipedes), Opiliones (harvestmen), Pseudoscorpiones (false scorpions) and land Isopoda (woodlice) on Clare Island. Recent records of these invertebrate taxa by other fieldworkers, together with the author's lists, are compared with those compiled in the original early twentieth-century faunal surveys on the island.

Introduction

The groups covered in this chapter were reported in multiple chapters in the original Clare Island Survey: Johnson (1912, 1915) studied the Diplopoda (millipedes) and Chilopoda (centipedes)—both groups are collectively referred to as 'myriopods'; Pack-Beresford (1911) studied the Opiliones (then referred to as Phalangida; harvestmen); Kew (1911) covered the Pseudoscorpiones (false scorpions) and the land Isopoda (woodlice) were part of Foster's (1912) study.

Methods

For the present survey, the author visited Clare Island on two occasions in 2002 and recorded millipedes (Diplopoda), centipedes (Chilopoda), harvestmen (Opiliones), false scorpions (Pseudoscorpiones) and woodlice (land Isopoda). On the first occasion (6–11 May 2002) collections were made in 24 of the 27 1km squares on the island and on the second visit (16–20 September 2002) collections were made in 25 1km squares. Thus, over the two visits, collections were made in all but one of the 1km squares, and most of the 1km squares were visited on both occasions. The 1km square not visited was L6586, which is a small patch of land difficult to access. Additional hand collected and pitfall trapped material was kindly provided by Stephen McCormack and this included additional records for millipedes (Diplopoda), which I have added to my own findings. The results of this fieldwork are detailed below. The numbers in brackets after each species indicates the number of 1km squares in which I recorded that species during the two visits. Species not recorded during the first Clare Island Survey are indicated using an asterisk (*).

Results and discussion

Diplopoda (millipedes)

Species recorded: *Glomeris marginata* (Villers) (3), **Nanogona polydesmoides* (Leach) (2), *Brachydesmus superus* Latzel (8), **Polydesmus angustus* Latzel (7), *Polydesmus inconstans* Latzel (4), **Macrosternodesmus palicola* Brölemann (1), **Proteroiulus fuscus* (Am Stein) (3), **Blaniulus guttulatus* (Fabricius) (2), **Boreoiulus tenuis* (Bigler) (1) and *Cylindroiulus latestriatus* (Curtis) (17).

The ten species collected in 2002 compares with just four species reported by Johnson (1912, 1915). Johnson's records of *Iulus luscus* Meinert are likely

Millipede *Cylindroiulus latestriatus*. Photo R. Anderson

to refer to *C. latestriatus*. Many of the additional species are likely to have been overlooked during the original survey. *N. polydesmoides, M. palicola, B. guttulatus* and *B. tenuis* were all associated with human habitation and are likely to have been brought to the island by human activity. *Polydesmus coriaceus* Porat, *Cylindroiulus punctatus* (Leach), *Ophyiulus pilosus* (Newport) and *Tachypodiulus niger* (Leach), which are amongst the commonest species on the mainland, were notable by their absence. The aliens *Leptoiulus belgicus* (Latzel) and *Cylindroiulus londinensis* (Leach) are expanding their range on the mainland and are among a number of species that might eventually colonise the island. The Irish millipede fauna has been reviewed by Doogue *et al.* (1993), and the British Isles fauna mapped by Lee (2006).

Chilopoda (centipedes)

Species recorded: *Stigmatogaster subterranea* (Shaw) (4), *Schendyla nemorensis* (C.L. Koch) (5), *Strigamia maritima* (Leach) (5), *Geophilus easoni* Arthur, Foddai, Lewis, Luczynski and Minelli (1), *Geophilus flavus* (De Geer) (2), *Geophilus insculptus* Attems (2), *Geophilus truncorum* (Bergsö and Meinert) (24), *Cryptops hortensis* Donovan (7), *Lithobius borealis* Meinert (13), *Lithobius forficatus* (Linnaeus) (23), *Lithobius melanops* Newport (10), *Lithobius variegatus* Leach (26), *Lithobius crassipes* L. Koch (1) and *Lamyctes emarginatus* (Newport) (2).

The fourteen species recorded compares with eight species reported by Johnson (1912). Two of the species on Johnson's list, *Hydroschendyla submarina* (Grube) and and *Lamyctes emarginatus*

Centipede *Lithobius variegatus*. Lecarrow, Clare Island. Photo J. Breen

(Newport) had not previously been recorded from Ireland. The first of these was not refound during the present survey, however this intertidal species is rare and difficult to collect. There is only one subsequent Irish record, from West Cork (Cawley 1998). Johnson's account also contained a summation of the scant amount of information on Irish myriapods available at that time. Among the additional species recorded in 2002 *C. hortensis* and especially *G. truncorum* were widespread on the island. The relatively inconspicuous *G. truncorum* was most likely overlooked by the naturalists who gathered myriapods during the first Survey, none of whom had much experience of these groups. However, *C. hortensis* is rather larger and more active, and is quite possibly a recent arrival on the island. It is a widespread centipede on southern offshore islands in Ireland (Cawley 2010). *S. subterranea* and *G. insculptus* were associated with human habitation on the island and likely arrived with the assistance of man. *Lithobius microps* Meinert, which is widespread on the mainland, was the only notable absentee. Pre-2000 Irish offshore island records for millipedes and centipedes were collated by Cawley (2000). Knowledge of the Irish centipedes and millipedes has increased greatly since the time of the original survey, however remarkably little additional information has been gathered on the symphylans. Specimens collected during the present survey remain to be identified.

Opiliones (harvestmen)

Species recorded: *Nemastoma bimaculatum* (Fabricius) (17), *Oligolophus tridens* (C.L. Koch) (4), *Paroligolophus agrestis* (Meade) (12), *Lacinius ephippiatus* (C.L. Koch) (2), *Mitopus morio* (Fabricius) (9), *Phalangium opilio* Linnaeus (7), *Megabunus diadema* (Fabricius) (6), *Rilaena triangularis* (Herbst) (8), *Leiobunum rotundum* (Latreille) (10) and *Nelima gothica* Lohmander (18).

Ten harvest-spider species were recorded compared to the seven species reported by Pack-Beresford (1911). *M. diadema* is a local species on the mainland, although less so in the north-west Cawley (2002). *Leiobunum blackwalli* Meade, which is widespread on the mainland, was the only notable absentee. The alien *Dicranopalpus ramosus* (Simon) was searched for without success. It has expanded rapidly on the mainland and is invading natural habitats such as woodland. It may only be a matter of time before it turns up on Clare Island.

Pseudoscorpiones (false scorpions)

Species recorded: *Neobisium carcinoides* (Hermann) (1) and *Neobisium maritimum* (Leach) (1).

Kew (1911) had reported three species from the island, including one, *Chthonius tetrachelatus* (Preyssler) which was not encountered during the present survey. The widespread *Chthonius ischnocheles* (Hermann) was not encountered during either survey. The synanthrophic *Dinocheirus panzeri* (C.L. Koch), which is widespread in cowsheds on the mainland, was searched for without success, however there were few suitable sites. Kew, who remains the most important worker on pseudoscorpions in the British Isles, carried out a thorough search of the island in 1910 and highlighted (Kew 1911) the great scarcity of these animals on the island at the time; that scarcity remains on the island today.

Land Isopoda (woodlice)

Species recorded: *Ligia oceanica* (Linnaeus) (12), *Androniscus dentiger* Verhoeff (1), *Haplophthalmus mengei* (Zaddach) (5), *Trichoniscoides saeroeensis* Lohmander (4), *Trichoniscus pusillus* Brandt (18), *Trichoniscus pygmaeus* Sars (3), *Philoscia muscorum* (Scopoli) (14), *Oniscus asellus* Linnaeus (25), *Cylisticus convexus* (De Geer) (1), *Porcellio dilatatus* Brandt (1), *Porcellio scaber* Latreille (22) and *Porcellio spinicornis* Say (1). Male specimens of *Oniscus* examined were referable to subspecies *asellus*. On present limited information subspecies *occidentalis* Bilton would appear to be confined to the south in Ireland and Britain.

Harvestman *Nelima gothica*. Lecarrow, Clare Island. Photo J. Breen

Woodlouse *Androniscus dentiger*. Photo R. Anderson

Twelve species of woodlice were recorded on the island, which compares with nine species reported by Foster (1912). *A. dentiger*, *C. convexus*, *P. dilatatus* and *P. spinicornis* are all likely to have been brought to the island by human activities. Foster's records of *Trichoniscoides albidus* (Budde-Lund) in all likelihood refer to *saeroeensis*. *Porcellionides cingendus* (Kinahan) remains unrecorded on Clare Island though present on nearby Inishbofin. The British Isles fauna has most recently been summarised by Gregory (2009).

REFERENCES

Cawley, M. 1998 New Irish vice-county records for centipedes (Chilopoda). *Bulletin of the British Myriapod Group* **14**, 10–17.

Cawley, M. 2000 Myriapod (Chilopoda and Diplopoda) notes from some Irish offshore islands. *Bulletin of the British Myriapod Group* **16**, 11–17.

Cawley, M. 2002 A review of the Irish harvestmen (Arachnida: Opiliones). *Bulletin of the Irish biogeographical Society* **26**, 106–37.

Cawley, M. 2010 A review of the Irish centipedes (Chilopoda). *Bulletin of the Irish biogeographical Society* **34**, 18–64.

Doogue, D., Fairhurst, C.P., Harding, P.T. and Jones, R.E. 1993 A review of Irish millipedes (Diplopoda). In M.J. Costello and K.S. Kelly (eds) *Biogeography of Ireland: past, present, and future. Occasional Publication of the Irish biogeographical Society* No. **2**, 83–97.

Foster, N.H. 1912 Clare Island Survey. Land and freshwater Isopoda. *Proceedings of the Royal Irish Academy* **31**(44), 1–4.

Gregory, S. 2009 *Woodlice and waterlice (Isopoda: Oniscoidea & Asellota) in Britain and Ireland*. Shrewsbury. FSC Publications.

Johnson, W.F. 1912 Clare Island Survey. Chilopoda and Diplopoda. *Proceedings of the Royal Irish Academy* **31**(33), 1–6.

Johnson, W.F. 1915 Addenda and corrigenda to section 2. 33-Chilopoda and Diplopoda. *Proceedings of the Royal Irish Academy* 31, 8–9.

Kew, H.W. 1911 Clare Island Survey. Pseudoscorpiones. *Proceedings of the Royal Irish Academy* **31**(38), 1–2.

Lee, P. 2006 *Atlas of the millipedes (Diplopoda) of Britain and Ireland*. Sofia. Pensoft.

Pack-Beresford, D.R. 1911 Clare Island Survey. Phalangida. *Proceedings of the Royal Irish Academy* **31**(36), 1–2.

CHAPTER 7

THE AQUATIC COLEOPTERA OF CLARE ISLAND

Stephen McCormack and T.K. McCarthy†

ABSTRACT

The aquatic beetle fauna of Clare Island (Coleoptera: Gyrinidae, Haliplidae, Noteridae, Dytiscidae, Hydraenidae, Helophoridae, Hydrophilidae, Scirtidae, Elmidae, Dryopidae, Chrysomelidae: Donaciinae) was surveyed. The composition of the aquatic beetle fauna of five islands—Britain, Ireland, Islay, Mull and Clare Island—were compared on the basis of flight ability and trophic role. The proportion of strong fliers was much greater on the three smaller islands studied. On Clare Island, 62% of the species are regarded as strong fliers or variable with regard to wing size or flight muscles, compared with 21% and 27% of those in Britain and Ireland respectively. A review of the trophic ecology of island aquatic Coleoptera assemblages suggested that a higher proportion of predators and scavengers are present in the offshore islands than in the faunas of Britain and Ireland. It is likely that the non-specialist habits of the majority of aquatic beetle species aids their ability to colonise new habitats.

Introduction

This study was initiated with a view to providing information on the present aquatic coleopteran fauna of Clare Island and to make comparisons with the earlier surveys of these insects on the island. Additional objectives were to analyse the variation in coleopteran species assemblages among sampling sites, with respect to environmental factors, and to compare the aquatic fauna of the island with those of some other islands, with respect to species composition, flight abilities and trophic roles.

Materials and methods

Samples of adult water beetles were collected between 9 June and 14 November 2002 from a representative series of aquatic habitats on Clare Island (Fig. 1). At each site a pond net and/or a small hand-held sieve (mesh size 1mm) was used depending on the size and depth of the water body. The net or sieve was swept through the water column while the substrate was disturbed to dislodge beetles. Particular attention was paid to searching at the water margins. In *Sphagnum* moss and floating vegetation an area was trampled, and beetles sieved from the puddle created. The contents of the sieve or net were sorted, and all adult beetles were kept. Where detritus was present this was shaken over a plastic sheet and beetles collected. Each site was searched until a full list of species for the site had been accumulated. This was deemed to have occurred when no more new species were found. Sampling time at each site varied from 30 minutes up to 2 hours depending on the size of the site and number of beetles and habitats present and the amount of vegetation or detritus present. Vegetation, water depth and substrate were noted. Beetles were killed with ethyl acetate vapour and preserved in 70% alcohol. Identifications were made using keys by Kevan (1962), Holmen (1987), Hansen (1987), Friday (1988), Drost *et al*. (1992), Nilsson and Holmen (1995) and Menzies and Cox (1996). Voucher specimens of each species recorded on Clare Island were sent to Garth Foster for

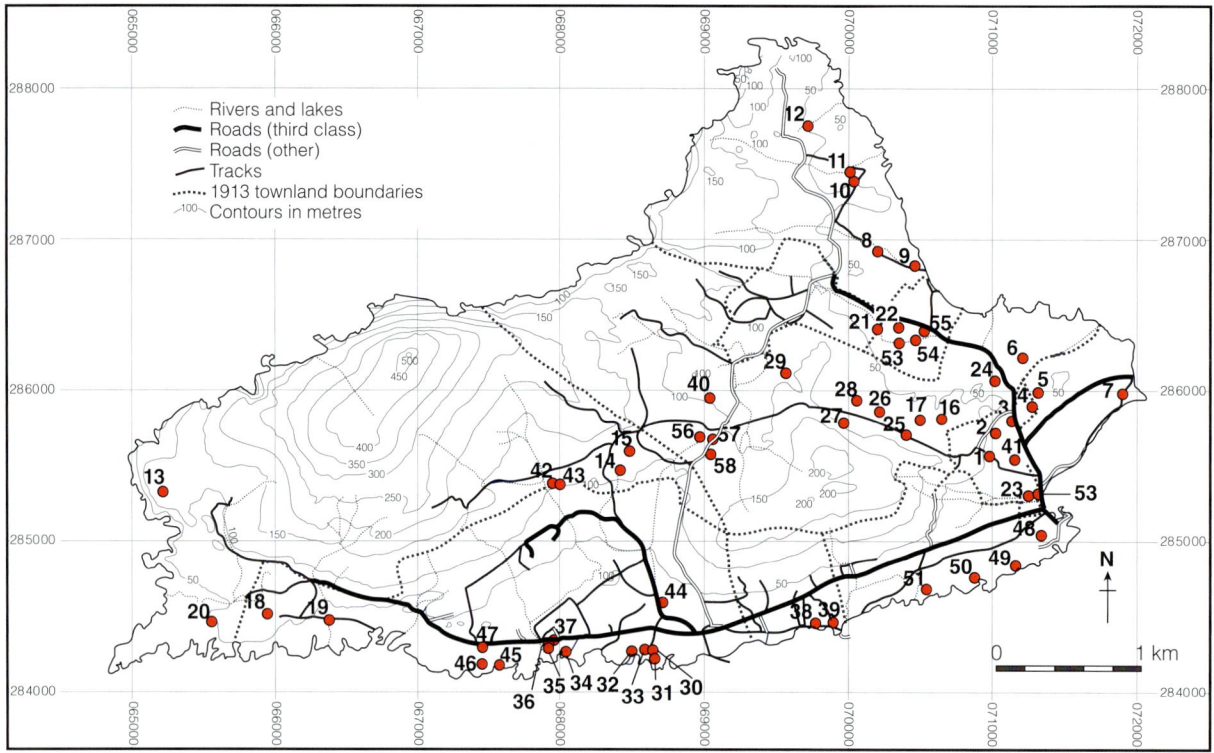

Fig. 1 Aquatic habitats on Clare Island where samples of adult water beetles were collected between 9 June and 14 November 2002

confirmation of identifications. Donaciinae were checked by Brian Nelson and Scirtidae by Jonty Denton. Voucher specimens were lodged with the National Museum of Ireland, accession code NMI. 36:2003. Nomenclature (Table 1) follows Ashe et al. (1998) except for more recent changes.

Results

A total of 87 species of aquatic Coleoptera have been recorded from Clare Island, 76 of which are recent records obtained in the present study or by Fahy (2003). Of these, 26 are new records for Clare Island but eleven species reported in the original survey were not re-recorded. A checklist is presented Table 1 summarising the species recorded in the present survey and those recorded by Balfour-Browne (1912) and Johnson and Halbert (1912). Balfour-Browne (1912) gave data on the number of sites in which each species occurred out a total of 45 sites sampled. In Table 1 the percentage occurrence of beetles at sites surveyed by Balfour-Browne (1912) and the 58 sites (Fig. 1) surveyed in the present work are compared.

The 87 species of aquatic Coleoptera so far recorded on Clare Island represents 38% (87/229) of the aquatic Coleoptera fauna of Ireland, excluding the families Heteroceridae and Curculionidae (Ashe et al. 1998). At 16km^2 the area of Clare Island is about 0.02% of the land area of Ireland. The composition of the Clare Island and Irish aquatic coleopteran faunas are compared, with respect to the relative importance of 16 families in Table 2. The dominant families are the Dytiscidae, Hydrophilidae and Hydraenidae with 82, 33 and 22 species in Ireland, respectively. The best represented family on Clare Island is the Hydrophilidae with 57.6% of the Irish species, which is well above the overall proportion of 38%.

A comparison of the composition of the aquatic beetle faunas was carried out using data from published lists of water beetles from five islands in the British Isles. Lists of beetles from Islay (Foster and Eyre 1988), Mull (Foster et al. 1991) and published checklists for Ireland (Ashe et al. 1998) and Britain (Friday 1988) were used. These lists were compared based on flight ability and trophic role of adult beetles. The relationship between land area and species richness was explored (McCormack 2003).

Table 3 shows the composition of island coleopteran faunas by reference to species richness of families. Data on four aquatic beetle families: Scirtidae, Chrysomelidae, Curculionidae and Heteroceridae were not available for all five

Table 1

Checklists of aquatic Coleoptera recorded in both Clare Island surveys, with data on the percentage occurrence of each species in sampling sites. (First Clare Island Survey: 45 sites. New Clare Island Survey: 58 sites. ✓ = species recorded but not at full survey sampling sites.

	First Survey (%)	New Survey (%)
Gyrinus marinus Gyllenhal	-	1.7
Gyrinus minutus Fabricius	6.7	8.6
Gyrinus substriatus Stephens	11.1	13.8
Haliplus confinis Stephens	2.2	1.7
Haliplus fulvus (Fabricius)	4.4	5.2
Haliplus lineatocollis (Marsham)	13.3	12.1
Haliplus ruficollis (De Geer)	26.7	5.2
Haliplus sibircus Motschulsky	-	1.7
Noterus clavicornis (De Geer)	-	8.6
Liopterus haemorrhoidalis (Fabricius)	-	3.4
Hygrotus inaequalis (Fabricius)	24.4	13.8
Hygrotus impressopunctatus (Schaller)	-	✓
Hydroporus discretus Fairmaire & Briscout	-	3.4
Hydroporus erythrocephalus (Linnaeus)	4.4	3.4
Hydroporus gyllenhalii (Schiödte)	44.4	43.1
Hydroporus incognitus Sharp	13.3	5.2
Hydroporus memnonius Nicolai	4.4	19
Hydroporus longulus Mulsant	2.2	-
Hydroporus nigrita (Fabricius)	22.2	22.4
Hydroporus obscurus Sturm	44.4	34.5
Hydroporus obsoletus Aubé	-	-
Hydroporus palustris (Linnaeus)	13.3	-
Hydroporus planus (Fabricius)	2.2	1.7
Hydroporus pubescens (Gyllenahal)	80	51.7
Hydroporus tessellatus Drapiez	4.4	6.9
Hydroporus tristis (Paykull)	35.6	34.5
Stictotarsus 12-pustulatus (Fabricius)	-	1.7
Stictonectes lepidus (Olivier)	6.7	3.4
Nebrioporus assimilis Paykull	2.2	✓
Agabus affinis (Paykull)	-	5.2
Agabus bipustulatus (Linnaeus)	66.7	44.8
Agabus nebulosus (Forster)	-	1.7
Agabus paludosus (Fabricius)	2.2	-
Agabus sturmii (Gyllenhal)	6.7	19
Ilybius fuliginosus (Fabricius)	4.4	5.2
Ilybius guttiger (Gyllenhal)	-	3.4
Ilybius montanus (Stephens)	-	15.5
Rhantus exsoletus (Forster)	-	✓
Rhantus suturellus (Harris)	4.4	8.6
Laccophilus minutus (Linnaeus)	13.3	5.2
Acilius sulcatus (Linnaeus)	-	1.7
Dytiscus marginalis Linnaeus	2.2	1.7

continued

Table 1. Cont.

	First Survey (%)	New Survey (%)
Dytiscus semisulcatus Müller	2.2	5.2
Limnebius truncatellus (Thunberg)	4.4	6.9
Ochthebius punctatus Stephens	-	1.7
Ochthebius lejolisii Mulsant & Rey	4.4	-
Ochthebius minimus (Fabricius)	2.2	-
Helophorus aequalis Thomson	11.1	1.7
Helophorus brevipalpis Bedel	28.9	13.8
Helophorus flavipes Fabricius	80	58.6
Helophorus grandis Illiger	-	3.4
Helophorus granularis Linnaeus	-	3.4
Helophorus obscurus Mulsant	-	6.9
Chaetarthria seminulum/simillima agg.	✓	12.1
Paracymus scutellaris (Rosenhauer)	20	17.2
Anacaena globulus (Paykull)	26.7	43.1
Anacaena limbata (Fabricius)	4.4	5.2
Anacaena lutescens (Stephens)	-	12.1
Laccobius atratus (Rottenburg)	-	3.4
Laccobius ytenensis Sharp	-	10.3
Laccobius bipunctatus (Fabricius)	4.4	1.7
Laccobius minutus (Linnaeus)	2.2	-
Helochares punctatus Sharp	13.3	19
Enochrus affinis (Thunberg)	6.7	8.6
Enochrus coarctatus (Gredler)	2.2	1.7
Enochrus fuscipennis (Thomson)	51.1	22.4
Hydrobius fuscipes (Linnaeus)	2.2	5.2
Coelostoma orbiculare (Fabricius)	2.2	6.9
Cercyon ustulatus (Preyssler)	-	✓
Cercyon littoralis (Gyllenhal)	✓	✓
Cercyon depressus Stephens	✓	-
Megasternum concinnum (Marsham)	✓	✓
Odeles marginata (Fabricius)	✓	-
Cyphon coarctatus Paykull	✓	✓
Cyphon hilaris Nyholm	-	✓
Cyphon laevipennis Tournier	-	✓
Cyphon ochraceus Stephens	✓	-
Cyphon palustris Thomson	✓	✓
Cyphon padi (Linnaeus)	-	✓
Cyphon variabilis (Thunberg)	✓	1.7
Esolus parallelepipedus (Müller)	✓	1.7
Dryops luridus (Erichson)	✓	20.7
Donacia simplex Fabricius	✓	-
Donacia thalassina Germar	✓	✓
Donacia versicolorea (Brahm)	-	1.7
Donacia vulgaris Zschach	-	✓
Plateumaris sericea Linnaeus	✓	3.4

Table 2
A comparison of the aquatic Coleoptera fauna of Clare Island, as reflected in species richness of 16 families, with that of Ireland, together with the percentage representation of species recorded on the island

	Ireland	Clare Island	%
Gyrinidae	10	3	30.0%
Haliplidae	13	5	38.5%
Hygrobiidae	1	0	0.0%
Noteridae	2	1	50.0%
Dytiscidae	82	34	41.5%
Hydraenidae	22	4	18.2%
Hydrociidae	3	0	0.0%
Helophoridae	16	6	37.5%
Georissidae	1	0	0.0%
Hydrophilidae	33	19	57.6%
Scirtidae	20	8	40.0%
Elmidae	4	1	25.0%
Dryopidae	3	1	33.3%
Heteroceridae	(5)		
Chrysomelidae	19	5	26.3%
Curculionidae	(20)		
	229	87	37.9%

islands and these are not included in Table 3. The G-test was used to test for significant differences in the faunal composition of the families in Table 3. No significant differences were found (p>0.05) indicating that the proportion of each island fauna is similar for all the islands.

Dispersal ability island faunas
Information on dispersal ability and trophic status for each of the species recorded on Clare Island was obtained from published sources by McCormack (2003) This information on life histories was obtained mainly from Holmen (1987), Nilsson and Holmen (1995), Hansen (1987), Balfour-Browne (1940, 1950, 1958). Information on flight ability is based mainly on Foster (1979) and Jackson (1952, 1956, 1956a, 1973). The flight capacity of the aquatic Coleoptera of five Islands in the British Isles is summarised in Figure 2. Of the water beetles recorded on Clare Island in the current survey, 63% are strong fliers or are polymorphic with regard to flight capacity. Of the Irish fauna, just over 39% of water beetles are strong fliers or with variable flight ability due to polymorphisms in wings and/or flight musculature. For the majority of species in Britain and Ireland there are no data on flight ability (Foster 1979). As can be seen in Figure 2, there is a marked increase in the proportion of strongly flying species with decrease in island size was noted, while the proportions of flightless species and species with variable flight ability were generally similar among the size range of islands.

Differences between islands in the proportion of each of the flight ability categories, strong fliers, variable fliers, flightless and unknown flight ability were investigated using the G-test. As Britain is the largest island, the proportion of each of the flight ability categories on that island were used as the expected values. Islay, Mull, and Clare Island all had highly significant differences in the flight capacity of their faunas (p<0.01), while Ireland did not (p>0.05).

Trophic niches
Adult aquatic beetles of the British Isles were assigned to one of four trophic categories: predators/scavengers, detritivores, herbivores or omnivores. The composition of the aquatic beetle faunas is illustrated in Figure 3. The greater part of the aquatic beetle fauna of the British Isles are predators or scavengers as a result of the predominance of the adephagan families. The Hydrophilidae are included here as detritivores

Table 3
Relative composition of the aquatic Coleoptera faunas, as indicated by checklists for 13 families, of five islands in the British Isles

	Clare Island		Islay		Mull		Ireland		Britain	
	species	%	species	%	species	%	species	%	species	%
Gyrinidae	3	4.1	5	5.6	3	6.4	10	5.3	11	4.0
Haliplidae	5	6.8	7	7.8	5	10.6	13	6.8	19	6.9
Hygrobiidae	0	0	0	0	0	0	1	0.5	1	0.4
Noteridae	1	1.4	0	0	0	0	2	1.1	2	0.7
Dytiscidae	34	45.9	46	51.1	23	48.9	82	43.2	114	
Hydraenidae	4	5.5	8	8.9	3	6.4	22	11.6	30	10.9
Helophoridae	6	8.2	7	7.8	4	8.5	16	8.4	20	7.3
Hydrochidae	0	0	0	0	0	0	3	1.6	6	2.2
Georissidae	0	0	0	0	0	0	1	0.5	1	0.4
Hydrophilidae	19	26.0	13	14.4	7	14.9	33	17.4	50	18.2
Elmidae	1	1.4	2	2.2	1	2.1	4	2.1	12	4.4
Dryopidae	1	1.4	2	2.2	1	2.1	3	1.6	8	2.9
Limnichidae	0	0	0	0	0	0	0	0	1	0.4
Number of species	74		90		47		190		275	

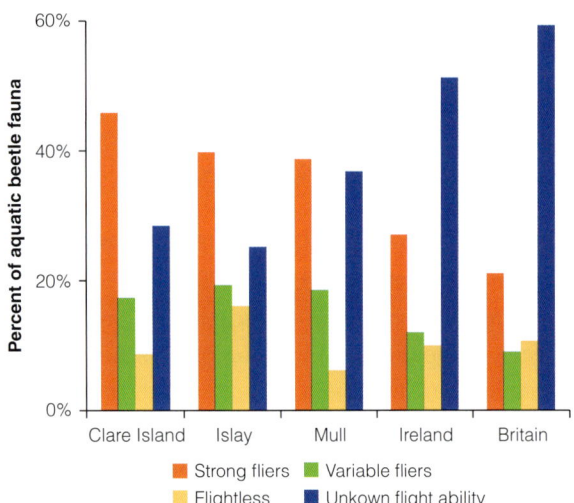

Fig. 2 The flight capacity of the aquatic Coleoptera of five Islands in the British Isles

although the larvae are predaceous (Balfour-Browne 1958). The Helophoridae, Hydraenidae, Elmidae and Dryopidae are also included as detritivores. The feeding habits of adult scirtid beetles are poorly known, although on the basis of the structure of the mouthparts it is thought that they are predatory (Klausnitzer 1996). The Haliplidae include omnivores, herbivores and carnivores. The sub-family Donaciinae is the only fully herbivorous taxon.

Significant differences between the observed proportions of each trophic group on the four smaller islands, and those expected from the proportions observed in the British fauna were assessed with a G-test. A significant difference (p<0.05) was found between the distribution of trophic roles in the Clare Island fauna and that of Britain. However, no significant differences were observed in respect of comparisons between the waterbeetle fauna of Britain and those of Islay, Mull and Ireland.

Discussion

Comparison of aquatic beetle checklists from the First and New Clare Island Surveys

Taxonomic changes, seasonality of sampling, relative rarity of many species, species turnover, habitat changes and differing sampling methods have contributed to the differences in the Clare Island fauna between 1909 and 2002 (Table 1).

Four additions to the species list of the First Survey are due to taxonomic changes in the intervening period. *Haliplus ruficollis* Brit. auct. was recorded by Balfour-Browne (1912). However, this taxon was split into several species by Edwards (1911). Unfortunately, no specimens of this species group from Clare Island were retained by Balfour-Browne (1912), who noted that more species may have been present on the island. In the current survey two species of this group were recorded: *Haliplus ruficollis* (DeGeer) and *Haliplus sibircus* Motschulsky (Table 1).

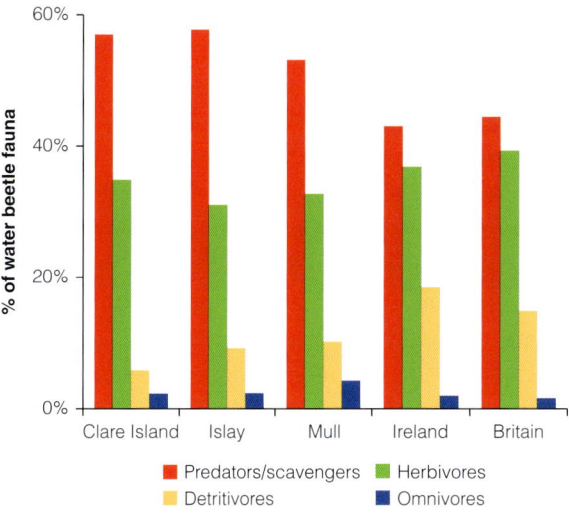

Fig. 3 The distribution of adult aquatic beetles across four trophic categories for each of the five islands

Anacaena limbata Fabricius, recorded by Balfour-Browne (1912), was later divided into two species: *Anacaena lutescens* (Stephens) and *Anacaena limbata* (Fabricius) by van Berge Henegouwen (1986). It is likely that both species, recorded in this study, along with *Anacaena globulus* (Paykull), were present on Clare Island in 1909. *Helophorus aquaticus* Linnaeus recorded by Balfour-Browne (1912) was also split into two species: *H. aequalis* Thomson and *H. grandis* Illiger (Kevan 1966), both of which were found on Clare Island (Table 1). *Cyphon variabilis sensu auct.* Brit. *partim* not Thunberg 1787, recorded by Johnson and Halbert (1912), is currently considered to be three species: *C. pubescens* (Fabricius), *C. variabilis* (Thunberg) and *C. laevipennis* Tournier (Kevan 1962). The latter two species were recorded from Clare Island in the current survey. *Cyphon laevipennis* is a recent addition to the Irish list from a site in Wexford (O'Connor 1999) and Clare Island was the third recorded location for the species in Ireland.

The species list of Balfour-Browne (1912) lacks some species that are currently quite widely distributed on Clare Island. Since it seems unlikely that a collector as experienced as Balfour-Browne would overlook such species, it must be concluded that there have been some changes in the water beetle fauna of Clare Island in the interim. Two species that are not easily overlooked, but not recorded in 1909, *Noterus clavicornis* and *Ilybius montanus*, now appear to be quite common. The former inhabits permanent stagnant pools with weeds, while the latter is found in a variety of habitats with acid water (particularly temporary waters). Further evidence for a change in the fauna of Clare Island are records of some relatively large Dytiscids: *Liopturus haemorrhoidalis, Agabus affinis, Agabus nebulosus, Ilybius guttiger, Rhantus exsoletus* and *Acilius sulcatus*. These species were found at only a few sites, but the large size of the beetles makes it unlikely that they were overlooked.

The specific habitat requirements, and general rarity, of some aquatic beetle species provides a possible explanation for their absence in either survey. Examples of such species absent from Balfour-Browne's list are: *Ochthebius punctatus, Helophorus granularis, Laccobius ytenensis* and *Laccobius atratus*. It is possible that the respective habitats of these species—brackish water, small grassy pools and muddy stream edges were overlooked by Balfour-Browne in the short time he spent on Clare Island. *Hydroporus longulus* and *Ochthebius lejolsii* recorded by Balfour-Browne were not re-recorded in this survey (Table 1), although their habitats—small trickling streams on high ground and coastal rock pools above high water—were searched. Differing collecting techniques may account for the absence of these species in the current survey. There may also be an element of luck in finding some of the small and cryptic species, which may occur only in low abundance.

A present-day absentee from Clare Island would appear to be *Hydroporus palustris*, a eurytopic species. Although it was formerly found in a wide range of freshwater habitats—it was common in 1909, occurring at six sites (Balfour-Browne 1912)—it was not recorded in the present survey.

Nebrioporus assimilis, a lake-dwelling species was not found at Creggan Lough in 2002. However, there is a recent record from 1999 (Fahy 2003). *Laccobius minutus*, and *Ochthebius minimus* were recorded by Balfour-Browne (1912) in one collection each and were probably overlooked in the current survey. Both species are reasonably common and widely distributed in Ireland. *Ochthebius minimus* is usually found in freshwater but can occur in brackish water. It was recorded from a marsh behind a boulder beach at Kinnacorra by Balfour-Browne (1912). *Agabus paludosus* also only occurred in one collection of Balfour-Browne and it is characteristic of small streams. As this habitat was sampled widely on Clare Island in the current survey, this may indicate that the species is still rare on Clare Island or absent.

Four species recorded by Johnson and Halbert (1912): *Cyphon ochraceus*, *Cercyon depressus*, *Donacia simplex* and *Donacia thalassina* were not re-recorded in current survey. The Clare Island record for the *Cyphon ochraceus* was the second record in Ireland for the species (Johnson and Halbert 1912) and its distribution in Ireland is still poorly known (Denton and Foster 2000). This species did not occur in Malaise trap samples or in sweep net samples, although other species of the genus were abundant. *Cercyon depressus* is found in shore drift and did not turn up in samples from this habitat whereas a species with similar ecological requirements, *Ceryon littoralis* was recorded on several occasions. *Donacia simplex* is one of more widespread members of the genus in Ireland. Its foodplant, *Sparganium* spp., is still found on Clare Island and this species could have been overlooked as the Donaciinae are more seasonal than many other aquatic beetles. *Donacia thalassina* feeds on *Carex* sp. and was recorded at Lough Merrignagh, in 1998 by Fahy (2003).

Flight abilities of Clare Island aquatic Coleoptera

The three smaller islands compared in Figure 2 (Clare Island, Islay and Mull) are separated from the adjacent mainland only by narrow stretches of sea. A relatively flat species-area curve (McArthur and Wilson 1967; McCormack 2003) is indicated by data for the five islands listed in Table 3. Dispersal of species from the mainland to island may be relatively frequent given that dispersal of aquatic Coleoptera is not severely hindered by separation by the sea. MacArthur and Wilson (1967) hypothesised that the number of species on an island would reach an equilibrium point but that the composition of the fauna would continue to change as new species colonised and established species became extinct. This phenomenon of species turnover may partly explain the observed change in the fauna of Clare Island between 1909 and 2002.

The data compiled for the five islands listed in Table 3 suggests that the proportion of strong flying aquatic Coleoptera increases as the island area decreases (Fig. 2). Most of the species regarded as common and widespread in Ireland are capable of flight and disperse readily. *Agabus bipustulatus*, *Hydrobius fuscipes*, *Helophorus brevipalpis* and *Dytiscus marginalis* are good examples of these species and they are often recorded in flight. These species are also regarded as eurytopic as they occur in an extremely wide range of habitats. Species characteristic of temporary habitats, including *Hydroporus planus*, *Agabus nebulosus* and *Ilybius montanus*, are also known to be active fliers. Members of the genus *Cyphon* are all strong fliers, as are the Donaciinae. The high proportion of strong flying aquatic beetles on Clare Island may indicate that on offshore islands the aquatic beetle populations frequently receive dispersing insects from the mainland.

Information on the flight ability of aquatic beetles is patchy and sometimes contradictory, as beetles may vary in their flight ability in different parts of their geographic ranges. Most of the research in this area was carried out by Jackson in Scotland. Many British and European specimens were dissected by Jackson (1952, 1956a, 1956b, 1973) but the majority of species in Britain and Ireland have not been studied in relation to flight ability. Species on Clare Island listed (McCormack 2003) as variable with regard to flight ability are *Haliplus fulvus*, *H. ruficollis*, *H. sibircus*, *Noterus clavicornis*, *Hygrotus inaequalis*, *Hydroporus discretus*, *H. erythrocephalus*, *H. gyllenhalii*, *H. memnonius*, *H. palustris*, *H. tristus*, *Stictotarsus 12-pustulatus*, *Helophorus granularis*, *Anacaena limbata* and *Laccophilus minutus*.

The differing composition of the aquatic beetle faunas of Scottish islands was noted by Balfour-Browne (1953). The differing compositions, even on neighbouring islands, was attributed to the different build up of species on an island over time and to the interactions between species, as

Dytiscus marginalis. Photo: Bernard Dupont. This item is distributed under the terms of the CC BY ShareAlike 2.0 Generic licence

Haliplus confinis. Photo: UR Schmidt. This item is distributed under the terms of the CC BY ShareAlike 4.0 licence

well as to the different habitats present. Changing environmental conditions are likely to cause movement of individuals or populations and isolated island faunas may be less stable than large mainland faunas (Balfour-Browne 1953).

The dispersal ability of *Noterus clavicornis* has been the subject of debate in the past. It is usually flightless but individuals capable of flight occur and it is found on some islands (Jackson 1956b). Its absence from Clare Island in 1909 and subsequent arrival and establishment illustrates the ongoing colonisation of the island. The flightless species recorded on Clare Island are *Anacaena globulus*, *Hydroporus obscurus*, *Hydroporus nigrita*, *Nebrioporus assimilis* and *Esolus parallelepipedus*; all are widely distributed throughout Ireland.

Trophic categories of aquatic Coleoptera on Clare Island

There appears to be a trend towards an increasing proportion of predators or scavengers with a consequent decline in the proportion of aquatic Coleoptera in the other trophic groups (Fig. 3) in the five islands listed in Table 3. On Clare Island however the Hydrophilidae are well represented and this is reflected in the larger proportion of detritivorous beetles on Clare Island than would be expected extrapolating the trend observed from Islay and Mull. Most of the aquatic beetles are non-specific in their feeding habits being predators or scavengers or detritivores. This non-specialisation increases their ability to establish themselves in new habitats. The trend seen in the decline of herbivorous beetles may reflect the difficulty that host specific herbivores have in colonising islands compared with non-specific predators and scavengers. Some of the Haliplidae including *Haliplus sibircus* are known to be facultative omnivores taking a wide range of food (Holmen 1987). Becker (1975) hypothesised that the greater trophic specialisation of herbivores makes it more difficult for them to establish populations in new environments. This would appear to be borne out by the present data. As aquatic habitats can be considered to be habitat islands, aquatic beetles are specialist island colonisers.

Overview

The present account provides an overview of our current knowledge of the aquatic Coleoptera of Clare Island and offers some new insights into the island biogeography of these interesting insects. However, it is clear that many new questions are also posed by the results. Future work could usefully be focused on the following areas: use of modern taxonomic tools for some difficult genera and larvae; studies on alary polymorphism and flight abilities; colonisation of experimental ponds and comparative studies between islands. Further analysis of the trophic interactions within the island's freshwater habitats would be useful. It could establish if frequent absence of fish predators or beetle parasites are important determinants of the composition of island aquatic Coleoptera species assemblages.

REFERENCES

Ashe, P., O'Connor, J.P. and Murray D.A. 1998 A checklist of Irish aquatic insects. *Occasional Publication of The Irish Biogeographical Society* **3**, 79.

Balfour-Browne, F. 1912 Aquatic coleoptera. Clare Island survey. *Proceedings of the Royal Irish Academy* **29**, 1–20.

Balfour-Browne, F. 1940 *British water beetles*. Volume 1. London. Ray Society.

Balfour-Browne, F. 1950 *British water beetles*. Volume 2. London. Ray Society.

Balfour-Browne, F. 1951 The aquatic Coleoptera of Ireland. *Entomologist's Gazette* **2**, 1–52.

Balfour-Browne, F. 1953 The aquatic Coleoptera of the western Scottish isles with a discussion on the sources of origin and means of arrival. *Entomologist's Gazette* **4**, 79–127.

Balfour-Browne, F. 1958 *British water beetles*. Volume 3. London. Ray Society.

Becker, P. 1975 Island colonisation by carnivorous and herbivorous Coleoptera. *Journal of Animal Ecology* **44**, 893–906.

Bilton, D.T. 1988 A survey of aquatic Coleoptera in central Ireland and the Burren. *Bulletin of the Irish Biogeographical Society* **11**, 77–94.

Bilton, D.T. and Lott, D.A. 1991 Further records of aquatic Coleoptera from Ireland. *Irish Naturalists' Journal* **23**, 389–97.

Denton, J. and Foster, G.N. 2000 Scirtid recording in Britain and Ireland - first progress report. *Latissimus* **12**:1.

Drost, M.B.P., Cuppen, H.P.J.J., van Nieukerken, E.J. and Schreijer, M. (eds) 1992 *De waterkevers van Nederland.* Utrecht. KNNV Uitgeverij.

Edwards, J. 1911 A revision of the British species of *Haliplus* Latrielle. *Entomologist's Monthly Magazine* **47**, 1–10.

Fahy, S. 2003 Studies of the aquatic insects of Clare Island, with particular reference to Chironomidae. Unpublished PhD thesis. National University of Ireland, Galway.

Foster, G.N. 1979 Flight and flightlessness. *Balfour-Browne Club Newsletter* **12**, 4–7.

Foster, G.N. 1995b Some records of aquatic Coleoptera in Kerry, including *Helophorus griseus* Herbst. (Coleoptera: Helophoridae) new for Ireland. *Irish Naturalists' Journal* **25**, 32–4.

Foster, G.N. and Eyre, M.D. 1988 The water beetles of Islay. *Glasgow Naturalist* **21**, 423–31.

Foster, G.N. and Eyre, M.D. 1992 *Classification and ranking of water beetle communities.* Peterborough. Joint Nature Conservaton Committee.

Foster, G.N., Nelson, B.H., Bilton, D.T., Lott, D.A., Merritt, R., Weyl, R.S. and Eyre, M.D. 1992 A classification and evaluation of Irish water beetle assemblages. *Aquatic conservation: marine and freshwater ecosystems* **2**, 185–202.

Foster, G.N., Spirit, M.G. and Counsell, D. 1991 A survey of water beetles in the western Highlands and on Mull. *Glasgow Naturalist* **22**, 21–9.

Friday, L.E. 1987 New records of aquatic Coleoptera from Cos Cork and Kerry. *Irish Naturalists' Journal* **22**, 343–45.

Friday, L.E. 1988 A key to the adults of British water beetles. *Field Studies* **7**, 1–151.

Holmen, M. 1987 The aquatic Adephaga (Coleoptera) of Fennoscandia and Denmark. I Gyrinidae, Haliplidae, Hygrobiidae and Noteridae. *Fauna Entomologica Scandinavica* **20**, 171.

Jackson, D.J. 1952 Observations on the capacity for flight of water beetles. *Physiological Entomology* **27**, 57–69.

Jackson, D.J. 1956a The capacity for flight of certain water beetles and its bearing on their origin in the western Scottish Isles. *Proceedings of the Linnean Society of London* **167**, 76–94.

Jackson, D.J. 1956b Observations on flying and flightless water beetles. *Journal of the Linnean Society of London* **43**, 18–42.

Jackson, D.J. 1958 Notes on some nematodes and trematodes infesting water beetles. *Entomologist's Monthly Magazine* **90**, 91–2.

Jackson, D.J. 1973 The influence of flight capacity on the distribution of aquatic Coleoptera in Fife and Kinross-shire. *Entomologist's Gazette* **24**, 247–93.

Johnson, W.F. and Halbert, J.N. 1912 Terrestrial Coleoptera. Clare Island Survey. *Proceedings of the Royal Irish Academy* **28**, 24.

Kevan, D.K. 1962 The British species of the genus *Cyphon* Paykull (Col., Helodidae), including three new to the British list. *Entomologist's Montly Magazine* **98**, 114–21.

Kevan, D.K. 1966 The British speices of *Helophorus* Illiger, subgenus *Helophorus* s. str. (Col., Hydrophilidae). *Entomologist's Monthly Magazine* **101**, 254–68.

Klausnitzer, B. 1996 Coleoptera Scirtidae, Marsh Beetles. In Nilsson, A.N. (ed.), *Aquatic Insects of North Europe. A taxonomic handbook,* 247. Stenstrup. Apollo Books.

MacArthur, R.H. and Wilson, E.O. 1967 *The theory of island biogeography.* Princeton. Princeton University Press.

McCormack, S. 2003 Studies on the aquatic coleoptera of Clare Island, Co. Mayo. Unpublished MSc thesis, National University of Ireland, Galway.

Menzies, I.S. and Cox, M.L. 1996 Notes on the natural history, distribution and identification of British reed beetles. *British Journal of Entomology and Natural History* **9**, 137–62.

Nelson, B., Foster, G.N., Weyl, R. and Anderson, R. 1998 The distribution of aquatic Coleoptera in Northern Ireland. Part 2: Families Hydraenidae, Helophoridae, Hydrochidae, Hydrophilidae, Elmidae and Dryopidae. *Bulletin of the Irish Biogeographical Society* **22**, 128–93.

Nilsson, A.N. and Holmen, M. 1995 The aquatic Adephaga (Coleoptera) of Fennoscandia and Denmark. II. Dytiscidae. *Fauna Entomologica Scandinavica* **32**, 188.

O'Connor, J.P. 1999 *Cyphon phragmiteticola* Nyholm (Scirtidae) new to Ireland. *The Coleopterist* **8**,116.

van Berge Henegouwen, A.L. 1986 Revision of the European species of *Anacaena* Thomson (Coleoptera: Hydrophilidae). *Entomologica Scandinavica* **17**, 393–407.

CHAPTER 8

AQUATIC BEETLE ASSEMBLAGES (INSECTA: COLEOPTERA) IN A CLARE ISLAND LOUGH

Stephen McCormack and T.K. McCarthy[†]

ABSTRACT

Intra-habitat spatial variations in aquatic beetle assemblages were studied in Clare Island with particular reference to contrasting vegetation types in an oligodystrophic lough. Distinctive aquatic beetle species assemblages were found in shallow water, in *Sphagnum*-rich habitats with low pH. Greatest aquatic beetle species richness was found in shallow water in vegetation composed of mosses and forbs. In contrast, both beetle abundance and species richness were lowest in open water with sparse vegetation.

Introduction

The aquatic beetle fauna of Clare Island, where a total of 87 species have so far been recorded, has been described by McCormack and McCarthy (present volume). A small shallow oligodystrophic waterbody, known locally as Poirtín Fhuinch Lough, was identified as having relatively high aquatic beetle species richness. This site had also been sampled during a number of other ecological studies. It was therefore selected for a detailed study of intra-habitat variations in aquatic beetle species abundances.

The specific aims of this study were to examine the composition of the aquatic beetle assemblages associated with four contrasting types of aquatic vegetation and to investigate other environmental factors that may influence the distribution and abundance patterns of aquatic beetle species within Poirtín Fhuinch Lough.

The role of littoral emergent and submerged macrophytes has been increasingly recognised in limnology and freshwater fish management Carpenter and Lodge 1986; Gasith and Hoyer 1998; Thomaz and Cunha 2010). Aquatic plants can influence the survival, reproductive success and population ecology of associated animal species in a variety of ways. They can provide refugia from large open water predators (Rennie and Jackson 2005); egg deposition substrates; shelter from extreme environmental conditions (Brönmark and Hansson 2005); and contribute to herbivore and detritivore food chains (Carpenter and Lodge 1986). Increased knowledge of aquatic plant-invertebrate associations may become progressively significant because of the changes occurring in many Irish mainland freshwater ecosystems. Studies of the relatively pristine and simple Clare Island aquatic habitats may provide valuable insights into natural species interactions no longer seen so clearly in inland waterbodies.

Study area

Poirtín Fhuinch Lough is the largest lough on Clare Island with an area of about 3000m^2. It was probably created by cutting turf in the early part of the 1900s (P. MacNamara, pers. comm.). It has a maximum depth of about 1m but is generally 60–70cm deep. The lough is situated in a natural hollow at about 70m altitude toward the eastern end of Clare Island and overlies Silurian bedrock (Graham 2001) The water level is very stable owing to the local topography and there are no fish present. The lough has a varied substrate and a variety of vegetation communities typical of oligodystrophic western Irish lakes. Much of the lough overlies a peaty or vegetated substratum. However, in exposed parts the substrate is rocky or gravely. *Littorella uniflora* and *Juncus bulbosus* are common in shallow exposed margins along the northeastern side of the lough. The vegetation is sparse in places as the area is subject to disturbance by wave action during the frequent southerly and westerly winds. The substrate is consequently stony or gravely or with some gritty peat. The absence of *Lobelia dortmanna* and presence of *Myriophyllum alterniflorum* and *Juncus bulbosus* likens this to the A22 *Littorella uniflora-Lobelia dortmanna* community of the National Vegetation Classification Scheme (Rodwell *et al.* 1995). This vegetation community is characteristic of barren stony shallows of clear infertile standing water particularly in western Britain (Rodwell *et al.* 1995).

Away from the exposed shores, the substrate is generally composed of bare or vegetated peat. *Potamogeton natans* was the only plant in much of the deeper parts of the lough, where it grows over bare peat substratum. In the lee of islands and around the southern and western areas of the lough where there is shelter from wave action, the charophyte *Nitella* sp. is dominant, growing to within 10–15cm of the water surface. Where there is shallow water over peat and around many of the islands, *Menyanthes trifoliata* is frequent as an emergent plant. The edges of Poirtín Fhuinch and the island coastline are in many places steep or overhanging where peat has been eroded and there is little opportunity for marginal aquatic plants to establish themselves. However, the indented outline of the lough provides sheltered areas with shallow water and sloping margins with floating mats of mosses and other aquatic plants.

Floating mats of *Hypericum elodes* and *Potamogeton polygonifolius* set in a carpet of *Sphagnum* mosses as well as *Ranunculus flammula* and *Hydrocotyle vulgaris* occur about the edges of

Looking to the northeast over Poirtín Fhuinch Lough, Clare Island (L707857). Photo: S. McCormack

the lough. This vegetation community corresponds to NVC community M29 *Hypericum elodes-Potamogeton polygonifolius* soakway (Elkington *et al.* 2002). This community is characteristic of heathland pools and soakways and is known from the western British Isles (Elkington *et al.* 2002). Another transitional community is composed of stands of *Sphagnum* and *Juncus effusus*, which are dominant to various degrees and can both form continuous cover. *Menyanthes trifoliata* and *Iris pseudacorus* were sometimes found growing through the *Sphagnum* carpet. This is a species-poor community corresponding to NVC M6c *Carex echinata-Sphagnum recurvum/auriculatum* mire (Elkington *et al.* 2002). A similar community occurred in wetter areas or shallow water where *Carex* sp. and *Eriophorum angustifolium* were common as well as *Sphagnum* sp. and *R. flammula*. This community corresponds to NVC community M6b *Carex nigra-Nardus stricta* subcommunity (Elkington *et al.* 2002).

Methods

A grid of 10m squares covering the lough was established. Vegetation types, substrate and water depth were noted and sketched onto graph paper at a scale of 1:100. Vegetation types were recorded on the basis of dominant species and these descriptions compared with Rodwell *et al.* (1995) and Elkington *et al.* (2002). This detailed map was redrawn at a scale of 1:200 and was presented by McCormack (2003).

Samples of water beetles were taken randomly in four contrasting vegetation types in the lough: 1. *Sphagnum*-dominated vegetation; 2. vegetation rich in forbs; 3. floating vegetation in deep water; and 4. sparse vegetation over exposed substrates. Sampling was confined to water less than 70cm deep and to areas that could safely be accessed from the shore. An open-ended cylinder of 56cm internal diameter was rapidly placed in the lough to isolate a column of water. Physical characteristics and vegetation structure of the 0.25m² area enclosed were recorded. Conductivity, pH and dissolved oxygen were measured with a Hach Sension 156 meter. Percentage cover of each vegetation category in Table 1 was estimated by eye, as was percentage cover of floating, submerged and emergent vegetation. Vegetation height was measured at five points in each quadrat (the centre and 15cm from the midpoint of each edge of each side). Vegetation height in the water column was measured using a ruler placed onto the solid substrate. A reading was taken at the highest point where vegetation touched the ruler. Where no piece of vegetation touched the ruler, the nearest tallest plant was measured. Where emergent vegetation was present the height recorded was equal

Table 1

Environmental variables for quadrats in each vegetation type (Median values in each category are presented. Differences in median pH, percentage cover of submerged vegetation and forb height were tested with Kruskal-Wallis test. Where significant difference was present the Mann-Whitney U-test was used *post hoc*. Superscript numbers indicate significant differences between the medians. U values and significant levels are given below. SD = standard deviation. % = 'percent cover'.

Vegetation type →		1	2	3	4
Number of samples →		n = 9	n = 9	n = 11	n = 9
pH	median	$4.49^{1,2,3}$	$6.14^{1,4,5}$	$7.08^{2,4}$	$7.13^{3,5}$
	mean	4.81	6.39	7.12	7.08
	SD	0.95	0.74	0.28	0.18
% cover submerged	median	10	60	50	30
	mean	23	45	53	38
	SD	25	28	43	28
forb height (cm)	median	$0.5^{6,7,8}$	$3^{6,9}$	$50^{7,9,10}$	$8^{8,10}$
	mean	0.72	5.09	39	8.22
	SD	0.71	5.66	27	5.91
Dissolved O_2		102.2	131.1	125.2	111.5
conductivity (µS cm^{-1})		204	210	173	176.8
depth (cm)		1	4	57	25
sedge height (cm)		0.5	3	0	0
Sphagnum height (cm)		0.5	1	0	0
Juncus height (cm)		0.5	1	0	0
Nitella height (cm)		0	0	12	0
% sedge		1	5	0	0
% forbs		5	45	10	20
% *Sphagnum*		95	1	0	0
% *Juncus*		1	0	0	1
% *Nitella*		0	0	20	0
% bare substrate		0	20	0	65
% emergent vegetation		50	30	0	0
% floating		10	5	4	0
distance from edge (m)		0.6	0.5	1	0.7
number of beetles		10	15	2	2
species richness		3	5	2	2

Mann-Whitney U- values: 1.U = 10, p<0.01; 2.U = 0, p<0.01; 3.U = 0, p<0.01; 4.U = 20, p<0.05; 5.U = 14, p<0.05; 6.U = 5, p<0.01; 7.U = 15, p<0.01; 8.U = 15, p<0.05; 9.U =19, p<0.05; 10.U = 20, p<0.05

to the depth of the water column. All beetles in the isolated column of water, vegetation and substrate were then collected using a pond net and by removing and sorting all vegetation. Beetles were identified and counted.

Ordination of the water beetle communities recorded was undertaken using detrended correspondence analysis (DCA), an unconstrained ordination technique used to show the major gradients in a data set (McGarigal *et al.* 2000). These analyses were carried out using Community Analysis Package Version 2.01 (Henderson and Seaby 2002). The non-parametric Kruska-Wallis test was used to assess differences between the medians of environmental parameters. The Mann-Whitney U-test was used *post hoc* where a significant difference was indicated by a Kruskal-Wallis test. Correlations between variables were assessed using Spearman rank-order correlation using SPSS for Windows (version 10.0).

Results

The location of each of the quadrats sampled for aquatic beetles is given elsewhere by McCormack (2003) who also gave an accurate representation of the lake area and a map of the vegetation communities present. Twenty-seven species of beetle were found, including two not elsewhere recorded on Clare Island: *Hygrotus impressopunctatus* and *Donacia vulgaris*. A total of 344 beetles were found in 35 of the 38 quadrats sampled. The most frequently occurring species was *Hydroporus obscurus*, which was present in nineteen quadrats. *Helochares punctatus* was the most abundant species with 99 individuals in thirteen quadrats. The highest species richness per quadrat was nine in quadrat 12. However, the median number of species per quadrat was 2.5. Quadrats 23 and 28 produced only one species each, *Acilius sulcatus* and *Helophorus flavipes* respectively, neither of which occurred in any other quadrat.

Water depth ranged from 0cm to 77cm and the median depth from all sites was 19cm. The median pH was 6.85 with an overall range of 4.04 to 7.65. Conductivity ranged from 111µS cm^{-1} to 295 µS cm^{-1}, with a median value of 179 µS cm^{-1}. Vegetation cover varied from total cover to almost none. Median values for each the environmental variables measured and number of beetles of each species in each quadrat are given in Table 1.

The detrended correspondence analysis was carried out without down-weighting rare species.

Two outliers (quadrats 23 and 28) were removed and the analysis was re-run with the remaining 33 quadrats. The relationship between ordination of the sites on DCA axes 1 and 2 (Fig. 1) and the environmental variables were analysed using Spearman rank-correlation. Correlations between each of the environmental variables were analysed similarly.

Axis 1 of the DCA plot was strongly correlated (p<0.01) with depth, pH, forb height, *Sphagnum* height, *Juncus* height. *Nitella* height, % cover sedges, % cover *Sphagnum*, % cover emergent vegetation and % bare substrate.

Axis 2 was strongly correlated (p<0.01) with pH, depth, forb height, *Sphagnum* height, *Juncus* height, % cover *Sphagnum*, % cover bare substrate and % cover emergent vegetation. pH was strongly (p< 0.01), correlated with conductivity, water depth, forb height, *sphagnum* height, *Juncus* height, % cover *Sphagnum*, % bare substrate and % cover of emergent vegetation. Water depth was correlated with pH (p<0.01), conductivity, forb height, *Sphagnum* height, *Juncus* height, *Nitella* height and with percentage cover of sedges, *Nitella* and emergent vegetation. Axis 3 was not significantly correlated with any of the environmental variables measured. Eigenvalues for the first three axes were 0.726, 0.547 and 0.321 respectively.

Quadrats with vegetation dominated by *Sphagnum* (vegetation type 1) are clustered on the left of the DCA plot, as is a cluster of *Sphagnum* moss dwelling beetles in the species plot (Fig. 1).

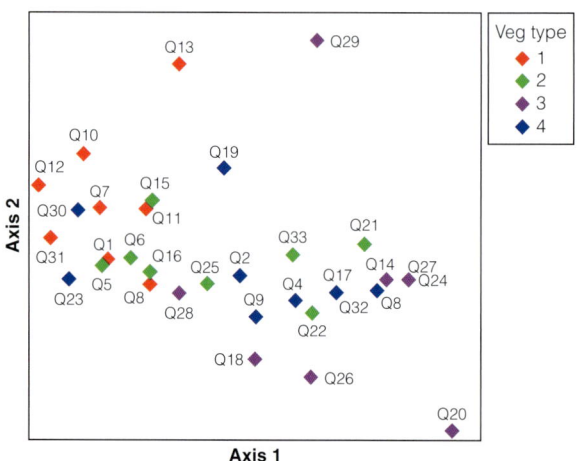

Fig. 1 DCA plot of quadrats from Poirtín Fhuinch Lough (Vegetation types of sites are coloured and ordination of species is plotted vegetation type: 1 *Sphagnum*, 2 Shallow margins rich forbs, 3 Floating vegetation in deep water, 4 sparsely vegetated bottom)

Species in this cluster are predominantly hydrophilid beetles such as: *Hydrobius fuscipes*, *Helochares punctatus*, *Coelostoma orbicualre* and *Enochrus fuscipennis*. However, three dytiscid beetles *Hydroporus gyllenhalii*, *Agabus affinis* and *Ilybius montanus* are also present. Also found in *Sphagnum* rich habitats was the chrysomelid *Plateumaris sericea*. *Hydroporus obscurus* was frequently found in *Sphagnum* rich habitats but it also occurred in less densely vegetated habitats. Forb rich margins and sparsely vegetated bottoms (vegetation types 2 and 4 respectively) are vaguely clustered about the centre of the plot and both are spread horizontally. Floating vegetation in deep water (vegetation type 1) is dispersed in the ordination plot with no discernible pattern.

The Kruskal-Wallis test was used to compare medians of three of the environmental variables across the vegetation types identified. Forb height, pH and percentage cover of submerged vegetation were taken as summarising the environmental data as there were significant correlations between these variables and Axis 1 of DCA. Forb height and pH were significantly correlated with Axis 2. Results are given in Table 1, as well as median values all the environmental variables measured. Median pH of *Sphagnum*-rich quadrats was significantly lower (p<0.01) than was found in other vegetation types. Median pH of vegetation type 2, forb-rich quadrats, was also significantly lower (p<0.05) than in vegetation types 3 and 4 respectively. Significant differences were also found in forb height across the vegetation types (Table 1).

Selected environmental variables are plotted in Figure 2 in order of increasing water depth. pH was generally lower where percentage cover of *Sphagnum* was high, which was in shallow water less than 10cm deep. pH values between 4.05 and 6.46 were measured in these quadrats. As water depth in quadrats increased, the proportion of *Sphagnum* present decreased and there was a corresponding rise in percentage cover of forbs. As the depth of water increased further past about 30cm, the percentage cover of forbs declined and the pH of the water, which was not directly influenced by *Sphagnum* is around neutral. Although not plotted in Figure 2, percentage cover of *Nitella* sp. in quadrats 33, 30, 31, 26 and 20 was as high as 100%, which accounts for the low percentage cover of forbs in these quadrats.

Discussion

A gradient from *Sphagnum* and forb-rich habitats in shallow water to deeper water with either little or some vegetation is evident in the DCA plot shown in Figure 1. *Sphagnum* was the dominant vegetation in the shallowest water from 0 to 5cm. This had an influence on the water chemistry, which was reflected in pH and conductivity. Water depth was strongly correlated (p<0.01) with overall vegetation height and percentage cover of

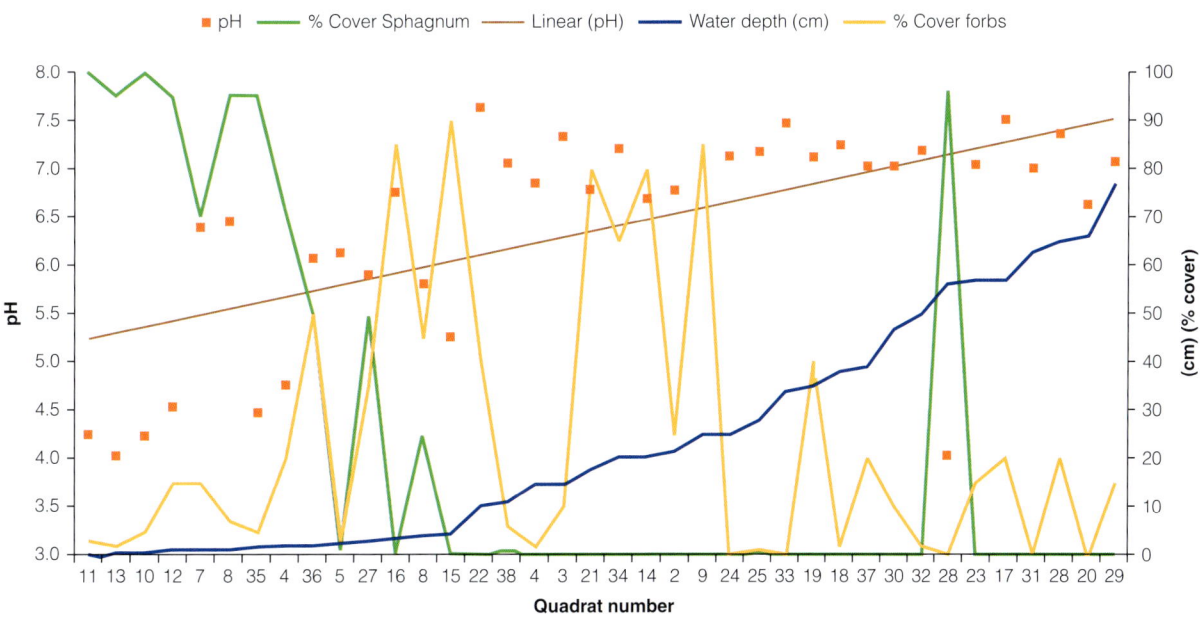

Fig. 2   Water depth, pH, % cover *Sphagnum*, % cover forbs in quadrats taken at Poirtín Fhuinch Lough, Clare Island

sedges, *Sphagnum, Nitella* sp. and bare substrate. This indicates that the nature of the habitat, particularly vegetation structure, changes markedly from marginal to sub-marginal parts of the lough. This difference is reflected in the ordination plot of vegetation types and species (Fig. 1). Strong correlations between Axis 1 of the plot and most of the measures of vegetation structure and composition, illustrates the overall change in the habitat across a gradient of water depth.

On the DCA plot, quadrats in *Sphagnum*-rich habitats are clustered on the left. *Sphagnum* moss dwelling groups had several beetle species that are poor swimmers and rarely occur away from *Sphagnum*. However, the dytiscid beetles present in the *Sphagnum*-dwelling group are strong swimmers and are less restricted in their habitats. Species richness and abundance were greatest in shallow margins rich in forbs (vegetation type 2) as species associated with open water habitats such as *Hygrotus inaequalis* occurred alongside the moss dwellers. *Sphagnum* often formed a carpet in which forbs such as *Potamogeton polygonifolius, Hypericum elodes* and *Ranunculus flammula* grew. The presence of *Sphagnum* in some of these quadrats is reflected in the pH values recorded in these quadrats.

On the DCA plot (Fig. 1) quadrats in vegetation types 1, 2 and 4 are generally more closely clustered than are those in vegetation type 3 (deep water, floating vegetation). This is in part due to the low abundance of beetles in some of these quadrats and the occurrence of species that did not occur in quadrats in other vegetation types. Species falling into this category are *Gyrinus minutus, Haliplus fulvus* and *Haliplus lineatocollis*. The former was present in abundance in Poirtín Fhuinch Lough as was *Gyrinus substriatus* but as these are extremely fast and evasive, only one individual occurred in quadrats. *Haliplus fulvus* and *Haliplus lineatocollis* were present in quadrats with *Nitella* sp. on which they may have been feeding (Holmen 1987). Quadrats in more open water had beetles representing a range of families including Dytiscidae, Gyrinidae, Haliplidae and Noteridae.

Species associated with open habitats—*Hygrotus inaequalis, Noterus clavicornis* and *Gyrinus minutus*—occurred on the right of the DCA plot. These species are stronger swimmers than the hydrophilid beetles associated with shallow water and *Sphagnum. Acilius sulcatus*, a species not included in the DCA plot was noted on several occasions swimming in open water. It only occurred in a single quadrat and was the only beetle present. However, the distinctive larvae of this species were seen swimming in schools, particularly in shallow water in the lough.

Quadrats in parts of Poirtín Fhuinch that were sparsely vegetated (vegetation type 4) were generally in shallower water than was present in the vegetation type 3 (Table 1). The stony substrate and exposure to wave action had a strong effect on the vegetation and beetles present. The percentage cover of bare substrate was highest in this vegetation type and consequently the number of beetles found was low. Plant species recorded from these quadrats such as *Juncus bulbosus, Littorella uniflora, Eleocharis multicaulis* and *Myriophyllum* sp. provide poor structural diversity. Beetle abundance and species richness were low in quadrats in this vegetation type. Dytiscid beetles predominated and *Noterus clavicornis* occurring in two quadrats. *Hydroporus obscurus* was the most frequent species, occurring in four quadrats. This species is reported by Nilsson and Holmen (1995) as occurring in *Sphagnum* pools and in bare shores of oligotrophic lakes.

Doyle and Foss (1986) recorded a change in the flora of Clare Island since the time of the first Clare Island Survey, mainly as a result of changing agricultural practices and land uses, particularly heavy grazing by sheep. A detailed account of the flora of the island, including references to Poirtín Fuinch was presented by Ryle (2013). However, it does not seem likely that the overall character of aquatic habitats such as peat pools, streams and ditches would have changed considerably in the intervening period. One freshwater habitat that has changed is Lough Avullin, which was partially drained and has infilled with reed beds. Likewise, Lough Merrignagh has infilled over time with floating mats of *Sphagnum* and has been reduced to a *Sphagnum* pool. This loss of open water habitat, however, has been partly offset by the creation of Poirtín Fhuinch Lough and various pools by peat cutting. The decline in peat cutting, which has all but ceased on Clare Island in recent years, may result in a decline in the number of small pools and a decline abundance of species that inhabit them.

In summary, the four vegetation types studied in Poirtín Fhuinch Lough provide very different habitats for aquatic beetles. Each of the vegetation types had different structural and chemical characteristics and this was reflected in the beetle assemblages found. Shallow water, rich in *Sphagnum*, had a distinctive fauna dominated by hydrophilid beetles. However, the highest density and diversity of beetles was found in water up to a few centimetres deep. In this area there were both *Sphagnum*-dwelling species and species of open water. In areas with over-exposed substrates beetle abundance was low but the composition of the species assemblages here were distinct from those in the shallow margins.

Eyre *et al.* (1986) suggested that habitats can be defined by beetle species assemblages similar to the way in which phytosociologists define habitats by the vegetation therein, as opposed to defining habitats on the basis of physical, chemical and other abiotic characteristics. The results of the Poirtín Fuinch survey indicate a link between the vegetation structure and aquatic beetles species, as was hypothesised prior to the fieldwork. These findings also suggest that results from comparable studies conducted elsewhere could prove useful. Thus, analyses of intra-habitat variation in assemblages of aquatic Coleoptera, and other insect taxonomic groups, can advance knowledge of various aspects of aquatic community composition and ecosystem processes.

REFERENCES

Boyd, C.E. 1971 *The limnological role of aquatic macrophytes and their relationship to reservoir management*. American Fisheries Society.

Brönmark, C. and Hansson, L.A. 2005 *The biology of lakes and ponds*. Oxford. Oxford University Press.

Carpenter, S.R. and Lodge, D.M. 1986 Effects of submersed macrophytes on ecosystem processes. *Aquatic botany* **26**, 341–70.

Doyle, G.J. and Foss, P.J. 1986 A resurvey of the Clare Island flora. *The Irish Naturalists' Journal* **22**(3), 85–9.

Dytham, C. 2011 *Choosing and using statistics: a biologist's guide*. New Jersey. John Wiley and Sons.

Elkington, T., Dayton, N., Jackson, D.L. and Strachan, I.M. 2002 *National Vegetation Classification: Field Guide to Mires and Heaths*. Peterborough Joint Nature Conservation Committee.

Eyre, M.D., Ball, S.G. and Foster, G.N. 1986 An initial classification of the habitats of aquatic Coleoptera in north-east England. *Journal of Applied Ecology* 23(3), 841–52.

Gasith, A. and Hoyer, M.V. 1998 Structuring role of macrophytes in lakes: changing influence along lake size and depth gradients. In E. Jeppesen, M. Søndergaard, M. Søndergaard and K. Christoffersen (eds) *The structuring role of submerged macrophytes in lakes*, 381–92. New York. Springer.

Graham, J. (ed.) 2001 *New Survey of Clare Island. Volume 2: Geology*, 75–86. Dublin. Royal Irish Academy.

Henderson, P.A. and Seaby, R.M.H. 2002 Community analysis package.

McCormack, S. 2003 Studies on the aquatic coleoptera of Clare Island, Co. Mayo. Unpublished M.Sc. Thesis. National University of Ireland, Galway, Department of Zoology.

McGarigal, K., Cushman, S., Stafford, S.G. and Stafford, S.G. 2000 *Multivariate statistics for wildlife and ecology research*. New York. Springer.

Nilsson, A.N. and Holmen, M. 1995 *The Aquatic Adephaga (Coleoptera) of the Fennoscandia and Denmark. Ii. Dytiscidae: II-Dytiscidea*. Leiden. Brill.

Rennie, M.D. and Jackson, L.J. 2005 The influence of habitat complexity on littoral invertebrate distributions: patterns differ in shallow prairie lakes with and without fish. *Canadian Journal of Fisheries and Aquatic Sciences* **62**(9), 2088–99.

Rodwell, J.S., Pignatti, S., Mucina, L. and Schaminée, J.H.J. 1995 European Vegetation Survey: update on progress. *Journal of Vegetation Science* **6**(5), 759–62.

Ryle, T. 2013 Vegetation–environment interactions on Clare Island, in D. Synott (ed.) *New survey of Clare Island. Volume 7: Plants and Fungi*. Dublin. Royal Irish Academy.

Schultz, R. and Dibble, E. 2012 Effects of invasive macrophytes on freshwater fish and macroinvertebrate communities: the role of invasive plant traits. *Hydrobiologia* **684**(1), 1–14.

Thomaz, S.M. and Cunha, E.R.D. 2010 The role of macrophytes in habitat structuring in aquatic ecosystems: methods of measurement, causes and consequences on animal assemblages' composition and biodiversity. *Acta Limnologica Brasiliensia* **22**(2), 218–36.

Zhu, B., Fitzgerald, D.G., Mayer, C.M., Rudstam, L.G. and Mills, E.L. 2006 Alteration of ecosystem function by zebra mussels in Oneida Lake: impacts on submerged macrophytes. *Ecosystems* **9**(6), 1017–28.

CHAPTER 9

THE CARABIDAE (COLEOPTERA) OF CLARE ISLAND

Roy Anderson and Stephen McCormack

ABSTRACT

The Coleoptera fauna of Clare Island is an impoverished subset of that found across Ireland, reflecting the isolation of the island ecosystem, its small size and restricted range of habitats. A reassessment of the fauna in 2002/3 indicated that changes in the past century have been random for the most part. Out of a total fauna of 66 species, 16 species were found in 2002/3 but not in 1912 and the same number in 1912 but not in 2002/3. These changes, however, do reflect a continuing decline in the woodland fauna and increase in species associated with acid, upland environments subject to moderate levels of grazing. Some of the apparent increase in the latter component may reflect the use of pitfall-trapping as a method in 2002/3, a technique not used in the earlier study. The main species of wider conservation concern on the island is *Carabus clatratus*. Fortunately, populations appear to be stable though localised to lower, poorly drained, areas with some open water and/or ongoing turf extraction.

Introduction

The original Clare Island Survey undertook to review the Coleoptera not just of Clare Island but of the surrounding area, including most of Clew Bay north to Achill (Johnson and Halbert 1912). The present report is restricted in scope as only Carabidae are presented here and samples were only collected on Clare Island. The plan was to revisit the island after a gap of about 90 years to see what changes, if any, had taken place in the fauna.

As in the original survey, surveying was mostly by hand, with beetles collected by turning stones, beating low vegetation, sieving moss and so on. But pitfall trapping was also used since the technique more efficiently samples ground-dwelling beetles in extensive terrain such as the grassland and dwarf shrub heath found on the island. Many Carabidae, even the larger beetles, are elusive in large areas of uniform vegetation and may be overlooked by hand searching. Over-reliance on pitfall-trapping was consciously avoided, however, as the technique is inefficient in areas where there is a sharp transition between habitats (i.e. ecotones), particularly on water margins. Methods were therefore adapted to the terrain being visited in an attempt to sample as wide an area as possible in the short period of time available.

Methodology

On 29–30 June 2002, a transect was followed by the first author across lower altitudes of the western, south-eastern, and eastern quadrants of the island. On 14–16 October 2002, a transect was followed across higher ground in the centre and north-west, e.g. across the top of Knockmore to the lighthouse at Lachnacranny. The second author similarly followed transects during 2002 and 2003. Sites where the collections were made are summarised in Figure 1.

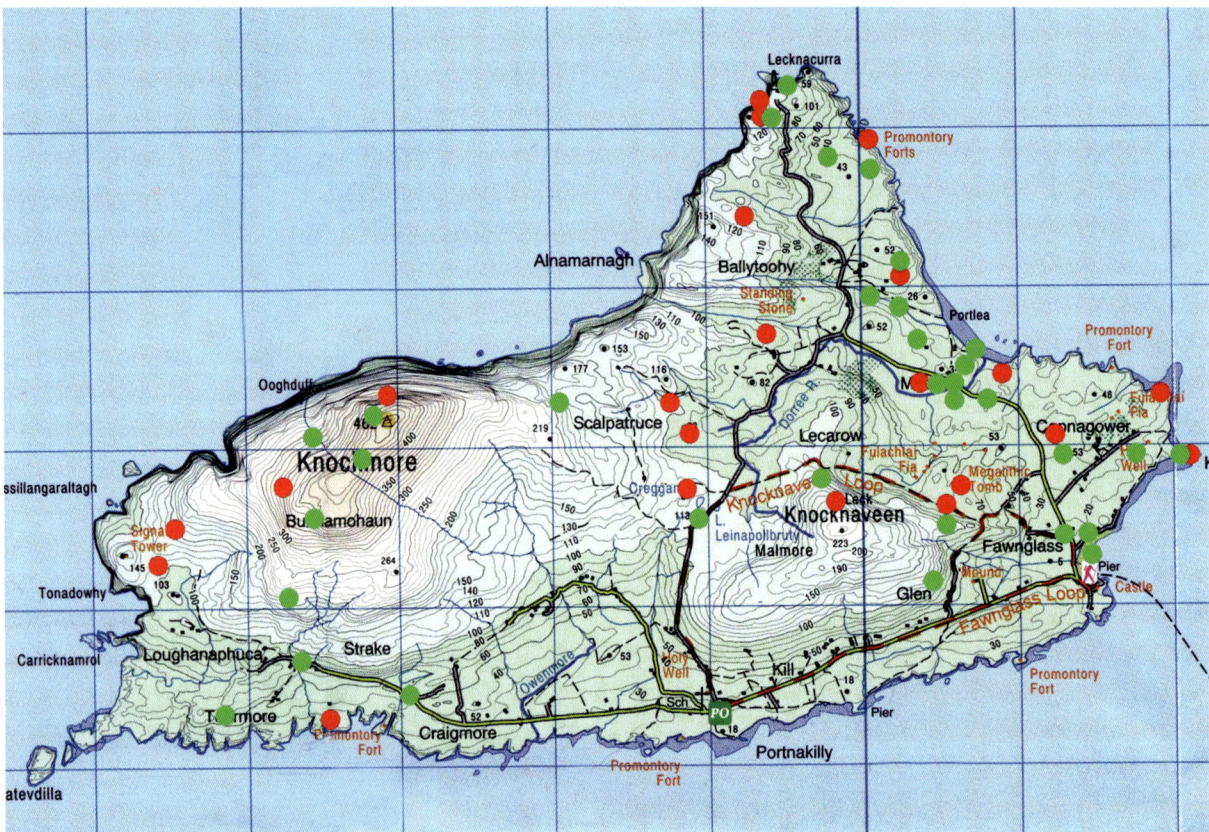

Fig. 1 Sampling sites, Clare Island 2002/3. Pitfall locations (•) and hand searching locations (•)

Pitfall traps were at the same time laid by the second author in selected vegetation types. A record of these is again indicated in Figure 1.

Transects were walked in clear weather and the following niches examined: undersides of stones and wood in pasture; moss under *Calluna*, and moss on tops of stone dykes (removed and beaten); grass or sedge tussocks (cut to beat into a sweep net); margins of freshwaters (splashed or examined by hand for riparian species, or vegetation and waterfall moss beaten into still water); clayey marine cliffs (splashed to flush out beetles).

Literature
Beetles were identified using Luff (2007).

Results
Species recorded and their site and status are summarised in Table 1 and compared with the results of the original survey given in Johnson and Halbert (1912). The nomenclature is that of Duff (2012).

Discussion
Sixteen species recorded in the 1912 survey were not seen in 2002/3. Some of these occur in niches that were not surveyed or could not be found in the recent survey. Examples are *Aepus robinii*, which lives in rock crevices and under deeply embedded stones on the marine foreshore (Thiele 1977). The site(s) where it was found in 1912 are unknown and suitable habitat was not detected in 2002.

The *Bembidion* species *aeneum*, *bipunctatum*, *bruxellense*, *doris* and *saxatile* are all missing from the 2002/3 list. *Bembidion aeneum*, *bipunctatum*, *bruxellense* and *saxatile* are essentially riparian species (Thiele 1977) with different degrees of reliance on specific habitats. *Bembidion saxatile* is restricted to coastal habitats in Ireland, particularly eroding soft cliffs (Anderson *et al.* 2000) (this is a habitat of very limited occurrence on the island, although *Bembidion stephensii* was found by the second author on soft low cliffs by the Harbour). The others are less stenotopic but could also occur in this habitat. One of the most interesting species is

Table 1
Carabidae on Clare Island: species recorded in 2002/3 compared to Johnson and Halbert (1912)

Species	2002, transects	2002, pitfalls	2002, all	1912
Abax parallelepipedus (Piller & Mitt., 1783)	+	+	+	+
Aepus robinii (Laboulbène, 1849)				+
Agonum fuliginosum (Panzer, 1809)		+	+	
Agonum gracile Sturm, 1814	+		+	
Agonum marginatum (Linnaeus, 1758)				+
Agonum muelleri (Herbst, 1784)	+	+	+	+
Agonum nigrum Dejean, 1828				+
Agonum thoreyi Dejean, 1828	+		+	
Amara communis (Panzer, 1797)		+	+	+
Amara lunicollis Schiødte, 1837	+	+	+	
Amara plebeia (Gyllenhal, 1810)				+
Anchomenus dorsalis (Pontoppidan, 1763)				+
Badister bullatus (Schrank, 1798)	+		+	
Bembidion aeneum Germar, 1824				+
Bembidion bipunctatum (Linnaeus, 1761)				+
Bembidion bruxellense Wesmael, 1835				+
Bembidion doris (Panzer, 1796)				+
Bembidion lampros (Herbst, 1784)	+	+	+	+
Bembidion mannerheimi Sahlberg, 1827	+	+	+	+
Bembidion pallidipenne (Illiger, 1802)	+		+	+
Bembidion saxatile Gyllenhal, 1827				+
Bembidion stephensii Crotch, 1866	+		+	
Bembidion tetracolum Say, 1825	+	+	+	+
Bradycellus caucasicus (Chaudoir, 1846)	+		+	
Bradycellus ruficollis (Stephens, 1828)	+		+	
Bradycellus verbasci (Duftschmid, 1812)	+	+	+	
Broscus cephalotes (Linnaeus, 1758)	+		+	+
Calathus fuscipes (Goeze, 1777)	+	+	+	+
Calathus melanocephalus (Linnaeus, 1758)	+		+	+
Calathus mollis (Marsham, 1802)	+		+	+
Carabus clatratus Linnaeus, 1761	+		+	+
Carabus granulatus Linnaeus, 1758	+	+	+	+
Carabus problematicus Herbst, 1786	+	+	+	+
Cicindela campestris Linnaeus, 1758	+		+	+
Clivina fossor (Linnaeus, 1758)		+	+	+
Curtonotus aulicus (Panzer, 1796)	+		+	+
Cychrus caraboides (Linnaeus, 1758)	+		+	+
Dyschirius globosus (Herbst, 1784)	+		+	
Elaphrus cupreus Duftschmid, 1812				+
Harpalus affinis (Schrank, 1781)				+
Harpalus latus (Linnaeus, 1758)	+		+	+
Harpalus rufipes (De Geer, 1774)	+	+	+	+
Leistus fulvibarbis Dejean, 1826				+
Loricera pilicornis (Fabricius, 1775)		+	+	+

continued

Table 1. Cont.

Species	2002, transects	2002, pitfalls	2002, all	1912
Nebria brevicollis (Fabricius, 1792)	+	+	+	+
Nebria salina Fairmaire & Laboulbène 1854	+	+	+	
Notiophilus aquaticus (Linnaeus, 1758)		+	+	+
Notiophilus biguttatus (Fabricius, 1779)		+	+	+
Notiophilus palustris (Duftschmid, 1812)	+	+	+	+
Notiophilus substriatus Waterhouse, 1833		+	+	
Olisthopus rotundatus (Paykull, 1790)	+	+	+	+
Paradromius linearis (Olivier, 1795)	+		+	
Paranchus albipes (Fabricius, 1796)	+	+	+	+
Patrobus assimilis Chaudoir, 1844				+
Poecilus cupreus (Linnaeus, 1758)				+
Poecilus versicolor (Sturm, 1824)		+	+	+
Pterostichus diligens (Sturm, 1824)	+	+	+	
Pterostichus melanarius (Illiger, 1798)	+	+	+	+
Pterostichus niger (Schaller, 1783)	+	+	+	+
Pterostichus nigrita* (Paykull, 1790)	+		+	+
Pterostichus rhaeticus Heer, 1837	+	+	+	
Pterostichus strenuus (Panzer, 1796)	+	+	+	+
Pterostichus vernalis (Panzer, 1795)	+	+	+	
Synuchus vivalis (Illiger, 1798)	+		+	+
Trechus obtusus Erichson, 1837	+	+	+	+
Trechus quadristriatus (Schrank, 1781)				+
Trichocellus placidus (Gyllenhal, 1827)	+		+	
Totals	43	31	51	50

**Pterostichus nigrita*, a species listed in the 1912 survey (Johnson and Halbert, 1912), has more recently been split into *P. nigrita* s.s. and *P. rhaeticus*. Both segregates were recorded in 2002/3 but which of the two were seen pre-1912 is unknown.

B. bipunctatum, which is declining in Britain but more stable in Ireland. It is associated with sandy/silty streambanks or shorelines and has a north-western and upland bias in range. Its absence from the current list suggests that is now rare or possibly extinct on the island.

Agonum nigrum is another species associated with seacoasts that was not recorded in 2002/3 (*Bembidion saxatile* has already been mentioned). It is a local and rather rare species in Ireland most often found in major river estuaries (Anderson *et al*. 2000) so it would not be expected on the island. Johnson and Halbert (1912) nevertheless report its presence although it could not be refound in the present survey.

Woodland species, not unexpectedly, are poorly represented in the island fauna. *Cychrus caraboides* is a mollusc predator feeding on slugs and shelled snails, usually in lowland broadleaf woods but also in dwarf shrub heath to moderate altitudes in Ireland (Anderson *et al*. 2000). While there are small areas of suitable heather habitat and low scrub at the eastern end of the island only a single specimen was seen, near Lecarrow in 2003. It is clearly now very rare, reflecting a decline that may have been going on since woodland disappeared and heather moor was reduced on the island. *Carabus problematicus* and *Leistus fulvibarbis* were also recorded from the island by Johnson and Halbert (1912). In the 2002/3 survey *Leistus fulvibarbis* was not seen at all and has clearly declined, perhaps to extinction. The appearance of patches of scrub in the eastern part of the island within the last century do not appear to have been helpful in this respect. *Carabus problematicus*, by contrast, was seen at a number of sites in 2002, mainly in bare, rocky areas. It appears not to be so dependent upon the shade and shelter of woods or indeed dwarf shrub heath and can colonise rocky places where heather has been reduced by overgrazing. Overall, however, the results suggest that historic declines in the woodland component of the island fauna are continuing.

Patrobus assimilis is generally found in rocky places or in dwarf shrub heath at altitude in Ireland. There is, however, some crossover with *Cychrus caraboides* and the two occasionally occur together. The fact that neither could be found either on the rocky summit and slopes of Knockmore is surprising. Another species that lives on well-drained upland soils is *Calathus melanocephalus*, the dark *nubigena* form being typical of Irish uplands. This was recorded in 1912 but only a single specimen could be found in 2002, by hand searching heather moor west of the Lighthouse, the same general area where a single *Cychrus* was found in 2003. An obvious mechanism for these changes in status is currently lacking.

There are sixteen species recorded in 2002 but not seen in 1912. Seven of these are typical of acid, exposed, upland environments and would be expected to be widespread on Clare Island: *Amara lunicollis, Bradycellus collaris, B. ruficollis, B. verbasci, Nebria salina, Pterostichus diligens*. What is remarkable is that none of these were seen during the 1912 survey. However, it seems likely that at least some of these were picked up on extensive grassland/heath by pitfall trapping, a technique not used in the earlier study. *Bradycellus ruficollis* and *B. collaris* were found on heather moor near The Mill and to the west of the Lighthouse. Both are rather strongly associated with *Calluna* (heather) in Ireland and *B. ruficollis* in particular would be absent only where heather had been removed by over-grazing or by turf removal. The apparent absence of these species on other parts of the island suggests that heather moorland has been scarce historically. In any event these *Bradycellus* species were not reported in 1912. It is just possible that that they colonised the island naturally post-1912, but more likely were overlooked.

Of the remaining additions, four are found in wetlands of one kind or another but not usually within peaty biotopes: *Agonum gracile, A. fuliginosum, A. thoreyi, Trichocellus placidus*. Areas of

Acidic scrub near The Mill. Photo: R. Anderson

Looking east across high ground at Knockmore.
Photo: R. Anderson

suitable wetland for these may have been scarce historically. Though they were not recorded in 1912 it is evident that a fair amount of suitable habitat exists today, for example in the marshy surround of the stream at Ballytoohy or at The Mill fen. Such areas may have developed after being wired off from grazing and a more diverse wetland biotope may have appeared subsequently. As with the previous group, whether the inhabitants of these areas have colonised the island recently or have recovered from a historical low, is impossible to say on present evidence.

Unlike the Mollusca, where a number of invasive aliens have appeared since 1912, there is little or no evidence of this kind of change in the Coleoptera between 1912 and 2002. This reflects wider experience in mainland Ireland where very few immigrant carabids are known. The integrity of the Clare Island Coleoptera has been more or less maintained with changes to the balance of species following longer-term trends in habitat disturbance and faunal decline. Although some recovery in the physical diversity of the grazing range and in the structural diversity of wetlands is apparent, this has led to only limited improvements in carabid diversity.

When comparing the island with the mainland, few distinctive differences appear. The most immediately obvious is the greater abundance, at least locally, of the larger predatory ground beetles, particularly *Carabus* spp., on the island. *Carabus clatratus* is a declining species in Europe because of land drainage and agricultural intensification. In Ireland it has fared rather better than in other areas of Europe and does well in peat areas where agricultural management is less intensive and more benign (Williams and Gormally 2010). There are other factors in its survival and persistence. It has been noted (R. Anderson, pers. obs.) that the species is very rare on The Mullet mainland in West Mayo but very common on offshore islands such as Inishkea North (D. Suddaby, pers. comm.). The only tangible difference between these areas is the absence of the fox *Vulpes vulpes* on the islands. Lower predation levels, particularly by foxes, may be an important factor in the current favourable environment on Clare Island for the larger ground beetles. Although *Carabus clatratus* was seen mainly in the vicinity of heather moor in the east of the island, and present in some numbers, it was also observed sparingly at the west end in an area of shallow peat cuttings and small pools (E. Regan, pers. comm.). This apparent localisation may simply reflect the distribution of suitably wet peatlands on the island, aided and abetted by turf cutting (pool creation) in some parts. *Carabus problematicus*, as already noted, is much more widely spread as it can adapt to bare rocky places in addition to areas with better cover. *Carabus granulatus* was the most restricted of the large carabids seen, probably again because of its habitat preferences—vegetated non-peat wetlands are favoured and these seem to be concentrated in the eastern parts of Clare Island.

Other species recorded are of lesser importance from a conservation viewpoint and occur in similar habitats on the mainland. Ireland has a rather poorly developed woodland fauna but those species found there, for example *Cychrus caraboides*, tend to be widespread and often common, especially in dwarf shrub heath ecosystems in the uplands. The rarity of *Cychrus* and inability to re-find *Leistus fulvibarbis* points to a continuing

Carabus clatratus. Photo: R. Anderson

Carabus problematicus. Photo: R. Anderson

decline in the relic woodland fauna of the island, even though land use post-1840 has become less intensive and scrub woodlands have sprung up in several areas.

Overall, the fauna is an impoverished subset of that found across Ireland, reflecting the isolation of the island ecosystem, its small size and restricted range of habitats. Colonisation by mainland species seems to be slow compared to, for example, terrestrial molluscs, and a number of mainland species remain unrecorded. Achill Island to the north, for example, has *Cymindis vaporariorum*, not seen on Clare Island. This is an important but stenotopic and rare species of well-drained, exposed western heath and would be expected on Clare Island but appears absent. Its reliance on areas of undisturbed heather moor may have led to its extinction during that part of the historical period when the human population was at or close to the island's maximum (pre-1840) level.

REFERENCES

Anderson, R., McFerran, D.M. and Cameron, A. 2000 *Atlas of the ground-beetles (Coleoptera: Carabidae) of Northern Ireland*, 246 pp. Belfast. Ulster Museum.

Duff, A.G. 2012 *Checklist of beetles of the British Isles, 2nd Edition,* 171 pp. United Kingdom. Pemberley Books (Publishing).

Johnson, W.F. and Halbert, J.N. 1912 Clare Island Survey. 28. Terrestrial Coleoptera. *Proceedings of the Royal Irish Academy* **31,** 1–24.

Luff, M.L. 2007 The Carabidae (ground beetles) of Britain and Ireland. *Handbooks for the Identification of British Insects* **4**(2) (2nd Edition) 247 pp. St. Albans. Royal Entomological Society.

Thiele, H.-U. 1977 *Carabid beetles in their environments*, 369 pp. Berlin. Springer Verlag.

Williams, C.D. and Gormally, M.J. 2010 The effects of blanket bog management on ground beetles (Carabidae) with particular reference to the threatened *Carabus clatratus* L. *Irish Wildlife Manuals, No. 47.* Dublin. National Parks and Wildlife Service, Department of the Environment, Heritage and Local Government.

CHAPTER 10

THE LEPIDOPTERA (BUTTERFLIES AND MOTHS) OF CLARE ISLAND

K.G.M. Bond

ABSTRACT

Fieldwork extending from late spring to early autumn 2000 to 2005, combined with extensive light-trapping, has produced a comprehensive list of Lepidoptera records for Clare Island, making it one of the best-worked sites in western Ireland. The checklist of the Lepidoptera of Clare Island indicates that 264 species have now been recorded on the island, compared with 92 in the original survey, and that these include a series of noteworthy records. One species, recorded in the recent survey, had not previously been recorded in Ireland and others are relatively rare.

Introduction

This report summarises the species of Lepidoptera recorded on Clare Island in the years 2000, 2002, 2003 and 2005 as part of the New Survey of Clare Island. A comparison is made between the species recorded during the original Clare Island Survey (Kane 1912) and the present study, and possible reasons for the loss of certain species are discussed. Species recorded in 2002 include the first Irish record of the tortrix moth, *Epinotia sordidana*.

Epinotia sordidana. Beutzen, Saxony, Germany. Photo: Friedmar Graf

Methods

During fieldwork carried out in 2000 to 2005, light-trapping was carried out in various parts of the island, using three types of light-trap: (1) Robinson mercury-vapour, used from mains electricity or a generator; (2) a Skinner type mercury-vapour trap, also used from mains or a generator; and (3) a much lighter Heath (actinic) trap, using a battery (the latter type of trap could easily be placed in the more remote and inaccessible parts of the island, but generally had a smaller catch due to the lower intensity of the light emitted). Specimens were also netted or recorded in flight at a wide range of localities, while larvae were recorded on their respective foodplants.

Identification of many of the more 'difficult' species, and of others that had been stored in ethanol was carried out by dissection. Permanent genitalia slides have been prepared, and some of these, along with the dissected specimens, and a range of other Clare Island specimens have been added to the collections of the National Museum of Ireland (Natural History).

Dates of fieldwork on Clare Island by the author

- 2000: 27–28 April; 17–18 June; 15–16 July; 10–11 August; 23–24 August
- 2002: 6–11 July; 23–24 July; 21–23 August; 27–29 September
- 2003: 8–9 July
- 2005: 13–14 July, 3 September

In addition, fieldwork was carried out on other dates in 2002 and 2003 by Stephen McCormack, and a smaller number of observations made by other recorders have been included to make this report as comprehensive as possible. In particular, the early season recording in 2003 provided records of several species which would otherwise have been overlooked. Where relevant, mention is also made to certain additional records as far back as 1990.

The original Clare Island Survey was spread over three years—the precise number of days is not known. Although Kane was the author of that report, some of the fieldwork was carried out by Greer or Johnson or others, especially on the Microlepidoptera. The months during which that survey was carried out were June 1909 (Kane), July 1909 (Kane), June 1910 (Greer and Kane); August 1910 (Johnson), May 1911 (Johnson) and July 1911 (Kane).

It is important to note that some species recorded during the original Clare Island Survey (Kane 1912), were in fact recorded from other locations in the vicinity of Clew Bay, County Mayo, and not from Clare Island itself. The species totals referred to in this report apply to only records specifically indicated by Kane as being recorded from the island itself. Clare Island Survey specimens taken during the original survey held in the National Museum of Ireland (NMINH) have been examined by the author.

The nomenclature and systematic arrangement used here follow that of *Fauna Europaea* (de Jong 2013). This employs a European numbering system, which is replacing the Bradley and Fletcher system used until now in the British literature. This newer system along with the corresponding Bradley and Fletcher (1979) numbering is used in Bond and O'Connor (2012). The botanical nomenclature follows Stace (1991).

Results

During the New Survey of Clare Island 264 species of Lepidoptera were recorded (Table 1; Appendix 1), in contrast with 92 during the original survey of 1912. It should be noted that the record of *Bryotropha politella* in the original survey is incorrect; dissection showed the specimen to be *B. terrella*, a species already recorded during that survey. Furthermore, a Clare Island

Table 1
Number of Lepidoptera species, by families, recorded during original survey (Kane 1912) and New Survey of Clare Island

Family	No. of spp. -original Survey	New Survey (2000-2005)	Change since original Survey
MICROPTERIGIDAE	1	2	+1
HEPIALIDAE	1	1	0
NEPTICULIDAE	0	4	+4
HELIOZELIDAE	0	1	+1
TINEIDAE	0	2	+2
GRACILLARIIDAE	0	6	+6
ARGYRESTHIIDAE	0	4	+4
YPSOLOPHIDAE	0	2	+2
PLUTELLIDAE	1	2	+1
GLYPHIPTERIGIDAE	1	1	0
LYONETIIDAE	0	1	+1
ELACHISTIDAE	1	11	+10
OECOPHORIDAE	1	3	+2
COLEOPHORIDAE	0	10	+10
BLASTOBASIDAE	0	1	+1
GELECHIIDAE	2	6	+4
ZYGAENIDAE	1	1	0
SESIIDAE	0	1	+1
TORTRICIDAE	14	36	+22
CHOREUTIDAE	0	1	+1
EPERMENIIDAE	0	1	+1
PTEROPHORIDAE	2	1	-1
PYRALIDAE	1	3	+2
CRAMBIDAE	5	19	+14
LASIOCAMPIDAE	3	2	-1
SATURNIIDAE	1	1	0
SPHINGIDAE	2	2	0
PIERIDAE	2	4	+2
LYCAENIDAE	3	3	0
NYMPHALIDAE	6	9	+3
GEOMETRIDAE	17	41	+24
NOTODONTIDAE	2	4	+2
NOCTUIDAE	20	72	+52
EREBIDAE	5	6	+1
Total	**92**	**264**	**+172**

specimen from the original survey, in the collections of the NMINH, of *Epiblema costipunctana*, taken by Bonaparte Wyse, has been identified following dissection. This represents an addition to the total of the original survey (Kane 1912). Of the 264 New Survey species, 72 were first recorded during 2002. By the completion of the 2000 report, 46 of the species recorded in the original survey had not been rediscovered, but by the end of 2002, this number had been reduced to 34, and it currently stands at 26 (Table 2). The total number of Lepidoptera species ever recorded from Clare Island is therefore 290.

As shown in Table 1, there have been substantial increases in the numbers of Microlepidoptera recorded in the New Survey. This is in part due to the availability of recently published guides to this group, such as the series *The moths and butterflies of Great Britain and Ireland* and various online sources such as *UKMoths* and *MothsIreland*. The substantial increase in the number of Noctuidae recorded is largely due to the increased efficiency of light-traps, in particular the mercury-vapour traps.

Important species recorded during New Survey
Clearly the number of species recorded from a site will increase when additional time and effort are included. The total number of species is therefore less useful as an indicator of the importance of the site than the proportion of rare or otherwise noteworthy species found. Seventeen of the species recorded are worth additional comments, and these are discussed individually below.

Table 2
Species of Lepidoptera recorded during the original survey (Kane 1912), but not during the New Survey. Information on foodplants is derived from *UKMoths*, Emmet (1991) and Skinner (2009)

Species	Foodplant(s) and habitat
Micropterix aureatella	Uncertain (ground flora?) – deciduous woodland
Hepialus humuli	Roots of grasses, etc. - generalist
Caryocolum marmoreum	*Cerastium fontanum* – mainly coastal, especially sandy areas
Archips rosana	Deciduous trees, *Myrica gale* – on bogs and heaths in Ireland
Aphelia viburnana	Mainly on heathland plants – bog and heaths
Ditula angustiorana	Various trees, also herbs – hedgerows, woodland
Eupoecilia angustana	Various herbs, incl. *Calluna* – mainly bogs and heaths
Aethes rubigana	*Arctium* spp. - generalist
Epinotia cruciana	*Salix* spp., prob. mainly *S. repens* – mainly boggy areas
Epiblema costipunctana	*Senecio jacobaea* - generalist
Dichrorampha plumbana	*Leucanthemum vulgare, Achillea millefolium* – calcareous grassland
Argyroploce palustrana	Mosses on bogs – wetlands, bogs
Argyroploce olivana	Herbs and mosses in boggy areas – wetlands, bogs
Platyptilia isodactylus	*Senecio aquaticus* – ditches, wetlands
Stenoptilia bipunctidactyla	Probably mainly *Succcisa pratensis* – bogs and heaths
Crambus uliginosellus	Probably on grasses – wetlands, bogs
Macrothylacia rubi	Heathers and heathland plants – heaths and moors
Cupido minimus	*Anthyllis vulneraria* – sandhills, cliffs, limestone quarries
Hipparchia semele	*Aira praecox, Ammophila arenaria* – coastal cliffs & sand-dunes
Xanthorhoe montanata	*Galium & Primula* spp. and other herbs – hedgerows, woodland
Nebula salicata latentaria	*Galium* spp. – mainly heaths, limestone pavement
Perizoma albulata	*Rhinanthus* – mainly coastal, sand-dunes, limestone
Diacrisia sannio (as *Nemeophila russula*)	Heathland species – bogs and heaths
Phragmatobia fuliginosa	Various herbs – mainly heathland and coastal grassland
Apamea sordens	*Dactylis glomerata* and other grasses - grassland
Apamea sublustris	Probably various grass species – mainly limestone grassland

Species of faunal interest recorded during the New Survey

1135 *Phyllonorycter quinqueguttella*. This species was first discovered in Ireland at Ballyconneely, Co. Galway (Emmet 1969). It has also been discovered by the author on Cape Clear Island, Co. Cork, and at St John's Point, Co. Donegal. It is a scarce and local species both in Britain and on mainland Europe. The larva mines the leaves of *Salix repens*.

1520 *Ochsenheimeria taurella*. This is a rather rare Irish species, with very few recent records. The larva feeds on *Poa* spp. and *Dactylis glomerata*, and its decline may be due to the loss of unimproved grasslands.

1535 *Rhigognostis annulatella*. This is a rather scarce cliff-dwelling species in Ireland. The larva feeds on *Cochlearia officinalis*.

2707a *Coleophora virgaureae*. This species is local in Ireland, being found in some coastal localities, and in the Burren. The larva is reported to feed on *Aster tripolium* and *Solidago virgaurea*.

2717 *Coleophora saxicolella*. Largely confined to coastal saltmarshes in Ireland, this species has been found mostly in the south and east. The larva feeds on the seeds of *Atriplex* and *Chenopodium* spp. As indicated elsewhere in this report (Appendix 1) the specimen was first misidentified as *Coleophora sternipennella*.

3324 *Monochroa lucidella*. There are only a few scattered Irish records of this species, from various parts of the south and west. The larva feeds on *Eleocharis palustris*.

3448 *Teleiopsis diffinis*. Nearly all the Irish records of this rather scarce and local species are either coastal or from the Burren. The larva feeds at the roots of *Rumex acetosa*.

4032 *Sesia bembeciformis*. There are relatively few recent records of the local Lunar Hornet Moth outside the northern counties. The larva burrows through, and feeds on the trunks of *Salix* spp.

4838 *Epinotia sordidana*. The Clare Island record is the earliest Irish record of this species. It is perhaps surprising that the first record is from an island where the reported main foodplant of the species, *Alnus glutinosa*, is quite scarce. In 2010 this species was also recorded from Cos Fermanagh and Leitrim (Langmaid and Young 2011).

5534 *Hellinsia tephradactyla*. This plume moth is very locally distributed over the south and west of Ireland; the Clare Island record is the most northerly Irish record. The larval foodplant is *Solidago virgaurea*.

7058 *Callophrys rubi*. The Green Hairstreak is a distinctly local butterfly in Ireland, with a predominantly western and north-western distribution. Although there is little direct evidence for it, the main Irish foodplant is believed to be *Vaccinium myrtillus*.

7309 *Lasiommata megera*. The Wall Brown butterfly is a local and declining species in Ireland, having disappeared from many northern and inland areas in recent decades. In Mayo it now appears to be confined to a few coastal sites. *Holcus lanatus* and *Dactylis glomerata* are reported as the main foodplants.

7334 *Coenonympha pamphilus*. The Small Heath, like the Wall Brown, is a local and declining butterfly in Ireland; in Mayo it is largely confined to a few local colonies. Various grasses including *Festuca* and *Poa* spp. are the foodplants.

7848 *Gnophos obfuscatus*. The Scotch Annulet is a very local Irish species, being almost totally confined to coastal cliffs from the Burren, Co. Clare northwards to northwest Donegal. The larva is reported to feed on heathers and *Saxifraga* spp.

Scotch Annulet, *Gnophos obfuscatus* coll. Mus. Zool. Oulu. Photo: M. Virtala

9650 *Aporophyla lueneburgensis*. The Northern Deep-brown Dart is a very local Irish species, being almost confined to the Burren, Co. Clare and some coastal sites in the northern half. The main foodplants are reported to be *Calluna* spp. and *Lotus corniculatus*.

10035 *Mythimna unipuncta*. This is a rather rare migrant to Ireland.

10347 *Agrotis trux*. This is a local cliff species, largely confined to the southern half of Ireland, with very few records north of a line from west Clare to Co. Dublin. The larva is reported to feed on various low-growing cliffs species.

Species found in the original Clare Island survey, but not during the New Survey

A total of 26 original survey species were not recorded during the New Survey (see Table 2). Although these species occur in a wide range of habitats, those that are characteristic of heathland habitats are well represented in this group, e.g. *Aphelia viburnana*, *Eupoecilia angustana*, *Macrothylacia rubi*, *Diacrisia sannio* and *Phragmatobia fuliginosa*. Other species that appear to have been lost include several of herb-rich grassland, whereas grass-feeders do not appear to have declined over the intervening period. The heathland species listed are still widespread, and in most cases common, over western Ireland, and this would seem to indicate a loss of heathland habitat on Clare Island during this period.

The Butterflies

Cupido minimus (Small Blue). The record of this species from the original survey of Clare Island may seem surprising in the light of its present Irish distribution. This species now appears to be absent from the entire coast from near Slyne Head to Killala, Co. Mayo, and there are very few inland sites in Mayo and Galway (apart from areas near the Burren). This species feeds only on Kidney Vetch (*Anthyllis vulneraria*), which Praeger (1911) described as 'frequent', but although still present on Clare Island (Tim Ryle, pers. comm.), the plant appears to have declined considerably since then.

Hipparchia semele (Grayling). This is a butterfly of dry grassland, largely coastal, but also occurring very locally inland on rocky slopes and eskers, or in quarries. Only one coastal 10km square in Mayo is indicated for the period 1995–2001 on the *ButterflyIreland* web page (www.geocities.com/butterflyireland/grayling.htm). A loss of undisturbed grassland on the coastal headlands on which Kane reported it may have caused the extinction of a few local colonies.

The Moths

Apamea sublustris (Reddish Light Arches). This is a local species in Ireland, with a north-westerly distribution. Little is known about the larva, but it is believed to feed on the roots of grasses.

Archips rosana. This species seems to have been widespread in the past, but there are very few recent records. Emmet (1969) recorded it as abundant, with the larva on *Myrica gale* (Bog Myrtle), in West Galway, but the larva also feeds on various fruit and deciduous trees (Emmet 1991).

White Ermine *Spilosoma lubricipeda* trapped by K. Bond, Capnagower. Photo: J. Breen

White Ermine *Spilosoma lubricipeda* larva. Crossing road at Fawnglass. Photo: J. Breen

Argyroploce olivana. Like the preceding, this is a very local wetland species in Ireland. The larva is reported to feed on 'herbaceous plants and mosses'.

Argyroploce palustrana. Another very local species, found on bogs and heaths, recorded only from Clare, Wicklow and Fermanagh. The larva feeds on mosses (Emmet 1991).

Crambus uliginosellus. The history of recording this species in Ireland is confused by misidentifications (Bond 1992). Although it was on earlier lists, the specimens available for checking indicated misidentifications. It has, however, now been confirmed as an Irish insect, with records from Sligo, Armagh and East Cork, so it is quite possible that the Clare Island record was correct, but no examples of it from the original survey (Kane 1912) have been found. Confirmed records indicate that it is a wetland insect of fens or saltmarsh margins.

Diacrisia sannio. This is a local species of boggy and heathy areas, largely confined to western parts of Ireland. The larva is reported to feed on heathers and *Hieracium* spp. (Emmet 1991)

Dichrorampha plumbana. A local species, which is found mainly in limestone areas and is absent from most of the south of Ireland. This is a species of dry pastures, whose larva feeds on *Leucanthemum vulgare* Lam. or *Achillea millefolium* (Emmet 1991).

Macrothylacia rubi (Fox Moth). This species is usually common on extensive tracts of heather (*Calluna vulgaris*), on which the larva feeds. It is probably most often detected as a larva, on or near heather, in the early spring, and its apparent absence could be due to relatively little fieldwork at this season.

Micropterix aureatella. This is another local species, mainly associated with deciduous woodland. It is a surprising find on Clare Island.

Platyptilia isodactyla (a plume moth). This is a local wetland species. While it may still occur on the island, it is possible that it is no longer present due to loss of habitat in which its foodplant, Marsh Ragwort (*Senecio aquatica*) occurs.

Stenoptilia bipunctidactyla (a plume moth). Like the plume moths in general, this species is a weak flyer, and easily overlooked. The larva

Buff-tip *Phalera bucephala* trapped by K. Bond, Capnagower. Photo: J. Breen

Caterpillar of the Knotgrass Moth, *Acronicta rumicis.* Capnagower. Photo: J. Breen

feeds mainly, possibly exclusively, on Devil's-bit Scabious (*Succisa pratensis*).

Conclusions

The Lepidoptera survey of Clare Island has provided a new level of insight into the faunal composition of an Irish west coast site. No other Irish island has received such a detailed survey of its lepidopteran fauna. While Irish west coast sites have probably undergone much less change due to human impacts than other Irish coasts, the changes shown by the failure to rediscover various first survey species suggest that there has been some loss of habits such as natural grassland, wildflower meadows and heath. The greatly increased number of Lepidoptera species recorded in the New Survey should not be taken to indicate an increase in Lepidoptera diversity or abundance. Powerful light-traps not available to the early twentieth-century entomologists are now available to moth recorders. Furthermore, identification of taxonomically difficult species is now aided by an extensive literature, backed up by readily accessible recourse to international experts.

Acknowledgements

I am grateful to Kieran McCarthy for making possible the fieldwork carried out in 2002 and later. A considerable amount of the information obtained in 2002 was due to the untiring efforts of Stephen McCormack in searching for insects at many sites using a variety of methods. Thanks are also due to Bob Cussen and Myles Nolan for providing interesting additional records.

REFERENCES

Bond, K.G.M. 1992 *Crambus uliginosellus* Zeller (Lepidoptera: Pyralidae), First Confirmed Irish Records. *The Irish Naturalist's Journal* **24**(4), 167–68.

Bond, K.G.M. 2000 Lepidoptera - Butterflies and Moths *Interim Report. New Clare Island Survey*. Dublin. Royal Irish Academy.

Bond, K.G.M. 2003 *Epinotia sordidana* (Lepidoptera: Tortricidae) new to Ireland. *Entomologist's Gazette* **54**, 142.

Bond, K.G.M. 2005 *Coleophora sternipennella* (Zetterstedt 1839) (Lepidoptera: Coleophoridae) new to Ireland. *Entomologist's Gazette* **56**, 65.

Bond, K.G.M. and O'Connor, J.P. 2012 Additions, deletions and corrections to *An annotated checklist of the Irish butterflies and moths (Lepidoptera)* with a concise checklist of Irish species and *Elachista biatomella* (Stainton 1848) new to Ireland. *Bulletin of the Irish Biogeographical Society* No. **36**, 60–179.

Bradley, J.D. and Fletcher, D.S. 1979 *A recorder's log book or label list of British Butterflies and Moths*. Colchester. Harley Books.

de Jong, Y.S.D.M. (ed.) 2013 *Fauna Europaea* version 2.6. Web Service available online at http://www.faunaeur.org

Emmet, A. M 1969 [Exhibits] Irish Microlepidoptera including *Lithocolletis quinqueguttella* Staint., *Ancylis inornatana* H.-S., *Stigmella hybnerella* Hubn., *Dechtiria pulverosella* Staint., *Nepticula ignobilella* Staint., *Coleophora adjunctella* Hodgk. *Proceedings and Transactions of the British Entomological and Natural History Society* **2**, 127–28.

Emmet, A.M. 1991 Life History and Habits of the British Lepidoptera. In A.M. Emmet and J. Heath (eds) *The Moths and Butterflies of Great Britain and Ireland. Vol. 7. Lasiocampidae - Thyatiridae*, 61–303. Colchester. Harley Books.

Kane, W.F. de V. 1912 Part 26. Lepidoptera. *Proceedings of the Royal Irish Academy* **31B**, 26.1–26.10.

Langmaid J.R. and Young, M.R. 2011 Microlepidoptera review of 2010. *Entomologist's Record* and *Journal of Variation* **123**, 249–77.

MothsIreland www.mothsireland.com/index.htm

Skinner, B. 2009 *Colour Identification Guide to Moths of the British Isles*. Stenstrup. Apollo Books.

Stace, C. 1991 *New Flora of the British Isles*. Cambridge. Cambridge University Press.

UKMoths 2013 at: http://ukmoths.org.uk/

APPENDIX 1

List of Lepidoptera recorded during the New Survey. Species numbering system is explained in the text. The number of species per family is in brackets after the family name.

MICROPTERIGIDAE (2)
7 *Micropterix aruncella* (Scopoli, 1763). Female on *Iris* (dissected), Fawnglass, L710853, 11.vii.2002; female from suction sample (dissected), Fawnglass, L712853, 1.viii.2002; two swept from heather, Lassu Wood, L703868, 8.v.2003.

10 *Micropterix calthella* (Linnaeus, 1761). One [male] swept from bog seepage (dissected), Portlea, L704866, 8.v.2003.

HEPIALIDAE (1)
69 *Pharmacis fusconebulosa* (DeGeer, 1778) MAP-WINGED SWIFT. One at m. v. trap, Kill, L698848, 18.vi.2000; six at m. v. trap, Ballytoohy More, L699870, 15.vii.2000; one at m. v. trap, Maum, L705864, 16.vii.2000; two at m. v. trap, Maum, L705864, 11.viii.2000; female (dissected) at m. v. trap, Fawnglass, L713854, 11.vii.2002; two at Skinner Trap, Creggan Lough, L689856, 14.vii.2005.

NEPTICULIDAE (4)
89 *Stigmella lapponica* (Wocke, 1862). Leaf-mines on *Betula pubescens*, Lassu Wood, L704866, 28.ix.2002; female swept (dissected), Lassu Wood, L703868, 8.v.2003.

139 *Stigmella salicis* (Stainton, 1854). Two vacated leaf-mines on *Salix*, Lassu Wood, L704866, 28.ix.2002.

151 *Stigmella continuella* (Stainton, 1856). Leaf-mines on *Betula pubescens*, Lassu Wood, L704866, 28.ix.2002.

307 *Ectoedemia occultella* (Linnaeus, 1767). Leaf-mines on *Betula pubescens*, Lassu Wood, L704866, 28.ix.2002.

HELIOZELIDAE (1)
331 *Heliozela hammoniella* Sorhagen, 1885. Vacated leaf-mines on *Betula pubescens*, Lassu Wood, L704866, 28.ix.2002.

TINEIDAE (2)
680 *Tinea pallescentella* Stainton, 1851 LARGE PALE CLOTHES MOTH. One indoors (dissected), Fawnglass, L712854, 8.v.2003.

700 *Monopis laevigella* ([Denis and Schiffermüller], 1775). Male (dissected), Capnagower (on willow), L715858, 10.vii.2002 (this specimen was first identified as *M. weaverella* (Scott, 1858)); female (dissected), Mill Cottage, L706864, 27.viii.2002.

GRACILLARIIDAE (6)
1116 *Caloptilia elongella* (Linnaeus, 1761). Male taken at m. v. trap (dissected), Maum, L705864, 16.vii.2000; larval leaf-mine and leaf-fold on *Alnus glutinosa*, Maum, L706864, 28.ix.2002.

1135 *Gracillaria syringella* (Fabricius, 1794). One at m. v. trap, Maum, L705864, 11.viii.2000.

1143 *Aspilapteryx tringipennella* (Zeller, 1839). One at m. v. trap, Maum, L705864, 16.vii.2000; male (dissected) at m. v. trap, Fawnglas, L713854, 9.vii.2002.

1293 *Phyllonorycter quinqueguttella* (Stainton, 1851). Leaf-mines on *Salix repens*, Toormore (clifftop), L662843, 28.ix.2002.

1300 *Phyllonorycter salicicolella* (Sircom, 1848). Leaf-mine on *Salix* (*S. aurita* or *S. auritaXcinerea*) (det. from larva), Lassu Wood, L704866, 28.ix.2002.

1326 *Phyllonorycter ulmifoliella* (Hübner, [1817]). Leaf-mines on *Betula pubescens* (det. from larva), Lassu Wood, L704866, 28.ix.2002.

ARGYRESTHIIDAE (4)
1402 *Swammerdamia pyrella* (Villers, 1789). Male (dissected) at m. v. trap, Lassu Wood, L705867, 22.viii.2002.

1453 *Argyresthia brockeella* (Hübner, [1813]). One, Lassu birchwood, L704867, 17.vi.2000.

1454 *Argyresthia goedartella* (Linnaeus, 1758). One, Maum, L663849, 15.vii.2000; one at m. v. trap, Maum, L705864, 16.vii.2000; one at m. v. trap, Maum L705864, 29.ix.2002.

1467 *Argyresthia bonnetella* (Linnaeus, 1758). One, Glen, L711851, 15.vii.2000.

YPSOLOPHIDAE (2)

1482 *Ypsolopha dentella* (Fabricius, 1775) HONEYSUCKLE MOTH. One at m. v. trap, Lassu Wood, L705867, 22.viii.2002.

1520 *Ochsenheimeria taurella* ([Denis and Schiffermüller], 1775). Female taken, Portnakilly, L690844, 22.viii.2002.

PLUTELLIDAE (2)

1525 *Plutella xylostella* (Linnaeus, 1758) DIAMOND-BACK MOTH. Two, Harbour, L712853, 17.vi.2000; two birchwood, L704867, 17.vi.2000; one at m. v. trap, Kill, L698848, 18.vi.2000; one, Glen, L706849, 17.vi.2000; one at m. v. trap, Maum, L705864, 11.viii.2000; one, Capnagower, L709861, 28.ix.2002.

1535 *Rhigognostis annulatella* (Curtis 1832). Male taken, Loughaunaphuca, L655843, 24.viii.2000.

GLYPHIPTERIGIDAE (1)

1580 *Glyphipterix thrasonella* (Scopoli, 1763). Two, Portacoolia, L692844, 17.vi.2000; two, Harbour, L712853, 17.vi.2000; two, male taken (dissected), birchwood, L702867, 17.vi.2000; two males, Church – Kill (on *Oenanthe crocata*), L689844, 9.vii.2002; five males (all dissected), swept from vegetation, Leck, L705858, 9.vii.2002; two, Lough Leinapollbruty, L691856, 13.vii.2005; one, Creggan Lough, L689857, 13.vii.2005.

LYONETIIDAE (1)

1627 *Lyonetia clerkella* (Linnaeus, 1758) APPLE LEAF MINER. Leaf-mines on *Malus domestica*, Maum, L705864, 11.viii.2000; leaf-mines on *Betula*, Lassu Wood, L704866, 28.ix.2002.

ELACHISTIDAE (11)

1722 *Agonopterix subpropinquella* (Stainton, 1849). Male at light (f. *rhodochrella* Herrich-Schäffer, 1854 - dissected), Maum, L705864, 27.ix.2002; male at light indoors (dissected), Fawnglass, L712854, 1.v.2003.

1736 *Agonopterix heracliana* (Linnaeus, 1758). One taken at light, hotel, L716855, 27.iv.2000; one at light, Maum, L705864, 28.iv.2000; one indoors, Fawnglass, L712854, 1.v.2003; one indoors, Fawnglass, L712854; 4.v.2003.

1763 *Agonopterix nervosa* (Haworth, 1811). Two, Lough Merrighnagh, L688852, 23.viii.2000; two males and female at Heath Trap, Capnagower, L712863, 23.vii.2002.

1776 *Depressaria radiella* (Goeze, 1783) (*heraclei* Retzius misidentification) PARSNIP MOTH, (*pastinacella* (Duponchel, 1838)). One at light, Harbour, L715855, 23.viii.2000.

1788 *Depressaria badiella* (Hübner, 1796). Male at m. v. trap (dissected), Toormore, L661844, 28.ix.2002.

1793 *Depressaria daucella* ([Denis and Schiffermüller], 1775). Female taken at light, hotel, L716855, 27.iv.2000; one at light, Fawnglass, L712854; 29.iv.2003; one indoors, Fawnglass, L712854, 4.v.2003; one at light, Fawnglass, L712854, 6.v.2003.

1828 *Elachista consortella* Stainton, 1851. Male (dissected), Alnamarnagh, L689875, 28.iv.2000.

1857 *Elachista alpinella* Stainton, 1854. Two (male dissected) at m. v. trap, Lassu Wood, L705867, 22.viii.2002; two males (both dissected), Kinnacorra saltmarsh, L720859, 22.viii.2002.

1863 *Elachista argentella* (Clerck, 1759). One, Ardal cliffs, L715864, 17.vi.2000; one, Glen, L706849, 17.vi.2000.

1865 *Elachista atricomella* Stainton, 1849. One at actinic trap, Maum, L706864, 14.vii.2005.

1883 *Elachista canapennella* (Hübner, [1813]). Male at m. v. trap (dissected), Maum, L705864, 23.viii.2002; male (dissected), Fawnglass, L712854, 30.iv.2003.

OECOPHORIDAE (3)

2282 *Endrosis sarcitrella* (Linnaeus, 1758) WHITE-SHOULDERED HOUSE-MOTH. One, Portacoolia, L692844, 17.vi.2000; one, indoors, Kill, L698847, 24.viii.2000; one indoors, Maum, L706864, 21.viii.2002; one, Fawnglass, L712854, 8.v.2003; one at actinic trap, Maum, L706864, 14.vii.2005.

2284 *Hofmannophila pseudospretella* (Stainton, 1849) BROWN HOUSE-MOTH. One found indoors, Fawnglass, L712854, 17.v.2003.

2348 *Pleurota bicostella* (Clerck, 1759). Two, Portlea, L704867, 17.vi.2000.

COLEOPHORIDAE (10)

2468 *Coleophora serratella* (Linnaeus, 1761). Case and leaf-mines on *Betula*, Lassu Wood, L703867, 17.vi.2000; larval cases and mines on *Betula pubescens*, Lassu Wood, L704866, 28.ix.2002.

2479 *Coleophora lusciniaepennella* (Treitschke, 1833). Leaf-mines on *Myrica gale*, Lassu Wood, L702867, 17.vi.2000; female taken (dissected), Lassu Wood, L704868, 17.vi.2000.

2572 *Coleophora discordella* Zeller, 1849. One at m. v. trap, Kinnacorra, L720862, 23.viii.2002.

2585 *Coleophora deauratella* Lienig and Zeller, 1846. Male at m. v. trap (dissected), Fawnglas, L713854, 11.vii.2002; male at Heath Trap (dissected), Capnagower, L712863, 23.vii.2002.

2593 *Coleophora albidella* ([Denis and Schiffermüller], 1775). Three at m. v. trap (two males taken and dissected), Lassu Wood, L705867, 22.viii.2002.

2688 *Coleophora tamesis* Waters, 1929. Female taken at m. v. trap (dissected), Maum, L705864, 16.vii.2000.

2692 *Coleophora alticolella* Zeller, 1849. Male taken (dissected), Portacoolia, L692844, 17.vi.2000; three, Glen, L706849, 17.vi.2000; one, Glen, L708850, 17.vi.2000; male taken (dissected), Portlea shore, L706867, 17.vi.2000; one, Kill (church), L689844, 9.vii.2002.

2693 *Coleophora taeniipennella* Herrich-Schäffer, 1855. Female taken (dissected), Ardal cliffs, L715864, 17.vi.2000.

2707a *Coleophora virgaureae* Stainton, 1857. Male and female (both dissected) at m. v. trap, Lassu Wood, L705867, 22.viii.2002; female at m. v. trap (dissected), Maum, L705864, 23.viii.2002; larval cases and leaf-mines on *Solidago virgaurea*, Capnagower, L712857, 28.ix.2002. (Note: this species was incorrectly named *obscenella* in some recent checklists).

2717 *Coleophora saxicolella* (Duponchel, [1843]). Female (dissected) at actinic trap, Harbour, L714849, 9.vii.2003. Note: this species was first determined as *C. sternipennella* (Zetterstedt, [1839]) and published as new to Ireland in Bond (2005); (the species was misidentified, the record refers to *C. saxicolella*) the record was corrected in Bond and O'Connor (2012).

BLASTOBASIDAE (1)

2905 *Blastobasis adustella* Walsingham, 1894 (*lignea* misidentification). One, Ballytoohy More, L699870, 15.vii.2000; one at m. v. trap, Maum, L705864, 16.vii.2000; three, Portlea, L706866, 10.viii.2000; five at m. v. trap, Maum, L705864, 11.viii.2000; two females at Heath Trap, Capnagower, L712863, 23.vii.2002; male and female at m. v. trap, Fawnglas, L712863, 23.vii.2002; fourteen at m. v. trap, Lassu Wood, L705867, 22.viii.2002; two at m. v trap, Maum, L705864, 23.viii.2002.

GELECHIIDAE (6)

3312 *Monochroa cytisella* (Curtis, 1837). Female taken at m. v. trap, Maum, L705864, 16.vii.2000.

3324 *Monochroa lucidella* (Stephens, 1834). Male taken at m. v. trap (dissected, det. O. Karsholt, who describes it as an 'unusually dark form'), Maum, L705864, 16.vii.2000; two males and female (all dissected) swept from vegetation, Lecarow, L705858, 9.vii.2002.

3373 *Bryotropha terrella* ([Denis and Schiffermüller], 1775). Three (male taken), cliffs opposite Glassilangaraltach, L654858, 15.vii.2000; male, Church – Kill (on *Oenanthe crocata*), L689844, 9.vii.2002; four males and two females (one of each dissected) at Heath Trap, Capnagower, L712863, 23.vii.2002; female (dissected), Beetle Head, L653843, 25.vii.2002; male at m. v trap (dissected), Maum, L705864, 23.viii.2002; two at actinic trap, Harbour, L714849, 9.vii.2003.

3404 *Exoteleia dodecella* (Linnaeus, 1758). Male at m. v. trap (dissected), Maum, L705864, 16.vii.2000.

3448 *Teleiopsis diffinis* (Haworth, 1828). Male at m. v. trap (dissected), Toormore, L661844, 28.ix.2002.

3833 *Neofaculta ericetella* (Geyer, [1832]). One, birchwood, L704867, 17.vi.2000; one, Ballytoohy, L693876, 22.v.2002.

ZYGAENIDAE (1)

3998 *Zygaena filipendulae* (Linnaeus, 1758) f. *stephensi* Dupont, 1900 SIX-SPOT BURNET. One, cliffs opposite Glassilangaraltach, L654858, 15.vii.2000; imago and two cocoons, Ooghaniska, L669846, 15.vii.2000; cocoon and pupa, Maum, L702865, 13.vii.2005; one, Kill, L695845, 13.vii.2005.

SESIIDAE (1)

4032 *Sesia bembeciformis* (Hübner, 1806) LUNAR HORNET MOTH. Female (dissected), Portlea, L705867, 27.vii.2002 (Myles Nolan).

TORTRICIDAE (36)

4268 *Agapeta hamana* (Linnaeus, 1758). Two at m. v. trap, Kill, L698848, 18.vi.2000.

4326 *Aethes cnicana* (Westwood, 1854). One at m. v. trap, Ballytoohy More, L699870, 15.vii.2000.

4385 *Acleris emargana* (Fabricius, 1775). One, Lassu Wood, L704866, 28.ix.2002.

4390 *Acleris variegana* ([Denis and Schiffermüller], 1775) GARDEN ROSE TORTRIX. One at m. v trap, Maum, L705864, 23.viii.2002; three at m. v. trap, Maum, L706864, 29.ix.2002.

4391 *Acleris aspersana* (Hübner, [1817]). Two at m. v. trap, Lassu Wood, L705867, 22.viii.2002.

4394 *Acleris hastiana* (Linnaeus, 1758). Male at m. v. trap, Lassu Wood, L705867, 22.viii.2002.

4397 *Acleris hyemana* (Haworth, 1811). Male taken, Alnamarnagh, L687872, 28.iv.2000.

4462 *Eana penziana* (Thunberg, 1791). Three (2 males, ssp. *colquhounana* (Barrett, 1884) taken, both dissected) at actinic trap, Harbour, L714849, 9.vii.2003.

4471 *Cnephasia incertana* (Treitschke, 1835) LIGHT GREY TORTRIX. Four at m. v. trap (male dissected), Ballytoohy More, L699870, 15.vii.2000 [although it is uncertain that all these were the same species, their similar appearance suggests that they were].

4491 *Cnephasia conspersana* Douglas, 1846. Male at m. v. trap (dissected), Lassu Wood, L705867, 22.viii.2002; four at m. v. trap (two males dissected), Fawnglas, L713854, 22.viii.2002.

4579 *Pandemis cerasana* (Hübner, 1786) BARRED FRUIT-TREE TORTRIX. Five at m. v. trap, Maum, L705864, 16.vii.2000.

4580 *Pandemis heparana* ([Denis and Schiffermüller], 1775) DARK FRUIT-TREE TORTRIX. One at m. v. trap, Maum, L705864, 16.vii.2000; one at m. v. trap, Maum, L705864, 10.viii.2000; two at m. v. trap, Maum, L705864, 11.viii.2000; nine at m. v. trap (male dissected), Lassu Wood, L705867, 22.viii.2002.

4655 *Bactra lancealana* (Hübner, [1799]). Four, Harbour, L712853, 17.vi.2000; one, Portacoolia, L692844, 17.vi.2000; one Lassu Wood, L699870, 15.vii.2000; three (small form), Toormore, L659854, 15.vii.2000; four at m. v. trap, Maum, L705864, 16.vii.2000; male, Church (Kill) on *Oenanthe crocata*, L689844, 9.vii.2002; one at m. v. trap, Lassu Wood, L705867, 22.viii.2002; 11 at actinic trap, Harbour, L714849, 9.vii.2003; two, Lough Leinapollbruty, L691856, 13.vii.2005; one at Skinner Trap, Creggan Lough, L689856, 14.vii.2005.

4656 *Bactra furfurana* (Haworth, 1811). Four (two taken, male dissected), Kinnacorra saltmarsh, L720860, 17.vi.2000; male swept from *Scirpus maritimus* (dissected), Kinnacorra, L721859, 10.vii.2002; male at actinic trap, Maum Marsh, L702863, 22.viii.2002.

4673 *Endothenia quadrimaculana* (Haworth, 1811). One at m. v. trap, Ballytoohy More, L699870, 15.vii.2000; one at m. v. trap, Maum, L705864, 16.vii.2000.

4692 *Apotomis semifasciana* (Haworth, 1811). Three at m. v. trap, Lassu Wood, L705867, 22.viii.2002; one at m. v trap, Maum, L705864, 23.viii.2002.

4711 *Orthotaenia undulana* ([Denis and Schiffermüller], 1775). Two, birchwood, L702867, 17.vi.2000; one at m. v. trap, Maum, L705864, 16.vii.2000.

4714 *Hedya nubiferana* (Haworth, 1811) MARBLED ORCHARD TORTRIX. Male taken, Maum, L663849, 15.vii.2000.

4715 *Hedya pruniana* (Hübner, [1799]) PLUM TORTRIX. Two at m. v. trap, Maum, L705864, 16.vii.2000; one at actinic trap, Maum, L706864, 14.vii.2005.

4728 *Celypha cespitana* (Hübner, 1817]). Male taken, Signal Tower, L658853, 15.vii.2000; two males at Heath Trap, Capnagower, L712863, 23.vii.2002; male, near Beetle Head, L654843, 23.vii.2002.

4731 *Celypha lacunana* ([Denis and Schiffermüller], 1775). One, Portlea shore, L706867, 17.vi.2000; one at m. v. trap, Ballytoohy More, L699870, 15.vii.2000; one at m. v. trap, Maum, L705864, 16.vii.2000; one at m. v. trap, Maum, L705864, 11.viii.2000; three at m. v. trap, Lassu Wood, L705867, 22.viii.2002; two at m. v. trap, Fawnglas, L713854, 22.viii.2002; one at actinic trap, Maum Marsh, L702863, 22.viii.2002; one at m. v. trap, Kinnacorra, L720862, 23.viii.2002; one at actinic trap, Maum, L706684, 14.vii.2005.

4806 *Lobesia littoralis* (Humphreys and Westwood, 1845). One, cliffs ENE of Signal Tower, L658853, 15.vii.2000; one, cliffs opposite Glassilangaraltach, L653859, 15.vii.2000; one at actinic trap, Harbour, L714849, 9.vii.2003.

4829 *Rhopobota naevana* (Hübner, [1817]) HOLLY TORTRIX. One at m. v. trap, Maum, L705864, 16.vii.2000; male taken at m. v. trap, Maum, L705864, 11.viii.2000; five at m. v. trap, Lassu Wood, L705867, 22.viii.2002; one at m. v trap, Maum, L705864, 23.viii.2002; one at m. v. trap, Maum, L706864, 29.ix.2002.

4838 *Epinotia sordidana* (Hübner, [1824]). Male at light (dissected), Maum, L705864, 27.ix.2002. NEW TO IRELAND (Bond 2003).

4839 *Epinotia caprana* (Fabricius, 1798). Female at actinic trap, Ballytoohy, L699869, 3.ix.2005.

4840 *Epinotia trigonella* (Linnaeus, 1758). Female at m. v. trap, Lassu Wood, L705867, 22.viii.2002.

4842 *Epinotia brunnichana* (Linnaeus, 1767). Nine at m. v. trap (two females dissected), Lassu Wood, L705867, 22.viii.2002; two at m. v. trap, Maum, L706864, 29.ix.2002.

4853 *Epinotia immundana* (Fischer von Röslerstamm, 1839) (*rhomboidella* misidentification). Four at m. v. trap (one taken), Maum, L705864, 10.viii.2000; eight at m. v. trap, Maum, L705864, 11.viii.2000; nine at m. v. trap, Lassu Wood, L705867, 22.viii.2002; nineteen at m. v trap, Maum, L705864, 23.viii.2002

4863 *Epinotia subocellana* (Donovan, 1806). One, Portlea shore, L706867, 17.vi.2000.

4932 *Eucosma cana* (Haworth, 1811). One at m. v. trap, Ballytoohy More, L699870, 15.vii.2000; four at m. v. trap, Maum, L705864, 16.vii.2000; one at m. v. trap, Maum, L705864, 11.viii.2000; male and female at Heath Trap, Toormore, L663847, 6.vii.2002; two females at m. v. trap, Fawnglas, L712863, 23.vii.2002; one at m. v. trap, Fawnglas, L713854, 22.viii.2002; one at m. v trap, Maum, L705864, 23.viii.2002; male (dissected), Capnagower, L712863, 23.vii.2003; one at actinic trap, Maum, L706864, 14.vii.2005.

4943 *Eucosma campoliliana* ([Denis and Schiffermüller], 1775). Two at m. v. trap, Maum, L705864, 16.vii.2000; one at actinic trap, Harbour, L714849, 9.vii.2003; one at Skinner trap, Creggan Lough, L689856, 14.vii.2005; one at actinic trap, Ooghcorragaun, L703875, 14.vii.2005.

4985 *Gypsonoma dealbana* (Frölich, 1828). Male at m. v. trap, Lassu Wood, L705867, 22.viii.2002.

4994 *Epiblema scutulana* ([Denis and Schiffermüller], 1775). One at m. v. trap, Kill, L698848, 18.vi.2000; four at m.v. trap (one taken), Kill, L698847, 24.viii.2000.

4995 *Epiblema cirsiana* (Zeller, 1843). Male taken, Glen, L708850, 17.vi.2000.

5057 *Ancylis unguicella* (Linnaeus, 1758). One taken, Portlea, L704867, 17.vi.2000; two, birchwood, L702867, 17.vi.2000.

5073 *Ancylis badiana* ([Denis and Schiffermüller], 1775). One at m. v. trap, Fawnglas, L713854, 22.viii.2002; two, Fawnglass, L712854, 30.iv.2003; one, Fawnglass, L712854, 6.v.2003; two, Ballytoohy, L699867, 7.v.2003.

CHOREUTIDAE (1)

5269 *Anthophila fabriciana* (Linnaeus, 1767). Two larvae on *Urtica dioica*., shore near hotel, L716856, 28.iv.2000; two, Glen, L706849, 17.vi.2000.

EPERMENIIDAE (1)

5303 *Epermenia chaerophyllella* (Goeze, 1783). Male taken, Alnamarnagh, L686869, 28.iv.2000.

PTEROPHORIDAE (1)

5534 *Hellinsia tephradactyla* (Hübner, [1813]). Larva on *Solidago virgaurea*, Capnagower, L712857, 28.ix.2002.

PYRALIDAE (3)

5569 *Aphomia sociella* (Linnaeus, 1758) BEE MOTH. One at m. v. trap, Lassu Wood, L705867, 22.viii.2002; two at m. v. trap, Kinnacorra, L720862, 23.viii.2002; one at actinic trap, Ooghcorragaun, L703875, 14.vii.2005.

5690 *Delplanqueia dilutella* ([Denis and Schiffermüller], 1775). Male taken at m. v. trap (dissected), Ballytoohy More, L699870, 15.vii.2000; male at m. v. trap (male taken), Maum, L705864, 16.vii.2000; five males at Heath Trap, Toormore, L663847, 6.vii.2002; two at actinic trap, Maum, L706864, 14.vii.2005; seven at actinic trap, Ooghcorragaun, L703875, 14.vii.2005.

6090 *Phycitodes saxicola* (Vaughan, 1870). Male taken at m. v. trap (dissected), Maum, L705864, 16.vii.2000; two at actinic trap (male dissected), Harbour, L714849, 9.vii.2003; one at actinic trap, Ooghcorragaun, L703875, 14.vii.2005.

CRAMBIDAE (19)

6168 *Scoparia ambigualis* (Treitschke, 1829). Male (dissected), shore opposite Carrickarelick, L701878, 4.vi.2000 (R. Cussen); one, Fawnglass, L711859, 17.vi.2000; two at actinic trap, Maum, L706864, 14.vii.2005.

6193 *Eudonia truncicolella* (Stainton, 1849). Two at m. v. trap (female dissected), Lassu Wood, L705867, 22.viii.2002; two at m. v. trap, Fawnglas, L713854, 22.viii.2002; four at m. v trap, Maum, L705864, 23.viii.2002; four at m. v. trap, Kinnacorra, L720862, 23.viii.2002.

6195 *Eudonia mercurella* (Linnaeus, 1758). Male taken at m. v. trap (dissected), Ballytoohy More, L699870, 15.vii.2000; one at m. v. trap, Maum, L705864, 10.viii.2000; two at m. v. trap, Maum, L705864, 11.viii.2000; two at m. v. trap, Lassu Wood, L705867, 22.viii.2002; four at m. v trap, Maum, L705864, 23.viii.2002; one at actinic trap, Maum, L706864, 14 July 2005.

6199 *Eudonia pallida* (Curtis, 1827). Two males at actinic trap (both dissected), Maum Marsh, L702863, 22.viii.2002; two at m. v trap, Maum, L705864, 23.viii.2002.

6241 *Chrysoteuchia culmella* (Linnaeus, 1758). One, Portlea shore, L706867, 17.vi.2000; one, Glen, L708850, 17.vi.2000; three at m. v. trap, Kill, L698848, 18.vi.2000; two at m. v. trap, Maum, L705864, 16.vii.2000; one at Heath Trap, Toormore, L663847, 6.vii.2002; one at m. v. trap, Fawnglas, L713854, 22.viii.2002; 15 at actinic trap, Harbour, L714849, 9.vii.2003; one at actinic trap, Creggan Lough, L690857, 9.vii.2003; one at Skinner trap, Creggan Lough, L689856, 14.vii.2005; one at actinic trap, Ooghcorragaun, L703875, 14.vii.2005.

6243 *Crambus pascuella* (Linnaeus, 1758). One, Toormore, L660853, 15.vii.2000; six, Toormore, L661853, 15.vii.2000; three, Toormore, L659854, 15.vii.2000; three at m. v. trap, Maum, L705864, 16.vii.2000; two, Portlea, L706866, 10.viii.2000; one at Heath Trap, Toormore, L663847, 6.vii.2002; one, Knockaveen ('*Nardus stricta*'), L702853, 10.vii.2002; two males at m. v. trap, Fawnglas, L712863, 23.vii.2002; one, Maum, L703866, 21.viii.2002; one at m. v. trap, Maum, L705864, 23.viii.2002; three, Lough Leinapollbruty, L691856, 13.vii.2005; two at actinic trap, Maum, L706964, 14.vii.2005; one at Creggan Lough, L689856, 14.vii.2005.

6251 *Crambus lathoniellus* (Zincken, 1817). One, Lassu Wood, L704867, 17.vi.2000; two, Lassu Wood, L702867, 17.vi.2000.

6253 *Crambus perlella* (Scopoli, 1763). Seven at m. v. trap, Ballytoohy More, L699870, 15.vii.2000; one, Toormore, L663849, 15.vii.2000; one, Toormore, L659854, 15.vii.2000; one at m. v. trap, Maum, L705864, 16.vii.2000; one at m. v. trap, Fawnglas, L713854, 9.vii.2002; twelve males at Heath Trap, Capnagower, L712863, 23.vii.2002; three at m. v. trap (male taken), Fawnglas, L713854, 22.viii.2002; two at actinic trap, Harbour, L714849, 9.vii.2003. **Note**: most of the 2002 specimens are of a golden-suffused form.

6258 *Agriphila tristella* ([Denis and Schiffermüller], 1775). One at m.v. trap, Kill, L698847, 24.viii.2000; seventeen at m. v. trap, Fawnglas, L713854, 22.viii.2002.

6260 *Agriphila inquinatella* ([Denis and Schiffermüller], 1775). One at m. v. trap, Fawnglas, L713854, 22.viii.2002.

6267 *Agriphila straminella* ([Denis and Schiffermüller], 1775). Three, Maum birchwood, L699870, 15.vii.2000; three, Maum birchwood, L701870, 15.vii.2000; three, cliffs opposite Glassilangaraltach, L653859, 15.vii.2000; two at m. v. trap, Ballytoohy More, L699870, 15.vii.2000; three, birchwood, L701870, 15.vii.2000; one, Toormore, L661853, 15.vii.2000; one, Toormore, L659854, 15.vii.2000; *circa* 50, Portlea, L706866, 10.viii.2000; five at m. v. trap, Maum, L705864, 11.viii.2000; *circa* 10, Creggan Lough, L688859, 23.viii.2000; one, Creggan Lough, L689857, 24.viii.2000; one, Portnakilly, L691844, 24.viii.2000; *circa* 10, NW slopes of Knockmore [*circa* 300m], L666863, 24.viii.2000; one, Knockmore [ENE of summit, alt. 400m], L671862, 24.viii.2000. (In 2000 *A. straminella* was common all around Knockmore and boggy areas to E and SE). Male in meadow, Fawnglas, L712863, 23.vii.2002; one, Maum, L703866, 21.viii.2002; two at m. v. trap, Lassu Wood, L705867, 22.viii.2002; five at m. v. trap, Fawnglas, L713854, 22.viii.2002; *circa* 20, Maum Stream, L705863 22.viii.2002; *circa* 20, Lough Leinapollbruty, L691856, 22.viii.2002; *circa* 20, Creggan Lough, L689856, 22.viii.2002; three, Portnakilly, L693842, 22.viii.2002; one, Kinnacorra saltmarsh, L720859, 22.viii.2002; eleven at m. v trap, Maum, L705864, 23.viii.2002; five at m. v. trap, Kinnacorra, L720862, 23.viii.2002; two (a small moorland race?), Capnagower, L709861, 28.ix.2002; one, Maum, L705864, 13.vii.2005; one at Creggan Lough, L689856, 14.vii.2005; one, Ooghcorragaun, L703875, 3.ix.2005. **Note**: the September 2002 record is by far the latest date this species has been recorded on the wing in Ireland; the flight period of this species ends in late August, except in certain mountain and moorland sites of the west and north-west).

6275 *Agriphila geniculea* (Haworth, 1811). Four, Kill, L696845, 23.viii.2000; four at m. v. trap, Kinnacorra, L720862, 23.viii.2002; one at actinic trap, L703875, 3.ix.2005.

6394 *Donacaula mucronella* ([Denis and Schiffermüller], 1775). Two at Skinner trap, Creggan Lough, L689856, 14.vii.2005.

6416 *Elophila nymphaeata* (Linnaeus, 1758) BROWN CHINA-MARK. One, Leck, L705858, 9.vii.2002; female (dissected), near Beetle Head, L654843, 25.vii.2002; one, Maum Stream, L705863 22.viii.2002; one, Lough Leinapollbruty, L691856, 22.viii.2002; one at m. v trap, Maum, L705864, 23.viii.2002; five, Lough Leinapollbruty, L691856, 13.vii.2005; one at Skinner trap, Creggan Lough, L689856, 14.vii.2005; one at actinic trap, Ooghcorragaun, L703875, 14.vii.2005.

6531 *Udea ferrugalis* (Hübner, 1796) RUSTY DOT PEARL. One, Kinnacorra saltmarsh, L720860, 17.vi.2000; one, Portacoolia, L692844, 17.vi.2000; one at m. v. trap, Kill, L698848, 18.vi.2000; one at m. v. trap, Ballytoohy More, L699870, 15.vii.2000; two at m. v. trap, Maum, L705864, 16.vii.2000; three, Kill, L696845, 23.viii.2000; one at m.v. trap, Kill, L698847, 24.viii.2000.

6538 *Udea lutealis* (Hübner, 1809). One at m. v. trap, Maum, L705864, 11.viii.2000; four, Kill, L696845, 23.viii.2000.

6563 *Anania fuscalis* ([Denis and Schiffermüller], 1775). One at actinic trap, Ooghcorragaun, L703875, 14.vii.2005.

6658 *Anania hortulata* (Linnaeus, 1758) SMALL MAGPIE. One at m. v. trap, Ballytoohy More, L699870, 15.vii.2000.

6719 *Nomophila noctuella* ([Denis and Schiffermüller], 1775) RUSH VENEER. Three, Harbour, L712853, 17.vi.2000; one, Lassu Wood, L702867, 17.vi.2000; two, Lassu Wood, L704867, 17.vi.2000; one, Ardal cliffs, L715864, 17.vi.2000; five, Kinnacorra saltmarsh, L720860, 17.vi.2000; four, Portacoolia, L692844, 17.vi.2000; one, Glen, L706849, 17.vi.2000; two at m. v. trap, Kill, L698848, 18.vi.2000; two at m. v. trap, Maum, L705864, 11.viii.2000; two, Creggan Lough, L688859, 23.viii.2000; four at m.v. trap, Kill, L698847, 24.viii.2000; one, Portnakilly, L691844, 24.viii.2000.

LASIOCAMPIDAE (2)
6752 *Lasiocampa quercus* (Linnaeus, 1758) OAK EGGAR/NORTHERN EGGAR. Larva feeding on *Calluna*, Lassu, L703869, 29.v.1999 (photo, R. Cussen).

6767 *Euthrix potatoria* (Linnaeus, 1758) DRINKER. Female at m. v. trap, Fawnglas, L713854, 11.vii.2002; two at Skinner trap, Creggan Lough, L689856, 14.vii.2005.

SATURNIIDAE (1)

6794 *Saturnia pavonia* (Linnaeus, 1758) EMPEROR. Female, Lassu, L703869, 4.vi.2000.

SPHINGIDAE (2)

6824 *Laothoe populi* (Linnaeus, 1758) POPLAR HAWK-MOTH. One at m. v. trap, Kill, L698848, 18.vi.2000; five at m. v. trap, Ballytoohy More, L699870, 15.vii.2000; one at m. v. trap, Fawnglas, L713854, 9.vii.2002; one at actinic trap, Maum, L706864, 14.vii.2005.

6862 *Deilephila elpenor* (Linnaeus, 1758) ELEPHANT HAWK-MOTH. One at m. v. trap, Kill, L698848, 18.vi.2000; one at actinic trap, Ooghcorragaun, L703875, 14.vii.2005.

PIERIDAE (4)

6995 *Pieris brassicae* (Linnaeus, 1758) LARGE WHITE. 2–9, Glen, L7084, 16.vi.1990 (Gordon D'Arcy); 2–9, Fawnglass, L710852, 1.vi.1999 (Dr Drewett); two, Kill, L694845, 24.viii.2000.

6998 *Pieris rapae* (Linnaeus, 1758) SMALL WHITE. 2–9, L7084, Glen, 11.vi.1990 (Gordon D'Arcy); three, Kill, L694845, 24.viii.2000.

7000 *Pieris napi* (Linnaeus, 1758) GREEN-VEINED WHITE. One, Glen, L7084, 15.vi.1990 (Gordon D'Arcy); 2–9, Glen, L7084, 16.vi.1990 (Gordon D'Arcy); 30–100, Fawnglass, L710852, 1.vi.1999 (Dr Drewett); 10–29, Portnakilly, L689846, 1.vi.1999 (Dr Drewett); 10–29, Maum, L700867, 1.vi.1999 (Dr Drewett); one, Toormore, L663847, 28.iv.2000; one, Glen, L708850, 17.vi.2000; one, Harbour, L712853, 17.vi.2000; six, Kill, L693844, 17.vi.2000; larva and 4 ova on *Cardamine pratensis*, Portacoolia, L692844, 17.vi.2000; two, Kill, L696845, 23.viii.2000; one, Portnakilly, L691844, 24.viii.2000; one, Maum Stream, L702864, 22.viii.2002.

7015 *Colias croceus* (Fourcroy, 1785) CLOUDED YELLOW. One, Portnakilly, L691844, 24.viii.2000; one, Strake, L665846, 24.viii.2000; one Kill, L690845, 24.viii.2000.

LYCAENIDAE (3)

7034 *Lycaena phlaeas* (Linnaeus, 1761) SMALL COPPER. One, Kill, L690845, 24.viii.2000. One specimen was also recorded *from an unspecified Clare Island locality* in May 1982 (A. Whilde, per Lloyd, 1984)

7058 *Callophrys rubi* (Linnaeus, 1758) GREEN HAIRSTREAK. One, Lassu Wood, L704867, 21.v.2002 (S. McCormack); two, Lassu Wood, L704867, 8.v.2003.

7163 *Polyommatus icarus* (Rottemburg, 1775) COMMON BLUE. Male, Kill, L693844, 17.vi.2000; male, birchwood, L699870, 15.vii.2000; three males, Portnakilly, L691844, 24.viii.2000; male, lane N. of church, L690845, 24.viii.2000; male, Strake, L665846, 24.viii.2000; two males, Harbour, L715849, 13.vii.2005; male, Ballytoohy, L698867, 13.vii.2005. An additional record is: 2–9, Glen, L7084, 16.vi.1990 (Gordon D'Arcy).

NYMPHALIDAE (9)

7243 *Vanessa atalanta* (Linnaeus, 1758) RED ADMIRAL. One, Portnakilly, L690844, 22.viii.2002; one, Portnakilly, L690843, 13.vii.2005; one on *Succisa pratensis* flowers, Capnagower, L713865, 3.ix.2005.

7245 *Vanessa cardui* (Linnaeus, 1758) PAINTED LADY. Two, Harbour, L715849, 13.vii.2005; one, Kill, L695845, 13.vii.2005.

7248 *Aglais io* (Linnaeus, 1758) PEACOCK. One, Lassu, L703870, 4.vi.2000 (R. Cussen); one, Kill, W698847, 24.viii.2000; three, Portnakilly, L691844, 24.viii.2000; one, Knockmore [ENE of summit, alt. 400m], L671862, 24.viii.2000; one on flowers of *Succisa pratensis*, Capnagower, L717863, 3.ix.2005

7250 *Aglais urticae* (Linnaeus, 1758) SMALL TORTOISESHELL. Four specimens were seen in June 1982 (Lloyd, 1984). One, Capnagower, L716859, 10.vii.2002; two, Kill (shop), L689844, 1.viii.2002; one, Kill, L695845, 13.vii.2005; one, Capnagower, L717863, 3.ix.2005; 12, Capnagower, L717863, 3.ix.2005.

7307 *Pararge aegeria tircis* (Godart, 1821) SPECKLED WOOD. Two, Maum Wood, L705867, 22.viii.2002. An additional record is: one, Glen, L7084, 16.vi.1990 (Gordon D'Arcy).

7309 *Lasiommata megera* (Linnaeus, 1767)
WALL BROWN. One basking in sunshine on *Rubus*, Fawnglass, W712854, 14.v.2003; one, Capnagower, L717863, 3.ix.2005.

7334 *Coenonympha pamphilus* (Linnaeus, 1758)
SMALL HEATH. Five, shore opposite Carrickarelick, L701878, 4.vi.2000 (R. Cussen). Additional records are: one, Glen, L7084, 15.vi.1990 (Gordon D'Arcy); 1029, Glen, L7084, 16.vi.1990 (Gordon D'Arcy).

7344 *Aphantopus hyperantus* (Linnaeus, 1758)
THE RINGLET. Three, birchwood, L699870, 15.vii.2000; one, birchwood, L701870, 15.vii.2000; two, Maum, L705864, 13.vii.2005; one, Kill, L695845, 13.vii.2005.

7350 *Maniola jurtina* (Linnaeus, 1758)
MEADOW BROWN. One, Maum, L704864, 15.vii.2000; one Toormore, L663849, 15.vii.2000; two, Toormore, L661853, 15.vii.2000; three, Toormore, L659854, 15.vii.2000; one, Ooghaniska, L669846, 15.vii.2000; four, Portnakilly, L691844, 24.viii.2000; one, Strake, L665846, 24.viii.2000; three, Kill, L690845, 24.viii.2000; one, Maum, L702865, 22.viii.2002; three, Maum Stream, L702864, 22.viii.2002; two, Maum Stream, L705863 22.viii.2002; one, Lassu Wood, L705865, 22.viii.2002; three, Portnakilly, L690844, 22.viii.2002; two, Maum, L705864, 13.vii.2005; one, Capnagower, L713865, 3.ix.2005; one, Ooghcorragaun, L702875, 3.ix.2005. An additional record is: one, Glen, L7084, 16.vi.1990 (Gordon D'Arcy).

GEOMETRIDAE (41)

7522 *Abraxas grossulariata* (Linnaeus, 1758)
THE MAGPIE. Five at m. v. trap, Ballytoohy More, L699870, 15.vii.2000; one, Harbour, L713854, 15.vii.2000; one at m. v. trap, Lassu Wood, L705867, 22.viii.2002; one at m. v. trap, Fawnglas, L713854, 22.viii.2002; one at actinic trap, Harbour, L714849, 9.vii.2003; one at Skinner trap, Creggan Lough, L689856, 14.vii.2005.

7527 *Lomaspilis marginata* (Linnaeus, 1758)
CLOUDED BORDER. One at m. v. trap, Ballytoohy More, L699870, 15.vii.2000; one at m. v. trap, Fawnglas, L713854, 22.viii.2002.

7596 *Petrophora chlorosata* (Scopoli, 1763)
BROWN SILVER-LINE. Shore opposite Carrickarelick, L701878, 4.vi.2000 (R. Cussen); one at m. v. trap, Ballytoohy, L699869, 7.v.2003.

7613 *Opisthograptis luteolata* (Linnaeus, 1758)
BRIMSTONE MOTH. One at m. v. trap, Ballytoohy More, L699870, 15.vii.2000.

7641 *Selenia dentaria* (Fabricius, 1775)
EARLY THORN. One at m. v. trap, Lassu Wood, L705867, 22.viii.2002; two at m. v. trap, Fawnglass, L712854, 30.iv.2003.

7654 *Crocallis elinguaria* (Linnaeus, 1758)
SCALLOPED OAK. One at m. v. trap, Kinnacorra, L720862, 23.viii.2002.

7777 *Alcis repandata* (Linnaeus, 1758)
MOTTLED BEAUTY. Two at m. v. trap, Ballytoohy More, L699870, 15.vii.2000.

7804 *Ematurga atomaria* (Linnaeus, 1758)
COMMON HEATH. Male, Pigeon Cove, L701878, 28.iv.2000 (R. Cussen); two, Ballytoohy More, L698874, 28.iv.2000.

7824 *Cabera pusaria* (Linnaeus, 1758)
COMMON WHITE WAVE. One at m. v. trap, Maum, L705864, 16.vii.2000; one at m. v. trap, Lassu Wood, L705867, 22.viii.2002.

7826 *Cabera exanthemata* (Scopoli, 1763)
COMMON WAVE. One at actinic trap, Maum, L706864, 14.vii.2005; one at actinic trap, Ooghcorragaun, L703874, 14.vii.2005.

7836 *Campaea margaritata* (Linnaeus, 1767)
LIGHT EMERALD. One at m. v. trap, Ballytoohy More, L699870, 15.vii.2000.

7848 *Gnophos obfuscatus* ([Denis and Schiffermüller], 1775) SCOTCH ANNULET. Two at actinic trap, Maum, L706864, 14.vii.2005; eleven at actinic trap, Ooghcorragaun, L703875, 14.vii.2005; two at actinic trap, Ooghcorragaun, L703875, 3.ix.2005.

7931 *Dyscia fagaria* (Thunberg, 1784)
GREY SCALLOPED BAR. One at m. v. trap, Ballytoohy, L698871, 5.v.2003; one at m. v. trap, Fawnglass, L712854, 6.v.2003; three at m. v. trap,

Ballytoohy, L699869, 7.v.2003; one at m. v. trap, Owenmore Bridge, L680851, 15.v.2003.

7965 *Pseudoterpna pruinata* (Hufnagel, 1767)
GRASS EMERALD. Male at Heath Trap, Capnagower, L712863, 23.vii.2002.

8064 *Scopula immutata* (Linnaeus, 1758)
LESSER CREAM WAVE. One at m. v. trap, Maum, L705864, 16.vii.2000.

8132 *Idaea biselata* (Hufnagel, 1767)
SMALL FAN-FOOTED WAVE. One at m. v. trap, Maum, L705864, 11.viii.2000; five at m. v. trap, Lassu Wood, L705867, 22.viii.2002; one at m. v trap, Maum, L705864, 23.viii.2002; one at m. v. trap, Kinnacorra, L720862, 23.viii.2002.

8161 *Idaea dimidiata* (Hufnagel, 1767)
SINGLE-DOTTED WAVE. One at m. v. trap, Kinnacorra, L720862, 23.viii.2002.

8184 *Idaea aversata* (Linnaeus, 1758) RIBAND WAVE. One at m. v. trap, Maum, L705864, 16.vii.2000; one at actinic trap, Maum, L706864, 14.vii.2005.

8239 *Scotopteryx chenopodiata* (Linnaeus, 1758)
SHADED BROAD-BAR. Two at actinic trap, Ooghcorragaun, L703875, 14.vii.2005.

8245 *Orthonama vittata* (Borkhausen, 1794)
OBLIQUE CARPET. One at m. v. trap, Fawnglas, L713854, 22.viii.2002; three at actinic trap, Maum Marsh, L702863, 22.viii.2002.

8249 *Xanthorhoe designata* (Hufnagel, 1767)
FLAME CARPET. One at m. v. trap, Maum, L705864, 11.viii.2000; one, Kill, L696845, 23.viii.2000; one at m. v trap, Maum, L705864, 23.viii.2002; one at m. v. trap, harbour, L714849, 9.vii.2003.

8253 *Xanthorhoe ferrugata* (Clerck, 1759)
DARK-BARRED TWIN-SPOT CARPET. One at m. v. trap, Lassu Wood, L705867, 22.viii.2002; one at m. v trap, Maum, L705864, 23.viii.2002; one at m. v. trap, Fawnglass, L712854, 30.iv.2003; one at m. v. trap, Ballytoohy, L699871, 5.v.2003; one at m. v. trap, Owenmore Bridge, L680851, 15.v.2003.

8275 *Epirrhoe alternata* (Müller, 1764)
COMMON CARPET. One, Ooghaniska, L669846, 15.vii.2000; one at actinic trap, Maum Marsh, L702863, 22.viii.2002; one at m. v. trap, Kinnacorra, L720862, 23.viii.2002; two at m. v. trap, Fawnglass, L712854, 30.iv.2003; one, Portlea, L705867, 8.vii.2003.

8289 *Camptogramma bilineata* (Linnaeus, 1758)
YELLOW SHELL. One, Glen, L711851, 15.vii.2000; eight, Leckascannelmore, L711865, 8.vii.2003; one at actinic trap, Ooghcorragaun, L703875, 14.vii.2005.

8332 *Eulithis populata* (Linnaeus, 1767)
NORTHERN SPINACH. One at m. v. trap, Maum, L705864, 16.vii.2000; one at m. v. trap, Lassu Wood, L705867, 22.viii.2002; one at m. v. trap, Fawnglas, L713854, 22.viii.2002; one at actinic trap, Ooghcorragaun, L703875, 14.vii.2005.

8338 *Ecliptopera silaceata* ([Denis and Schiffermüller], 1775) SMALL PHOENIX. One at m. v. trap, Maum, L705864, 11.viii.2000; five at m. v. trap, Kinnacorra, L720862, 23.viii.2002; three at m. v trap, Maum, L705864, 23.viii.2002; one at m. v. trap, Fawnglass, L712854, 30.iv.2003; one at actinic trap, Maum, L706864, 14.vii.2005.

8348 *Dysstroma truncata* (Hufnagel, 1767)
COMMON MARBLED CARPET. Two, Maum birchwood, L704867, 17.vi.2000; one, Glen, L710851, 17.vi.2000; one at m. v. trap, Lassu Wood, L705867, 22.viii.2002; two at m. v trap, Maum, L705864, 23.viii.2002.

8356 *Thera obeliscata* (Hübner, 1787)
GREY PINE CARPET. Four at m. v. trap, Maum, L706864, 29.ix.2002; one at actinic trap, Maum, L706864, 14.vii.2005.

8391 *Hydriomena furcata* (Thunberg, 1784)
JULY HIGHFLYER. One at m. v. trap, Ballytoohy More, L699870, 15.vii.2000; one at m. v. trap, Maum, L705864, 16.vii.2000; thirteen at m. v. trap, Lassu Wood, L705867, 22.viii.2002; one at actinic trap, Ballytoohy, L699869, 3.ix.2005.

8393 *Hydriomena ruberata* (Freyer, 1831). One at m. v. trap, Fawnglass, L712854, 30.iv.2003; one at m. v. trap, Fawnglass, L712854, 1.v.2003; one at m. v. trap, Ballytoohy, L699869, 7.v.2003.

8436 *Euphyia unangulata* (Haworth, 1809)
SHARP-ANGLED CARPET. One at actinic trap, Ooghcorragaun, L703875, 14.vii.2005.

8456 *Perizoma alchemillata* (Linnaeus, 1758) SMALL RIVULET. Two at m. v. trap, Ballytoohy More, L699870, 15.vii.2000; two at m. v. trap, Maum, L705864, 16.vii.2000; one at actinic trap, Maum, L7068645, 14.vii.2005.

8462 *Perizoma blandiata* ([Denis and Schiffermüller], 1775) PRETTY PINION. One at m. v. trap, Ballytoohy More, L699870, 15.vii.2000; one at m. v. trap, Maum, L705864, 16.vii.2000; fourteen at actinic trap, Maum, L706864, 14.vii.2005; three at actinic trap, Ooghcorragaun, L703875, 14.vii.2005.

8465 *Mesotype didymata* (Linnaeus, 1758) TWIN-SPOT CARPET. One, NW slopes of Knockmore [circa 300m], L666863, 24.viii.2000 (what appeared to be several more flying lower down north-facing cliff, out of sunlight).

8484 *Eupithecia pulchellata* Stephens, 1831 FOXGLOVE PUG. One f. *hebudium* Sheldon, 1899 (dissected) at m. v. trap, Kill, L698848, 18.vi.2000; one at m. v. trap, Ballytoohy More, L699870, 15.vii.2000; female f. *hebudium* (dissected) at Heath Trap, Capnagower, L712863, 23.vii.2002.

8509 *Eupithecia centaureata* ([Denis and Schiffermüller], 1775) LIME-SPECK PUG. One at actinic trap, Creggan Lough, L690857, 9.vii.2003; one at actinic trap, Ooghcorragaun, L703875, 14.vii.2005.

8527 *Eupithecia absinthiata* (Clerck, 1759) WORMWOOD PUG. Female taken at light (dissected), Harbour, L715855, 23.viii.2000; female at m. v. trap (dissected), Kill, L698847, 24.viii.2000; larvae on *Solidago virgaurea*, Capnagower, L712857, 28.ix.2002; female taken at m. v. trap (dissected), Fawnglas, L713854, 22.viii.2002.

8537 *Eupithecia subfuscata* (Haworth, 1809) GREY PUG. Male at m. v. trap (dissected), Fawnglass, L713854, 18.v.2003.

8570 *Eupithecia nanata* (Hübner, [1813]) NARROW-WINGED PUG. Two at m. v. trap, Ballytoohy More, L699870, 15.vii.2000; male at Heath Trap, Capnagower, L712863, 23.vii.2002; female at m. v. trap, Fawnglas, L712863, 23.vii.2002; one at m. v. trap, Lassu Wood, L705867, 22.viii.2002; two at m. v trap, Maum, L705864, 23.viii.2002; three at m. v. trap, Kinnacorra, L720862, 23.viii.2002; one at m. v. trap, Fawnglass, L712854, 6.v.2003; one at domestic light, Fawnglass, L712854, 6.v.2003; one at actinic trap, Maum, L706964, 14.vii.2005; four at actinic trap, Ooghcorragaun, L703875, 14.vii.2005.

8599 *Gymnoscelis rufifasciata* (Haworth, 1809) DOUBLE-STRIPED PUG. One at light, hotel, L716855, 28.iv.2000; three males at Heath Trap, Capnagower, L712863, 23.vii.2002; one at actinic trap, Maum, L706864, 14.vii.2005.

8681 *Acasis viretata* (Hübner, [1799]) YELLOW-BARRED BRINDLE. One at m. v. trap, Maum, L705864, 10.viii.2000.

NOTODONTIDAE (4)

8704 *Cerura vinula* (Linnaeus, 1758) PUSS MOTH. One at Heath Trap, Toormore, L663847, 6.vii.2002; one at actinic trap, Ooghcorragaun, L703875, 14.vii.2005.

8719 *Notodonta ziczac* (Linnaeus, 1758) PEBBLE PROMINENT. One at m. v. trap, Ballytoohy More, L699870, 15.vii.2000; one at m. v. trap, Lassu Wood, L705867, 22.viii.2002 (as *Eligmodonta ziczac* in 2000 report); one at actinic trap, Ooghcorragaun, L703875, 14.vii.2005.

8728, *Pheosia gnoma* (Fabricius, 1776) LESSER SWALLOW PROMINENT. Two at m. v. trap, Ballytoohy More, L699870, 15.vii.2000; two at m. v. trap, Lassu Wood, L705867, 22.viii.2002.

8750 *Phalera bucephala* (Linnaeus, 1758) BUFF-TIP. One at actinic trap, Maum, L706864, 14.vii.2005; one at Skinner trap, Creggan Lough, L689856, 14.vii.2005.

NOCTUIDAE (72)

8780 *Subacronicta megacephala* ([Denis and Schiffermüller], 1775) POPLAR GREY. One, Craigmore, L674842, 20.v.1997.

8787 *Acronicta rumicis* (Linnaeus, 1758) KNOT GRASS. One at m. v. trap, Maum, L705864, 16.vii.2000; male at m. v. trap (dissected),

Fawnglass, L713854, 11.vii.2002; one at m. v. trap, Lassu Wood, L705867, 22.viii.2002.

8863 *Hypenodes humidalis* Doubleday, 1850
MARSH OBLIQUE-BARRED. Female at m. v. trap (dissected), Lassu Wood, L705867, 22.viii.2002; one at actinic trap, Maum, L706864, 14.vii.2005.

8994 *Hypena proboscidalis* (Linnaeus, 1758) SNOUT. Four at m. v. trap, Ballytoohy More, L699870, 15.vii.2000.

9008 *Rivula sericealis* (Scopoli, 1763) STRAW DOT. One, birchwood, L699870, 15.vii.2000; one, birchwood, L701870, 15.vii.2000; four at m. v. trap, Maum, L705864, 16.vii.2000; one, Portlea, L706866, 10.viii.2000; one at m. v. trap, Lassu Wood, L705867, 22.viii.2002; two at m. v. trap, Fawnglas, L713854, 22.viii.2002; six at actinic trap, Maum, L706864, 14.vii.2005.

9045 *Diachrysia chrysitis* (Linnaeus, 1758) BURNISHED BRASS. Two at m. v. trap, Maum, L705864, 16.vii.2000; one at m. v. trap, Fawnglas, L713854, 9.vii.2002; three males at m. v. trap, Fawnglas, L712863, 23.vii.2002; one at m. v trap, Maum, L705864, 23.viii.2002; two at m. v. trap, Fawnglas, L713854, 22.viii.2002; one at m. v. trap, Kinnacorra, L720862, 23.viii.2002.

9053 *Plusia festucae* (Linnaeus, 1758) GOLD SPOT. Four at m. v. trap (male dissected), Maum, L705864, 16.vii.2000; male at m. v. trap (dissected), Maum, L705864, 11.viii.2000; one at m. v. trap, Fawnglas, L713854, 9.vii.2002; female, 'Bogbean', L654844, 24.vii.2002; one at m. v trap, Maum, L705864, 23.viii.2002; one at Skinner trap, Creggan Lough, L689856, 14.vii.2005.

9056 *Autographa gamma* (Linnaeus, 1758) SILVER Y. One, Lackwee, L650842, 24.viii.2000; one at m. v. trap, Fawnglas, L713854, 22.viii.2002; one at m. v. trap, Maum, L706864, 29.ix.2002.

9059 *Autographa pulchrina* (Haworth, 1809) BEAUTIFUL GOLDEN Y. One at m. v. trap, Fawnglas, L713854, 9.vii.2002; male at m. v. trap, Fawnglas, L712863, 23.vii.2002.

9061 *Autographa jota* (Linnaeus, 1758) PLAIN GOLDEN Y. One at m. v. trap, Fawnglas, L713854, 9.vii.2002.

9091 *Abrostola tripartita* (Hufnagel, 1766) THE SPECTACLE. One at m. v. trap, Ballytoohy More, L699870, 15.vii.2000; two at m. v. trap, Maum, L705864, 16.vii.2000; one at Heath Trap, Toormore, L663847, 6.vii.2002; male at Heath Trap, Capnagower, L712863, 23.vii.2002; two males at m. v. trap, Fawnglas, L712863, 23.vii.2002; one at m. v. trap, Fawnglass, L712854, 30.iv.2003.

9093 *Abrostola triplasia* (Linnaeus, 1758) DARK SPECTACLE. One at m. v. trap, Ballytoohy More, L699870, 15.vii.2000; one at m. v. trap, Fawnglas, L713854, 9.vii.2002.

9199 *Cucullia umbratica* (Linnaeus, 1758) SHARK. One at m. v. trap, Fawnglas, L713854, 9.vii.2002.

9407 *Stilbia anomala* (Haworth, 1812) ANOMALOUS. One at m. v trap, Maum, L705864, 23.viii.2002; one at light, Maum, L705864, 27.ix.2002; two at actinic trap, Ooghcorragaun, L703875, 3.ix.2005.

9433 *Caradrina clavipalpis* (Scopoli, 1763) PALE MOTTLED WILLOW. One, Fawnglas, L714852, 6.vii.2002; one at m. v. trap, Lassu Wood, L705867, 22.viii.2002; one at actinic trap, Ooghcorragaun, L703875, 14.vii.2005.

9449 *Hoplodrina octogenaria* (Goeze, 1781) THE UNCERTAIN. Three at m. v. trap, Kinnacorra, L720862, 23.viii.2002.

9450 *Hoplodrina blanda* ([Denis and Schiffermüller], 1775) THE RUSTIC. Three at m. v. trap (male dissected), Ballytoohy More, L699870, 15.vii.2000; ten at m. v. trap, Maum, L705864, 16.vii.2000; two males at Heath Trap, Capnagower, L712863, 23.vii.2002; female at actinic trap (dissected), Maum Marsh, L702863, 22.viii.2002; three at m. v trap, Maum, L705864, 23.viii.2002; two at m. v. trap, Kinnacorra, L720862, 23.viii.2002; one at actinic trap, Maum, L706864, 14.vii.2005; two at Skinner trap, Greggan Lough, L689856, 14.vii.2005; two at actinic trap, Ooghcorragaun, L703875, 14.vii.2005.

9503 *Euplexia lucipara* (Linnaeus, 1758) SMALL ANGLE SHADES. One at m. v. trap, Kill, L698848, 18.vi.2000; one at m. v. trap, Ballytoohy More, L699870, 15.vii.2000.

9505 *Phlogophora meticulosa* (Linnaeus, 1758) ANGLE SHADES. One at m. v trap, Maum, L705864, 23.viii.2002; four at m. v. trap, Toormore, L661844, 28.ix.2002; two at m. v. trap, Maum, L706864, 29.ix.2002; two at m. v. trap, Fawnglas, L712854, 30.iv.2003; two at m. v. trap, Ballytoohy, L698871, 5.v.2003; one at m. v. trap, Fawnglass, L712854, 6.v.2003; three at m. v. trap, Ballytoohy, L699869, 7.v.2003; four at m. v. trap, Owenmore Bridge, L680851, 15.v.2003; one at actinic trap, Ballytoohy, L699869, 3.ix.2005.

9556 *Xanthia togata* (Esper, 1788) PINK-BARRED SALLOW. Two at actinic trap, Ballytoohy, L699869, 3.ix.2005.

9591 *Agrochola lunosa* (Haworth, 1809) LUNAR UNDERWING. One at light, Bayview Hotel (Capnagower), L716856, 27.ix.2002; five at m. v. trap, Toormore, L661844, 28.ix.2002; one at m. v. trap, Maum, L706864, 29.ix.2002.

9650 *Aporophyla lueneburgensis* (Freyer, 1848) NORTHERN DEEP-BROWN DART. Male taken at m. v. trap, Lassu Wood, L705867, 22.viii.2002.

9741 *Mniotype adusta* (Esper, 1790) DARK BROCADE. One at actinic trap, Maum, L706864, 14.v.2005.

9748 *Apamea monoglypha* (Hufnagel, 1766) DARK ARCHES. 26 at m. v. trap, Ballytoohy More, L699870, 15.vii.2000; seven at m. v. trap, Maum, L705864, 16.vii.2000; one at m. v. trap, Maum, L705864, 10.viii.2000; three at m. v. trap, Maum, L705864, 11.viii.2000; five at m. v. trap, Kill, L698847, 24.viii.2000; one at m. v. trap, Fawnglas, L713854, 9.vii.2002; six males and two females at Heath Trap, Capnagower, L712863, 23.vii.2002; 48 at m. v. trap, Lassu Wood, L705867, 22.viii.2002; eight at actinic trap, Maum Marsh, L702863, 22.viii.2002; fifteen at m. v. trap, Fawnglas, L713854, 22.viii.2002; 26 at m v trap, Maum, L705864, 23.viii.2002; four at m. v. trap, Kinnacorra, L720862, 23.viii.2002; four at m. v. trap, Toormore, L661844, 28.ix.2002; two at m. v. trap, Maum, L706864, 29.ix.2002; two at actinic trap, Creggan Lough, L690857, 9.vii.2003; one at actinic trap, Harbour cliffs, L714849, 9.vii.2003; nine at actinic trap, Maum, L706864, 14.vii.2005; twelve at Skinner trap, Creggan Lough, L689856, 14.vii.2005; two at actinic trap, Ooghcorragaun, L703875, 14.vii.2005; two at actinic trap, Ballytoohy, L699869, 3.ix.2005; one at actinic trap, Ooghcorragaun, L703875, 3.ix.2005.

9752 *Apamea lithoxylaea* ([Denis and Schiffermüller], 1775) LIGHT ARCHES. One at m. v. trap, Maum, L705864, 16.vii.2000; two males at Heath Trap, Capnagower, L712863, 23.vii.2002.

9755 *Apamea crenata* (Hufnagel, 1766) CLOUDED-BORDERED BRINDLE. Male at m. v. trap, Fawnglas, L713854, 9.vii.2002.

9766 *Apamea remissa* (Hübner, [1809]) DUSKY BROCADE. One at m. v. trap, Kill, L698848, 18.vi.2000; three at m. v. trap, Ballytoohy More, L699870, 15.vii.2000; eleven at m. v. trap, Maum, L705864, 16.vii.2000; male (dissected) at Heath Trap, Toormore, L663847, 6.vii.2002; male at Heath Trap, Toormore, L663847, 6.vii.2002; female at m. v. trap, Fawnglas, L713854, 9.vii.2002; two males at Heath Trap, Capnagower, L712863, 23.vii.2002.

9784 *Oligia fasciuncula* (Haworth, 1809) MIDDLE-BARRED MINOR. One at m. v. trap, Maum, L705864, 16.vii.2000.

9787 *Litoligia literosa* (Haworth, 1809) ROSY MINOR. One at m. v trap, Maum, L705864, 23.viii.2002.

9789 *Mesapamea secalis* (Linnaeus, 1758) COMMON RUSTIC. Male (dissected) at Heath Trap, Capnagower, L712863, 23.vii.2002 (similar specimens were seen in numbers at light-traps at Maum, Fawnglas, Lassu Wood, Capnagower, and Kinnacorra in 2000 and 2002, but not retained).

9790 *Mesapamea secalella* Remm, 1983 LESSER COMMON RUSTIC. Two at m. v. trap (male dissected), Maum, L705864, 11.viii.2000 (similar specimens were seen at light-traps at Maum Marsh and Lassu Wood in 2002, but not retained).

9795 *Photedes minima* (Haworth, 1809) SMALL DOTTED BUFF. Seven at actinic trap, Maum Marsh, L702863, 22.viii.2002.

9801 *Luperina testacea* ([Denis and Schiffermüller], 1775) FLOUNCED RUSTIC. One at light, Harbour,

L715855, 23.viii.2000; two at m. v. trap, Kill, L698847, 24.viii.2000; four at m. v. trap, Fawnglas, L713854, 22.viii.2002; one at m. v. trap, Kinnacorra, L720862, 23.viii.2002; one at m. v. trap, Toormore, L661844, 28.ix.2002; one at actinic trap, Ballytoohy, L699869, 3.ix.2005.

9831 *Amphipoea lucens* (Freyer, 1845) LARGE EAR. Female at m. v. trap (dissected), Maum, L705864, 11.viii.2000; two at m. v. trap (two taken – male and female dissected, 15 *Amphipoea* at trap), Lassu Wood, L705867, 22.viii.2002 (three possible further specimens at light at Maum, 29.ix.2002); one at actinic trap, Ballytoohy, L699869, 3.ix 2005 (female, dissected, two other *Amphipoea* at trap).

9832 *Amphipoea crinanensis* (Burrows, 1908) CRINAN EAR. Male at m. v. trap (dissected), Kill, L698847, 24.viii.2000. [Other records of *Amphipoea* – dissection is required for identification: Fifteen at m. v. trap, Fawnglas, L713854, 22.viii.2002; five at m. v trap, Maum, L705864, 23.viii.2002; two at m. v. trap, Kinnacorra, L720862, 23.viii.2002; one at actinic trap, Maum Marsh, L702863, 22.viii.2002.]

9834 *Hydraecia micacea* (Esper, 1789) ROSY RUSTIC. Seven at m. v. trap, Kill, L698847, 24.viii.2000; four at m. v. trap, Fawnglas, L713854, 22.viii.2002; one at m. v trap, Maum, L705864, 23.viii.2002; one at m. v. trap, Toormore, L661844, 28.ix.2002; two at actinic trap, Ballytoohy, L699869, 3.ix.2005; one at actinic trap, Kinnacorra, L720859, 3.ix.2005.

9857 *Helotropha leucostigma* (Hübner, [1808]) THE CRESCENT. One at m. v. trap, Maum, L705864, 11.viii.2000; one at m. v. trap, Lassu Wood, L705867, 22.viii.2002; one at m. v. trap, Fawnglas, L713854, 22.viii.2002; one at m. v trap, Maum, L705864, 23.viii.2002; one at m. v. trap, Kinnacorra, L720862, 23.viii.2002.

9876 *Denticucullus pygmina* (Haworth, 1809) SMALL WAINSCOT. One at m. v. trap, Maum, L705864, 11.viii.2000; seven at m. v. trap, Fawnglas, L713854, 22.viii.2002; four at m. v. trap, Kinnacorra, L720862, 23.viii.2002; one at m. v trap, Maum, L705864, 23.viii.2002; one at m. v. trap, Maum, L706864, 29.ix.2002; thirteen at actinic trap, Ballytoohy, L699869, 3.ix.2005.

9917 *Lacanobia oleracea* (Linnaeus, 1758) BRIGHT-LINE BROWN-EYE. Two at m. v. trap, Kill, L698848, 18.vi.2000; 18 at m. v. trap, Ballytoohy More, L699870, 15.vii.2000; 26 at m. v. trap, Maum, L705864, 16.vii.2000; two males at Heath Trap, Capnagower, L712863, 23.vii.2002; one at m. v. trap, Lassu Wood, L705867, 22.viii.2002; four at m. v. trap, Fawnglas, L713854, 22.viii.2002; four at m. v. trap, Kinnacorra, L720862, 23.viii.2002; nine at m. v trap, Maum, L705864, 23.viii.2002; one at actinic trap, Maum, L706864, 14.vii.2005; 29 at Skinner trap, Creggan Lough, L689856, 14.vii.2005.

9925 *Hada plebeja* (Linnaeus, 1761) SHEARS. Three at m. v. trap, Kill, L698848, 18.vi.2000; two at m. v. trap, Maum, L705864, 16.vii.2000; one at Heath Trap, Ruinte River (Strake), L686846, 7.vii.2002; one at actinic trap, Ooghcorragaun, L703875, 14.vii.2005.

9955 *Sideridis rivularis* (Fabricius, 1775) CAMPION. One at actinic trap, Harbour cliffs, L714849, 9.vii.2003; one at actinic trap, Mau, L699869, 3.ix.2005.

9957 *Hadena perplexa capsophila* (Duponchel, 1842) POD LOVER. One, at m. v. trap, Kill, L698848, 18.vi.2000; one at m. v. trap, Maum, L705864, 16.vii.2000; male (dissected) at Heath Trap, Ruinte River (Strake), L686846, 7.vii.2002; female (dissected) at m. v. trap, Fawnglas, L712863, 23.vii.2002; one at actinic trap, Ooghcorragaun, L703875, 14.vii.2005.

9984 *Melanchra persicariae* (Linnaeus, 1761) DOT MOTH. One at m. v. trap, Kill, L698848, 18.vi.2000; female at m. v. trap, Fawnglas, L712854, 9.vii.2002; one at m. v trap, Maum, L705864, 23.viii.2002.

9985 *Ceramica pisi* (Linnaeus, 1758) BROOM MOTH. One at m. v. trap, Kill, L698848, 18.vi.2000; one at Skinner trap, Creggan Lough, L689856, 14.vii.2005.

9987 *Mamestra brassicae* (Linnaeus, 1758) CABBAGE MOTH. One at m. v. trap, Maum, L705864, 16.vii.2000.

10000 *Mythimna conigera* ([Denis and Schiffermüller], 1775) BROWN-LINE BRIGHT-EYE. One at m. v. trap, Fawnglas, L713854, 22.viii.2002.

10006 *Mythimna impura* (Hübner, 1808) SMOKY WAINSCOT. Three at m. v. trap, Maum, L705864, 16.vii.2000; two at m. v. trap, Maum, L705864, 11.viii.2000; four at m. v. trap, Fawnglas, L713854, 22.viii.2002; one at actinic trap, Maum Marsh, L702863, 22.viii.2002; two at m. v trap, Maum, L705864, 23.viii.2002; one at actinic trap, Maum, L706864, 14.vii.2005; one at actinic trap, Ballytoohy, L699869, 3.ix.2005.

10035 *Mythimna unipuncta* (Haworth, 1809) WHITE-SPECK. Two at m. v. trap (male taken), Toormore, L661844, 28.ix.2002.

10037 *Orthosia incerta* (Hufnagel, 1766) CLOUDED DRAB. One at m. v. trap, Fawnglass, L712854, 30.iv.2003; two at m. v. trap, Ballytoohy. L699869, 7.v.2003.

10038 *Orthosia gothica* (Linnaeus, 1758) HEBREW CHARACTER. Nine at m. v. trap, Fawnglass, L712854, 30.iv.2003; one at m. v. trap, Fawnglass, L712854, 1.v.2003; one at m. v. trap, Fawnglass, L712854, 2.v.2003; one at m. v. trap, Fawnglass, L712854, 6.v.2003; five at m. v. trap, Ballytoohy, L699869, 7.v.2003.

10044 *Orthosia cerasi* (Fabricius, 1775) COMMON QUAKER. One at m. v. trap, Ballytoohy, L699869, 7.v.2003.

10048 *Orthosia gracilis* ([Denis and Schiffermüller], 1775) POWDERED QUAKER. [At or near] Maum, L705864, iv-v.1994 (A.A. Myers).

10062 *Cerapteryx graminis* (Linnaeus, 1758) ANTLER MOTH. Eighteen at m. v. trap, Maum, L705864, 11.viii.2000; six at m. v. trap, (male taken), Kill, L698847, 24.viii.2000; six at m. v. trap, Kinnacorra, L720862, 23.viii.2002; three at m. v trap, Maum, L705864, 23.viii.2002; one at actinic trap, Ooghcorragaun, L703875, 3.ix.2005.

10082 *Axylia putris* (Linnaeus, 1761) FLAME. Seven at m. v. trap, Kill, L698848, 18.vi.2000; eleven at m. v. trap, Ballytoohy More, L699870, 15.vii.2000; twelve at m. v. trap, Maum, L705864, 16.vii.2000; one at Heath Trap, Toormore, L663847, 6.vii.2002; one at actinic trap, Harbour cliffs, L714849, 9.vii.2003; one at actinic trap, Maum, L706864, 14.vii.2005; one at actinic trap, Ooghcorragaun, L703875, 14.vii.2005.

10086 *Ochropleura plecta* (Linnaeus, 1761) FLAME-SHOULDER. Three at m. v. trap, Kill, L698848, 18.vi.2000; two at m. v. trap, Ballytoohy More, L699870, 15.vii.2000; six at m. v. trap, Maum, L705864, 16.vii.2000; two at m. v. trap, Kill, L698847, 24.viii.2000; two males at Heath Trap, Toormore, L663847, 6.vii.2002; one at m. v. trap, Fawnglas, L713854, 9.vii.2002; two males at Heath Trap, Capnagower, L712863, 23.vii.2002; one at actinic trap, Maum Marsh, L702863, 22.viii.2002; one at m. v. trap, Fawnglass, L712854, 1.v.2003; one at m. v. trap, Fawnglass, L712854, 2.v.2003; one at m. v. trap, Fawnglass, L712854, 6.v.2003; one at m. v. trap, Ballytoohy, L699869, 7.v.2003; one at m. v. trap, Owenmore Bridge, L680851, 9.v.2003; one at m. v. trap, Owenmore Bridge, L680851, 15.v.2003; two at actinic trap, Creggan Lough, L690587, 9.vii.2003; one at actinic trap, Harbour cliffs, L714849, 9.vii.2003; two at m. v. trap, Maum, L706864, 14.vii.2005; one at actinic trap, Ooghcorragaun, L703875, 14.vii.2005.

10089 *Diarsia mendica* (Fabricius, 1775) INGRAILED CLAY. One at m. v. trap, Maum, L705864, 16.vii.2000.

10093 *Diarsia rubi* (Vieweg, 1790) SMALL SQUARE-SPOT. One at m. v. trap, Kill, L698847, 24.viii.2000; two at actinic trap, Maum Marsh, L702863, 22.viii.2002; two at m. v. trap, Fawnglas, L713854, 22.viii.2002; one at m. v. trap, Kinnacorra, L720862, 23.viii.2002; one at m. v. trap, Ballytoohy, L699869, 7.v.2003; one at m. v. trap, Owenmore Bridge, L680851, 15.v.2003; four at m. v. trap, Fawnglass, L712854, 18.v.2003; one at actinic trap, Ballytoohy, L699869, 3.ix.2005; one at actinic trap, Kinnacorra, L720859, 3.ix.2005; one at actinic trap, Ooghcorragaun, L703875, 3.ix.2005.

10096 *Noctua pronuba* (Linnaeus, 1758) LARGE YELLOW UNDERWING. Two at m. v. trap, Ballytoohy More, L699870, 15.vii.2000; three at m. v. trap, Maum, L705864, 16.vii.2000; nine at m. v. trap, Maum, L705864, 11.viii.2000; one at m. v. trap, Kill, L698847, 24.viii.2000; male and female at Heath Trap, Capnagower, L712863, 23.vii.2002; six at m. v. trap, Lassu Wood, L705867, 22.viii.2002; one at m. v. trap, Fawnglas, L713854, 22.viii.2002; one at actinic trap, Maum Marsh, L702863, 22.viii.2002; four at m. v trap, Maum, L705864, 23.viii.2002; one at m. v. trap, Toormore, L661844, 28.ix.2002; two at actinic trap, Maum, L706864, 14.vii.2005; one at actinic trap, Maum, L689856, 14.vii.2005.

10099 *Noctua comes* Hübner, [1813]
LESSER YELLOW UNDERWING. Four at m. v trap, Maum, L705864, 23.vii.2002; three at m. v. trap, Lassu Wood, L705867, 22.viii.2002; four at actinic trap, Maum Marsh, L702863, 22.viii.2002; four at m. v. trap, Fawnglas, L713854, 22.viii.2002; male and female at Heath Trap, Capnagower, L712863, 23.viii.2002; two at m. v. trap, Kinnacorra, L720862, 23.viii.2002; two at m. v. trap, Toormore, L661844, 28.ix.2002; two at m. v. trap, Maum, L706864, 29.ix.2002.

10103 *Noctua janthe* (Borkhausen, 1792)
LESSER BROAD-BORDERED YELLOW UNDERWING. Six at m. v. trap, Maum, L705864, 11.viii.2000; nine at m. v. trap, Lassu Wood, L705867, 22.viii.2002; one at m. v. trap, Fawnglas, L713854, 22.viii.2002; two at m. v trap, Maum, L705864, 23.viii.2002.

10105 *Noctua interjecta* Hübner, 1803
LEAST YELLOW UNDERWING. One at m. v. trap, Maum, L705864, 11.viii.2000.

10113 *Lycophotia porphyrea* ([Denis and Schiffermüller], 1775) TRUE LOVER'S KNOT. Six at m. v. trap, Ballytoohy More, L699870, 15.vii.2000; ten at m. v. trap, Maum, L705864, 16.vii.2000; one at m. v. trap, Maum, L705864, 11.viii.2000; two males and female (dissected) at Heath Trap, Capnagower, L712863, 23.vii.2002; one at m. v. trap, Fawnglas, L713854, 22.viii.2002; seven at m. v. trap, Lassu Wood, L705867, 22.viii.2002; two at actinic trap, Maum Marsh, L702863, 22.viii.2002; six at m. v trap, Maum, L705864, 23.viii.2002; two at m. v. trap, Kinnacorra, L720862, 23.viii.2002; two at actinic trap, Creggan Lough, L690857, 9.vii.2003; 26 at actinic trap, Maum, 14.vii.2005; three at actinic trap, Ooghcorragaun, L703875, 14.vii.2005.

10153 *Standfussiana lucernea* (Linnaeus, 1758)
NORTHERN RUSTIC. Two at m. v. trap (male taken), Ballytoohy More, L699870, 15.vii.2000; one at Heath Trap, Toormore, L663847, 6.vii.2002; two males at Heath Trap, Capnagower, L712863, 23.vii.2002; two at actinic trap, Creggan Lough, L690857, 9.vii.2003; one at actinic trap, Harbour cliffs, L714849, 9.vii.2003; one at actinic trap, Maum, L706864, 14.vii.2005; one at actinic trap, Ooghcorragaun, 14.vii.2005.

10204 *Xestia baja* ([Denis and Schiffermüller], 1775) DOTTED CLAY. One at m. v. trap, Ballytoohy More, L699870, 15.vii.2000; male at Heath Trap, Capnagower, L712863, 23.vii.2002; one at m. v. trap, Fawnglas, L713854, 22.viii.2002; one at m. v. trap, Lassu Wood, L705867, 22.viii.2002; one at actinic trap, Maum, L706864, 14.vii.2005.

10207 *Xestia castanea* (Esper, 1798)
NEGLECTED RUSTIC. Two at m. v. trap (male taken), Lassu Wood, L705867, 22.viii.200; two at actinic trap, Ballytoohy, L699869, 3.ix.2005.

10212 *Xestia xanthographa* ([Denis and Schiffermüller], 1775) SQUARE-SPOT RUSTIC. Four at m. v. trap, Maum, L705864, 10.viii.2000; fifteen at m. v. trap, Maum, L705864, 1.viii.2000; eight at m. v. trap, Kill, L698847, 24.viii.2000; eighteen at m. v. trap, Fawnglas, L713854, 22.viii.2002; four at m. v. trap, Lassu Wood, L705867, 22.viii.2002; fifteen at m. v trap, Maum, L705864, 23.viii.2002; one at m. v. trap, Kinnacorra, L720862, 23.viii.2002; four at m. v. trap, Toormore, L661844, 28.ix.2002; two at m. v. trap, Maum, L706864, 29.ix.2002; four at actinic trap, Ballytoohy, L699869, 3.ix.2005; two at actinic trap, Ooghcorragaun, L703875, 3.ix.2005.

10224 *Cerastis rubricosa* ([Denis and Schiffermüller], 1775) RED CHESTNUT. Two at m. v. trap, Fawnglass, L712854, 30.ix.2003; one at m. v. trap, Fawnglass, L712854, 2.v.2003; one at m. v. trap, Ballytoohy, L688871, 5.v.2003; one at m. v. trap, Fawnglass, L712854, 6.v.2003; one at m. v. trap, Ballytoohy, L699869, 7.v.2003.

10280 *Euxoa tritici* (Linnaeus, 1761)
WHITE-LINE DART. Female at m. v. trap (dissected), Fawnglas, L713854, 22.viii.2002.

10346 *Agrotis ipsilon* (Hufnagel, 1766)
DARK SWORD-GRASS. Two at m. v. trap (male taken), Toormore, L661844, 28.ix.2002.

10347 *Agrotis trux* (Hübner, 1824).
CRESCENT DART. Male at Heath Trap, Capnagower, L712863, 23.vii.2002; male at m. v. trap, Fawnglas, L712863, 23.vii.2002; one at actinic trap, Harbour cliffs, L714849, 9.vii.2003.

10348 *Agrotis exclamationis* (Linnaeus, 1758)
Heart and Dart. 16 at m. v. trap, Kill, L698848, 18.vi.2000; three at m. v. trap, Maum, L705864, 16.vii.2000; male (dissected) at Heath Trap, Ruinte River (Strake), L686846, 7.vii.2002.

10356 *Agrotis vestigialis* (Hufnagel, 1766)
ARCHER'S DART. Male m. v. trap, Fawnglas, L712863, 23.vii.2002.

EREBIDAE (6)
[this includes the former ARCTIIDAE and LYMANTRIIDAE]
10397 *Orgyia antiqua* (Linnaeus, 1758) The Vapourer. Vacated cocoon on *Crataegus monogyna*, Maum, L706864, 28.ix.2002; vacated cocoon on *Betula pubescens*, Lassu Wood, L704866, 28.ix.2002.

10464 *Nudaria mundana* (Linnaeus, 1761)
MUSLIN FOOTMAN. Two at m. v. trap, Ballytoohy More, L699870, 15.vii.2000; two at m. v. trap, Maum, L705864, 16.vii.2000; two at actinic trap, Maum, L706864, 14.vii.2005.

10566 *Spilosoma lutea* (Hufnagel, 1766)
BUFF ERMINE. Nine at m. v. trap, Kill, L698848, 18.vi.2000; 14 at m. v. trap, Ballytoohy More, L699870, 15.vii.2000; four at m. v. trap, Maum, L705864, 16.vii.2000; one at Heath Trap, Toormore, L663847, 6.vii.2002; three at actinic trap, Maum, L706864, 14.vii.2005; one at Skinner trap, Creggan Lough, L689856, 14.vii.2005; one at actinic trap, Ooghcorragaun, L703875, 14.vii.2005.

10567 *Spilosoma lubricipeda* (Linnaeus, 1758)
WHITE ERMINE. One, Portlea shore, L706867, 25.vi.1999 (B. Cussen); seven at m. v. trap, Kill, L698848, 18.vi.2000; two at m. v. trap, Ballytoohy More, L699870, 15.vii.2000; one at Heath Trap, Toormore, L663847, 6.vii.2002.; one at m. v. trap, Fawnglass, L712854, 30.iv.2000; one at m. v. trap, Ballytoohy, L699869, 7.v.2003; one at actinic trap, Maum, L706864, 14.vii.2005.

10598 *Arctia caja* (Linnaeus, 1758) GARDEN TIGER. Larva on *Urtica dioica*, shore near hotel, L716856, 28.iv.2000; larva on *Urtica dioica*, SW side of harbour, L7185, 28.iv.2000; nine at m. v. trap, Ballytoohy More, L699870, 15.vii.2000; two at m. v. trap, Maum, L705864, 16.vii.2000; two at m. v. trap, Maum, L705864, 11.viii.2000; one at m. v trap, Maum, L705864, 23.viii.2002; one at actinic trap, Maum, L706864. 14.vii.2005.

10607 *Tyria jacobaeae* (Linnaeus, 1758)
THE CINNABAR. Three at m. v. trap, Kill, L698848, 18.vi.2000.

CHAPTER 11

CHIRONOMIDAE (INSECTA: DIPTERA) OF CLARE ISLAND AND THE ADJACENT WEST MAYO MAINLAND, IRELAND

Declan A. Murray

ABSTRACT

Studies of two-winged insects in the Order Diptera during the Clare Island survey between 1910 and 1911 reported 8 species-level taxa of Chironomidae (as Tendipedidae) on Clare Island and 38 from the nearby mainland of west Co. Mayo. Qualitative studies of the Chironomidae between 1998 and 2002 for the New Survey of Clare Island yielded 132 species while collections on the mainland between 2000 and 2018 yielded records of 253 species-level taxa. The marked difference in numbers of species recorded between the previous and the New Survey is due to a more intensive and extended collection effort. One hundred and sixty of the species recorded were unknown at the time of the first survey and were described as new to science since 1911. Eleven species in the present survey—two on Clare Island and nine on the mainland of west Co. Mayo—were first records for the Irish faunal inventory. None of the species recorded are unique to Clare Island but not all were recorded from west Mayo in this study. Five of the ten saline-tolerant halophilous species of Chironomidae on record from the British Isles, belonging to four of the five genera known from European marine waters, occupy coastal habitats of Clare Island. The chironomid fauna of Clare Island is an unsurprising subset of the mainland's fauna with approximately 24% of the known Irish Chironomidae on record for the island. The impression from the first Clare Island Survey that there 'was little of note regarding the dipterous fauna' could also apply to knowledge of the island's chironomid fauna today—in spite of the fact that sixteen times as many species have been recorded during the New Survey of Clare Island compared to one hundred years ago.

Introduction

Studies of two-winged insects (Diptera) in the Clare Island survey were reported by Grimshaw (1912) in his account of insects collected *'during a few days' visit to the island in July, 1910, and a sojourn of about a fortnight in the neighbourhood of Westport exactly a year after this date'*. In addition to specimens Grimshaw collected personally, his colleague J.N. Halbert also provided material from Clare Island while other entomologists involved in that survey, F. Balfour-Brown, Rev. W.F. Johnson and C. Morley, made specimens available to him from their collections on Achill Island, the environs of Westport, Louisburg and Roonagh as well as from locations further inland at Laghta, Clogher and Castlebar Lough. Grimshaw (1912) examined over 4,000 specimens and documented 519 species from 45 insect families in the Order Diptera, including 44 species-level taxa in the Family 'Tendipedidae (Chironomidae)'.

Grimshaw's use of the family name Tendipedidae, with (Chironomidae) in parentheses, reveals his awareness of an ongoing taxonomic debate at that time. The family name Chironomidae, based on the genus name *Chironomus* from Meigen (1803), had been in widespread use throughout the nineteenth century. However, the discovery by Hendel (1908) of an earlier Meigen publication (Meigen 1800) in which the name *Tendipes* was applied to the same genus, resulted in considerable nomenclatural confusion as many (but not all) authors adopted Tendipedidae as the family name rather than Chironomidae. In October 1910, as Grimshaw and his colleagues completed the first year of their fieldwork for the Clare Island survey, the International Commission on Zoological Nomenclature ruled (ICZN Opinion 28, October 1910) that the Meigen (1800) nomenclature should take precedence over the later work (Meigen 1803). Grimshaw (1912) was aware of this situation and in his report for the Clare Island survey, read before the Academy in November 1912, he accepted that ICZN ruling and opted *'to use such names without question'* (Grimshaw 1912, 25 p.3) using Tendipedidae rather than Chironomidae in his paper. In spite of the 1910 ICZN ruling, use of nomenclature from both of Meigen's 1800 and 1803 publications continued, with many European taxonomists adhering to the 1803 work while American and Russian taxonomists based their nomenclature on Meigen (1800). Confusion, debate and conflicting use of both family names ensued for half a century until a second ICZN opinion in 1963 (Opinion 678, October 1963) overturned the October 1910 ruling and placed the Meigen (1800) publication, and nomenclature, on the index of invalid works. This action suppressed all taxonomic names based on Meigen (1800) and validated the Meigen (1803) nomenclature—and use of the now accepted family name Chironomidae.

The Chironomidae, commonly known as non-biting midges, are holometabolous nematoceran Diptera that typically have a short-lived adult aerial phase. The juvenile larval and pupal life-history stages of the majority of species occur in freshwater habitats although some are found in moist and semi-aquatic environments and a small minority occur in marine coastal environments. Although Grimshaw (1912) reported 44 species-level taxa for the 'Family Tendipedidae (Chironomidae)', four of these were biting midges, that are now placed in a separate family that was not formally recognised until several years after the Clare Island survey when Malloch (1917) assigned biting midges to the Family Ceratopogonidae. Thus, 40 species-level taxa of Chironomidae were recorded during the first Clare Island survey. Disappointingly, only eight were reported from Clare Island, identified from no more than sixteen specimens collected on the wing personally by Grimshaw or by his colleague J.N. Halbert. The other species documented were collected from sites on mainland west Co. Mayo.

At the time of the first survey entomological studies were based exclusively on adult insects that were caught in flight by net. Little attention was then given to the juvenile insect stages when Miall and Hammond (1900) commented that the chironomid pupa was *'hardly more than the fly enclosed in a temporary skin, and details of its structure cannot be understood without constant reference to the structure of the fly'*. However, the value of the external structural details of the pupa and particularly of the pupal exuviae—the chitinous skin cast off by the adult insect as it emerges (Pl. I)—was recognised by the German taxonomist August Thienemann who promoted their use in taxonomic and ecological investigations (Thienemann 1910). Pupal exuviae exhibit a wealth of morphological features that readily allows identification to genus and frequently to species level. In comparison with net collection of adult Chironomidae on the wing, chironomid pupal exuviae are more readily obtained from the water surface of lakes and ponds, or from drift-nets in flowing waters, and their examination provides a methodology for assessment of faunal composition and distribution. The study by Carmel F. Humphries MRIA (former Professor

Pl. I Adult female *Psectrocladius sordidellus* emerging and leaving its pupal skin floating on the water surface. Photo: James Lindsey

of Zoology at University College Dublin and PhD mentor of the author), under the guidance of August Thienemann, on the seasonal periodicity of chironomid emergence from the Großer Plöner See, Germany, that was based on collections of pupal exuviae, was a classic early ecological application of the methodology (Humphries 1938).

The prime aim of the present study was to provide an inventory of the Chironomidae of Clare Island and the adjacent mainland of west Co. Mayo in the general area in which collections had been undertaken during the original Clare Island survey (CIS). This qualitative study of Chironomidae for the New Survey of Clare Island (NSCI) is based not only on collections of adult Chironomidae but also on pupal exuviae collected from diverse aquatic habitats. Collections were undertaken on Clare Island in April 2000 and June and August 2002 and on mainland west Mayo in May, August and September 2004. Results from Clare Island presented here include records of adult Chironomidae from an independent study on aquatic insects of Clare Island in 1998 and 1999 by Fahy (2002). The bulk of the author's fieldwork was completed by 2004. However, unforeseen delays in compilation of this volume in the *New Survey of Clare Island* series provided opportunities for the inclusion of valuable additional species records from west Mayo in collections of chironomid pupal exuviae provided by the Environmental Protection Agency (EPA) from lakes around Westport, Castlebar and Delphi between 2006 and 2008. Records are also included from more recent collections by the author in August 2017 on Achill Island (Murray 2017a,b) and in March 2018 from a site previously visited near Westport in 2004.

Methods
Field collections

Adult Chironomidae were obtained by aerial sweep netting but many records were acquired by skimming surface waters with a fine mesh net (Pl. II) and by use of drift-nets placed overnight in flowing water habitats to collect pupal exuviae (Pl. III) as described by Langton (1991). Collections by EPA field research staff were made by skimming a fine mesh net along lake margins, where prevailing onshore winds generate accumulations of floating exuviae in foam, following procedures outlined in Wilson and Ruse (2005). Such samples frequently also included mature pupae and occasionally partially emerged

Pl. II Author using skim net to collect pupal exuviae at a) Loughnapucar stream (Site 31) and b) Creggan Lough, (Site 20). Photos: D. Murray

Pl. III Drift net in position on the River Doree at Maum, Site 6. Photo: D. Murray

and intact adult specimens (Pl. IV). All specimens were preserved in the field in 70% alcohol.

Preparations and specimen identifications

Slide preparations of pupal exuviae and adult chironomids were made, using Euparal as mountant,

Pl. IV Skim net collection of chironomid pupal exuviae, Lough Acorrymore (Site 36). Photo: D. Murray

following Pinder (1986; 1989). Taxonomic treatment of results follows current nomenclature (Andersen et al. 2013; Ashe and O'Connor 2012; Murray et al. 2013; 2014; 2015; 2018; Spies 2005 and Spies and Sæther 2004; 2013). Identifications of adult male Chironomidae were based on Pinder (1978), Murray and Fittkau (1989) and Langton and Pinder (2007). Pupal exuviae were determined from Langton (1991), Langton and Visser (2003) and Langton et al. (2015). Reference was also made to relevant taxonomic revisions, including Silva and Ekrem (2015), Ekrem (2004), Fittkau (1962), Fittkau and Murray (1986), Hirvenoja (1973), Reiss and Fittkau (1971), Sæther (1990; 1995; 2004; 2005), Sæther and Sublette (1983), Sæther and Wang (1995), Sæther et al. (2000), Spies and Dettinger-Klem (2015), Spies and Bolton (2013), Strenzke (1959), Stur and Ekrem (2006), Wiederholm (1986; 1989) and by use of extensive reference specimens in the personal collections of the author.

Slide voucher preparations from Clare Island are deposited in the National Museum of Ireland, Dublin (NMI) where a number of voucher specimens have also been incorporated into the Heritage Council Collection of Irish Chironomidae (HCCIC), (Murray 2005). Some additional voucher slide preparations from Clare Island by Fahy (2002) are also deposited in NMI.

Sampling sites and collections
Clare Island
Collections were taken at a range of locations and habitats on the island. Thirty-three sites were visited between 27–29 April 2000 and between 4–5 June and 21–22 August 2002. Some sites were visited on more than one occasion with a resulting total of 46 collections. Results from the study on aquatic insects of Clare Island from April 1998 to November 1999 by Fahy (2002) provide additional data from one site, Creggan Lough, as well as original data from L. Poirtin Fuinch. Chironomidae were thus obtained from 34 sites on the island. A list of sites, location details and collection dates is given in Appendix Ia and approximate positions are indicated in Figure 1. Photographs of Sites 6, 19, 20 and 32 are given in John (2007): Site 6 - Plate II; Site 19 - Plate IX; Site 20 (20F) - Plate X; Site 32 - Plate XVI. Photographs of Sites 7, 8 and 14 are given in Whitton (2007): Site 7 - Plate XI; Site 8 - Plate XII; Site 14 - Plate X. Photos of Site 17, Portnakilly and Site 33, Leckacanny, are given as Plates I and II in Myers and McGrath (2002).

Mainland
Between 2000 and 2008, collections were made at 41 sites from a variety of aquatic habitats in west Co. Mayo. Some samples were obtained in 2000 and 2003 but the majority was obtained from fieldwork in 2004 supported by a grant from the Praeger Committee of The Royal Irish Academy. Collections of chironomid pupal exuviae provided by the Environmental Protection Agency were obtained from 11 lakes, sampled in 2006 and 2008, three of which had also been sampled by the author in 2004. A delay in compilation and publication of the present volume fortuitously facilitated collections in August 2017 from nine additional sites from the northeast of Achill Island (Murray 2017a). With the inclusion of a collection in March 2018 from a site previously visited at Westport in 2004, specimens from a total of 94 collections were examined from 53 sites in west Mayo. A list of these mainland sites, with location data and collection dates, is given in Appendix 1b and corresponding information for sites on Achill Island is given in Appendix 1c. Approximate positions of these sites are indicated in Figures 2 and 3.

Clare Island survey sites
Precise details of locations where collections were made during the first Clare Island survey were not given by Grimshaw (1912) whose records were reported simply as 'Clare Island' or from mainland districts as, for example, 'Achill', 'Westport House Gardens' or 'Louisburg'. From the many collections of Diptera by Grimshaw and his colleagues, species records of Chironomidae were provided from

CHAPTER 11: CHIRONOMIDAE (INSECTA: DIPTERA) OF CLARE ISLAND AND THE ADJACENT WEST MAYO MAINLAND, IRELAND

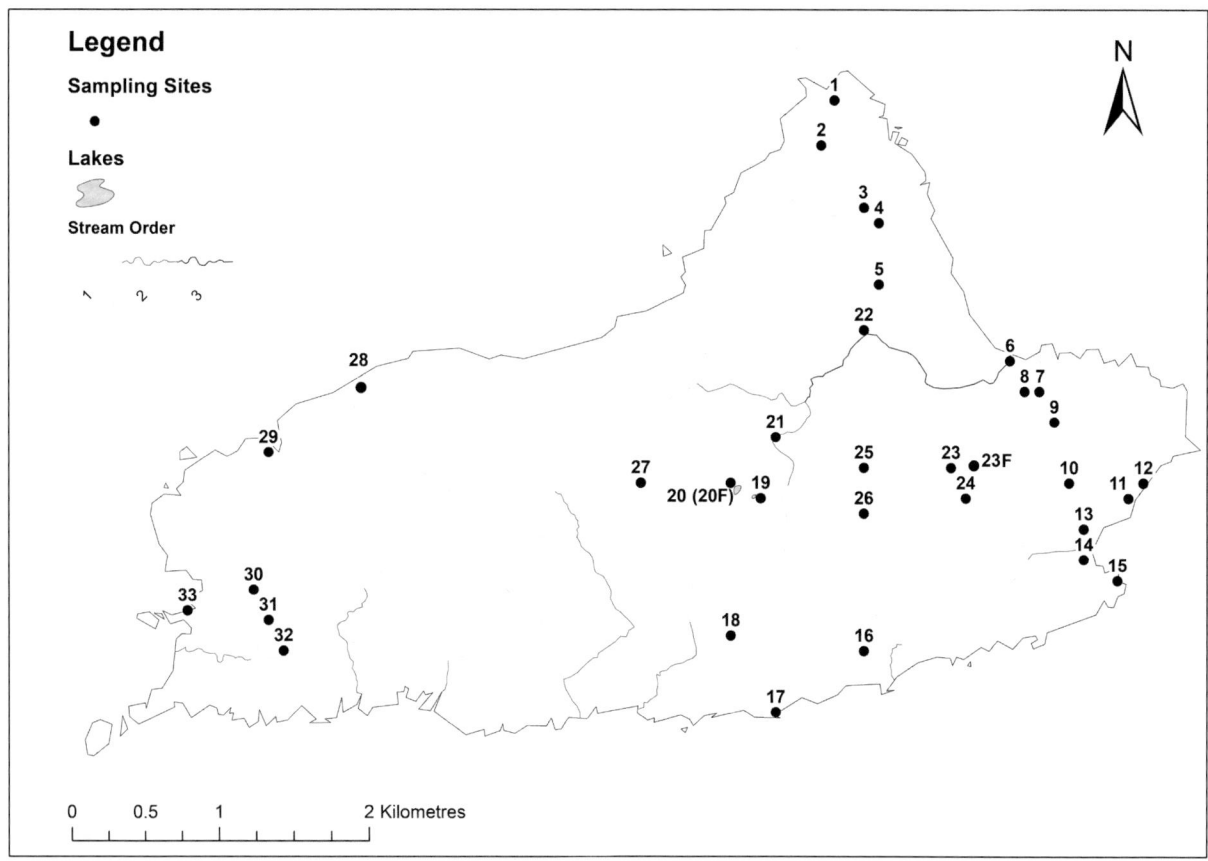

Fig. 1 Map showing locations of sampling sites on Clare Island. See Appendix Ia for details of individual sites.

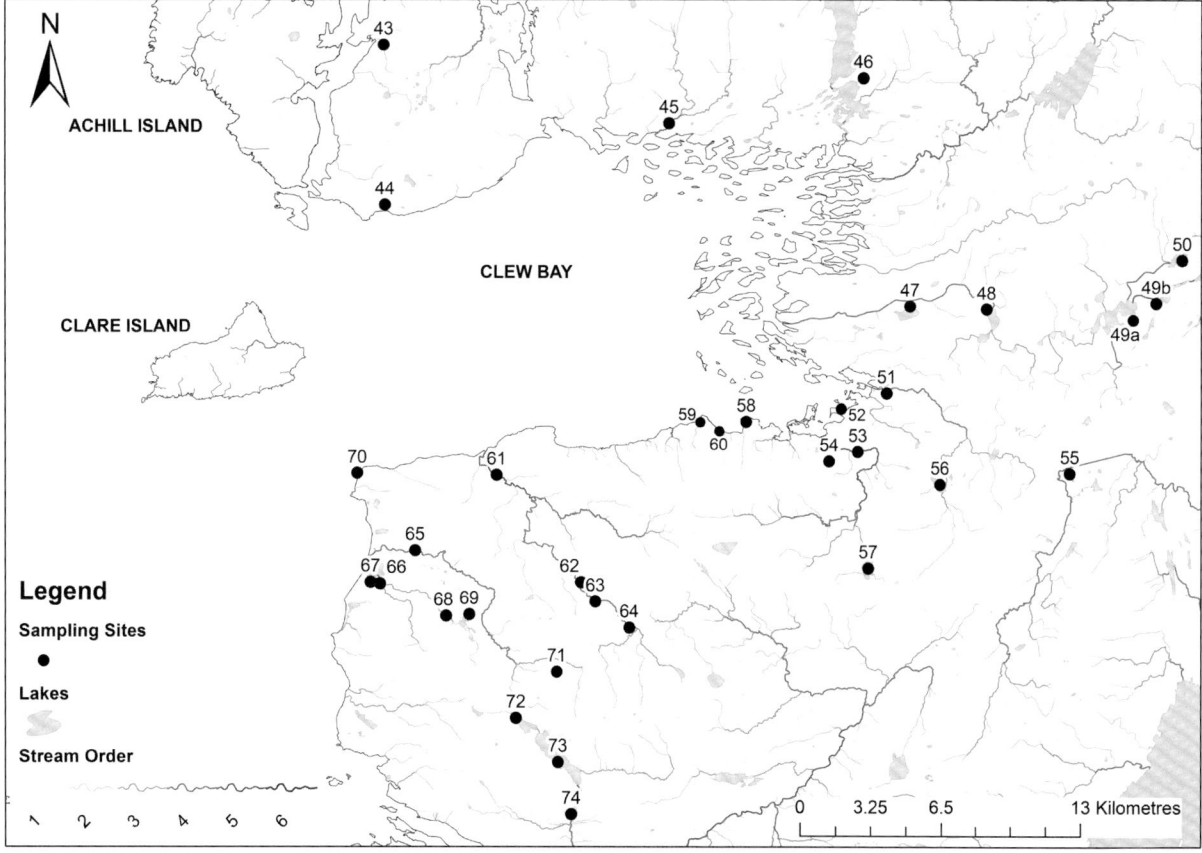

Fig. 2 Map showing locations of sampling sites on mainland west Mayo. See Appendix 1b for details of individual sites.

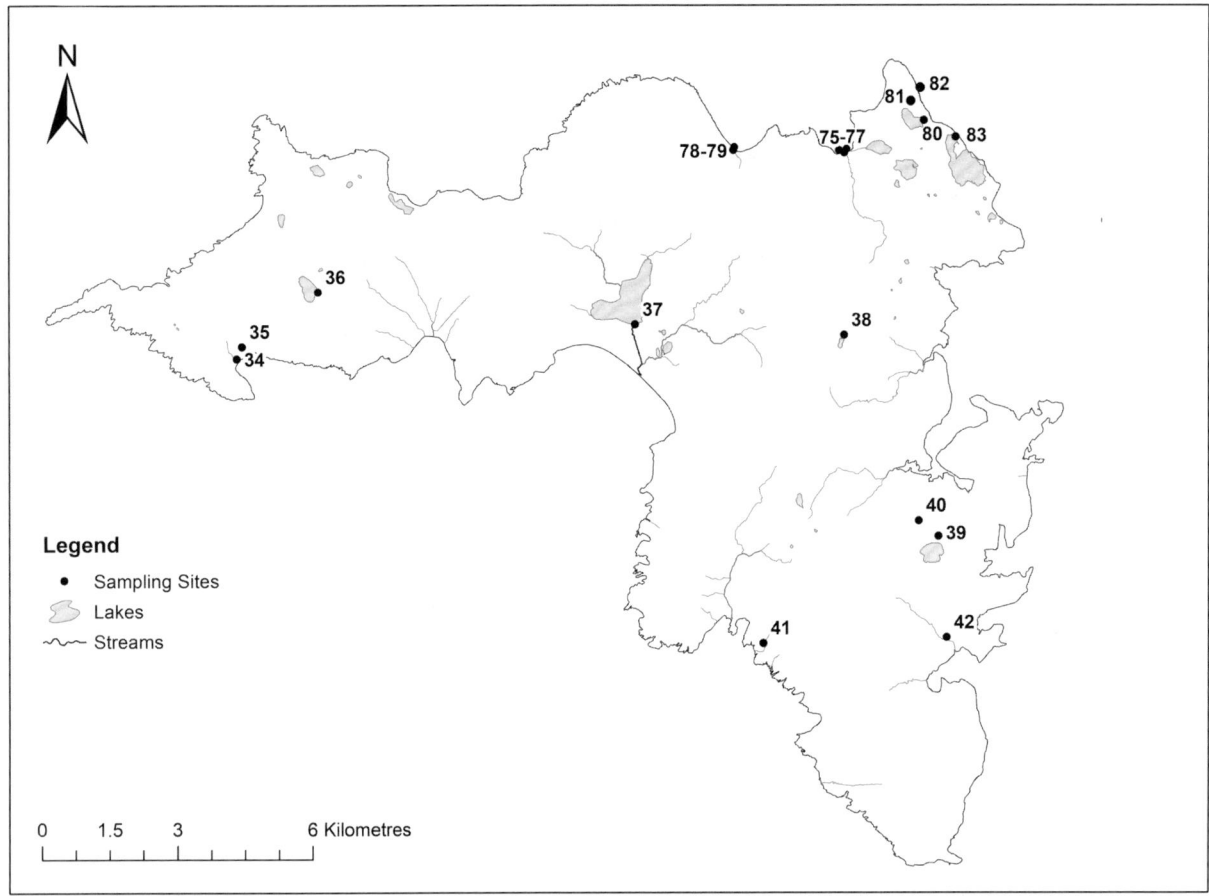

Fig. 3 Map showing locations of sampling sites on Achill Island. See Appendix 1c for details.

Clare Island and ten locations in west Mayo. A list of these locations / districts is given in Appendix Id in which each location is assigned an individual code. Records by Grimshaw from Clare Island are prefixed by the letter 'G' while the ten mainland locations are cited in sequence from G1–G10 with corresponding site numbers from the present survey.

Results and comments

A broad generic concept prevailed at the time of the first Clare Island survey. Since then enhanced characterisation of taxa and taxonomic revision has resulted in significant nomenclatural change providing a more uniform and ecologically consistent taxonomic concept of genera in the Family Chironomidae. The nominal 40 species-level taxa in ten genera that were documented by Grimshaw (1912) are now placed in 27 genera and some subgeneric categories (Table 1). A total of 300 species-level taxa have been documented during the present survey (Table 2). A list of these taxa, including records from Grimshaw (1912) and Fahy (2002) is given in Appendix II. Some of the Grimshaw records are now uncertain, either from a misidentification of species that are now more clearly defined or from an inability to verify some records in the absence of intact voucher material. An inventory of all species found in this survey, giving collection dates and distribution information and additional comments on some records, is provided in Appendix III. Details of the known distribution in Ireland of the species reported here are also included in Murray (2017a) and Murray *et al.* (2013; 2014; 2015; 2018).

The extant Chironomidae of Clare Island

The collections by Fahy (2002) yielded 56 species of Chironomidae from identification of adult males captured in floating emergence traps on Creggan Lough (Site 20F) and Lough Poirtin Fuinch (Site 23F) and in Malaise trap collections of flying insects adjacent to Creggan Lough. Murray and Murray (2003) reported 103 species, now reduced to 100 from recent taxonomic revisions, 28 of which had also been recorded by Fahy (*loc. cit.*). One hundred and thirty-two species for the island are documented here (Table 3). None of these are unique to

Table 1
Species of Chironomidae in the taxonomic sequence listed by Grimshaw (1912) on Clare Island (I) and mainland (M) Mayo with their current nomenclature

1912 record	D	+ Current nomenclature
Tendipes (Chironomus) annularis Deg	M	*Chironomus (Chironomus) annularius* auctt.
T. aprilinus Mg.	M	*Chironomus (Chironomus) aprilinus* Meigen, 1818
T. brevitibialis Ztt.	M	*Dicrotendipes nervosus* (Staeger, 1839)
T. dispar Mg.	M	*Synendotendipes dispar* (Meigen, 1830)
T. dorsalis Mg.	I, M	*Chironomus alpestris* (Goetghebuer, 1934) = *C. dorsalis* auct.
T. ferrugineovittatus Ztt.	M	*Chironomus (Chironomus) plumosus* (Linnaeus, 1758)
T. nigrimanus Staeg.	M	*Demicryptochironomus (s. str.) vulneratus* (Zetterstedt, 1838)
T. nubeculosus Mg.	M	*Polypedilum (Polypedilum) nubeculosum* (Meigen, 1804)
T. pedellus Deg.	M	*Microtendipes pedellus* (De Geer, 1776)
T. pictulus Mg.	M	*Stictochironomus pictulus* (Meigen, 1830)
T. plumosus L.	M	*Chironomus (Chironomus) plumosus* (Linnaeus, 1758)
T. prasinatus Staeg.	M	*Pseudochironomus prasinatus* (Staeger, 1839)
T. psittacinus Mg.	M	*Cryptochironomus psittacinus* (Meigen, 1830)
T. pusillus L.	M	*Kloosia pusilla* (Linnaeus, 1767)
T. riparius Mg.	M	*Chironomus (Chironomus) riparius* Meigen, 1804
T. rufipes L.	M	*Demeijerea rufipes* (Linnaeus, 1761)
T. tentans Fab.	I, M	*Chironomus (Chironomus) tentans* Fabricius, 1805
T. viridis Mcq.	M	*Glyptotendipes (Caulochironomus) viridis* (Macquart 1834)
Cricotopus bicinctus Mg.	M	*Cricotopus (Cricotopus) bicinctus* (Meigen, 1818)
C. motitator L.	M	*Cricotopus (Cricotopus) annulator* Goetghebuer, 1927
C. silvestris Fab.	M	*Cricotopus (Isocladius) sylvestris* (Fabricius, 1794)
C. tibialis Mg.	I, M	*Cricotopus (Cricotopus) tibialis* (Meigen, 1804)
C. trifasciatus Panz.	M	*Cricotopus (Isocladius) trifasciatus* (Meigen, 1810, in Panzer)
C. trifasciatus var *tricinctus* Mg.	M	*Cricotopus (Isocladius) tricinctus* (Meigen, 1818)
Cricotopus sp.	M	? unknown
Orthocladius sordidellus Ztt.	I, M	*Psectrocladius (P.) sordidellus* (Zetterstedt, 1838)
Tanytarsus flavipes Mg.	M	*Phaenopsectra flavipes* (Meigen, 1818)
T. punctipes Wied.	M	*Phaenopsectra punctipes* (Wiedemann, 1817)
T. pusio Mg.	M	nomen nudum
T. sp (? *gmundensis*)	I	*Micropsectra junci* (Meigen, 1818)
Metriocnemus fuscipes Mg.	M	*Metriocnemus (Metriocnemus) fuscipes* (Meigen, 1818)
Metriocnemus modestus Mg.	M	*Polypedilum (Pentapedilum) sordens* (van der Wulp, 1874)
Diamesa ammon Hal.	I	*Potthastia gaedii* (Meigen, 1838)
Prodiamesa obscurimana Mg.	I, M	A likely misidentification of *Prodiamesa olivacea* (Meigen, 1818)
Procladius nervosus Mg.	M	*Clinotanypus (Clinotanypus) nervosus* (Meigen, 1818)
Pelopia (Tanypus) culiciformis L.	M	*Procladius (Holotanypus) culiciformis* (Linnaeus, 1767)
Pelopia sp.	M	*Procladius (Psilotanypus) rufovittatus* (van der Wulp, 1874)
Ablabesmyia nebulosa Mg.	I, M	*Macropelopia (Macropelopia) nebulosa* (Meigen, 1804)
A. phatta Egger.	M	*Ablabesmyia (Ablabesmyia) phatta* (Egger, 1863)
A. pygmaea V.d Wulp.	M	*Zavrelimyia (Paramerina) cingulata* (Walker, 1856)

+following Ashe and O'Connor (2009, 2012), Silva and Ekrem (2015), Murray *et al.* (2018).
D = Distribution: I = Clare Island, M = Mainland - west County Mayo.

Table 2
Numbers of species in subfamilies of Chironomidae on Clare Island, West Mayo and combined survey total

Subfamily	Clare Island	West Mayo	Survey total
Tanypodinae	16	35	37
Telmatogetoninae	1	-	1
Diamesinae	1	3	4
Prodiamesinae	1	1	1
Orthocladiinae	63	111	136
Chironominae	50	103	121
Total	132	253	300

Table 3
Chironomidae on record from Clare Island, 2018

Subfamily Tanypodinae

Ablabesmyia longistyla - F
Ablabesmyia monilis - F, M
Ablabesmyia phatta - F
Arctopelopia barbitarsis - F
Arctopelopia griseipennis - M
Krenopelopia nigropunctata - M
*?*Larsia atrocincta* - F
Macropelopia adaucta - *M*
*[*Macropelopia nebulosa*] - G
Macropelopia notata - F, M
Monopelopia tenuicalcar - M
Natarsia punctata - F, M
Nilotanypus dubius - M
Procladius (Holotanypus) choreus - F, M
Procladius (H.) saggitalis - F
Zavrelimyia (Paramerina) cingulata - F, M
Zavrelimyia (Zavrelimyia) barbatipes - M

Subfamily Telmatogetoninae

Thalassomya frauenfeldi - M

Subfamily Diamesinae

Diamesa insignipes - M
*[*Potthastia gaedii?*] - G

Subfamily Prodiamesinae

**Prodiamesa olivacea* - G, M

Subfamily Orthocladiinae

Acamptocladius reissi - M
Acricotopus lucens - M
Brillia bifida - M
Camptocladius stercorarius - M
**Cardiocladius fuscus* - F

continued

Table 3. Cont.

Subfamily Orthocladiinae (Cont.)

Chaetocladius dentiforceps - F
Chaetocladius melaleucas - F, M
Chaetocladius perennis - F, M
Clunio marinus - M
Corynoneura edwardsi - F, M
Corynoneura lobata - M
Corynoneurella paludosa - M
Cricotopus (Cricotopus) albiforceps - M
Cricotopus (C.) annulator - F
Cricotopus (C.) ephippium - M
Cricotopus (C.) festivellus - F
Cricotopus (C.) pilosellus - F, M
Cricotopus (C.) pulchripes - M
[*Cricotopus (C.) tibialis*] - G
Cricotopus (Isocladius) ornatus - M
Cricotopus (I.) sylvestris - F
Cricotopus (Paratrichocladius) rufiventris -M
Eukiefferiella claripennis - M
Eukiefferiella cyanea - M
Eukiefferiella devonica - M
Eukiefferiella minor / fittkaui - M
Halocladius fucicola - M
Halocladius variabilis - F, M
Heterotanytarsus apicalis - M
Heterotrissocladius marcidus - M
Limnophyes angelicae - F, M
Limnophyes gurgicola - M
Limnophyes habilis - F
Limnophyes minimus - M
Limnophyes natalensis - M
Limnophyes pentaplastus - M
Metriocnemus (M.) eurynotus - F, M
Metriocnemus (M.) fuscipes - F, M
Metriocnemus (M.) picipes - M
Orthocladius (Eudactylocladius) fuscimanus - M
Orthocladius (Orthocladius) dentifer - M
Orthocladius (O.) oblidens - F, M
Parametriocnemus stylatus - M
Paraphaenocladius impensus - M
Paraphaenocladius irritus - M
Paraphaenocladius pseudirritus - M
Psectrocladius (Allopsectrocladius) obvius -F, M
Psectrocladius (A.) platypus - F, M
Psectrocladius (Psectrocladius) fennicus - F
Psectrocladius (P.) limbatellus - M
Psectrocladius (P.) oligosetus - M
Psectrocladius (P.) sordidellus - G, M

continued

Table 3. Cont.

Subfamily Orthocladiinae (Cont.)

Psectrocladius (P.) ventricosus - M
Pseudorthocladius curtistylus - F
Pseudorthocladius filiformis - M
Pseudorthocladius macrovirgatus - M
Pseudosmittia oxoniana - M
Pseudosmittia trilobata - *M*
Rheocricotopus (Psilocricotopus) chalybeatus - M
Smittia pratorum - M
Synorthocladius semivirens - M
Thalassosmittia thalassophila - M
Thienemannia gracilis - M
Thienemanniella majuscula - M
Thienemanniella Pe 2a - M

Subfamily Chironominae Tribe Chironomini

[Chironomus (Chironomus) alpestris] – G (as "*Tendipes dorsalis* Mg.")
Chironomus (C.) aprilinus - F, M
Chironomus (C.) commutatus - F, M
Chironomus (C.) nuditarsis - M
Chironomus (C.) pseudothummi - F, M
Chironomus (C.) riparius - F, M
Chironomus (C.) tentans - F
Cladopelma krusemani - F, M
Demeijerea rufipes - F
Dicrotendipes nervosus - F
Endochironomus tendens - F
Glyptotendipes cauliginellus - F, M
Glyptotendipes pallens - M
Glyptotendipes paripes - M
Lauterborniella agrayloides - M
Microtendipes chloris - M
Microtendipes diffinis - F
Microtendipes pedellus - F
Pagastiella orophila - F, M
Parachironomus cinctellus - M
**Parachironomus mauricii*- F
Parachironomus parilis - F
Parachironomus tenuicaudatus - F
Parachironomus vitiosus - M
**?Parachironomus* sp Pe - M
Phaenopsectra flavipes - M
Phaenopsectra punctipes - F
**Polypedilum (Pentapedilum) sordens* - G, M
Polypedilum (Pen.) uncinatum - M
Polypedilum (Polypedilum) acutum -F, M
Polypedilum (Pol.) arundineti - F
Polypedilum (Pol.) pedestre - F

continued

Table 3. Cont.

Subfamily Chironominae Tribe Chironomini (Cont.)

Polypedilum (Tripodura) pullum - F

Tribe Pseudochironomini

Pseudochironomus prasinatus - F, M

Tribe Tanytarsini

Cladotanytarsus atridorsum - F
Cladotanytarsus nigrovittatus - F
Micropsectra atrofasciata - M
**Micropsectra junci* - G, M
Micropsectra lindrothi - M
Micropsectra pallidula – M
Micropsectra roseiventris - M
**Paratanytarsus brevicalcar* - M
Rheotanytarsus curtistylus - M
Stempellinella brevis - M
Tanytarsus buchonius - M
Tanytarsus gracilentus - M
Tanytarsus gregarius - F, M
Tanytarsus lestagei - M
Tanytarsus pallidicornis - F
Tanytarsus signatus - F, M
Zavrelia pentatoma - M

Records from the original Clare Island survey and not found in this survey are enclosed in square brackets [ ]; Abbreviations used for sources of records: G = Grimshaw (1912), F= Fahy (2002), M = Murray (this paper)
* See text and Appendix III for further comment

the island but 46 were not found amongst the 253 species recorded in collections from west Mayo. The 63 species of Orthocladiinae and 50 species of Chironominae together account for the majority of taxa recorded. Sixteen species of Tanypodinae are recognised and one each in the subfamilies Diamesinae, Prodiamesinae and Telmatogetoninae.

Grimshaw (1912) recorded just one species of Tanypodinae, '*Ablabesmyia nebulosa* Mg.', that was transferred to the genus *Macropelopia* by Thienemann and Kieffer (1916). *Macropelopia nebulosa* was not recorded on Clare Island in the current study but two other species in the genus, *M. adaucta* and *M. notata*, occurred at a number of sites. Fittkau (1962) drew attention to the uncertain validity of some species records in the genus *Macropelopia* by early researchers and was aware of frequent mistaken determinations of *M. nebulosa* with the closely related *M. notata*. In the absence of reference material for Clare Island, the identity of the specimen identified by Grimshaw (1912) as '*Ablabesmyia nebulosa*' is unconfirmed, but its presence on the island is possible. All three *Macropelopia* species are common on the mainland and occurred in collections from west Co. Mayo in the present survey.

The subfamily Telmatogetoniinae was established by Wirth (1949) for marine coastal dwelling Chironomidae including *Thalassomya frauenfeldi* which occurred in collections at Capnagower (Pl. V) in April 2000. This remains as only the third record of the species from Irish coastal waters (Murray *et al.* 2018). Previous records, also from Irish Atlantic coastal sites, are from Inishtearaght Island, Co. Kerry (Murray and Ashe 1982) and Kilkee, Co. Clare (Murray 2000). One species of Diamesinae, *Potthastia gaedii*, was previously reported by Grimshaw (1912) under its synonym *Diamesa ammon*; it was not found on the island in the present study, but it did occur in several mainland collections. However, adults of another Diamesinae species, *D. insignipes,* were obtained in April 2000 at Capnagower (Site 11). The subfamily Prodiamesinae is represented by *Prodiamesa olivacea* from a specimen collected on the wing adjacent to the River Doree (Site 22) in June 2002. Murray and Murray (2003) considered the strong likelihood that the record of *Prodiamesa obscurimana* by Grimshaw (1912) was a misidentification of *P. olivacea* (Appendix III).

Grimshaw (1912) reported only two species of Orthocladiinae from Clare Island: *Cricotopus tibialis* Mg. and *Psectrocladius sordidellus* Zett. In the present study 63 species are documented, including *P. sordidellus.* (Pl V). While *Cricotopus tibialis* was not found on the island, it was recorded in collections from Achill Island and it is known from other locations further inland in Co. Mayo (Murray *et al.* 2014).

Two species, *Acamptocladius reissi* and *Limnophyes angelicae*, were reported as new to

Pl. V Marine rock pool at Capnagower (Site 12). Photo: D. Murray

the Irish fauna from Clare Island by Murray and Murray (2003). Both have since been recorded on the Irish mainland but not, as yet, from west Mayo (Langton 2004; Murray 2005; Murray *et al.* 2014; 2018). The record by Fahy (2002) of a single adult male of *Cardiocladius fuscus* in collections from a floating emergence trap on Lough Poirtin Fuinch is doubtful and is regarded here as a misidentification. Larvae of *Cardiocladius* are characteristically found in fast-flowing well-oxygenated water in rivers and streams, living in a phoretic association with and preying on larvae and pupae of Simuliidae (Andersen *et al.* 2013). Grimshaw (1912) recorded one species of Simuliidae, *Simulium reptans* Linnaeus, from Clare Island but Fahy (*loc. Cit*) provides no current record of Simuliidae from the lentic waters of Lough Poirtin Fuinch (or elsewhere on Clare Island). *Cardiocladius* larvae are unlikely to be present in the lough without a host species and, in the absence of a voucher specimen in NMI, that record of *Cardiocladius fuscus* is considered unreliable.

Four saline-tolerant halobiontic Orthocladiinae—*Clunio marinus*, *Halocladius fucicola*, *Halocladius variabilis* and *Thalassosmittia thalassophila*—that are typical inhabitants of marine coastal littoral habitats and rock pools, were recorded at three sites (11, 12 and 15) between Capnagower and Glen on the east coast of the island. *H. fucicola* and *T. thalassophila* were also present in collections at Portnakilly Harbour (Site 17). Unusually, Fahy (2002) reported a total of 25 adult *Halocladius variabilis* in Malaise trap collections around Creggan Lough, 90 metres above sea level (Site 20F), in 1998 and 1999. It is possible that adults, emerging from marine coastal waters were carried inland by wind and caught in the Malaise traps adjacent to the lough. Surprisingly, and without comment, Fahy (2002) also reported six adults of *H. variabilis,* four in October 1998 and two in August 1999, from emergence trap collections on the freshwater Lough Poirtin Fuinch that lies in a peaty basin surrounded by blanket bog at 60 metres above sea level (Site 23F).

Grimshaw (1912) recorded three species of Chironominae on the island: *Tendipes dorsalis* (= *Chironomus alpestris* Goetghebuer, 1934), *Tendipes* (= *Chironomus*) *tentans* and an unidentified *Tanytarsus* sp., about which Grimshaw (1912) commented that 'it could be *Tanytarsus gmundensis*'. In this New survey 50 species-level taxa of Chironominae (32 in the Tribe Chironomini, 1 Pseudochironomini and 17 Tanytarsini) were present in the collections, including 2 of the species reported by Grimshaw. *Chironomus tentans* was found in Creggan Lough by Fahy (2002). The taxon *T. gmundensis*, tentatively recognised by Grimshaw, is now a synonym of *Micropsectra junci* that was found at two locations in June and August 2004. In spite of Grimshaw's reservation it is likely that the record of *T. gmundensis* from the first survey was correct. There has been taxonomic uncertainty about the status of early records of '*Tendipes dorsalis*', currently considered a synonym of *Chironomus alpestris*. Murray and Murray (2003) reported *C. dorsalis* but that record was incorrect and the specimen has since been identified as *C. commutatus* (see Appendix III). *C. alpestris* is known from several locations in Ireland including a site at Letterfrack Co. Galway, approximately 30km south of the study area of west Mayo and Clare Island.

Recent re-examination of the collections in April 2000 from a small pool at Maum (Site 7) have revealed exuviae of *Chironomus nuditarsis* that constitutes an additional species record that had been previously overlooked and not included in the listing by Murray and Murray (2003). At least five species of *Parachironomus* are known from the island. Murray and Murray (2003) recorded *Parachironomus vitiosus* from Creggan Lough (Site 20). Fahy (2002) recorded *P. parilis* from Creggan Lough and L. Poirtin Fuinch and *P. tenuicaudatus* from the L. Poirtin Fuinch. However, examination of the single FCIC voucher specimen labelled as *P. paralis* in NMI revealed that specimen to be a misidentified adult male of *P. mauricii* (Goegthebuer, 1921). Following a recent review and revised interpretation of pupal exuviae of some *Parachironomus* species by Spies and Bolton (2013), re-examination of specimens in the author's collections from Creggan Lough, subsequent to the account reported in Murray and Murray (2003), has yielded an adult male of a fifth species *P. cinctellus* as well as pupal exuviae that run to *P. mauricii* in the key by Langton and Visser (2003). Spies and Bolton (2013) revealed incorrect associations between some adult *Parachironomus* species and pupal exuviae in Langton (1991) and Langton and Visser (2003) and have shown that exuviae identified as *P. mauricii* from the latter work could also belong to at least two other species—*P. biannulatus* Goetghebuer and *P. kuzini* Shilova—whose exuviae are as yet unknown. In the absence of associated specimens (i.e. pharate male or reared adult male) from Creggan Lough, the pupal exuviae from this site are reported here as *Parachironomus* sp. Pe sensu Langton (1991) since there is no direct evidence that they belong to the adult *P. mauricii* that was misidentified as *P. parilis* by Fahy (2002). The FCIC voucher adult male *P. mauricii* from Creggan Lough in NMI is the sole valid record of the species for Ireland. It is not yet confirmed from the mainland, but it is likely to occur since Murray *et al*. (2015; 2018) document occurrences of the pupal morphotype from 33 locations in Ireland. The identities of the pupal morphotypes *Polypedilum* Pe 7 and *Polypedilum* Pe 8, determined from Langton (1991), that were reported in Murray and Murray (2003) have been updated to *Polypedilum* (*Polypedilum*) *acutum* and *P*. (*P*.) *uncinatum* respectively, based on identification from Langton and Visser (2003).

Pseudochironomus prasinatus was found at three locations on Clare Island. Grimshaw (1912) recorded this species as *Tendipes prasinatus* from two mainland sites but did not cite a record from Clare Island. Nevertheless, Edwards (1929, p377) attributed a record from Clare Island to Grimshaw, as well as from three sites in west Mayo. It is not certain if this citation by Edwards was in reference to a record from the island itself or if it was a general allusion to records from the Clare Island survey as has occurred with records of some marine taxa (Myers and McGrath 2002). Further comments on other species of Chironominae are given in Appendix III.

Chironomid fauna in mainland collections

Two hundred and fifty-three species-level taxa in five subfamilies have been identified in collections from the mainland sites (Table 2, Appendices II and III). As on Clare Island the subfamilies Orthocladiinae and Chironominae collectively constitute the greater overall proportion of taxa recorded with 111 and 103 species respectively. Thirty-five species of Tanypodinae are documented along with three species in the subfamily Diamesinae and one species of Prodiamesinae.

Of the 38 taxa listed from the mainland by Grimshaw (1912), *Tendipes pusio* Mg. is a nomen nudum and no longer recognised, *T. ferrugineovittatus*, is a synonym of *Chironomus plumosus* while a species of *Cricotopus* was not identified (Table 1). Grimshaw (1912) drew particular attention to specimens from Belclare (Site G7) and Louisburg (Site G10) that he had identified as *Metriocnemus modestus*. Since he regarded this species as 'new to the British list' he gave a brief description and illustration of the male genitalia. However, this was regarded as a misidentification by F.W. Edwards, British Museum, London, who examined Grimshaw's specimen of '*M. modestus*' and identified it as *Pentapedilum sordens* van der Wulp. (Edwards 1929, p.310). A specimen from Castlebar Lough reported by Grimshaw as *Pelopia* sp. was also reidentified by Edwards (1929, p.302) as *Procladius (Psilotanypus) rufovittatus* (van der Wulp, 1874). Nine taxa reported by Grimshaw did not occur in the recent collections on the mainland. Five, *Cricotopus trifasciatus* var *tricinctus* Mg, [= *Cricotopus (Isocladius) tricinctus*], *Metriocnemus fuscipes*, *Tendipes brevitibialis* Ztt. [= *Dicrotendipes nervosus* (Staeger)], *Tendipes dispar* [= *Synendotendipes dispar*] and *Tendipes dorsalis* [=*Chironomus (Chironomus) alpestris*], are common in Irish freshwater habitats and no particular significance is attributed to their absence in the collections. Two of the aforementioned species, *Metriocnemus fuscipes* and *Dicrotendipes nervosus* were present in the collections on Clare Island. Further comments on these, and the remaining taxa cited by Grimshaw, *Pelopia (Tanypus) culiciformis*, [=*Procladius (Holotanypus) crassinervis*], *Tendipes (Chironomus) annularis*, [=*Chironomus (Chironomus) annularius*], *Tendipes viridis* [=*Glyptotendipes (Caulochironomus) viridis*] and *Tendipes pusilla* [=*Kloosia pusilla?*], are given in the species accounts (Appendix III). Grimshaw (1912) reported *Pseudochironomus prasinatus* from two mainland sites as specimens collected by J.N. Halbert on Achill and in his own collection from 'a Lough near Westport' (= Ballin Lough). However, Edwards (1929, p. 377) in his examination of Grimshaw's material noted specimens of *P. prasinatus* from three locations: Castlebar, Ballin and Coolbarren Lough. The record from Coolbarren Lough (grid reference L981869) approximately 1km southwest of Ballin Lough had not been cited by Grimshaw (1912). During the present study this species was recorded at Ballin Lough as well as from Achill (Keel Lough) and from three other loughs in the Westport region, L. Aille, L. Knappabeg and L. Moher (Appendix III).

Nine species found in the mainland collections were recorded for the first time from Ireland. Five species—*Limnophyes asquamatus*, *Orthocladius (Eudactylocladius) olivaceus*, *O. (Symposiocladius) holsatus*, *Psectrocladius (Psectrocladius) ventricosus* and *Demicryptochironomus (Irmakia) neglectus*—were obtained during the Praeger Committee supported fieldwork in 2004 and reported by Murray and Murray (2006). The remaining four species—*Tanypus kraatzi*, *Cladotanytarsus iucundus*, *Neozavrelia cuneipennis* and *Tanytarsus lactescens*—were identified from pupal exuviae in collections by the Environmental Protection Agency (Murray 2010).

The record of *Metriocnemus ephemerus* in August 2017 from a shallow pond on a small area of machair at Dugort on Achill Island is only the second known record of the species that had been described as new to science just two years earlier from adult male specimens collected in a tidal bay on the mouth of the River Bann, Co. Derry (Langton 2015b). The collections at Dugort in 2017 included pharate adult males and females as well as pupal exuviae. A description of the pupal exuviae has been given in Langton and Murray (2018) while a description of the adult female is pending. The machair is a flat sandy plain with wet and dry grasslands that develops when sand (mostly fragmented calcareous shells) is blown inland by prevailing winds from beaches and sand dunes. Machair plains are unique to the west of Ireland and parts of west and north Scotland and are areas of special scientific interest that have been designated priority habitats under the EU Habitats Directive. It remains to be seen if *M. ephemerus* is restricted to such habitats. Langton (2015b) noted that *M. ephemerus* appears closely related to *M. sibericus* (Lundstrom, 1915), only known from ponds in coastal regions of the New Siberian Islands of north-east Russia.

Six species characteristic of saline habitats occurred in the coastal collections of west Mayo, four of which, *Clunio marinus*, *Halocladius fucicola*, *H. variabilis* and *Thalassosmittia thalassophila* also occurred in the Clare Island collections. The remaining two, *Halocladius varians* and *Chironomus salinarius* are documented from collections in 2017 on Achill Island from a coastal rock pool at Dugort (Site 78) and from the saline inundated Lough Sruhill (Site 83).

Discussion

There have been significant advances in chironomid systematics and biodiversity assessment techniques since the first Clare Island survey. Taxonomic revisions have led to synonymy of some species and new species have been recognised and described. One hundred and sixty-four species, 54% of those reported in the present study, were unknown at the time of the first survey and have been described as new to science since 1911, including *Metriocnemus ephemerus* described by Langton (2015b) that was collected on Achill Island in 2017 (Murray 2017a). The dynamic aspect of recent studies on the Irish Chironomidae is evident from the increase in the documented species inventory during the time span of the work on which this paper is based. In the twenty-year period between 1998 and 2018 the inventory of Chironomidae of Ireland increased from the 396 species documented at the end of the twentieth century by Ashe *et al.* (1998) to 540 species level taxa documented (up to December 2017) in Murray *et al.* (2018). Since then an additional two species have been reported, one each by Langton and Ruse (2018) and Murray and Langton (2018) bringing the known Irish inventory to 542 species of Chironomidae (Murray 2019). The increase of 146 species in the period between 1998 and 2019 includes 11 species records derived from collections undertaken during the present study.

Grimshaw (1912) demonstrated a broad taxonomic expertise by reporting on his identification of 519 species from the 45 diverse dipteran insect families he studied during the Clare Island survey. His inventory of only eight taxa from the family Chironomidae (as Tendepedidae) on Clare Island could not have fully represented the true situation. Undoubtedly other chironomid species were present at that time but were not caught by the chance aerial sweep-net collection of flying insects. In this New Survey, concentrating on one dipteran family, a single day's collecting of adult Chironomidae and pupal exuviae on Clare Island in April 2000 yielded 34 species. The eight species reported from Clare Island by Grimshaw (1912) represented approximately 21% of the species he documented from the adjacent mainland and about 16% of the then known Irish chironomid fauna.

The floral and faunal composition of any region is determined by a range of factors including geographic location, climate, landmass area, topography, availability of suitable habitats as well as opportunities for dispersal. Proximity to the Irish mainland is a determining factor influencing the composition of Clare Island's aquatic dwelling chironomid fauna. However, habitats typical of larger rivers and lakes are not available on the island but several small streams, some large ponds (small lakes) and numerous bog pools provide suitable habitats. Species belonging to six of the eight subfamilies of Chironomidae known from Ireland were obtained in this survey (Table 4). Not surprisingly, in the absence of suitable habitats, species in the subfamilies Podonominae and Buchonomyiinae were not encountered. The

Table 4
Total number and *percentage representation of species in subfamilies of Chironomidae known from Clare Island, west Mayo (this survey) and Ireland

Subfamily	Clare Island %	West Mayo %	Ireland %
Tanypodinae	12	14	12
Buchonomyiinae	0	0	<1
Podonomiinae	0	0	<1
Telmatogetoninae	<1	0	<1
Diamesinae	<1	1	2.0
Prodiamesinae	<1	<1	<1
Orthocladiinae	48	44	45
Chironominae	38	41	40
Total species number	132	253	542
As % of the known Irish species	24	47	

* rounded to nearest whole percent

132 species reported from Clare Island account for approximately 24% of the currently known Irish Chironomid fauna while approximately 47% of the known Irish species are documented from the collections at mainland sites in west Mayo.

The less species-rich subfamilies Telmatogetoninae, Diamesinae and Prodiamesinae have a low representation while species in the subfamilies Orthocladiinae and Chironominae comprise the majority of records. The Tanypodinae, with predominantly carnivorous larvae, have a smaller representation in comparison to the Orthocladiinae and Chironominae, whose larvae have a more diverse range of feeding habitats. While the proportion of Tanypodinae in the west Mayo collections is slightly greater than on Clare Island (which is similar to the all-Ireland composition) the relative proportions of species within the Orthocladiinae and Chironominae in the west Mayo collections compare favourably with the overall Irish chironomid fauna at 44% versus 45% and 41% versus 40% respectively. In contrast, the Orthocladiinae form a relatively larger component of 48% on Clare Island compared to the 38% for the Chironominae. Larval Orthocladiinae favour oxygen-rich waters whereas the lentic, standing water habitats with organic enrichment, in which oxygen may sometimes be lacking, are favoured by Chironominae (Pinder 1995; Wilson and Ruse 2005). The slightly greater occurrence of Orthocladiinae on Clare Island may be an indication of greater availability of suitable oxygen-rich aquatic habitats or a proportional lesser availability of standing water habitats for Chironominae species.

Sixteen times as many species have been recorded in the Family Chironomidae during this New Survey of Clare Island compared with the first survey, undoubtedly due to the more intense sampling effort on one Family of Diptera, in contrast to the broader range of insect families reviewed in the first survey of Clare Island. Grimshaw's (1912) impression that there 'was little of note regarding the dipterous fauna' could, generally speaking, apply to knowledge on the Chironomidae of Clare Island today. The record of one species, *Parachironomus mauricii*, is thus far documented as unique to the island, although it is likely to occur on the mainland (Murray *et al.* 2015). The existence of five halophilous chironomid species on the island's marine coastal habitats is noteworthy. Sixteen halophilous species in five genera are known from European coastal waters, fifteen documented by Legakis and Murray (2001) and one, *Telmatogeton murrayi*, thus far only known from Iceland, Britain and Ireland (Murray 1999; Sæther 2009; Langton and Handcock 2013; Langton 2015a). Ten species are on record from the British Isles (Murray 2000; Langton and Handcock 2013; Langton 2015) five of which, from four of the five genera recorded in European marine waters, occupy coastal habitats of Clare Island.

Grimshaw (1912) recorded a total of 519 species of Diptera, from the mainland and Clare Island, during the first survey. The 211 species he reported from the island corresponded to approximately 17% of the 1,210 species of Diptera known from Ireland at that time. A century later some 3,397 dipteran species are on record in Ireland (Chandler *et al.* 2008; Chandler 2016) but comparative data is not available from the New Survey for species in the 45 dipteran families reported by Grimshaw (1912). The species of Chironomidae recorded during the first survey accounted for about 3% of the then known Irish Diptera, while the eight species on Clare Island represented less than one percent (0.66%). Chironomidae documented in the New Survey account for approximately 7.0 % of the currently known Irish Diptera and the extant Clare Island Chironomidae comprise approximately 4% of the total number of dipteran species—an almost six-fold increase compared to the 0.66% representation at the time of the original survey.

It is unfortunate that studies on other dipteran families were not undertaken in the New Survey, a fact largely attributable to the lack of relevant taxonomic expertise in Ireland. In March 1987, O'Connor *et al.* (1988), at a Royal Irish Academy seminar on *The State of Taxonomy in Ireland*, highlighted the dwindling status of taxonomy in Ireland at that time. They recommended that taxonomic studies be promoted in third-level institutions and that, on their retirement, existing taxonomists be replaced by other taxonomists, though not necessarily in the same field. In recent decades there has been an increase in global awareness of natural heritage and biodiversity issues but the situation outlined by O'Connor *et al.* (loc. cit.) remains largely unchanged. Ireland is not unique in this position as Agnarsson and Kuntner (2007) highlighted the persistent attrition of taxonomists and regarded taxonomy as 'a science in crisis'. While Costello *et al.* (2013) maintained that on a global scale, taxonomists were 'not in danger of extinction', in Europe Fontaine *et al.* (2012) had earlier noted that non-professional taxonomists (individuals who are not remunerated for taxonomic studies—skilled amateurs and retired professionals), acting on a volunteer basis had been responsible for over 60% of the formal taxonomic publications in which new species were described in the decade between 1998 and 2007.

The drive to explore nature and the diversity of life is an inherent human trait. The multidisciplinary First survey of Clare Island was a demonstration of such ambition to acquire knowledge and understanding on a broad range of topics. The entomological contributions by Grimshaw and his colleagues at the beginning of the twentieth century were an integral part of that goal. One hundred years later Courchamp *et al.* (2014) highlighted a need to foster such fundamental research, to advance knowledge and understanding, rather than solely pursuing applied research with a defined objective. More recently, partly because of a shift in priorities away from taxonomy towards applied research, Thomson *et al.* (2018) have also attributed the global decline in taxonomic expertise to the dwindling instruction in fundamental taxonomic principles and nomenclature at many universities.

Assessment of biodiversity, the act of exploring diversity in nature to find out 'what's there' is not possible without appropriate expertise. Promotion of the scientific discipline of taxonomy in academic curricula, leading to a proficiency in formal identification of various forms of life, is necessary to provide the expertise for future biodiversity assessment, perhaps to facilitate another survey of Clare Island in the next century.

Acknowledgements

Financial assistance for fieldwork on Clare Island in 2002 through a Heritage Council of Ireland grant to Dr T.K. McCarthy[†], National University of Ireland, Galway, is gratefully acknowledged as is the financial support from the Praeger Committee of the Royal Irish Academy towards fieldwork on Clare Island and in west Mayo in 2004. Some records in west Mayo were obtained during the tenure of grants from the Heritage Council of Ireland under the 2005 and 2006 Wildlife Grant Scheme (projects WLD/2005/13985 and WLD/2006/14748) that are gratefully acknowledged. I am grateful to research staff of the Environmental Protection Agency—Catherine Bradley, Gary Free, Bryan Kennedy, Ruth Little, Patricia McCreesh, Caroline Plant and Wayne Trodd—who collected samples of pupal exuviae from lakes in the Westport area and to the Agency's Senior Research Officer, Dr Deirdre Tierney, for permission to publish records from those collections. I acknowledge the inputs of Dr Patrick Ashe[†] while confirming identification of some voucher specimens in NMI. I thank Siobhan Atkinson, School of Biology and Environmental Science, University College Dublin for assistance with preparation of the maps. I thank my son Finbarr for some collections in west Mayo and my wife, Freddie (W.A. Murray), for technical assistance with fieldwork, constructive and practical comments on the manuscript and for her support and understanding.

REFERENCES

Agnarsson, I. and Kuntner, M. 2007 Taxonomy in a changing world: seeking solutions for science in a crisis. *Systematic Biology* **56**, 531–39.

Andersen. T., Sæther, O.A., Cranston, P.S. and Epler, J.H. 2013 The larvae of Orthocladiinae (Diptera: Chironomidae) of the Holarctic Region – Keys and daignoses. In T. Andersen, P.S. Cranston and J.H. Epler (eds) *The larvae of Chironomidae (Diptera) of the Holarctic region – Keys and diagnoses.* Insect Systematics and Evolution, Supplement **66**, 189–386.

Ashe, P. and O'Connor, J.P. 2009 *A World Catalogue of Chironomidae (Diptera) Part 1, Buchonomyiinae, Podonominae, Aphroteniinae, Tanypodinae, Usambaromyiinae, Diamesinae, Prodiamesinae and Telmatogetoninae.* Irish Biogeographical Society and National Museum of Ireland, Dublin. 455 pp.

Ashe, P. and O'Connor, J.P. 2012 *A World Catalogue of Chironomidae, (Diptera) Part 2 Orthocladiinae,* Irish Biogeographical Society and National Museum of Ireland, Dublin, 968 pp.

Ashe, P., O'Connor, J.P. and Murray, D.A. 1998 A checklist of Irish aquatic insects. *Occasional Publication of the Irish Biogeographical Society* **3**, 1- 80.

Chandler, P.J. 2016 Changes to the Irish Diptera Checklist (30). *Dipterists Digest (Second Series)*, **22**, 212.

Chandler, P.J., O'Connor, J.P. and Nash, R. 2008 *An annotated checklist of the Irish two-winged flies (Diptera).* Irish Biogeographical Society and National Museum of Ireland. Dublin. 261 pp.

Cranston, P.S., Dillon, M.E., Pinder, L.C.V. and Reiss, F. 1989 The adult males of Chironominae (Diptera: Chironomidae) of the Holarctic region - Keys and diagnoses In T. Wiederholm (ed.), *Chironomidae of the Holarctic region - Keys and diagnosis*. Part 3. Adult males. *Entomologica scandinavica* Supplement **34**, 353–502.

Coe, R.L. 1950 Family Chironomidae. Handbooks for the Identification of British Insects. *Royal Entomological Society of London* **9** (2), 121–206.

Costello, M.J., May, R.M. and Stork, N.E. 2013 Can we name Earth's species before they go extinct? *Science* **339**, 413–16.

Courchamp, F. Dunne, J.A., Le Maho, Y., May, R.M. Thébaud, C. and Hochberg, M.E. 2014 Fundamental ecology is fundamental. *Trends in Ecology and Evolution* **1886** 1–8.

Edwards, F.W. 1929 British nonbiting midges. *Transactions of the Royal Entomological Society, London* **77**, 279–364.

Ekrem, T. 2004 Immature stages of European Tanytarsus species 1. The *eminulus-*, *gregarius-*, *lugens-* and *mendax* species groups (Diptera, Chironomidae). *Deutsche Entomologische Zeitschrift* **51**, 97–146.

Fahy, S. 2002 Studies on the Aquatic Insects of Clare Island, with particular reference to Chironomidae. Unpublished PhD Thesis, National University of Ireland, Galway.

Fittkau, E.J. 1962 Die Tanypodinae (Diptera: Chironomidae) Die Tribus Tanypodiini, Macropelopiini, und Pentaneurini. *Abhandlung zur Larvalsystematik der Insekten* **6**, 1–453. Akademie Verlag, Berlin.

Fittkau, E.J. and Murray, D.A. 1986 The pupae of Tanypodinae (Diptera: Chironomidae) of the Holarctic region -Keys and diagnoses. In: T. Wiederholm (ed.), *Chironomidae of the Holarctic region - Keys and diagnosis*. Part 2. Pupae. *Entomologica scandinavica* Supplement **28**, 31–113.

Fontaine, B., van Achterberg, K., Alonzo-Zarazaga, M.A., Araujo, R., Asche, M. *et al.* 2012 New Species in the Old World: Europe as a frontier in Biodiversity Exploration, a test bed for twenty-first Century Taxonomy. *PLoS ONE* 7:e36881. doi:101371/journal. pone.0036881. 1–7 (accessed June 2014).

Grimshaw, P.H. 1912 Clare Island Survey: Diptera. *Proceedings of the Royal Irish Academy* **31** (1911–15), Section 25, 1–9.

Hendel, F. 1908 Nouvelle classification des mouches à deux ailes (Diptera L.) D'aprés un plan tout nouveay par J.G. Meigen, Paris, an VIII (1800 v.s.) Mit einem Kommentar herausgegeben von Friedrich Hendel (Wien). *Verhandlungen der Kaiserlich-Koniglichen Zoologisch-Botanischen Gesellschaft in Wien (Abhandlungen)* **58**, 43–69.

Hirvenoja, M. 1973 Revision der Gattung *Cricotopus* van der Wulp und ihrer Verwandten (Diptera, Chironomidae). *Annales Zoologici Fennici* **10**, 1–363.

Humphries, C.F. 1938 The Chironomid Fauna of the Grosser Plöner See, the Relative Density of its Members and their Emergence Period. *Archiv für Hydrobiologie* **33**, 535–48.

ICZN 1910 International Commission on Zoological Nomenclature, Opinion 28, Shall the Nouvelle Classification of Meigen 1800 be given precedence over Meigen's Versuch 1803. *Science* 1910, **32**(820) 380–82.

ICZN 1963 International Commission on Zoological Nomenclature, Opinion 678. The suppression under plenary powers of the pamphlet published by Meigen 1800. *Bulletin of Zoological Nomenclature* **20**, 339–42.

John, D. 2007 Freshwater and terrestrial habitats. In M.D. Guiry, D.M. John, F. Rindi and T.K. McCarthy (eds), *New Survey of Clare Island*. Volume 6: *The Freshwater and Terrestrial Algae*, 13–19. Dublin. Royal Irish Academy.

Krusemann, G. 1933 Tendepedidae Neerlandicae. Pars I. genus Tendipes cum generibus finitimis. *Tijdschrift Entomologiae* **76**, 119–216.

Langton, P.H. 1980 The genus *Psectrocladius* Kieffer (Diptera: Chironomidae) in Britain. *Entomologist's Gazette* **31**, 75–88.

Langton, P.H. 1991 *A key to pupal exuviae of West Palaearctic Chironomidae*. Privately published by the author, Cambridgshire.

Langton, P.H. 2004 Additions to the Irish List of Chironomidae (Diptera), including the first species of the millennium new to the British Isles. *Dipterists Digest (Second Series)*, **10**, 131–4.

Lanton, P.H. 2015a *Telmatogeton murrayi* Sæther (Diptera, Chironomidae) new to Ireland. *Dipterists Digest (Second Series)*, **22**, 27.

Langton, P.H. 2015b *Metriocnemus ephemerus* sp. nov. (Diptera, Chironomidae) from Northern Ireland. *Dipterists Digest (Second Series)*, **22**, 35–42.

Langton, P.H and Handcock, G. 2013 *Telmatogeton murrayi* Sæther and *T. japonicus* Tokunaga (Diptera, Chironomidae) new to Britain. *Dipterists Digest (Second Series)*, **20**, 157–60.

Langton, P.H. and Murray, D.A. 2018 The pupa of *Metriocnemus* (*Metriocnemus*) *ephemerus* Langton (Diptera, Chironomidae). Dipterists Digest (Second Series) **25**: 188–92.

Langton, P.H. and Pinder, L.C.V. 2007 Keys to the adult male Chironomidae of Britain and Ireland. *Freshwater Biological Association Scientific Publication* **64**, Volumes 1 and 2.

Langton, P.H. and Ruse, L.P. 2018 *Nanocladius* (*Nanocladius*) *distinctus* (Malloch, 1915) (Diptera, Chironomidae) new to Britain and Ireland. *Dipterists Digest (Second Series)* **25**: 17–19.

Langton P.H. and Visser, H. 2003 *Chironomidae exuviae - A key to pupal exuviae of the West Palaearctic Region*. Interactive System for the European Limnofauna Biodiversity Centre of ETI. UNESCO Publishing, Paris.

Langton, P.H., Bitusik, P. and Mitterova, J. 2015 A contribution towards a revision of West Palaearctic *Procladius* Skuse (Diptera: Chironomidae). *Chironomus Journal of Chironomidae Research* **26**: 41–44.

Legakis, A. and Murray, D. 2001 Insecta. In M.J. Costello, C.S. Emblow and R. White (eds), *European Register of Marine Species, a check-list of the marine species in Europe and a bibliography of guides to their identification*. *Patrimoines naturels* **50**, 323–24.

Lindeberg, B. 1968 Population differences in *Tanytarsus gracilentus* (Holmg.) (Dipt., Chironomidae). *Annales Zoologici Fennici* **5**, 88–91.

Malloch, J.R. 1917 A preliminary classification of Diptera, exclusive of Pupipara, based on larval and pupal characters, with keys to the imagines in certain families. *Bulletin of the Illinois State Laboratory of Natural History* (1918) **12**, 161–409.

Meigen, J.W. 1800 *Nouvelle classification des mouches à deux ailes (Diptera L.) d'apres un plan tout nouveaux*. J.J. Fuchs, Paris, 40 pp.

Meigen, J.W. 1803 Versuch einer neuen GattungsEinteilung der europäischen zweiflügligen Insekten. (*Magazin fur Insektenkunde (Illiger)* **2**, 259–81.

Miall, L.C. and Hammond, A.R. 1900 *The Structure and Life-history of the Harlequin Fly (Chironomus)*. Clarendon Press, Oxford.

Murray, D.A. 1972 A list of the Chironomidae (Diptera) known to occur in Ireland with notes on their distribution. *Proceedings of the Royal Irish Academy* **72B**: 275–93.

Murray, D.A. 1999 Two marine coastal-dwelling Chironomidae (Diptera) new to the fauna of Iceland: *Telmatogeton japonicus* Tokunaga (Telmatogetoninae) and *Clunio marinus* Haliday (Orthocladiinae). *Bulletin of the Irish biogeographical Society* **23**, 89–91.

Murray, D.A. 2000 First record of *Telmatogeton japonicus* Tokunaga (Dipt., Chironomidae) from the British Isles and additional records of halobiontic Chironomidae from Ireland. *Entomologist's Monthly Magazine* **136** (2000), 157–9.

Murray, D.A. 2005 *Irish Chironomidae (Diptera: Insecta) Reference Collection and Catalogue*. Report to The Heritage Council WLD/2005/13985.

Murray, D.A. 2010 Records of Chironomidae (Diptera) in Ireland - twenty additions and notes on four morphotypes. *Bulletin of the Irish Biogeographical Society* **34**, 85–96.

Murray, D.A. 2017a The Chironomidae (Insecta, Diptera) of Achill Island, Co. Mayo – new records and checklist. *Bulletin of the Irish biogeographical Society* **41**, 90–102.

Murray, D.A. 2017b A second record of *Metriocnemus ephemerus* Langton 2015 from Ireland. *Dipterists Digest* **25**, 155.

Murray, D.A. 2019 Records and a checklist of Chironomidae (Insecta: Diptera) in County Louth and an updated summary of species distribution in Ireland. *Bulletin of the Irish biogeographical Society* **43**, 42–56.

Murray, D.A. and Ashe, P. 1982 First records of the sub-families Podonominae and Telmatogetoninae (Diptera: Chironomidae) from Ireland. *Irish Naturalists' Journal* **20**, 546–7.

Murray D.A. and Fittkau, E.J. 1989 The adult males of Tanypodinae (Diptera: Chironomidae) of the Holarctic region - Keys and diagnoses. In T. Wiederholm (ed.), *Chironomidae of the Holarctic region - Keys and diagnosis*. Part 3. Adult males. *Entomologica scandinavica* Supplement **34**, 37–123.

Murray, D.A. and Langton, P.H. 2018 Doubt and certainty in records of *Micropsectra recurvata* Goetghebuer (Insecta: Diptera: Chironomidae) in Ireland. *Bulletin of the Irish biogeographical Society* **42**, 155–58.

Murray, D.A. and Murray, W.A. 2003 A reassessment of Chironomidae (Diptera) of Clare Island, Co. Mayo, with first records of *Acamptocladius reissi* Cranston and Sæther and *Limnophyes angelicae* Sæther (Orthocladiinae) for the Irish faunal checklist. *Bulletin of the Irish Biogeographical Society* **27**, 255–69.

Murray, D.A. and Murray, W.A. 2006 Notes on six Chironomidae (Diptera, Insecta) new to Ireland from Co. Mayo. *Bulletin of the Irish Biogeographical Society* **30**, 294–98.

Murray, D.A., O'Connor, J.P. and Ashe, P.J. 2018 *Chironomidae (Diptera) of Ireland – a review, checklist and their distribution in Europe*. Occasional publication of the Irish Biogeographical Society **Number 12**, 412p.

Murray, D.A., Langton, P.H., O'Connor, J.P. and Ashe, P. 2013 Distribution records of Irish Chironomidae (Diptera): Part 1 – Buchonomyiinae, Podonominae, Tanypodinae, Telmatogetoniinae, Diamesinae and Prodiamesinae. *Bulletin of the Irish biogeographical Society* **37**, 208–336.

Murray, D.A., Langton, P.H., O'Connor, J.P. and Ashe, P. 2014 Distribution records of Irish Chironomidae (Diptera): Part 2 – Orthocladiinae. *Bulletin of the Irish biogeographical Society* **38**, 72–318.

Murray, D.A., Langton, P.H., O'Connor, J.P. and Ashe, P. 2015 Distribution records of Irish Chironomidae (Diptera): Part 3 – Chironominae. *Bulletin of the Irish biogeographical Society* **39**, 7–192.

Myers, A. and McGrath, D. 2002 General Introduction to Clare Island Intertidal Marine Studies. In A. Myers (ed.) *New Survey of Clare Island*. Volume 3: *Marine Intertidal Ecology,* 1–5. Dublin. Royal Irish Academy.

O'Connor, J.P., Scannell, M.J.F. and Speight, M.C.D. 1988 The state of taxonomy in Ireland. In C. Moriarty (ed.), *Taxonomy, putting plants and animals in their place,* 202–8. Proceedings of a seminar, Royal Irish Academy, Dublin.

Pagast, F. 1947 Systematik und Verbreitung der um die Gattung *Diamesa* grupppierten Chironomiden. *Archiv für Hydrobiologie* **41**, 435–596.

Pinder, L.C.V. 1978 A key to the adult males of British Chironomidae. *Freshwater Biological Association Scientific Publication* **37**, volumes 1 and 2.

Pinder, L.C.V. 1986 The Pupae of Chironomidae - Introduction. In T. Wiederholm (ed.) *Chironomidae of the Holarctic region - Keys and diagnoses*. Part 2. Pupae. *Entomologica Scandinavica* Supplement **28**, 1–7.

Pinder, L.C.V. 1989 The Adult males of Chironomidae - Introduction. In T. Wiederholm (ed.) *Chironomidae of the Holarctic region - Keys and diagnoses. Part 3. Adult males*. Entomologica Scandinavica Supplement **34**, 1–9.

Pinder, L.C.V. 1995 The habitats of chironomid larvae. In P. Armitage, P.S. Cranston and L.C.V. Pinder (eds) *The Chironomidae: The biology and ecology of non-biting Midges*, 107–35. Chapman and Hall, London.

Reiss, F. 1969 Revision der Gattung *Micropsectra* Kieff., 1909 (Diptera, Chironomidae) 1. die *attenuata* Gruppe der Gattung *Micropsectra*. Beschreibung 5 neuer Arten aus Mitteleuropa und Nordafrika. *Deutsche Entomologische Zeitschrift* **16**, 431–49.

Reiss, F. 1988 Die Gattung *Kloosia* Kruseman, 1913 mit der Neubeschreibung zweier Arten (Diptera, Chironomidae). *Spixiana* Supplement **14**, 35–44.

Reiss, F. and Fittkau, E.J. 1971 Taxonomie und Ökologie europäisch verbreiteter *Tanytarsus*-Arten (Chironomidae, Diptera). *Archiv für Hydrobiologie* Supplement **40**, 75–200.

Sæther, O.A. 1990 A review of the genus *Limnophyes* Eaton from the Holarctic and Afrotropical regions (Diptera: Chironomidae, Orthocladiinae). *Entomologica Scandinavica* **35**, 1–135.

Sæther, O.A. 1995 *Metriocnemus* van der Wulp: seven new species, revision of species and new records (Diptera: Chironomidae) *Annales de Limnologie* **31**(1), 35–64.

Sæther, O.A. 2004 Three new species of *Orthocladius* subgenus *Eudactylocladius* (Diptera: Chironomidae) from Norway. *Zootaxa* **508**, 1–12.

Sæther, O.A. 2005 A new subgenus and new species of *Orthocladius* van der Wulp with a phylogenetic evaluation of the validity of the subgenera of the genus (Diptera: Chironomidae) *Zootaxa* **974**, 1–56.

Sæther, O.A. 2009 *Telmatogeton murrayi* sp. n. from Iceland and *T. japonicus* Tokunaga from Madeira (Diptera: Chironomidae). *Aquatic Insects* **31**, 31–44.

Sæther, O.A. and Sublette, J.E. 1983 A review of the genera *Doithrix* n.gen., *Georthocladius* Strenzke, *Parachaetocladius* (Wülker) and *Pseudorthocladius* Goetghebuer. (Diptera: Chironomidae Orthocladiinae). *Entomologica Scandinavica* Supplement **20**, 1–100.

Sæther O.A. and Wang, X. 1995 Revision of the genus *Paraphaenocladius* Thienemann 1924 of the world (Diptera: Chironomidae, Orthocladiinae). *Entomologica scandinavica* Supplement **48**, 1–69.

Sæther, O.A., Ashe, P. and Murray, D.A. 2000 Family Chironomidae. In L. Papp and B. Darvas (eds), *Contributions to a manual of Palaearctic Diptera* (with special reference to the flies of economic importance) **A 6**, 113–334. Budapest. Science Herald.

Silva, F.L. and Ekrem, T. 2015 Phylogenetic relationships of nonbiting midges in the aubfamily Tanypodinae (Diptera: Chironomidae) inferred from morphology. *Systematic Entomology* (2015) DOI: 10.1111/syen.12141.

Spies, M. 2005 On selected family-group names in Chironomidae (Insecta, Diptera), and related nomenclature. *Zootaxa* **894**, 1–12. Auckland, New Zealand. Magnolia Press. www.mapress.com/zootaxa/.

Spies, M. and Bolton, S.J. 2013 On the first record from Britain of *Parachironomus elodeae* (Townes) (Diptera, Chironomidae). *Dipterists Digest* **20**, 1–7.

Spies, M. and Dettinger-Klemm, A. 2015 Diagnoses for *Nubensia*, n. gen. (Diptera, Chironomidae, Chironomini), with the first full descriptions of the adult female and larva of *N. nubens* (Edwards, 1929). *Zootaxa* **3994**(1), 109–21.

Spies, M. and Sæther, O.A. 2004 Notes and recommendations on taxonomy and nomenclature of Chironomidae (Diptera). *Zootaxa* **752**, 1–90. Auckland, New Zealand. Magnolia Press. www.mapress.com/zootaxa/.

Spies, M. and Sæther, O.A. 2013 Chironomidae. Fauna Europaea: Chironomidae. In P. Beuk and T. Pape (eds) *Diptera Nematocera. Fauna Europaea version* **2.6**. Internet database: <http://www.faunaeur.org/>

Strenzke, K. 1959 Revision der Gattung *Chironomus* Meig. I. Die Imagines von 15 norddeutschen Arten und Unterarten. *Archiv für Hydrobiologie* **56**, 1–42.

Stur, E. and Ekrem, T. 2006 A revision of West Palaearctic species of the *Micropsectra atrofasciata* species group (Diptera: Chironomidae). *Zoological Journal of the Linnaean Society* **146**, 165–225.

Svensson, B.S. 1986 *Eukiefferiella ancyla* sp. n. (Diptera: Chironomidae) a commensalistic midge on *Ancylus fluviatilis* Müller (Gastropoda: Ancylidae). *Entomologica scandinavica* **17**, 292–8.

Thienemann, A. 1910 Das Sammeln von Puppenhäuten der Chironomiden. *Archiv für Hydrobiologie* **6**, 213–4.

Thienemann, A. and Kieffer, J.J. 1916 Schwedische Chironomiden. *Archiv für Hydrobiologie* Supplement **2**, 483–554.

Thomson, S.A., Pyle, R.L. Ahyong, S.T. Alono-Zaragoza, M., Ammirati, J., Araya, J.F. et al. 2018 Taxonomy based on science is necessary for global conservation. *PLoS Biology* **16**(3), e2005075. https://doi.org/10.1371/journal.pbio.2005075.

Whitton, B.A. 2007 *The Blue-Green Algae (Cyanophyta)*. In M.D. Guiry, D.M. John, F. Rindi and T.K. McCarthy (eds), *New Survey of Clare Island. Volume 6: The Freshwater and Terrestrial Algae*, 141–204. Dublin. Royal Irish Academy.

Wiederholm, T. (ed.) 1986 *Chironomidae of the Holarctic region - Keys and diagnosis*. Part 2. Pupae. *Entomologica scandinavica* Supplement **28**, 1–481

Wiederholm, T. (ed.) 1989 *Chironomidae of the Holarctic region - Keys and diagnosis*. Part 3. Adult males. *Entomologica scandinavica* Supplement **34**, 1–532.

Wilson, R.S. and Ruse, L.P. 2005 A Guide to the Identification of Genera of Chironomid Pupal Exuviae occurring in Britain and Ireland (including common genera from northern Europe) and their use in monitoring lotic and lentic waters. *Freshwater Biological Association,* Special Publication **13**, 1–176.

Wirth, W.W. 1949 A revision of clunionine midges with descriptions of a new genus and four new species (Diptera: Tendepididae). *University of California Publications in Entomology* **8**, 151–82.

APPENDIX Ia

NSCI sites, Irish Grid Reference (IGR), location data and sampling dates on Clare Island

Site No	IGR	Townland	Locality	Sampling dates
1	L696882	Ballytoohy	Wetland near lighthouse	20-Aug-02
2	L695879	Ballytoohy	Seepage near lighthouse	20-Aug-02
3	L698875	Ballytoohy	Artificial tank, lighthouse road	20-Aug-02
4	L699874	Ballytoohy	1st order stream, lighthouse road	20-Aug-02
5	L699870	Ballytoohy	Roadside, Ballytoughey	20-Aug-02
6	L708865	Maum	River Doree, 100m from sea	28-Apr-00
7	L710863	Maum	Pond east of Fawnglass road	28-Apr-00, 5-Jun-02, 20-Aug-02
8	L709863	Maum	Pool west of Fawnglass road	28-Apr-00, 5-Jun-02
9	L711861	Maum	1st order stream Fawnglass road	28-Apr-00
10	L712857	Fawnglass	Fawnglass to Maum road	04-Jun-02, 20-Aug-02
11	L716856	Capnagower	Roadside at hotel	28-Apr-00, 5-Jun-02, 21-Aug-02
12	L717857	Capnagower	Marine pools, below hotel	29-Apr-00, 5-Jun-02, 20-Aug-02
13	L713854	Fawnglass	Stream at Community Centre	28-Apr-00, 21-Aug-02
14	L713852	Fawnglass	Glen stream, 50m above sea	28-Apr-00, 5-Jun-02
15	L716851	Glen	Marine pools, Granuaile's Castle	05-Jun-02
16	L698846	Kill	Portnakilly road	05-Jun-02
17	L692842	Strake	Portnakilly harbour	05-Jun-02
18	L689847	Strake	Roadside, Abbey-Knocknaveen	05-Jun-02
19	L691856	Lecarrow	Leinapollbauty Lough	05-Jun-02
20	L689857	Lecarrow	Creggan Lough	21-Aug-02
20F	L689857	Lecarrow	Creggan Lough (Fahy, 2002)	Apr–Nov, 1998 and 1999
21	L692860	Lecarrow	Dorree River	05-Jun-02
22	L698867	Lecarrow	Dorree River, 1km d/s site 21	05-Jun-02
23	L704858	Lecarrow	Pollabrandy stream Knocknaveen	04-Jun-02
23F	L706857	Lecarrow	Poirtin Fuinch (Fahy, 2002)	Apr–Nov 1998 and 1999
24	L705856	Lecarrow	Artificial tank, Knocknaveen	04-Jun-02, 21-Aug-02
25	L698858	Lecarrow	Knocknaveen	28-Apr-00, 04-Jun-02, 21-Aug-02
26	L698855	Lecarrow	Knocknaveen, seepage on north hillside	04-Jun-02
27	L683857	Scalpatruce	Bog pool, Knockmore	21-Aug-02
28	L664864	Bunnamohaun	Seepage, Knockmore	21-Aug-02
29	L658859	Bunnamohaun	Small pool east of Signal tower	21-Aug-02
30	L657850	Bunnamohaun	Loughnapucar stream, S.E. Signal Tower	21-Aug-02
31	L658848	Bunnamohaun	Loughnapucar stream, d/s site 30	21-Aug-02
32	L659846	Bunnamohaun	Humic pond	21-Aug-02
33	L652849	Bunnamohaun	Over rock pool, Leckacanny	21-Aug-02

Photographs of Sites 6, 19, 20 and 32 are given in John (2007): Site 6 - Plate II; Site 19 - Plate IX; Site 20, (20F) - Plate X; Site 32 - Plate XVI

Photographs of Sites 7, 8 and 14 are given in Whitton (2007): Site 7 - Plate XI; Site 8 - Plate XII; Site 14 -Plate X

APPENDIX Ib

NSCI sites, Irish Grid Reference (IGR), location data and sampling dates on mainland west Mayo				
Site no	IGR	Townland	Location	Sampling dates
43	L755997	Achill	Pollranny Sweeney, stream	06-May-04
44	L756926	Achill	Bolinglanna Stream	06-May-04
45	L886962	Mulranny	Owengarve River, Rosgalliv	04-Aug-04
46	L975983	Furnace, Newport	L. Feeagh (EPA samples)	11-Apr-08, 25-Jun-08, 24-Jul-08, 17-Sep-08
47	L996882	Westport	L. Ballin (EPA samples)	22-Apr-08, 25-Jun-08, 18-Sep-08
48	M032880	Clogher	Clogher Lough, North shore	30-Sep-04
49a	M106879	Castlebar	Castlebar Lough, North shore	07-May-04
49b	M109883	Castlebar	Castlebar Lough (L. Lannagh, EPA samples)	23-Apr-08, 30-Jun-08, 21-Aug-08, 26-Sep-08
50	M121902	Castlebar	Lough Mallard, Castlebar	30-Sep-04
51	L985843	Westport	Westport Woods Hotel grounds	07-May-02, 30-Sep-04, 07-Mar-2018
52	L959822	Westport	Owenwee River, Belclare	07-May-02, 30-Sep-04
53	L972817	Westport	Owenwee River, Aghamore	05-Aug-00, 08-May-03
54	L958813	Knappagh	Roadside, Prospect	07-May-04
55	M070807	Westport	L. Aille (EPA samples)	22-Apr-08, 24-Jun-08, 8-Aug-08, 18-Sep-08
56	M010803	Aghagower, Westport	L. Knappabeg (EPA samples)	21-Apr-08, 24-Jun-08, 23-Jul-08, 18-Sep-08
57	L977766	Liscarney, Westport	L. Moher (EPA samples)	21-Apr-08, 24-Jun-08, 23-Jul-08, 18-Sep-08
58	L921830	Murrisk	Carrokeel Lake - coastal	05-Aug-04
59	L908824	Murrisk	Archeological site 9, Murrisk	07-May-04
60	L908825	Murrisk	Archeological site 9, coastal	07-May-04
61	L807807	Louisburg	Bunowen River	07-May-04, 05-Aug-04
62	L845759	Laghta	Bog marsh pool, west of road	07-May-04
63	L852751	Laghta	Bunowen River	30-Sep-04
64	L867739	Laghta	Bunowen tributary	07-May-04
65	L770773	Killeen	Carrowinsky River, NE Killeen	05-Aug-04
66	L754759	Cloonlaur	Clapper Bridge Ford	05-Aug-04
67	L749759	Roonagh	Roonagh Lough, SE shore	30-Sep-04
68	L784744	Derrygarve	1st road bridge d/s L. Nahaltora	06-Jun-04
69	L794745	Cregganbaun	L. Nahaltora, Cregganbaun	30-Sep-04
70	L744808	Roonagh	South of Roonagh Pier	27-Apr-00
71	L834720	Glenkeen	Carrowinskey River, SE Cregganbaun	09-May-03
72	L815699	Doolough Pass	L. Glencullin (EPA samples)	03-Apr-08, 26-Jun-08, 25-Aug-08, 22-Sep-08
73	L835680	Doolough Pass	Doo Lough (+EPA samples from 2006 and 2008)	06-Jun-02, 22-Aug-02, 5-Apr-06, 9-May-06, 28-Jun-06, 12-Jul-06, 16-Aug-06, 3-Apr-08, 25-Jun-08, 27-Aug-08
74	L841657	Delphi	L. Fin, Delphi (EPA samples)	19-May-05, 5-Apr-06, 9-May-06, 28-Jun-06, 12-Jul-06, 11-Sep-06

APPENDIX Ic

NSCI sites, Irish Grid Reference (IGR), location data and sampling dates on Achill Island

Site no	IGR	Townland	Location	Sampling dates
34	F560043	Achill Island	Keem Bay west - stream 1	05-Aug-04
35	F516045	Achill Island	Keem Bay east - stream 2	05-Aug-04
36	F578057	Achill Island	L. Acorrymor (+EPA samples 2008)	05-Aug-04, 29-Sep-04, 07-Apr-08, 17-Jun-08, 16-Jul-08, 17-Sep-08
37	F649050	Achill Island	Keel lake (+EPA samples 2008)	07-Apr-08, 17-Jun-08, 16-Jul-08, 17-Sep-08
38	F695048	Achill Island	Bunacurry lake	29-Sep-04
39	F716005	Achill Island	Bog pool, Sraheene	06-May-04
40	F712008	Achill Island	Bog pool, Glendarary	05-Aug-04
41	L677982	Achill Island	Pond and outflow, Camport	29-Sep-04
42	L718983	Achill Island	Sraheens Bridge, stream	06-May-04
75	F695087	Achill Island	Dugort – at window	28-Aug-2017
76	F697088	Achill Island	River Ballynagappul, Doogort	28-Aug_2017
77	F694886	Achill Island	Pond on machair flatland, Doogort	28-Aug_2017
78	F671089	Achill Island	Rock pools on Pollawaddy strand	29-Aug-2017
79	F671088	Achill Island	Stream at bridge, Pollawaddy strand	29-Aug-2017
80	F713094	Achill Island	Lough Doo	29-Aug-2017
81	F711097	Achill Island	Bog pool, west on Ridge Point road	29-Aug-2017
82	F717099	Achill Island	Rock pool, east on Ridge Point road	29-Aug-2017
83	F720091	Achill Island	Lough Sruhill	29-Aug-2017

APPENDIX Id

Locations from the Clare Island Survey (CIS) and allocated alpha-numeric code (CIS site) and corresponding numbers for NSCI sites (Appendices Ib and Ic)

CIS Location	CIS site	NSCI site
*Clare Island	G	1-33
Mainland Mayo (including Achill Island)		
Achill NE (Bunacurry)	G1	38, 75–83
Achill	G1	34–37, 39–42
Glendarary	G2	40
Mulranny	G3	45
Clogher	G4	48
Castlebar Lough	G5	49
Westport Demesne	G6	51
Westport Demesne garden	G6	51
Westport	G6	51
Westport riverside	G6	51
Westport environs	G6	51
Belclare	G7	52
Knappagh	G8	53
Lough near Westport (Ballin Lough)	G9	47
Louisburg	G10	61

*Records from the CIS for Clare Island were given without site location data

APPENDIX II

Records of Chironomidae from Clare Island and the west Mayo mainland by Grimshaw (G), Fahy (F) and Murray (M). Clare Island Survey (CIS); New Survey of Clare Island (NSCI)

	Clare Island		Mainland	
	CIS	NSCI	CIS	NSCI
Subfamily Tanypodinae				
Ablabesmyia (Ablabesmyia) longistyla Fittkau, 1962		F		M
Ablabesmyia (A.) monilis (Linnaeus, 1758)		F, M		M
Ablabesmyia (A.) phatta (Egger, 1863)		F	G	M
Apsectrotanypus trifascipennis (Zetterstedt, 1838)				M
Arctopelopia barbitarsis (Zetterstedt, 1850)		F		M
Arctopelopia griseipennis (van der Wulp, 1859)		M		M
Clinotanypus nervosus (Meigen, 1818)			G	M
Conchapelopia melanops (Meigen, 1818)				M
Conchapelopia pallidula (Meigen, 1818)				M
Conchapelopia viator (Kieffer, 1911)				M
Krenopelopia nigropunctata (Staeger, 1839)		M		M
Larsia sp. ? *atrocincta* (Goetghebuer, 1942)		F		
Larsia curticalcar (Kieffer, 1818)				M
Macropelopia adaucta Kieffer, 1916		M		M
Macropelopia nebulosa (Meigen, 1804)	G		G	M
Macropelopia notata (Meigen, 1818)		F, M		M
Monopelopia tenuicalcar (Kieffer, 1918)		M		M
Natarsia punctata (Fabricius, 1805)		F, M		
Nilotanypus dubius (Meigen, 1923)		M		M
Procladius (Holotanypus) choreus (Meigen, 1804)		F, M		M
Procladius (H.) crassinervis (Zetterstedt, 1838)				M
*[*Procladius (H.) culiciformis* (Linnaeus, 1767)]			G	
Procladius (H.) sagittalis (Kieffer, 1909)		F		M
Procladius (H.) signatus (Zetterstedt, 1850)				M
Procladius (H.) simplicistilus Freeman, 1948				M
Procladius (Psilotanypus) flavifrons Edwards, 1929				M
Procladius (P.) lugens Kieffer, 1915				M
Procladius (P.) rufovittatus (van der Wulp, 1874)			G	M
Rheopelopia maculipennis (Zetterstedt, 1848)				M
Tanypus (Tanypus) kraatzi (Kieffer, 1912)				M
Tanypus (T.) vilipennis (Kieffer, 1918)				M
Thienemannimyia (Thienemannimyia) laeta (Meigen, 1818)				M
Thienemannimyia (T.) northumbrica (Edwards, 1929)				M
Trissopelopia longimana (Staeger, 1839)				M
Zavrelimyia (Paramerina) cingulata (Walker, 1856)		F, M	G	M
Zavrelimyia (P.) divisa (Walker, 1856)				M
Zavrelimyia (Zavrelimyia) barbatipes (Kieffer, 1911)		M		M
Zavrelimyia (Z.) melanura (Meigen, 1804)				M
Subfamily Telmatogetoninae				
Thalassomya frauenfeldi Schiner, 1856		M		
Subfamily Diamesinae				
Diamesa (Diamesa) insignipes Kieffer, 1908		M		

	Clare Island		Mainland	
	CIS	NSCI	CIS	NSCI
Potthastia gaedii (Meigen, 1838)	G			M
Potthastia longimanus Kieffer, 1922				M
Protanypus morio (Zetterstedt, 1838)				M
Subfamily Prodiamesinae				
Prodiamesa olivacea (Meigen, 1818) [? As *P. obscurimana*]	?G	M	?G	M
Subfamily Orthocladiinae				
Acamptocladius reissi Cranston & Sæther, 1981		M		
Acamptocladius submontanus (Edwards, 1932)				M
Acricotopus lucens (Zetterstedt, 1850)		M		M
Brillia bifida (Kieffer, 1909)		M		M
Brillia longifurca Kieffer, 1921				M
Bryophaenocladius subvernalis (Edwards, 1929)				M
Camptocladius stercorarius (De Geer, 1776)		M		
?*Cardiocladius fuscus* Kieffer, 1924		?F		
Chaetocladius (*Chaetocladius*) *dentiforceps* (Edwards, 1929)		F		
Chaetocladius (*C.*) *melaleucas* (Meigen, 1818)		F, M		M
Chaetocladius (*C.*) *perennis* (Meigen, 1830)		F, M		M
Clunio marinus Haliday, 1855		M		M
Corynoneura arctica Kieffer, 1923				M
Corynoneura carriana Edwards, 1924				M
Corynoneura celtica Edwards, 1924				M
Corynoneura edwardsi Brundin, 1949		F, M		M
Corynoneura gratias Schlee, 1968				M
Corynoneura lobata Edwards, 1924		M		M
Corynoneura Pe2a sensu Langton & Viser, 2003				M
Corynoneurella paludosa Brundin, 1949		M		
Cricotopus (*Cricotopus*) *albiforceps* (Kieffer, 1916)		M		
Cricotopus (*C.*) *annulator* Goetghebuer, 1927		F	G	M
Cricotopus (*C.*) *bicinctus* (Meigen, 1818)			G	M
Cricotopus (*C.*) *ephippium* (Zetterstedt, 1838)		M		
Cricotopus (*C.*) *festivellus* (Kieffer, 1906)		F		
Cricotopus (*C.*) *flavocinctus* (Kieffer, 1924)				M
Cricotopus (*C.*) *fuscus* (Kieffer, 1909)				M
Cricotopus (*C.*) *pallidipes* Edwards, 1929				M
Cricotopus (*C.*) *pilosellus* Brundin, 1956		F, M		M
Cricotopus (*C.*) *polaris* (Kieffer, 1926)				M
Cricotopus (*C.*) *pulchripes* Verrall, 1912		M		M
Cricotopus (*C.*) *similis* Goetghebuer, 1921				M
[*Cricotopus* (*C.*) *tibialis* (Meigen, 1804)]	G		G	M
Cricotopus (*C.*) *tremulus* (Linnaeus, 1758)				M
Cricotopus (*C.*) *trifascia* (Edwards, 1929)				M
Cricotopus (*C.*) *tristis* Hirvenoja, 1973				M
Cricotopus (*Isocladius*) *brevipalpis* Kieffer, 1909				M
Cricotopus (*I.*) *intersectus* (Staeger, 1839)				M
Cricotopus (*I.*) *laricomalis* (Edwards, 1932)				M
Cricotopus (*I.*) *ornatus* (Meigen, 1818)		M		

	Clare Island		Mainland	
	CIS	NSCI	CIS	NSCI
Cricotopus (I.) sylvestris (Fabricius,1794)		F	G	M
[*Cricotopus (I.) tricinctus* (Meigen, 1818)]			G	
Cricotopus (I.) trifasciatus (Meigen, 1810)			G	M
Cricotopus (I.) Pe 2 sensu Langton 1991				M
Cricotopus (I.) Pe 5 sensu Langton 1991				M
Cricotopus (Paratrichocladius) rufiventris (Meigen, 1830)		M		M
Cricotopus (P.) skirwithensis (Edwards, 1929)				M
Eukiefferiella ancyla Svensson, 1986				M
Eukiefferiella brevicalcar (Kieffer, 1911)				M
Eukiefferiella claripennis (Lundbeck, 1898)		M		M
Eukiefferiella clypaeata (Kieffer, 1923)				M
Eukiefferiella coerulescens (Kieffer, 1926)				M
Eukiefferiella cyanea Thienemann, 1936		M		
Eukiefferiella devonica (Edwards, 1929)		M		M
Eukiefferiella dittmari Lehmann, 1972				M
Eukiefferiella ilkleyensis (Edwards, 1929)				M
Eukiefferiella sp. *Pe minor* (Edwards, 1929) / *fittkaui* Lehman, 1972		M		M
Eukiefferiella tirolensis Goetghebuer, 1938				M
Georthocladius (Georthocladius) luteicornis (Goetghebuer, 1941)				M
Halocladius (Halocladius) fucicola (Edwards, 1926)		M		M
Halocladius (H.) variabilis (Staeger, 1839)		F, M		M
Halocladius (H.) varians (Staeger, 1839)				M
Heleniella ornaticollis (Edwards, 1929)				M
Heterotanytarsus apicalis (Kieffer, 1921)		M		M
Heterotrissocladius grimshawi (Edwards, 1929)				M
Heterotrissocladius marcidus (Walker, 1856)		M		M
Krenosmittia camptophleps (Edwards, 1929)				M
Limnophyes angelicae Sæther, 1990		F, M		
Limnophyes asquamatus Søgaard Andersen, 1937				M
Limnophyes gurgicola (Edwards, 1929)		M		
Limnophyes habilis (Walker, 1856)		F		
Limnophyes minimus (Meigen, 1818)		M		
Limnophyes natalensis (Kieffer, 1914)		M		
Limnophyes pentaplastus (Kieffer, 1921)		M		
Limnophyes pumilio (Holmgren, 1869)				M
Metriocnemus (Metriocnemus) ephemerus Langton, 2015				M
Metriocnemus (M.) eurynotus (Holmgren, 1883)		F, M		
Metriocnemus (M.) fuscipes (Meigen, 1818)		F, M	G	
Metriocnemus (M.) picipes (Meigen, 1818)		M		M
Nanocladius (Nanocladius) balticus (Palmén, 1959)				M
Nanocladius (N.) dichromus (Kieffer, 1906)				M
Nanocladius (N.) rectinervis (Kieffer, 1911)				M
Orthocladius (Eudactylocladius) fuscimanus (Kieffer, 1908)		M		
Orthocladius (Eud.) olivaceus (Kieffer 1911)				M
Orthocladius (Euorthocladius) ashei Soponis, 1990				M
Orthocladius (Euo.) rivicola Kieffer, 1911				M

	Clare Island		Mainland	
	CIS	NSCI	CIS	NSCI
Orthocladius (Mesorthocladius) frigidus (Zetterstedt, 1838)				M
Orthocladius (Orthocladius) dentifer Brundin, 1947		M		M
Orthocladius (O.) glabripennis Goetghebuer, 1921				M
Orthocladius (O.) oblidens (Walker, 1856)		F, M		M
Orthocladius (O.) rivinus Potthast, 1914				M
Orthocladius (O.) rubicundus (Meigen, 1818)				M
Orthocladius (O.) ruffoi Rossaro & Prato, 1991				M
Orthocladius (O.) wetterensis Brundin, 1956				M
Orthocladius (Pogonocladius) consobrinus (Holmgren, 1869)				M
Orthocladius (Symposiocladius) holsatus Goetghebuer, 1937				M
Paracladius conversus (Walker, 1856)				M
Parakiefferiella bathophila (Kieffer, 1912)				M
Parakiefferiella coronata (Edwards, 1929)				M
Parakiefferiella scandica (Brundin, 1947)				M
Parakiefferiella smolandica (Brundin, 1947)				M
Parametriocnemus stylatus (Spärck, 1923)		M		M
Paraphaenocladius impensus impensus (Walker, 1856)		M		M
Paraphaenocladius irritus irritus (Walker, 1856)		M		
Paraphaenocladius pseudirritus pseudirritus Strenzke, 1950		M		M
Psectrocladius (Allopsectrocladius) obvius (Walker, 1856)		F, M		M
Psectrocladius (A.) platypus (Edwards, 1929)		F, M		M
Psectrocladius (Mesopsectrocladius) barbatipes Kieffer, 1923				M
Psectrocladius (Psectrocladius) fennicus Storå, 1939		F		M
Psectrocladius (P.) limbatellus (Holmgren, 1869)		M		M
Psectrocladius (P.) octomaculatus Wülker, 1956				M
Psectrocladius (P.) oligosetus Wülker, 1956		M		M
Psectrocladius (P.) oxyura Langton, 1985				M
Psectrocladius (P.) psilopterus (Kieffer, 1906)				M
Psectrocladius (P.) sordidellus (Zetterstedt, 1838)	G	M	G	M
Psectrocladius (P.) ventricosus Kieffer, 1925		M		M
Psectrocladius (P.) sp. A sensu Langton, 1980				M
Pseudorthocladius (Pseudorthocladius) curtistylus (Goetghebuer, 1921)		F		
Pseudorthocladius (P.) filiformis (Kieffer, 1908)		M		M
Pseudorthocladius (P.) macrovirgatus Sæther & Sublette, 1983		M		
Pseudosmittia oxoniana (Edwards, 1922)		M		M
Pseudosmittia trilobata (Edwards, 1929)		M		
Rheocricotopus (Psilocricotopus) chalybeatus (Edwards, 1929)		M		M
Rheocricotopus (Rheocricotopus) fuscipes (Kieffer, 1909)				M
Smittia pratorum (Goetghebuer, 1927)		M		
Synorthocladius semivirens (Kieffer, 1909)		M		M
Thalassosmittia thalassophila (Bequaert & Goetghebuer, 1913)		M		M
Thienemannia gracilis Kieffer, 1909		M		
Thienemanniella acuticornis (Kieffer, 1912)				M
Thienemanniella clavicornis (Kieffer, 1911)				M
Thienemanniella majuscula (Edwards, 1924)		M		
Thienemanniella vittata (Edwards, 1924)				M

	Clare Island		Mainland	
	CIS	NSCI	CIS	NSCI
Thienemanniella sp Pe 2a sensu Langton & Visser, 2003		M		M
Tvetenia bavarica (Goetghebuer, 1934)				M
Tvetenia calvescens (Edwards, 1929)				M
Tvetenia discoloripes (Goetghebuer & Thienemann, 1936)				M
Tvetenia verralli (Edwards, 1929)				M
Subfamily Chironominae – Tribe Chironomini				
Benthalia carbonaria (Meigen, 1804)				M
*[*Chironomus alpestris* Goetghebuer, 1934 = ?*dorsalis* Meigen, 1818]	G		G	
[*Chironomus* (*Chironomus*) *annularius* Meigen, 1818]			G	
Chironomus (*C.*) *anthracinus* Zetterstedt, 1860				M
Chironomus (*C.*) *aprilinus* Meigen, 1818		F, M	G	M
Chironomus (*C.*) *commutatus* Keyl, 1960		F		
Chironomus (*C.*) *nuditarsis* Keyl, 1961		M		
Chironomus (*C.*) *piger* Strenzke, 1959				M
Chironomus (*C.*) *plumosus* (Linnaeus, 1758)			G	M
Chironomus (*C.*) *pseudothummi* Strenzke, 1959		F, M		
Chironomus (*C.*) *riparius* Meigen, 1804		F	G	M
Chironomus (*C.*) *salinarius* Kieffer, 1915				M
Chironomus (*C.*) *tentans* Fabricius, 1805	G	F	G	M
Cladopelma krusemani (Goetghebuer, 1935)		F, M		
Cladopelma virescens (Meigen, 1818)				M
Cladopelma viridulum (Linnaeus, 1767)				M
Cryptochironomus albofasciatus (Staeger, 1839)				M
Cryptochironomus obreptans (Walker, 1856)				M
Cryptochironomus psittacinus (Meigen, 1830)			G	M
Cryptochironomus supplicans (Meigen, 1830)				M
Demeijerea rufipes (Linnaeus, 1761)		F	G	M
Demicryptochironomus (*Demicryptochironomus*) *vulneratus* (Zetterstedt, 1838)			G	M
Demicryptochironomus (*Irmakia*) *neglectus* Reiss, 1988				M
Dicrotendipes lobiger (Kieffer, 1921)				M
[*Dicrotendipes nervosus* (Stæger, 1839)]		F	G	
Dicrotendipes notatus (Meigen, 1818)				M
Dicrotendipes pulsus (Walker, 1856)				M
Dicrotendipes tritomus (Kieffer, 1916)				M
Endochironomus (*Endochironomus*) *albipennis* (Meigen, 1830)				M
Endochironomus (*E.*) *tendens* (Fabricius, 1775)		F		
Glyptotendipes (*Caulochironomus*) *foliicola* sensu Pinder, 1978				M
Glyptotendipes (*C.*) *scirpi* (Kieffer, 1915)				M
[*Glyptotendipes* (*C.*) *viridis* (Macquart, 1834)]			G	
Glyptotendipes (*Glyptotendipes*) *barbipes* (Staeger, 1839)				M
Glyptotendipes (*G.*) *cauliginellus* (Kieffer, 1913)		F, M		M
Glyptotendipes (*G.*) *pallens* (Meigen, 1804)		M		M
Glyptotendipes (*G.*) *paripes* (Edwards, 1929)		M		M
Harnischia curtilamellata (Malloch, 1915)				M
? [*Kloosia pusilla* (Linné, 1767)]			[G]	

	Clare Island		Mainland	
	CIS	NSCI	CIS	NSCI
Lauterborniella agrayloides (Kieffer, 1911)		M		
Microtendipes chloris (Meigen, 1818)		M		M
Microtendipes diffinis (Edwards, 1929)		F		M
Microtendipes pedellus (De Geer, 1776)		F	G	M
Microtendipes rydalensis (Edwards, 1929)				M
Nubensia nubens (Edwards, 1929)				M
Pagastiella orophila (Edwards, 1929)		F, M		M
Parachironomus cinctellus (Goetghebuer, 1921)		M		M
Parachironomus frequens (Johannsen, 1905)				M
Parachironomus gracilor Kieffer (1918)				M
Parachironomus mauricii (Kruseman, 1933)		F		
Parachironomus parilis (Walker, 1856)		F		
Parachironomus tenuicaudatus (Malloch, 1915)		F		
Parachironomus vitiosus (Goetghebuer, 1921)				M
Parachironomus sp. Pe 2, sensu Langton & Viser 2003		M		M
Parachironomus sp. Pe sensu Langton, 1991		M		M
Paracladopelma camptolabis (Kieffer, 1913)				M
Paracladopelma laminatum (Kieffer, 19210				M
Paracladopelma nigritulum (Goetghebuer, 1942)				M
Paralauterborniella nigrohalteralis (Malloch, 1915)				M
Phaenopsectra flavipes (Meigen, 1818)		M	G	M
Phaenopsectra punctipes (Wiedemann, 1817)		F	G	
Polypedilum (*Pentapedilum*) *sordens* (van der Wulp, 1874)		M	G	M
Polypedilum (*Pen.*) *uncinatum* (Goetghebuer, 1921)		M		M
Polypedilum (*Polypedilum*) *acutum* Kieffer, 1915		F, M		
Polypedilum (*Pol.*) *arundineti* (Goetghebuer, 1921)		F		M
Polypedilum (*Polypedilum*) *nubeculosum* (Meigen, 1804)			G	M
Polypedilum (*Polypedilum*) *pedestre* (Meigen, 1830)		F		
Polypedilum (*Tripodura*) *pullum* (Zetterstedt, 1838)		F		M
Polypedilum (*Uresipedilum*) *convictum* (Walker, 1856)				M
Polypedilum (*Uresipedilum*) *cultellatum* Goetghebuer, 1931				M
Sergentia coracina (Zetterstedt, 1850)				M
Stictochironomus pictulus (Meigen, 1830)			G	M
Stictochironomus sticticus (Fabricius, 1781)				M
[*Synendotendipes dispar* (Meigen, 1830)]			G	
Xenochironomus xenolabis (Kieffer, 1916)				M
Subfamily Chironominae - Tribe Pseudochironomini				
Pseudochironomus prasinatus (Stæger, 1839)		F, M	G	M
Subfamily Chironominae - Tribe Tanytarsini				
Cladotanytarsus atridorsum Kieffer, 1924		F		M
Cladotanytarsus iucundus Hirvenoja, 1962				M
Cladotanytarsus lepidocalcar Kieffer, 1938				M
Cladotanytarsus mancus (Walker, 1856)				M
Cladotanytarsus nigrovittatus (Goetghebuer, 1922)		F		M
Cladotanytarsus pallidus Kieffer, 1922				M
Cladotanytarsus vanderwulpi (Edwards)				M

	Clare Island		Mainland	
	CIS	NSCI	CIS	NSCI
Micropsectra atrofasciata (Kieffer, 1911)		M		M
Micropsectra attenuata Reiss, 1969				M
Micropsectra junci (Meigen, 1818)	G	M		
Micropsectra lindebergi Säwedal, 1976				M
Micropsectra lindrothi Goetghebuer, 1931		M		M
Micropsectra notescens (Walker, 1856)				M
Micropsectra pallidula (Meigen, 1813)		M		M
Micropsectra roseiventris (Kieffer, 1909)		F, M		M
Neozavrelia cuneipennis (Edwards, 1929)				M
Neozavrelia luteola Goetghebuer, 1941				M
Paratanytarsus bituberculatus (Edwards)				M
Paratanytarsus brevicalcar (Kieffer, 1909)		M		
Paratanytarsus dissimilis (Johannsen, 1905)				M
Paratanytarsus inopertus (Walker, 1856)				M
Paratanytarsus laccophilus (Edwards, 1929)				M
Paratanytarsus laetipes (Zetterstedt, 1850)				M
Paratanytarsus lauterborni (Kieffer, 1909)				M
Paratanytarsus penicillatus (Goetghebuer, 1928)				M
Paratanytarsus tenuis (Meigen, 1813)				M
Rheotanytarsus curtistylus (Goetghebuer, 1921)		M		
Rheotanytarsus pellucidus (Walker, 1848)				M
Rheotanytarsus pentapoda (Kieffer, 1909)				M
Stempellina bausei (Kieffer, 1911)				M
Stempellinella brevis (Edwards, 1929)		M		M
Stempellinella edwardsi Spies & Sæther, 2004				M
Tanytarsus brundini Lindeberg, 1963				M
Tanytarsus buchonius Reiss & Fittkau, 1971		M		M
Tanytarsus ejuncidus (Walker, 1856)				M
Tanytarsus eminulus (Walker, 1856)				M
Tanytarsus gracilentus (Holmgren, 1883)		M		
Tanytarsus gregarius Kieffer, 1909		F, M		M
Tanytarsus lactescens Edwards, 1929				M
Tanytarsus lestagei Goetghebuer, 1922		M		M
Tanytarsus medius Reiss & Fittkau, 1971				M
Tanytarsus pallidicornis (Walker, 1856)		F		
Tanytarsus signatus (van der Wulp, 1859)		F, M		
[*T. pusio* Mg.]			[G]	
Tanytarsus striatulus Lindeberg, 1976				M
Tanytarsus sylvaticus van der Wulp, 1859				M
Tanytarsus telmaticus Lindeberg, 1959				M
Tanytarsus usmaensis Pagast, 1931				M
Virgatanytarsus triangularis (Goetghebuer, 1928)				M
Zavrelia pentatoma Kieffer, 1913		M		M
Total	8	132	39	253

* [ ] Square brackets denote records from CIS not found in NSCI. Comments on records are given in Appendix III.

APPENDIX III
ANNOTATED INVENTORY OF CHIRONOMIDAE OF CLARE ISLAND AND WEST COUNTY MAYO

Chironomidae from Clare Island and the west County Mayo mainland are listed by subfamily and alphabetically by genus and species following the taxonomic sequence in Murray *et al.* (2018). Species binomials in square brackets '[binomial]' denote those given by Grimshaw (1912). Doubtful records are enclosed in angle brackets, with a question mark preceeding the species binomial, *viz.* <?binomial>. Records for each taxon / species are given by site number followed by date(s) of collection(s); abbreviations are used for months. Precise information on the collection sites in the first Clare Island survey were not given by Grimshaw (1912) and his records from Clare Island are denoted here by the letter 'G', while his records from the mainland have been assigned site numbers with the prefix 'G'. Details of the sites on Clare Island (Sites 1-33) and on the west Mayo mainland (Sites 34-83) are given in Appendices Ia, Ib, Ic and 1d. Records by Fahy (2002) on Clare Island from Creggan Lough (Site 20) are indicated as '20F' while records from Lough Poirtin Fuinch are cited '23F'. Records from the mainland are from personal collections of the author, augmented by records from examination of chironomid pupal exuviae from small lakes in the general area of investigation of the original survey in west Mayo that were provided by the Environmental Protection Agency (EPA). Delays in publication of the present volume afforded opportunities to undertake further collections in August 2017 from nine additional sites (75 to 83) on Achill Island (Murray, 2017) and from one site (51) in March 2018 at Westport that had been visited previously. Records from those collections are included. Comments are given for individual species records where appropriate.

Representative slide preparations have been incorporated as voucher specimens into the Heritage Council Collection of Irish Chironomidae (HCCIC) deposited in the National Museum of Ireland (NMI) (Murray, 2005). In the following list the reference number of specimens in the HCCIC is indicated by the symbol '#' followed by the species/specimen number in the collection. Slide mounted voucher specimens from Fahy (2002) deposited in NMI are indicated here as 'FCIC' (Fahy Clare Island Chironomidae). Some FCIC specimens were identified by P.H. Langton and the abbreviation 'PHL' is used to distinguish these. Additional slide-mounted and alcohol preserved specimens from the survey will be deposited in the NMI as the 'New Survey of Clare Island Chironomidae Collection - D. A. Murray'.

The inventory
Subfamily Tanypodinae

Ablabesmyia (Ablabesmyia) longistyla Fittkau, 1962
Clare Island: 20F; 11-Jun-1998 to 05-Oct-1998 and 10-Jun-1999 to 12-Oct-1999. 23F; 10-Aug-1998.

Mainland: 36; 05-Aug-2004. 45; 04-Aug-2004. 46; 24-Jul-2008. 49b; 21-Aug-2008. 55; 24-Jun, 08-Aug-2008. 57, 24-Jun and 21-Jul-2008. 61; 05-Aug-2004. 66; 05-Aug-2004. 72; 26-Jun-2008. 73; 22-Aug-2002, 12-Jul-2006. 74; 28-Jun and 12-Jul-2006.

Comment: Although Fahy (2002) recorded *A. longistyla* from Clare Island, the single voucher adult male in FCIC, NMI was a misidentified *Ablabesmyia monilis*. That slide has been appropriately correctly relabelled.

Ablabesmyia (Ablabesmyia) monilis (Linnaeus, 1758)
Clare Island: 5; 20-Aug-2002. 7; 28-Apr-2000, 05-Jun-2002. 19; 05-Jun-2002. 20; 21-Aug-2002. 20F; 10-Jun, 07-Jul-1999. 23F; 07-Jul-1999.

Mainland: 36; 29-Sep-2004, 17-Jun, 16-Jul, 17-Sep-2008. 37; 17-Jun, 16-Jul, 17-Sep-2008. 38; 29-Sep-2004. 40; 05-Aug-2004. 45; 04-Aug-2004. 46; 25-Jun, 24-Jul, 17-Sep-2008. 47; 25-Jun-2008; 49b; 21-Aug-2008. 51; 07-May-2004. 55; 24-Jun, 08-Aug, 18-Sep-2008. 57; 24-Jun, 21-Jul, 18-Sep-2008. 66; 05-Aug-2004. 72; 26-Jun, 25-Aug-2008. 73; 12-Jul-2006. 74; 19-May-2005, 28-Jun, 12-Jul-2006. 81; 29-Aug-2017.

Comment: Voucher male in FCIC, NMI.

Ablabesmyia (Ablabesmyia) phatta (Egger, 1863)
[*A. phatta* Egger. in Grimshaw (1912)]

Mainland: 37; 17-Jun and 16-Jul-2008. 40; 05-Aug-2004. 47; 25-Jun-2008. 49b; 21-Aug-2008. 73; 22-Aug-2002. G7; Jul-1911.

Apsectrotanypus trifascipennis (Zetterstedt, 1838)
Mainland: 46; 24-Jul-2008. 74; 12-Jul-2006.

Arctopelopia barbitarsis (Zetterstedt, 1850)
Clare Island: <20F; 11-Jun-1998 to 03-Nov-1998 and 10-Jun-1999 to 02-Nov-1999. 23F; 06-Jul-1998 to 15-Oct-1998 and 02-Jun-1999 to 02-Nov-1999>
Mainland: 57; 21-Apr, 21-Jul, 18-Sep-2008. 72; 25-Aug-2008.

Comment: The voucher FCIC, NMI adult male from Creggan Lough on Clare Island is a misidentified *Arctopelopia griseipennis*.

Arctopelopia griseipennis (van der Wulp, 1859)
Clare Island: 8; 28-Apr-2000. 19; 05-Jun-2002. 20; 21-Aug-2002;
Mainland: 36; 29-Sep-2004. 55; 18-Sep-2008. 69; 30-Sep-2004.

Clinotanypus (*Clinotanypus*) *nervosus* (Meigen, 1818)
[*Procladius nervosus* Mg. in Grimshaw (1912)]
Mainland: 49b, 30-Jun-2008. 55; 24-Jun-2008. 57; 24-Jun-2008. G7; 1910-1911.

Conchapelopia (*Conchapelopia*) *melanops* (Meigen, 1818)
Mainland: 56; 23-Jul-2008. 57; 21-Jul-2008. 66; 05-Aug-2004. 72; 25-Aug-2008. 73; 12-Jul-2006.

Conchapelopia (*Conchapelopia*) *pallidula* (Meigen, 1818)
Mainland: 39, 06-May-2004. 51; 07-May-2004.

Conchapelopia (*Conchapelopia*) *viator* (Kieffer, 1911)
Mainland: 61; 05-Aug-2004. 65; 05-Aug-2004. 73; 12-Jul-2006.

Krenopelopia nigropunctata (Stæger, 1839)
Clare Island: 12; 20-Aug-2002.
Mainland: 35; 05-Aug-2004.

Larsia sp. ? *atrocincta* (Goetghebuer, 1942)
Clare Island: 20F; 13-Jul-1998 to 05-Oct-1998 and 10-Jun-1999 to 20-Aug-1999. 23F; 10-Jun-1999.

Comment: Fahy (2002) was uncertain of the identification of this taxon and queried it as *Larsia atrocincta*?. The condition of the slide mounted FCIC voucher specimen in NMI labelled as '*Larsia atrocincta* ?' is such that specific determination is not possible. It is, however, a *Larsia* species. Only two species of *Larsia* are known from Ireland (Murray *et al.*, 2013). Pupal exuviae of the other species, *L. curticalcar*, have been found on the mainland at Cregganbawn, Site 69.

Larsia curticalcar (Kieffer, 1918)
Mainland: 69; 30-Sep-2004.
Comment: See above under *L. atrocincta*.

Macropelopia (*Macropelopia*) *adaucta* Kieffer, 1916
Clare Island: 2; 20-Aug-2002. 7; 28-Apr-2000. 9; 05-Jun-2002. 10; 20-Aug-2002. 11; 28-Apr-2000. 27; 21-Aug-2002. 29; 21-Aug-2002
Mainland: 36; 07-Apr-2008, 17-Jun-2008. 38; 29-Sep-2004.39; 06-May-2004. 40; 05-Aug-2004. 41; 29-Sep-2004. 57; 21-Apr-2008. 62; 07-May-2004. 69; 30-Sep-2004.70; 27-Apr-2000, 27-Jul-2000. 72; 26-Jun-2008. 74; 28-Jun-2006.

Macropelopia (*Macropelopia*) *nebulosa* (Meigen, 1804)
[?*Ablabesmyia nebulosa* Mg in Grimshaw (1912).]
Clare Island: G, Jul-1911
Mainland: 37; 17-Sep-2008. 40; 05-Aug-2004. 55; 18-Sep-2008. 69; 30-Sep-2004. 72; 22-Sep-2008. 73; 22-Aug-2002. G1; Jul-1910. G2; Jul-1910. G3; Jul-1911.

Comment: Grimshaw (1912) remarked that '*Ablabesmyia nebulosa*' was 'a well-marked species'. Now in the genus *Macropelopia* it was not found on Clare Island in the NSCI whereas two other *Macropelopia* species, *M. adaucta* and *M. notata*, were found. It is possible that Grimshaw observed *M. nebulosa*, characterised by usually distinct wing markings, but Fittkau (1962) cautioned that early records of *M. nebulosa* and *M. notata* were frequently confused. Both species have distinct wing markings, *M. nebulosa* more so than *M. notata*, and are superficially similar. Adult males are readily differentiated by the absence of a gonocoxite lobe in *M. nebulosa* (present in *M. notata*), a feature clearly visible in modern slide preparations but difficult to observe in pinned specimens that Grimshaw examined. The third species, *M. adaucta*, which has been recorded at seven locations on the Island during this study, was not described until 1916. Nevertheless it is unlikely that Grimshaw had examples of *M. adaucta* since the adult male is clearly distinguished by the absence of wing markings. Separation of all three species is unambiguous on features of the pupal exuviae. Adults and pupal exuviae of *M. adaucta* and *M. notata* only were found on Clare Island in this study. All three species were noted in mainland samples and it is possible that *M. nebulosa* also occurs on Clare Island.

Macropelopia (*Macropelopia*) *notata* (Meigen, 1818)
Clare Island: 7; 20-Aug-2002. 24; 21-Aug-2002. 20F; 11-Jun, 05-Oct-1998, 10-Jun, 07-Jul, 03-Aug-1999.

Mainland: 61; 05-Aug-2004.

Comment: Voucher adult male in FCIC, NMI and in HCCIC #43 from site 61.

Pl. VI *Macropelopia notata* adult. Photo: James Lindsey

Monopelopia (*Monopelopia*) *tenuicalcar* (Kieffer, 1918)
Clare Island: 7; 05-Jun-2002
Mainland: 62; 07-May-2004.

Natarsia punctata (Fabricius, 1805)
Clare Island: 7; 05-Jun-2002. 20F; 10-Jun-1999.
Comment: Voucher specimen in FCIC, NMI.

Nilotanypus dubius (Meigen, 1804)
Clare Island: 31; 21-Aug-2002

Mainland: 45; 04-Aug-2004. 53; 05-Aug-2004. 61; 07-May, 05-Aug-2004. 63; 30-Sep-2004. 65; 05-Aug-2004. 68; 06-Jun-2004. 74; 12-Jul-2006.

Comment: Adult male voucher specimen #51 from Site 31 HCCIC, NMI.

Procladius (*Holotanypus*) *choreus* (Meigen, 1804)
Clare Island: 1, 20-Aug-2002. 7; 20-Aug-2002. 9; 05-Jun-2002. 10; 20-Aug-2002. 20; 21-Aug-2002. 29; 21-Aug-2002. 32; 21-Aug-2002. 20F; 11-Jun-1998 to 01-Sep-1998 and 10-Jun-1999 to 02-Nov-1999. 23F; 02-Jun-1998 to 07-Sep-1999

Mainland: 36; 07-Apr, 05-Aug-2004. 38; 29-Sep-2004. 41; 29-Sep-2004. 43; 06-May-2004. 46; 25-Jun-2008. 47; 22-Apr-2008. 49a; 07-May-2004, 30-Jun, 21-Aug-2008. 51; 07-May-2004. 55; 08-Aug, 18-Sep-2008. 56; 18-Sep-2008. 57; 21-Apr, 24-Jun-2008. 58; 05-Aug-2004. 61; 05-Aug-2004. 66; 05-Aug-2004. 67; 30-Sep-2004. 74; 05-Apr-2006.

Comment: Voucher specimen in FCIC, NMI.

Procladius (*Holotanypus*) *crassinervis* (Zetterstedt, 1838).
Mainland: 46; 11-Apr-2008.

Comment: This is a is widespread species in Ireland easily identified as adult male from Pinder (1978) and Langton and Pinder (2007) and as pupa, that has globose thoracic horns with a reduced plastron, that key readily in Langton (1991), Langton and Visser (2003) and Langton *et al.* (2015). However, there is conflict with the interpretation of this taxon and *P. culiciformis* (see below). Further discussion on the issue is given in Murray *et al.* (2018).

<?*Pelopia* (*Tanypus*) *culiciformis* L. (Grimshaw, 1912)> = *Procladius* (*Holotanypus*) *culiciformis* (Linnaeus, 1767).
Mainland: G4; Jul-1911, G6; Jul-1911, G7; Jul-1911, G7; Jul-1911, G9, Jul-1911.

Comment: Grimshaw (1912) gave the the only published records of this taxon, from eight male and seven female adult Chironomidae, in Ireland. It was included in the inventories of Murray, 1972, Ashe *et al.* (1998), Ashe and O'Connor (2009) and Murray *et al.* 2013). However these records of *Procladius* (*Holotanypus*) *culiciformis* (L.) remain doubtful because of taxonomic uncertainty. The taxon was considered a variety of *P. choreus* by Edwards (1929) and also later by Coe (1950) who, nevertheless, considered it as 'possibly a distinct species'. Pinder (1978) and Langton and Pinder (2007) were unable to separate adults of *P. choreus* (Meig.) from *P. culiciformis* (L.) sensu Coe (1950). More recently Spies and Sæther (2013) recognise *P. culiciformis* as a valid taxon and cite *P. crassinervis* (see above) as a junior synonym. This synonymy was not accepted by Murray *et al.* (2018).

Procladius (*Holotanypus*) *sagittalis* (Kieffer, 1909)
Clare Island: 20F; 11-Jun-1998 to 13-Oct-1998 and 10-Jun-1999 to 13-Aug-1999. 23F; 10-Jun-1999 to 20-Aug-1999

Mainland: 38; 29-Sep-2004. 56; 23-Jul-2008, 23-Jul-2008, 18-Sep-2008. 57; 21-Jul-2008, 18-Sep-2008.

Comment: Voucher specimen in FCIC, NMI.

Procladius (*Holotanypus*) *signatus* (Zetterstedt, 1850)
Mainland: 37; 17-Jun, 16-Jul-2008. 46; 25-Jun, 24-Jul, 17-Sep-2008. 72; 26-Jun, 25-Aug, 22-Sep-2008. 73; 22-Aug-2002, 12-Jul-2006.

Procladius (*Holotanypus*) *simplicistilus* Freeman, 1948
Mainland: 57; 21-Apr-2008.

Procladius (*Psilotanypus*) *flavifrons* Edwards, 1929
Mainland: 47; 25-Jun, 18-Sep-2008. 49b; 23-Apr-2008. 57; 24-Jun, 21-Jul, 18-Sep-2008.

Pl. VII *Procladius* (*Psilotanypus*) lugens adult male. Photo: James Lindsey

Procladius (*Psilotanypus*) *lugens* Kieffer, 1915
Mainland: 56; 21-Apr, 24-Jun-2008. 80; 29-Aug-2017.

Procladius (*Psilotanypus*) *rufovittatus* (van der Wulp, 1874)
[*Pelopia* sp. in Grimshaw (1912)]
Mainland: 56; 23-Jul-2008. G5.

Comment: Grimshaw (1912) reported a single female '*Pelopia* sp' from Castlebar Lough as a 'pretty little species which I cannot refer to any described form'. His brief reference to the specimen's very small size and body colour pattern suggests it could have been a *Psilotanypus* species. Edwards (1929, p302) examined Grimshaw's specimen and identified it as *Psilotanypus rufovittatus*.

Rheopelopia maculipennis (Zetterstedt, 1838)
Mainland: 53; 05-May-2004.

Tanypus (*Tanypus*) *kraatzi* (Kieffer, 1912)
Mainland: 56; 24-Jun-2008.

Comment: The record from Site 56, L. Knappabeg, was the first record of the species from Ireland (Murray, 2010).

Tanypus (*Tanypus*) *vilipennis* (Kieffer)
Mainland: 56; 24-Jun, 23-Jul-2008.

Thienemannimyia (*Thienemannimyia*) *laeta* (Meigen, 1818)
Mainland: 66; 05-Aug-2004.

Thienemannimyia (*Thienemannimyia*) *northumbrica* (Edwards, 1929)
Mainland: 46; 24-Jul-2008. 47; 25-Jun-2008. 73; 06-Jun-2002, 12-Jul-2006.

Trissopelopia longimana (Staeger, 1839)
Mainland: 51; 07-May-2004. 53; 05-Aug-2004. 63; 30-Sep-2004. 66; 05-Aug-2004.

Zavrelimyia (*Paramerina*) *cingulata* (Walker, 1856)
[*Ablabesmyia pygmaea* V.d Wulp in Grimshaw (1912)]
Clare Island: 7; 20-Aug-2002. 19; 05-Jun-2002. 20; 21-Aug-2002. 20F; 11-Jun-1998 to 13-Oct-1998 and 10-Jun-1999 to 14-Sep-1999.

Mainland: 36; 29-Sep-2004, 17-Jun, 16-Jul, 17-Sep-2008. 41; 29-Sep-2004. 55; 24-Jun, 08-Aug, 18-Sep-2008. 57; 21-Jul-2008. 61; 05-08-2004. 69; 30-Sep-2004. 72; 25-Aug, 22-Sep-2008. 73; 05-Jun-2002. G7; Jul-1911.

Comment: Pupal exuviae voucher specimen #54 from site 73 in HCCIC, NMI. Adult male voucher specimen in FCIC, NMI.

Zavrelimyia (*Paramerina*) *divisa* (Walker, 1856)
Mainland: 53; 08-May-2003

Comment: Adult male voucher specimen #55 in HCCIC, NMI.

Zavrelimyia (*Zavrelimyia*) *barbatipes* (Kieffer, 1911)
Clare Island: 8; 05-Jun-2002

Mainland: 35; 05-Aug-2004. 70; 27-Apr, 27-Jul-2000.

Zavrelimyia (*Zavrelimyia*) *melanura* (Meigen, 1804)
Mainland: 69; 30-Sep-2004.

Subfamily Telmatogetoninae

Thalassomya frauenfeldi Schiner, 1856
Clare Island: 12; 29-Apr-2000.

Comment: Adult female voucher specimen #125 in HCCIC, NMI. This is only the third record of this coastal marine species from Ireland. Until now it is only known from Inistearaght Island, Co. Kerry (Murray and Ashe, 1982) and Kilkee, Co. Clare (Murray, 2000).

Subfamily Diamesinae

Diamesa insignipes Kieffer, 1908
Clare Island: 11; 28-Apr-2000.

Comment: See *Potthastia gaedii*.

Potthastia gaedii (Meigen, 1838)
[?*Diamesa ammon* Hal in Grimshaw (1912)]

Clare Island: G, Jul-1910

Mainland: 42; 06-May-2004. 46; 25-Jun, 24-Jul, 17-Sep-2008. 51; 07-May-2004. 52; 07-May-2004. 61; 05-Aug-2004. 65; 05-Aug-2004. 66; 05-Aug-2004. 74; 05-Apr, 09-May, 12-Jul-2006.

Comment: Grimshaw recorded *Diamesa ammon*, Hal (= Haliday, 1856) now a synonym of *P. gaedii*, from Clare Island. Grimshaw (1912, p. 8) referred to the characteristic 'heart-shaped fourth tarsal segment' of the single female specimen he examined - a morphological feature now known to be common to species of *Potthastia* and *Diamesa* and remarked 'the wings are milky white notwithstanding the fact that Haliday described them as hyaline in this sex' indicating that perhaps the specimen Grimshaw examined was not *Diamesa ammon* described by Haliday but could have been *P. gaedii* which is widely distributed in Ireland (Murray *et al*. 2013) and found at seven mainland sites in this study. If correct, the 1912 record would be the sole account of *P. gaedii* from Clare Island. However in the present study male and female imagines of *Diamesa insignipes*, which also have a characteristic cordiform tarsomere 4, were collected at Site 11 on Clare Island.

Potthastia longimanus Kieffer, 1922
Mainland: 37; 07-Apr-2008. 47; 22-Apr-2008. 69; 30-Sep-2004. 72; 25-Aug-2008. 73; 22-Aug-2002; 05-Apr, 12-Jul-2006. 74; 05-Apr, 09-May-2006.

Protanypus morio (Zetterstedt, 1838)
Mainland: 36; 29-Sep-2004, 17-Sep-2008. 38; 29-Sep-2004. 46; 11-Apr-2008. 57; 18-Sep-2008. 72; 22-Sep-2008.

Subfamily Prodiamesinae

Prodiamesa olivacea (Meigen, 1818)
[?*Prodiamesa obscurimana* Mg in Grimshaw (1912)]

Clare Island: 22; 05-Jun-2002. ?G, Jul-1910.

Mainland: 37; 17-Sep-2008. 38; 29-Sept-2004. 51; 07-May-2004. 56; 21-Apr, 23-Jul-2008. 61; 07-May, 05-Aug-2004. ?G10, Jul-1910.

Comment: Grimshaw (1912) identified three adult male specimens from Clare Island, and two female specimens collected by C. Morley from the mainland at Louisburg, as *Prodiamesa obscurimana* Mg. a species originally described in the genus *Prodiamesa* but subsequently transferred to *Odontomesa* on designation of generic status by Pagast (1947). However, according to Ashe and O'Connor (2009) *Prodiamesa obscurima*na is a questionable synonym of *Odontomesa fulva* (Kieffer), a species not known from Ireland in extant collections but which is recorded in Britain. Murray and Murray (2003) consider the possibility that Grimshaw may have misidentified *P. obscurimana* for the rather common *P. olivacea* which is here reported from Site 22 on the River Doree on Clare Island and from six locations on the mainland, including Louisburg, where specimens reported by Grimshaw were also collected.

Subfamily Orthocladiinae

Acamptocladius reissi Cranston and Sæther, 1982
Clare Island: 7; 20-Aug-2002

Comment: This was the first record of the species from Ireland (Murray and Murray, 2003). It has since been recorded in samples collected from the remote mountain lake L. Ouler, County Wicklow, in August 2005 (Murray *et al*. 2014).

Acamptocladius submontanus (Edwards, 1932)
Mainland: 74; 09-May-2006.

Acricotopus lucens (Zetterstedt, 1850)
Clare Island: 29; 21-Aug-2002

Mainland: 72; 26-Jun-2008. 81; 29-Aug-2017.

Comment: Voucher specimen, (Pe) #162, from Clare Island Site 29 in HCCIC, NMI.

Brillia bifida (Kieffer, 1909)
Clare Island: 13; 28-Apr-2000. 14; 28-Apr-2000, 05-Jun-2002. 25; 28-Apr-2002
Mainland: 59; 07-May-2004.

Brillia longifurca Kieffer, 1921
Mainland: 61; 30-Sep-2004. 63; 30-Sep-2004. 66; 05-Aug-2004.

Bryophaenocladius subvernalis (Edwards, 1929)
Mainland: 74; 05-Apr-2006.

Camptocladius stercorarius (De Geer, 1776)
Clare Island: 11; 28-Apr-2000.

Comment: Larvae of this species are coprophilic living in, and feeding on, dung of farm animals mammals such as cattle and sheep.

<?*Cardiocladius fuscus*> Kieffer, 1924
Clare Island: 23F; 13-Jul-1999.

Comment: This record by Fahy (2002), based on a single adult male from an emergence trap on Lough Poirtin Fuinch is doubtful. *Cardiocladius* larvae are typical denizens of fast flowing water and live in association with, and prey on, larval Simuliidae. Since the record is based on a specimen from a lentic waterbody, from which there are no records of Simuliidae and in the absence of a voucher specimen in NMI this is considered a likely misidentification (see main text).

Chaetocladius (*Chaetocladius*) *dentiforceps* (Edwards, 1929)
Clare Island: 20F; 10-Aug, 13-Oct-1998.
Comment: Voucher specimen in FCIC, NMI.

Chaetocladius (*Chaetocladius*) *melaleucus* (Meigen, 1818)
Clare Island: 1, 20-Aug-2002. 10; 04-Jun-2002. 18; 05-Jun-2002. 20F; 11-Jun-1998 to 05-Oct-1998 and 10-Jun-1999 to 14-Sep-1999. 26; 28-Apr-2002. 29; 21-Aug-2002.
Mainland: 38; 29-Sep-2004. 42; 06-May-2004. 63; 30-Sep-2004. 69; 30-Sep-2004.

Comment: Adult male voucher specimen #195 from site 26 in HCCIC, NMI. Additional voucher specimen in FCIC, NMI.

Chaetocladius (*Chaetocladius*) *perennis* (Meigen, 1830)
Clare Island: 11; 28-Apr-2000. 12; 20-Aug-2002. 13; 28-Apr-2000. 20F; 07-Jul-1999
Mainland: 70; 27-Apr-2000. 79; 29-Aug-17.
Comment: Voucher specimen in FCIC, NMI.

Clunio marinus Haliday, 1855
Clare Island: 12; 29-Apr-2000, 05-Jun-2002; 20-Aug-2002
Mainland: 70; 27-Apr-2000. 83; 29-Aug-2017.

Corynoneura arctica Kieffer, 1923
Mainland: 67; 30-Sep-2004. 73; 22-Aug-2002.

Corynoneura carriana Edwards, 1924
Mainland: 47; 18-Sep-2008. 61; 30-Sep-2004.

Corynoneura celeripes Winnertz, 1852
Mainland: 81; 29-Aug-2017.

Corynoneura celtica Edwards
Mainland: 61; 05-Aug-2004. 62; 07-May-2004.

Corynoneura edwardsi Brundin, 1949
Clare Island: 29; 21-Aug-2002. 20F; 06-Jul-1998 to 03-Nov-1998 and 05-May-1999 to 12-Oct-1999. 23F; 13-Aug, 20-Aug-1999
Mainland: 50; 30-Sep-2004. 73; 22-Aug-2002.
Comment: Voucher specimen in FCIC, NMI (det PHL).

Corynoneura gratias Schlee, 1968
Mainland: 42; 06-May-2004.

Corynoneura lobata Edwards, 1924
Clare Island: 25. 28-Apr-2002
Mainland: 41; 29-Sep-2004. 61; 05-Aug-2004.

Corynoneura Pe2a sensu Langton and Visser 2003
Mainland: 49a; 07-May-2004. 53; 05-Aug-2004. 59; 07-May-2004. 61; 07-May-2004.

Comment: Exuviae of this pupal morphotype, not yet associated with the adult, are common in Irish rivers and streams (Murray *et al.* 2018).

Corynoneurella paludosa Brundin, 1949
Clare Island: 31; 21-Aug-2002.
Comment: Voucher specimen (Pe) #224 in HCCIC, NMI.

Cricotopus (*Cricotopus*) *albiforceps* (Kieffer, 1916)
Clare Island: 20; 21-Aug-2002.
Comment: Voucher specimen (Pe) #226 in HCCIC, NMI.

Cricotopus (*Cricotopus*) *annulator* Goetghebuer, 1927
[*Cricotopus motitator* L. in Grimshaw (1912)]
Clare Island: 20F; 11-Jun-1998 to 03-Nov-1998 and 10-Jun-1999 to 02-Nov-1999
Mainland: 53; 07-May-2004. G10; Jul-1910 (collected by J.N. Halbert)
Comment: Adult male voucher specimen #227 from site 53 in HCCIC, NMI. Additional voucher specimen in FCIC, NMI (det. PHL).

Cricotopus (*Cricotopus*) *bicinctus*
(Meigen, 1818)
[*Cricotopus bicinctus* Mg in Grimshaw (1912)]
Mainland: 61; 30-Sep-2004. 74; 12-Jul-2006; 19-May-2005; 19-May-2005. G6 and G9; Jul-1911.

Cricotopus (*Cricotopus*) *ephippium* (Zetterstedt, 1838)
Clare Island: 19; 05-Jun-2002.

Cricotopus (*Cricotopus*) *festivellus* (Kieffer, 1906)
Clare Island: 20F; 13-Jul-1998 to 05-Oct-1998 and 07-Jul-1999 to 02-Nov-1999. 23F; 04-Aug to 14-Sep-1998 and 13-Jul-1999 to 07-Sep-1999.
Comment: Voucher specimen in FCIC, NMI (det. PHL).

Cricotopus (*Cricotopus*) *flavocinctus* (Kieffer, 1924)
Mainland: 77; 28-Aug-2017.

Cricotopus (*Cricotopus*) *fuscus* (Kieffer, 1909)
Mainland: 61; 05-Aug-2004.

Cricotopus (*Cricotopus*) *pallidipes* Edwards, 1929
Mainland: 34; 05-Aug-2004. 45; 04-Aug-2004. 50; 30-Sep-2004. 74; 28-Jun, 12-Jul-2006.

Cricotopus (*Cricotopus*) *pilosellus* Brundin, 1956
Clare Island: 20; 21-Aug-2002. 20F; 11-Jun-1998 to 05-Oct-1998 and 10-Jun-1999 to 02-Nov-1999. 23F; 13-Aug-1999
Mainland: 50; 30-Sep-2004.
Comment: Voucher specimen in FCIC, NMI (det. PHL).

Cricotopus (*Cricotopus*) *polaris* Kieffer
Mainland: 50; 30-Sep-2004.

Cricotopus (*Cricotopus*) *pulchripes* Verrall, 1912
Clare Island: 6; 28-Apr-2000
Mainland: 73; 25-Aug-2008.

Cricotopus (*Cricotopus*) *similis* Goetghebuer, 1921
Mainland: 61; 07-May, 05-Aug, 30-Sep-2004.

Cricotopus (*Cricotopus*) *tibialis* (Meigen, 1804)
Clare Island: G, Jul-1910
Mainland: G4, G7 and G9, Jul-1911. 77; 28-Aug-2017.
Comment: *Cricotopus tibialis* is a common species on the Irish mainland (Murray *et al.* 2013). It was not found on Clare Island during this study but did occur in a collection on Achill Island and is a possible component of the extant fauna of Clare Island.

Cricotopus (*Cricotopus*) *tremulus* (Linnaeus, 1758)
Mainland: 61; 07-May-2004.

Cricotopus (*Cricotopus*) *trifascia* Edwards, 1929
Mainland: 66; 05-Aug-2004.

Cricotopus (*Cricotopus*) *tristis* Hirvenoja, 1973
Mainland: 73; 06-Jun, 22-Aug-2002.

Cricotopus (*Isocladius*) *brevipalpis* Kieffer, 1909
Mainland: 57; 18-Sep-2008.

Cricotopus (*Isocladius*) *intersectus* (Stæger, 1839)
Mainland: 50; 30-Sep-2004. 77; 28-Aug-2017.

Cricotopus (*Isocladius*) *laricomalis* Edwards, 1932
Mainland: 42; 06-May-2004. 49a; 07-May-2004.

Cricotopus (*Isocladius*) *ornatus* (Meigen, 1818)
Clare Island 19; 05-Jun-2002.

Cricotopus (Isocladius) sylvestris (Fabricius, 1794)
[*Cricotopus silvestris* Fab in Grimshaw (1912)]
Clare Island: 20F; 04-Aug, 01-Sep-1998, 10-Jun, 14-Sep-1999, 23F; 10-Aug-1998
Mainland: 51; 30-Sep-2004. 67; 30-Sep-2004. G6, G7 and G9, Jul-1911.
Comment: Voucher specimen in FCIC, NMI.

Cricotopus (Isocladius) tricinctus (Meigen, 1818)
[*Cricotopus trifasciatus* var. *tricinctus* Mg in Grimshaw (1912)]
Mainland: G4; G9, Jul-1911.
Comment: Although it was not recorded in the present study records of this species exist from the coastal Lough Arusbeg, near Cleggan, County Galway (Murray *et al.* 2014)

Cricotopus (Isocladius) trifasciatus (Meigen, in Panzer, 1813)
[*Cricotopus trifasciatus* Panz. in Grimshaw (1912)]
Mainland: 37; 17-Sep-2008. 38; 29-Sep-2004. G3; Jul-1911.

Cricotopus (Isocladius) Pe 2 sensu Langton and Visser, 2003
Mainland: 48; 30-Sep-2004.
Comment: Langton and Viser (2003) suggest that exuviae of this morphotype might belong to *C. relucens* Hirvenoja, 1972, a species not yet known from the British Isles.

Cricotopus (Isocladius) Pe 5 sensu Langton and Visser, 2003
Mainland: 41, 29-Sep-2004.
Comment: The characteristic exuviae collected from a pool at Site 41 on the southern face of Knockmore, Achill was first described by Langton (1991) from a peat pool near Lough Staing, Scotland. It remains to be linked with a described species.

Cricotopus (Paratrichocladius) rufiventris (Meigen, 1830)
Clare Island: 6; 28-Apr-2000
Mainland: 46; 25-Jun-2008. 57; 24-Jun-2008. 61; 05-Aug-2004. 66; 05-Aug-2004. 74; 19-May-2005.

Cricotopus (Paratrichocladius) skirwithensis (Edwards, 1929)
Mainland: 61; 07-May-2004. 63; 30-Sep-2004. 66; 05-Aug-2004. 71, 09-May-2003.

Comment: Adult male voucher specimen #461from site 71 in HCCIC, NMI.

Eukiefferiella ancyla Svensson, 1986
Mainland: 52; 07-May-2004. 61; 07-May-2004.
Comment: Larvae of this species, described by Svensson (1986), are phoretic and live along the inner rim of the mantel cavity of the gastropod mollusc *Ancylus fluviatilis* (Müller). This molluscan limpet is common in fast flowing waters in Ireland where other records of *E. anclya* are documented (Murray *et al.* 2014, 2018).

Eukiefferiella brevicalcar (Kieffer, 1911)
Mainland: 73; 09-May-2006. 74; 05-Apr-2006.

Eukiefferiella claripennis (Lundbeck, 1898)
Clare Island: 6; 28-Apr-2000
Mainland: 44; 06-May-2004. 59; 07-May-2004. 61; 30-Sep-2004. 66; 05-Aug-2004.

Eukiefferiella clypeata (Thienemann, 1919)
Mainland: 53; 05-Aug-2004. 61; 05-Aug-2004. 66; 05-Aug-2004. 71; 09-May-2003.

Eukiefferiella coerulescens (Kieffer, 1926)
Mainland: 34; 05-Aug-2004. 36; 29-Sep-2004, 07-Apr, 16-Jul, 17-Sep-2008. 45; 04-Aug-2004. 46; 24-Jul, 17-Sep-2008. 52; 30-Sep-2004. 53; 05-Aug-2004. 55; 08-Aug-2008. 63; 30-Sep-2004. 72; 25-Aug, 22-Sep-2008. 73; 05-Apr, 09-May, 12-Jul-2006.

Eukiefferiella cyanea Thienemann, 1936
Clare Island: 6; 28-Apr-2000.

Eukiefferiella devonica (Edwards, 1929)
Clare Island: 6; 28-Apr-2000.
Mainland: 34; 05-Aug-2004. 51; 07-May-2004. 61; 05-Aug, 30-Sep-2004. 74; 19-May-2005.

Eukiefferiella dittmari Lehmann, 1972
Mainland: 45; 04-Aug-2004. 61; 30-Sep-2004.

Eukiefferiella ilkleyensis (Edwards, 1929)
Mainland: 34; 05-Aug-2004.

Eukiefferiella tirolensis Goetghebuer, 1938
Mainland: 71; 09-May-2003. 74; 19-May-2005.

Eukiefferiella sp. Pe *minor* (Edwards, 1929) / *fittkaui* Lehman, 1972

Clare Island: 6; 28-Apr-2000

Mainland: 73; 22-Aug-2002.

Comment: It is not possible to discriminate between pupal exuviae of the two closely related species (Langton and Visser, 2003) but these records are most likely of *E. minor*.

Georthocladius (Georthocladius) luteicornis (Goetghebuer, 1941)

Mainland: 72; 26-Jun-2008.

Halocladius (Halocladius) fucicola (Edwards, 1926)
Clare Island: 15; 05-Jun-2002. 17; 05-Jun-2002.

Mainland: 82, 83, 29-Aug-2017.

Comment: Records of this halobiontic species are from eastern, southern and western coastal regions of Ireland.

Halocladius (Halocladius) variabilis (Staeger, 1839)
Clare Island: 11; 28-Apr-2000. 12; 29-Apr-2000. [20F; 11-Jun, 13-Jul, 15-Oct-1998; 14-Jul, 13-Aug, 07, 14-Sept and 06-Oct-1999. <23F; 05-Oct-1998, 13-Aug-1999>].

Mainland: 70; 27-Apr-2000.

Comment: An adult male from Site 11, Clare Island, is voucher specimen #325 in HCCIC, NMI. There is as additional voucher specimen from Malaise Trap collections (Site 20F) in FCIC, NMI (det PHL). This is a halobiontic coastal species whose juvenile stages are typical of marine rock pools (Hirvenoja, 1973). Records of this species from emergence trap collections on the freshwater Lough Poirtin Fuinch (Site 23F) by Fahy (2002) are doubtful.

Halocladius varians (Staeger, 1839)
Mainland: 78, 83; 29-Aug-2017.

Comment: Existing records of *H. varians* in Ireland are from intertidal zones from south, west and northern coasts.

Heleniella ornaticollis (Edwards, 1929)
Mainland: 34; 05-Aug-2004. 41; 29-Sep-2004. 59; 07-May-2004. 61; 07-May-2004.

Heterotanytarsus apicalis (Kieffer, 1921)
Clare Island: 6; 28-Apr-2000.

Mainland: 36; 05-Aug, 29-Sep-2004, 17-Jun, 16-Jul, 17-Sep-2008. 37; 07-Apr-2008. 38; 29-Sep-2004. 41; 29-Sep-2004. 46; 11-Apr-2008. 52; 07-May-2004. 55; 22-Apr-2008. 72; 26-Jun, 25-Aug, 22-Sep-2008. 73; 05-Apr, 09-May-2006. 74; 05-Apr, 09-May-2006.

Heterotrissocladius grimshawi (Edwards, 1929)
Mainland: 36; 07-Apr, 16-Jul, 17-Sep-2008. 51; 07-May-2004. 52; 07-May-2004. 57; 21-Apr-2008. 73; 06-Jun-2002; 12-Jul-2006. 74; 05-Apr-2006.

Heterotrissocladius marcidus (Walker, 1856)
Clare Island: 7; 28-Apr-2000

Mainland: 73; 12-Jul-2006.

Comment: Adult male voucher specimen #337 from site 7 in HCCIC, NMI.

Krenosmittia camptophleps (Edwards, 1929)
Mainland: 61; 07-May-2004. 73; 06-Jun, 22-Aug-2002. 74; 28-Jun,12-Jul-2006.

Limnophyes angelicae Sæther, 1990
Clare Island: 20; 21-Aug-2002. 20F; 13-Jul, 05-Oct, 13-Oct-1998, 10-Jun, 07-Jul, 14-Jul, 14-Sep, 06-Oct-1999.

Comment: Adult male voucher specimen #341 from Creggan Lough, 21 Aug 2002, in HCCIC, NMI was the first published record of the species from Ireland (Murray and Murray, 2003). Additional voucher specimen in FCIC, NMI.

Limnophyes asquamatus Søgaard Andersen, 1937
Mainland: 73; 22-Aug-2002.

Comment: Adult male voucher specimen #343 in HCCIC, NMI. This record from Lough Doo was the first record of the species from the Ireland (Murray and Murray, 2006).

Limnophyes gurgicola (Edwards, 1929)
Clare Island: 10; 20-Aug-2002. 11; 28-Apr-2000

Comment: Adult male voucher specimen #351 from site 11 in HCCIC, NMI.

Limnophyes habilis (Walker, 1856)
Clare Island: 20F; 11-Jun-1998.

Comment: Voucher specimen in FCIC, NMI.

Limnophyes minimus (Meigen, 1818)
Clare Island: 11; 28-Apr-2000. 12; 20-Aug-2002

Comment: Adult male voucher specimen #355 from site 11 in HCCIC, NMI.

Limnophyes natalensis (Kieffer, 1914)
Clare Island: 1; 20-Aug-2002. 2; 20-Aug-2002. 10; 04-Jun-2002. 18; 05-Jun-2002.

Limnophyes pentaplastus (Kieffer, 1921)
Clare Island: 10; 04-Jun-2002. 22; 05-Jun-2002.

Limnophyes pumilio (Holmgren, 1869)
Mainland: 36; 29-Sep-2004. 38; 29-Sep-2004.

Metriocnemus ephemerus Langton, 2015
Mainland: 77; 28-Aug-2017.

Comment: Adult male and females, pupae and pupal exuviae were collected in a surface skim sample from a pond in the *machair* (a plain of wind-blown sand in level ground behind the sand dunes) at Barrynagappul Beach (Golden Strand), Dugort, Achill Island. *M. ephemerus* was only recently described as new to science from Ireland based on adult males collected from a tidal bay on the lower reach of the River Bann, County Derry (Langton 2015b). The pond at Dugort is a second location for the species which is thus far unique to Ireland (Murray 2017a, b). A description of the pupa based on specimens collected on Achill Island in 2017 is given by Langton and Murray (2018).

Metriocnemus (*Metriocnemus*) *eurynotus* (Holmgren, 1883)
Clare Island: 10; 04-Jun-2002. 13; 04-Jun-2002. 18; 05-Jun-2002. 20F; 10-Jun-1999.

Comment: recorded as *M. hygropetricus* by Fahy (2002). Voucher specimen in FCIC, NMI.

Metriocnemus (*Metriocnemus*) *fuscipes* (Meigen, 1818).
[*Metriocnemus fuscipes* Mg in Grimshaw (1912)]
Clare Island: 1; 20-Aug-2002. 2; 20-Aug-2002. 5; 20-Aug-2002. 10; 20-Aug-2002. 12; 20-Aug-2002. 18; 05-Jun-2002. 27; 21-Aug-2002. 28; 21-Aug-2002. 20F; 11-Jun-1998 to 05-Oct-1998 and 07-Jul-1999 to 02-Nov-1999. 23F; 13-May-1999

Mainland: G4, G7, G8, G9, Jul-1911.

Comment: Adult male voucher specimen #379 from site 28 in HCCIC, NMI. Additional voucher specimen in FCIC, NMI.

Pl. VIII *Metriocnemus fuscipes* adult male. Photo: James Lindsey

[*Metriocnemus modestus* Mg in Grimshaw (1912)]
Misidentification, see *Polypedilum* (*Pentapedilum*) *sordens*. (Subfamily Chironominae)

Metriocnemus (*Metriocnemus*) *picipes* (Meigen, 1818)
Clare Island: 11; 28-Apr-2000. 24; 21-Aug-2002.
Mainland: 75; 28-Aug-2017.

Comment: An adult male specimen from Clare Island, voucher #381 from site 24, is deposited in HCCIC, NMI. Several adult males were collected at Site 75, Dugort on Achill Island in 2017.

Nanocladius (*Nanocladius*) *balticus* (Palmén, 1959)
Mainland: 46; 24-Jul-2008. 73; 12-Jul-2006.

Nanocladius (*Nanocladius*) *dichromus* (Kieffer, 1906)
Mainland: 47; 22-Apr-2008, 25-Jun-2008. 49b; 30-Jun, 21-Aug-2008. 55; 08-Aug-2008. 57; 21-Apr, 24-Jun, 18-Sep-2008.

Nanocladius (*Nanocladius*) *rectinervis* (Kieffer, 1911)
Mainland: 46; 24-Jul-2008. 46; 17-Sep-2008. 50; 30-Sep-2004. 53; 05-Aug-2004. 74; 28-Jun-2006.

Orthocladius (*Eudactylocladius*) *fuscimanus* (Kieffer, 1908)
Clare Island: 11; 28-Apr-2000.

Orthocladius (*Eudactylocladius*) *olivaceus* (Kieffer, 1911)
Mainland: 46; 24-Jul-2008. 73; 22-Aug-2002

Comment: Voucher specimen (Pe) #396 from Site 73 in HCCIC, NMI. The record from Lough Doo was the first report of the species from Ireland (Murray and Murray 2006).

Orthocladius (*Euorthocladius*) *ashei* Soponis, 1990
Mainland: 52; 07-May-2004.

Orthocladius (*Euorthocladius*) *rivicola* Kieffer, 1911
Mainland: 71, 09-May-2003.
Comment: Adult male (#399) and pupal exuviae (#400) voucher specimens in HCCIC, NMI.

Orthocladius (*Mesorthocladius*) *frigidus* (Zetterstedt, 1838)
Mainland: 43; 06-May-2004. 59; 07-May-2004. 74; 09-May-2006.

Orthocladius (*Orthocladius*) *dentifer* Brundin, 1947
Clare Island: 6; 28-Apr-2000
Mainland: 74; 09-May-2006.
Comment: Voucher specimen (Pe) #408 from Site 6 in HCCIC, NMI. The record from Clare Island in 2000 was cited as *Orthocladius* Pe4 sensu Langton 1991 in Murray and Murray (2003). That pupal morphotype has since been associated with *O. dentifer* in the revised key of Langton and Visser (2003).

Orthocladius (*Orthocladius*) *glabripennis* (Goetghebuer, 1921)
Mainland: 73; 22-Aug-2002.

Orthocladius (*Orthocladius*) *oblidens* (Walker, 1856)
Clare Island: 20F; 11-Jun-1998. 31; 21-Aug-2002
Mainland: 56; 23-Jul-2008.
Comment: A voucher slide in FCIC-NMI, from Site 20F, labelled '*Orthocladius* sp.' was examined and determined as *Orthocladius* (*O.*) *oblidens*. It has been appropriately re-labelled.

Orthocladius (*Orthocladius*) *rivinus* Potthast, 1914
Mainland: 47; 22-Apr-2008. 73; 12-Jul-2006. 74; 05-Apr, 09-May-2006.

Orthocladius (*Orthocladius*) *rubicundus* (Meigen, 1818)
Mainland: 46; 25-Jun-2008. 51; 07-May-2004. 59; 07-May-2004. 61; 07-May, 05-Aug, 30-Sep-2004. 65; 05-Aug-2004. 66; 05-Aug-2004. 73; 22-Aug-2002.

Orthocladius (*Orthocladius*) *ruffoi* Rossaro and Prato, 1991
Mainland: 72; 25-Aug-2008.

Orthocladius (*Orthocladius*) *wetterensis* Brundin, 1956.
Mainland: 77; 28-Aug-2017.

Orthocladius (*Pogonocladius*) *consobrinus* (Holmgren, 1869).
Mainland: 37; 17-Jun-2008. 37; 16-Jul, 17-Sep-2008. 57; 21-Apr, 24-Jun-2008. 57; 18-Sep-2008. 69; 30-Sep-2004. 72; 26-Jun-2008.
Comment: Voucher specimen (Pe) #428 from site 69 in HCCIC, NMI.

Orthocladius (*Symposiocladius*) *holsatus* Goetghebuer, 1937
Mainland: 50; 30-Sep-2004
Comment: Voucher specimen (Pe) #430 in HCCIC, NMI. The record from Lough Mallard was the first occurrence of the species from Ireland (Murray and Murray 2006).

Paracladius conversus (Walker, 1856)
Mainland: 53; 08-May-2003.

Parakiefferiella bathophila (Kieffer, 1912)
Mainland: 36; 16-Jul-2008, 17-Jun-2008. 37; 16-Jul-2008. 45; 04-Aug-2004. 72; 25-Aug-2008. 73; 06-Jun-2002.

Parakiefferiella coronata (Edwards, 1929)
Mainland: 36; 16-Jul-2008. 55; 24-Jun-2008. 72; 26-Jun-2008. 73; 28-Jun-2006. 74; 28-Jun-2006, 12-Jul-2006.

Parakiefferiella scandica (Brundin, 1947)
Mainland: 73; 09-May-2006.

Parakiefferiella smolandica (Brundin, 1947)
Mainland: 57; 21-Apr-2008. 58; 05-Aug-2004. 74; 9-May-2006.
Comment: Voucher specimen (Pe) #442 from site 58 in HCCIC, NMI.

Parametriocnemus stylatus (Spärck, 1923)
Clare Island: 26; 04-Jun-2002. 31; 21-Aug-2002
Mainland: 50; 30-Sep-2004. 51; 07-May-2004. 52; 30-Sep-2004. 53; 05-Aug-2004. 61; 07-May, 05-Aug-2004. 63; 30-Sep-2004. 66; 05-Aug-2004. 68; 06-Jun-2004.

Paraphaenocladius impensus impensus (Walker, 1856)
Clare Island: 10; 27-Apr-2000. 11; 28-Apr-2000
Mainland: 49b, 26-Sep-2008. 67; 30-Sep-2004.

Paraphaenocladius irritus irritus (Walker, 1856)
Clare Island: 12; 29-Apr-2000.

Paraphaenocladius pseudirritus subsp. *pseudirritus* Strenzke, 1950
Clare Island: 1; 20-Aug-2002. 2; 20-Aug-2002.
Mainland: 52; 30-Sep-2004
Comment: Adult male voucher specimen #455 from site 1 in HCCIC, NMI.

Psectrocladius (Allopsectrocladius) obvius (Walker, 1856)
Clare Island: 1; 20-Aug-2002. 7; 05-Jun-2002. 20; 21-Aug-2002. 20F; 11-Jun, 05-Oct-1998 and 10-Jun-1999 to 06-Oct-1999
Mainland: 38; 29-Sep-2004. 40, 05-Aug-2004.
Comment: Voucher specimen in FCIC, NMI.

Psectrocladius (Allopsectrocladius) platypus (Edwards, 1929)
Clare Island: 7; 05-Jun-2002, 20-Aug-2002. 27; 21-Aug-2002. 29; 21-Aug-2002. 20F; 11-Jun-1998 to 05-Oct-1998 and 10-Jun-1999 to 06-Oct-1999. 23F; 02-Jun-1998 to 03-Nov-1998 and 05-May-1999 to 14-Sep-1999
Mainland: 30; 06-May-2004
Comment: Voucher specimen (Pe) #468 from site 7 in HCCIC, NMI. Additional voucher specimen in FCIC, NMI.

Psectrocladius (Mesopsectrocladius) barbatipes Kieffer, 1923.
Mainland: 81; 29-Aug-2017.

Psectrocladius (Psectrocladius) fennicus Storå, 1939
Clare Island: 20F; 06-Jul-1998, 13-Aug-1999. 23F; 11-Jun-1998 to 01-Dec-1998 and 07-Jul-1999 to 21-Oct-1999
Mainland: 36; 07-Apr-2008.
Comment: Voucher specimen in FCIC, NMI.

Psectrocladius (Psectrocladius) limbatellus (Holmgren, 1869)
Clare Island: 29; 21-Aug-2002
Mainland: 69; 30-Sep-2004. 73; 22-Aug-2002. 81; 29-Aug-2017.

Psectrocladius (Psectrocladius) octomaculatus Wülker, 1956
Mainland: 62; 07-May-2004. 64; 07-May-2004. 74; 12-Jul-2006.
Comment: Voucher specimen (Pe) #482 from site 64 in HCCIC, NMI.

Psectrocladius (Psectrocladius) oligosetus Wülker, 1956
Clare Island: 7; 28-Apr-2000, 20-Aug-2002.
Mainland: 36; 05-Aug-2004. 40, 05-Aug-2004. 41, 29-Sep-2004.
Comment: Voucher specimens adult male (#483) and pupal exuviae (#484) from site 7 in HCCIC, NMI.

Psectrocladius (Psectrocladius) oxyura Langton, 1985
Mainland: 55; 24-Jun-2008.

Psectrocladius (Psectrocladius) psilopterus (Kieffer, 1906)
Mainland: 36; 05-Aug-2004. 07-Apr, 17-Jun, 17-Sep-2008. 38; 29-Sep-2004. 39, 06-May-2004. 41, 29-Sep-2004. 45; 04-Aug-2004. 49a, 07-May-2004, 21-Aug-2008. 55; 24-Jun, 18-Sep-2008. 57; 21-Apr-2008, 24-Jun-2008. 74; 12-Jul-2006.
Comment: Voucher specimen (Pe) #488 from site 45 in HCCIC, NMI.

Psectrocladius (Psectrocladius) sordidellus (See Pl. I) (Zetterstedt, 1838)
[*Orthocladius sordidellus* Ztt. In Grimshaw (1912)]
Clare Island: 13; 04-Jun-2002. G; Jul-1910.
Mainland: 36; 29-Sep-2004. 16-Jul, 17-Sep-2008, 48; 30-Sep-2004. 50; 30-Sep-2004. 53; 05-Aug-2004. 57; 21-Apr, 24-Jun, 18-Sep-2008. 67; 30-Sep-2004. G2, G3, G5, G6, G7, G9, Jul-1911.

Psectrocladius (Psectrocladius) ventricosus Kieffer, 1925
Clare Island: 12; 20-Aug-2002
Mainland: 37; 17-Jun, 16-Jul, 17-Sep-2008. 58; 05-Aug-2004.
Comment: Pupal exuviae voucher specimen #494 from site 58 in HCCIC, NMI. *P. ventricosus* is a brackish water species. On Clare Island pupal exuviae were collected from a marine pool on the foreshore opposite the Bayview Hotel. The Mainland sites at the outflow from L. Keel

(Achill Island, site 37) and Carrokeel Lough (site 58) are shallow coastal pools formed behind sand barriers through which sea water may percolate giving rise to slightly saline conditions. The record from Carrokeel was reported as new to the Irish fauna in Murray and Murray (2006).

Psectrocladius sp A, sensu Langton, 1980
Mainland: 36; 29-Sep-2004.

Comment: Pupal exuviae of this distinct morphotype, were first noted by Langton (1980) from Lochs in northern Scotland. The adult is known (Langton and Pinder, 2007) but not yet described. This morphotype is widespread in Ireland (Murray *et al.* 2014, 2018) including two locations in Co. Mayo - L. Alone, a remote lake in the Ox Mountains, and Kylemore Lough (specimens in the authors collection). Voucher specimens of the morphotype, #491 (adult male) and # 492 (pe), from L. Ouler, Co. Wicklow are in the HCCIC, NMI (Murray 2005).

Pseudorthocladius (Pseudorthocladius) curtistylus (Goetghebuer, 1921)
Clare Island: 23F; 13-Jul, 20-Aug-1999.

Comment: Voucher specimen in FCIC, NMI.

Pseudorthocladius (Pseudorthocladius) filiformis (Kieffer, 1908)
Clare Island: 3; 20-Aug-2002. 8; 28-Apr-2000, 05-Jun-2002. 13; 04-Jun-2002. 18; 05-Jun-2002. 20; 21-Aug-2002. 27; 21-Aug-2002
Mainland: 71, 09-May-2003. 72; 26-Jun-2008.
Comment: Adult male voucher specimen #497 from Site 8 in HCCIC, NMI.

Pseudorthocladius (Pseudorthocladius) macrovirgatus Sæther and Sublette, 1983
Clare Island: 10; 20-Aug-2002. 28; 21-Aug-2002.

Pseudosmittia oxoniana (Edwards, 1922)
Clare Island: 10; 20-Aug-2002
Mainland: 36; 29-Sep-2004. 38; 29-Sep-2004. 61; 30-Sep-2004. 73; 22-Aug-2002.

Pseudosmittia trilobata (Edwards, 1929)
Clare Island: 2; 20-Aug-2002.

Rheocricotopus (Psilocricotopus) chalybeatus (Edwards, 1929)
Clare Island: 23; 04-Jun-2002
Mainland: 61; 05-Aug-2004.

Rheocricotopus (Rheocricotopus) fuscipes (Kieffer, 1909)
Mainland: 56; 21-Apr-2008.

Smittia pratorum (Goetghebuer, 1927)
Clare Island: 11; 28-Apr-2000. 22; 05-Jun-2002. 31; 21-Aug-2002

Synorthocladius semivirens (Kieffer, 1909)
Clare Island: 11; 21-Aug-2002. 23; 04-Jun-2002. 30; 21-Aug-2002
Mainland: 36; 05-Aug-2004. 37; 29-Sep-2004, 07-Apr, 17-Jun, 16-Jul, 17-Sep-2008. 38; 29-Sep-2004. 46; 25-Jun-2008. 46; 24-Jul-2008. 47; 22-Apr, 18-Sep-2008. 49b; 23-Apr, 21-Aug, 26-Sep-2008. 50; 30-Sep-2004. 51; 07-May, 30-Sep-2004. 55; 22-Apr, 08-Aug, 18-Sep-2008. 56; 21-Apr, 23-Jul-2008. 57; 21-Apr, 24-Jun, 18-Sep-2008. 61; 30-Sep-2004. 66; 05-Aug-2004. 68; 06-Jun-2004. 72; 26-Jun, 22-Sep-2008. 73; 06-Jun, 22-Aug-2002, 05-Apr, 09-May, 28-Jun, 12-Jul, 16-Aug-2006.

Comment: Adult male voucher specimen #541 from Site 73 in HCCIC, NMI.

Thalassosmittia thalassophila (Bequaert and Goetghebuer, 1913)
Clare Island: 11; 22-Aug-2002. 12; 28-Apr-2000, 05-Jun, 20-Aug-2002. 15; 05-Jun-2002. 17; 05-Jun-2002
Mainland: 60; 07-May-2004. 71; 27-Jul-2000. 82; 29-Aug-2017.

Comment: An adult male voucher slide preparation of this marine coastal dwelling species from Clare Island, Site 12 is #543 in HCCIC, NMI.

Thienemannia gracilis Kieffer, 1909
Clare Island: 12; 29-Apr-2000.

Thienemanniella acuticornis (Kieffer, 1912)
Mainland: 53; 05-Aug-2004.

Thienemanniella clavicornis (Kieffer, 1911)
Mainland: 53; 08-May-2003.

Comment: Adult male voucher specimen #549 in HCCIC, NMI.

Thienemanniella majuscula (Edwards, 1924)
Clare Island: 9; 05-Jun-2002. 26; 28-Apr-2002. 31, 21-Aug-2002.

Thienemanniella vittata (Edwards, 1924)
Mainland: 45; 04-Aug-2004. 49a; 07-May-2004. 52; 30-Sep-2004. 61; 30-Sep-2004.

Comment: Voucher specimen (Pe) #588 from Site 52 in HCCIC, NMI.

Thienemanniella sp Pe2a sensu Langton and Visser 2003
Clare Island: 26; 04-Jun-2002.
Mainland: 34; 05-Aug-2004. 59; 07-May-2004.

Tvetenia bavarica (Goetghebuer, 1934)
Mainland: 35; 05-Aug-2004.

Tvetenia calvescens (Edwards, 1929)
Mainland: 42; 06-May-2004. 43; 06-May-2004. 44; 06-May-2004. 45; 04-Aug-2004. 51; 30-Sep-2004. 52; 30-Sep-2004. 53; 05-Aug-2004. 59; 07-May-2004. 61; 30-Sep-2004. 66; 05-Aug-2004. 68; 06-Jun-2004. 74; 19-May-2005, 05-Apr, 11-Sep-2006,

Comment: Voucher specimen (Pe) #566 from Site 44 in HCCIC, NMI.

Tvetenia discoloripes (Goetghebuer and Thienemann, 1936)
Mainland: 61; 30-Sep-2004.

Tvetenia verralli (Edwards, 1929)
Mainland: 52; 30-Sep-2004. 53; 05-Aug-2004. 59, 07-May-2004. 68; 06-Jun-2004.

Comment: Voucher specimen (Pe) #570 from Site 59 in HCCIC, NMI.

Subfamily Chironominae
Tribe Chironomini

Benthalia carbonaria (Meigen, 1804)
Mainland: 55; 24-Jun-2008.
Comment: Previous records of this taxon in Ireland are under its synonym *Lobochironomus dissidens*.

Chironomus (*Chironomus*) *alpestris*
Goetghebuer, 1934
[?*Tendipes dorsalis* Mg in Grimshaw (1912)]
C. dorsalis Auctt., *Chironomus dorsalis* sensu Strenzke, 1959 and Langton and Visser (2003)
Clare Island: G, Jul-1910
Mainland: G6; Jul-1911.
Comment: Grimshaw (1912) noted that the two males, collected by J.N. Halbert on Clare Island, and one female that he himself collected in the gardens of Westport Demesne and identified as *Tendipes dorsalis* Mg., belonged to 'a variable but common species' and remarked that 'the transversely banded abdomen serves to distinguish it from its allies'. However, abdominal colour patterns, while useful, are not absolutely reliable for taxonomic discrimination. Twentieth century taxonomic revisions assigned species-level status to several *Chironomus* species previously recognised as 'variable'. Some 25 species of *Chironomus* are currently known from Ireland (Murray *et al.* 2018), including *Chironomus alpestris*. A number of these have transverse abdominal bands but are distinguished by critical morphological features of the male hypopygium. According to Spies and Sæther (2004) many early determinations and citations of *C. dorsalis* are unreliable and thus the records of *T. dorsalis* by Grimshaw (1912) are uncertain given that ten species of *Chironomus* are recognised in the NSCI, five on Clare Island and eight in west Mayo. The specimens reported as *C. dorsalis* in Murray and Murray (2003) were incorrectly determined - see *C. commutatus* below.

Chironomus (*Chironomus*) *annularius* Meigen, 1818
[*Tendipes* (*Chironomus*) *annularis* Deg. in Grimshaw (1912)]

Mainland: G1, 1910-1912

Comments: The record of *Tendipes* (*Chironomus*) *annularis* Deg. from the original survey was based on three specimens (two male, one female) collected by J.N. Halbert from 'Achill, N.E.' (Grimshaw, 1912, p. 4). Spies and Sæther (2004) noted that many publications since the eighteenth century have cited at least four different biological species variously spelled '*annularis*', '*annularia*', '*annularius*' and '*annularius*'. Since a comprehensive taxonomic review is necessary to resolve this confusion, the record by Grimshaw is tentatively retained here as *Chironomus annularius* Meigen. Although it did not occur in collections in this survey this species is known from several locations in Ireland including a record from Hurney, on the west coast of Lough Corrib, County Galway (Murray *et al.* 2015).

Chironomus (*Chironomus*) *anthracinus*
Zetterstedt, 1860
Mainland: 56; 18-Sep-2008.

Comment: *Chironomus anthracinus* is a widespread and common species in Ireland whose larvae are found in enriched lakes and ponds.

Chironomus (*Chironomus*) *aprilinus* Meigen, 1818
[Reported by Grimshaw (1912) as *Tendipes aprilinus* Mg.]
Clare Island: 8; 28-Apr-2000. 11; 27-Apr-2000. 14; 05-Jun-2002. 20F; 13-Aug-1999, 20-Aug-1999.
Mainland: 51; 07-May-2004, 30-Sep-2004. 67; 30-Sep-2004. G6, G9, Jul-1911.
Comment: Voucher specimen (Pe) #582 from Site 51 in HCCIC, NMI. Additional voucher specimen in FCIC, NMI (det. PHL).

Chironomus (*Chironomus*) *commutatus* Keyl, 1960
Clare Island: 7; 28-Apr-2000. 20F; 11-Jun, 13-Jul-1998 and 10-Jun to 13-Aug-1999.
Comment: Voucher specimen in FCIC, NMI (det. PHL). Pupal exuviae reported from Site 7 on Clare Island as *C. dorsalis* in Murray and Murray (2003) determined from Langton (1991) were incorrectly identified and are reported here as *C. commutatus* from Langton and Visser (2003). See comments for *C. alpestris* above.

Chironomus (*Chironomus*) *nuditarsis* Keyl, 1961
Clare Island: 7; 05-Jun-2002.
Comment: The record of this species derives from a recent re-examination of the bulk sample from Site 7 in which pupal exuviae were observed that had been overlooked in Murray and Murray (2003). The omission was discovered too late for inclusion in Murray *et al.* (2018).

Chironomus (*Chironomus*) *piger* Strenzke, 1959
Mainland: 41; 29-Sep-2004. 62; 07-May-2004.

Chironomus (*Chironomus*) *plumosus*
(Linnaeus, 1758)
[*Tendipes plumosus* L. and *T. ferrugineovittatus* Ztt. in Grimshaw (1912)]
Mainland: 58; 05-Aug-2004. G6, G7, G10 1910-1912.
Comment: Grimshaw examined a single female from Westport (G6) commenting 'it is very similar to the common *T. plumosus* L'.

Chironomus (*Chironomus*) *pseudothummi*
Strenzke, 1959
Clare Island: 1, 20-Aug-2002. 7; 28-Apr-2000. 13; 04-Jun-2002. 32; 21-Aug-2002. 20F; 02-Jun, 11-Jun, 01-Sep-1998, 10-Jun, 06-Oct-1999.
Comment: Voucher specimen in FCIC, NMI (det. PHL)

Chironomus (*Chironomus*) *riparius* Meigen, 1804
[Reported by Grimshaw (1912) as *Tendipes riparius* Mg.]
Clare Island: 7; 5-Jun-2002. 20F; 11-Jun-1998 to 13-Oct-1998 and 10-Jun-1999 to 14-Sep-1999. 23F; 14-Sep-1999
Mainland: 53; 08-May-2003. G2, G6, Jul-1911.
Comment: Voucher specimen in FCIC, NMI (det. PHL). The record from Site 7 on Clare Island had been overlooked in Murray and Murray (2003).

Chironomus (*Chironomus*) *salinarius* (Kieffer, 1915)
Mainland: 83; 29-Aug-2017.
Comment: This record from the brackish Sruhill (Dooniver) Lough on the northeast coastline of Achill Island is the first record of this saline-tolerant species from County Mayo (Murray 2017a).

Chironomus (*Chironomus*) *tentans* Fabricius, 1805
[*Tendipes tentans* Fab in Grimshaw (1912)]
Clare Island: 20F; 11-Jun-1998 to 01-Sep-1998 and 10-Jun-1999 to 20-Aug-1999. G; Jul-1910
Mainland: 47; 25-Jun-2008. G9, Jul-1911.
Comment: Voucher specimen in FCIC, NMI. Labelled as '*Metriocnemus tentans*'.

Cladopelma krusemani (Goetghebuer, 1935)
Clare Island 19; 05-Jun-2002. 20; 21-Aug-2002. 20F; 11-Jun to 01-Sep-1998 and 10-Jun to 06-Oct-1999. 23F; 06-Jul-1998, 07-Jul, 13-Aug-1999.
Comment: Voucher specimen in FCIC, NMI.

Cladopelma virescens (Meigen, 1818)
Mainland: 55; 18-Sep-2008.

Cladopelma viridulum (Linnaeus, 1767)
Mainland: 37; 17-Jun, 16-Jul-2008. 46; 24-Jul-2008. 56; 23-Jul-2008. 72; 25-Aug-2008. 80; 29-Aug-2017.

Cryptochironomus albofasciatus (Staeger, 1839)
Mainland: 46; 25-Jun-2008.

Cryptochironomus obreptans (Walker, 1856)
Mainland: 46; 25-Jun, 24-Jul, 17-Sep-2008. 56; 23-Jul-2008. 58; 05-Aug-2004.

Cryptochironomus psittacinus (Meigen, 1830)
Mainland: 56; 23-Jul-2008, 24-Jun-2008. G2, G4 Jul-1911.

Cryptochironomus supplicans (Meigen, 1830)
Mainland: 47; 25-Jun-2008. 73; 22-Aug-2002.

Demeijerea rufipes (Linnaeus, 1761)
[*Tendipes rufipes* L. in Grimshaw (1912)]
Clare Island: 23F; 07-Sep-1999
Mainland: 57; 18-Sep-2008. G2, Jul-1910 (collected by J.N. Halbert).
Comment: Voucher specimen from Site 23F in FCIC, NMI.

Demicryptochironomus (*Demicryptochironomus*) *vulneratus* (Zetterstedt, 1838)
[as *Tendipes nigrimanus* Staeg. in Grimshaw (1912)]
Mainland: 37; 17-Jun, 17-Sep-2008. 47; 25-Jun-2008. 56; 23-Jul-2008. 72; 26-Jun-2008. G6; Jul-1911.

Demicryptochironomus (*Irmakia*) *neglectus* Reiss, 1988
Mainland: 61; 05-Aug-2004.
Comment: First record for Ireland by Murray (2010).

Dicrotendipes lobiger (Kieffer, 1921)
Mainland: 73; 12-Jul-2006.
Comment: Pupal exuviae voucher specimen #656 in HCCIC, NMI.

Dicrotendipes nervosus (Staeger, 1839)
[*Tendipes brevitibialis* Ztt. in Grimshaw (1912)]
Clare Island: 23F; 02-Jun-1998, 11-Jun-1998, 06-Jul-1998, 10-Jun-1999
Mainland: G2, G5, G6; Jul-1911.
Comment: Voucher specimen in FCIC, NMI.

Dicrotendipes notatus (Meigen, 1818)
Mainland: 51; 07-May-2004.
Comment: Adult male voucher specimen #659 from site 51 in HCCIC, NMI.

Dicrotendipes pulsus (Walker, 1856)
Mainland: 36; 17-Jun-2008. 47; 22-Apr-2008. 50; 30-Sep-2004. 56; 24-Jun, 23-Jul-2008.
Comment: Voucher specimen (Pe) #664 from Site 73 in HCCIC, NMI.

Dicrotendipes tritomus (Kieffer, 1916)
Mainland: 74; 28-Jun-2006.

Endochironomus albipennis (Meigen, 1830)
Mainland: 37; 17-Jun-2008, 16-Jul-2008. 47; 22-Apr, 18-Sep-2008. 49a, 07-May-2004. 50; 30-Sep-2004. 55; 08-Aug-2008. 56; 23-Jul-2008.

Endochironomus tendens (Fabricius, 1775)
Clare Island: 20F; 04-Aug-1998. 23F; 10-Aug-1998
Mainland: 37; 17-Jun-2008. 37; 16-Jul-2008. 49a; 21-Aug-2008. 57; 24-Jun-2008. 58; 05-Aug-2004. 74; 28-Jun-2006.
Comment: Voucher specimen (Pe) #672 from Site 58 in HCCIC, NMI. Additional voucher specimen in FCIC, NMI.

Glyptotendipes (*Caulochironomus*) *foliicola* sensu Pinder, 1978 (adult male) and Langton and Visser, 2003 (pupal exuviae).
Mainland: 48; 30-Sep-2004. 49b; 23-Apr-2008, 30-Jun-2008. 57; 21-Apr-2008.
Comment: Voucher specimen (Pe) #674 from Site 48 in HCCIC, NMI. Because of taxonomic uncertainty (Spies and Saether, 2004, 2013) the status of this taxon is unclear although pupal exuviae and adult male are recognised. To avoid uncertainty the source determination keys for the records of this species of *Glyptotendipes* from west Mayo are cited above.

Glyptotendipes (*Caulochironomus*) *scirpi* (Kieffer, 1915)
Mainland: 55; 24-Jun, 23-Jul-2008. 57; 21-Jul-2008.

Glyptotendipes (*Caulochironomus*) *viridis* (Macquart, 1834)
[*Tendipes viridis* Mcq. in Grimshaw (1912)]
Mainland: G4; Jul-1911.
Comment: Grimshaw (1912) recorded two males of *Tendipes viridis,* now recognised in the genus *Glyptotendipes*. According to Spies and Sæther (2004) the status of *Glyptotendipes* (*Caulochironomus*) *viridis* Macquart is uncertain. Confusion persists between the taxa *G. viridis* and *Glyptotendipes* (*Caulochironomus*) *imbicilis* (Walker, 1856) since the latter has been 'a name of varied usage' that 'different authors had interpreted quite differently'. Pending taxonomic resolution to this issue Grimshaw's record is retained. While it has not been found in the present survey there are records in Ireland from seven other locations (Murray *et al.* 2018).

Glyptotendipes (Glyptotendipes) barbipes
(Staeger, 1839)
Mainland: 47; 25-Jun-2008.

Glyptotendipes (Glyptotendipes) cauliginellus
(Kieffer, 1913)
Clare Island 19; 05-Jun-2002. 20F; 11-Jun-1998, 10-Jun, 07-Jul, 13-Aug-1999. 23F; 13-Aug-1999
Mainland: 49b, 21-Aug-2008. 56; 23-Jul-2008. 57; 24-Jun, 21-Jul-2008. 74; 28-Jun-2006

Comment: cited under the synonym *Glyptotendipes gripkoveni* (Kieffer) in Fahy (2002) with voucher specimen in FCIC, NMI.

Glyptotendipes (Glyptotendipes) pallens
(Meigen, 1804)
Clare Island: 20; 21-Aug-2002
Mainland: 49b, 07-May-2004, 30-Jun-2008. 55; 08-Aug-2008. 56; 23-Jul-2008.

Glyptotendipes (Glyptotendipes) paripes
(Edwards, 1929)
Clare Island: 9, 05-Jun-2002. 32; 21-Aug-2002
Mainland: 47; 22-Apr-2008. 55; 08-Aug-2008. 81; 29-Aug-2017.

Harnischia curtilamellata (Malloch, 1915)
Mainland: 49b, 21-Aug-2008. 56; 24-Jun, 23-Jul-2008.

<?*Kloosia pusilla* (Linnaeus, 1767)>
[*Tendipes pusillus* L. in Grimshaw (1912)]
Mainland: G8?

Comment: This is an uncertain record. Grimshaw remarked on the very small size (2mm) of the four specimens he examined and the overall bright green colour 'with three shining black stripes on the thorax the middle one of which is distinctly double behind'. He drew attention to the fact that it was 'only reputed British' but stated he had seen specimens taken in Edinburgh. *Tendipes pusillus* was transferred to a new genus *Kloosia*, erected by Krusemann (1933). According to Cranston et al. (1989) adult males of *Kloosia* are small with wing length of 1.1 to 1.4mm and have a gold-brown body with separate gold-brown vittae (the 'stripes' on the thorax to which Grimshaw referred). According to Reiss (1988) larvae of *K. pusilla* are members of the sand dwelling community in the beds of large rivers. The species is not yet known from Britain or Ireland (Spies and Sæther 2013). It is not possible to confirm Grimshaw's record.

Lauterborniella agrayloides (Kieffer, 1911)
Clare Island: 20; 21-Aug-2002.

Comment: Pupal exuviae voucher specimen #696 in HCCIC, NMI.

Microtendipes chloris (Meigen, 1818)
Clare Island: 20; 21-Aug-2002. 25; 28-Apr-2000. 32; 21-Aug-2002
Mainland: 73; 06-Jun-2002, 22-Aug-2002.

Microtendipes diffinis (Edwards, 1929)
Clare Island: 20F; 11-Jun-1998, 07-Jul, 02-Aug-1999. 23F; 07-Jul-1999
Mainland: 61; 07-May, 05-Aug-2004. 66; 05-Aug-2004.
Comment: Voucher specimen in FCIC, NMI.

Microtendipes pedellus (De Geer, 1776)
[as *T. pedellus* Deg in Grimshaw (1912).]
Clare Island: 20F; 11-Jun-1998 to 05-Oct-1998 and 10-Jun-1999 to 07-Sep-2009. 23F; 06-Jul-1998 to 14-Sep-1998 and 19-Apr-1999 to 07-Sep-1999
Mainland: 61; 05-Aug-2004. G5, G6, G7, G10 Jul-1911.
Comment: Voucher specimen in FCIC, NMI.

Microtendipes rydalensis (Edwards, 1929)
Mainland: 66; 05-Aug-2004.

Nubensia nubens (Edwards, 1929) syn *Polypedilum (Pentapedilum) nubens* (in Spies and Dettinger-Klemm 2015).
Mainland: 46; 24-Jul, 17-Sep-2008. 49b, 21-Aug-2008. 57; 24-Jun-2008. 73; 12-Jul-2006. 74; 12-Jul-2006.

Pagastiella orophila (Edwards, 1929)
Clare Island 19; 05-Jun-2002. 20F; 11-Jun, 13-Jul-1998, 02-Jun, 10-Jun-1999. 23F; 02-Jun-1998 to 14-Sep-1998 and 02-Jun-1999 to 13-Aug-1999
Mainland: 37; 17-Jun-2008. 72; 26-Jun-2008. 74; 28-Jun-2006.

Comment: Voucher specimen in FCIC, NMI.

Parachironomus cinctellus (Goetghebuer, 1921)
Clare Island: 20; 21-Aug-2002
Mainland: 57; 18-Sep-2008; 72; 25-Aug-2008

Parachironomus frequens (Johannsen, 1905)
Mainland: 37; 17-Jun-2008, 17-Sep-2008.

Parachironomus gracilor Kieffer (1918)
syn *P. arcuatus* (Goetghebuer, 1919)
Mainland: 37; 16-Jul-2008. 47; 25-Jun-2008. 49b; 26-Sep-2008. 57; 18-Sep-2008.

Parachironomus mauricii (Kruseman, 1933)
Clare Island: 20F

Comment: The adult male voucher specimen in FCIC, NMI was misidentified and the slide was originally labelled by Fahy (2002) as *Parachironomus parilis*. This is the only confirmed record of the species from Ireland. There are two labels on the slide, one placed directly over the other. The top label shows the handwritten species name *Parachironomus paralis* and an NMI reference number 'NMI 31:2003'. The writing on the label underneath is mostly obscured but the word 'Malaise' is visible, indicating the specimen was taken in Malaise trap collections at Creggan Lough. There is no visible indication of the date or year (1998 or 1999) of capture.

Parachironomus paralis (Walker, 1856)
Clare Island: 20F; 11-Jun-1998 to 13-Oct-1998 and 10-Jun-1999 to 2-Nov-1999. 23F; 11-Jun-1998.

Comment: Fahy recorded *P. paralis* at Creggan Lough and L. Poirtin Fuinch but the single voucher specimen from the Malaise trap collections at Creggan Lough in NMI is a misidentified *P. mauricii* (see above). *P. paralis* did not occur in the author's collections on Clare Island, or the mainland, but the species is known from Lough Pollacappul, Kylemore, about 30km south of Clare Island in County Galway (Murray *et al.* 2015). It is possible that some of the more than 200 specimens that Fahy (2002) obtained in Malaise Trap collections on Clare Island are *P. parilis*.

Parachironomus tenuicaudatus (Malloch, 1915)
Clare Island: 23F; 01-Sep-1998, 02-Aug-1999.
Comment: Voucher specimen in FCIC, NMI.

Parachironomus vitiosus (Goetghebuer, 1921)
Clare Island: 20; 21-Aug-2002
Mainland: 57;18-Sep-2008. 72; 25-Aug-2008.

Comment: Adult male voucher specimen #743 from site 20, in HCCIC, NMI.

Parachironomus Pe 2a sensu Langton and Visser 2003
Mainland: 37; 17-Jun, 16-Jul-2008.

Comment: Pupal exuviae of this distinct morphotype, found in collections from Keel lake, Achill, have not yet been associated with the adult.

? *Parachironomus* sp. Pe sensu Langton, 1991
Clare Island: 20; 21-Aug-2002
Mainland: 37; 17-Sep-2008. 61; 05-Aug-2004.

Comment: According to Spies and Bolton (2013) pupal exuviae determined as '? *Parachironomus* sp. Pe.' from Langton (1991) or as *P. mauricii* from Langton and Viser (2003) may also be linked with other adult *Parachironomus* species. The morphotype is also known from nearby Co. Sligo and Co. Roscommon (Murray *et al.* 2015). A voucher pupal exuviae #732 from Site 61 is deposited in HCCIC, NMI.

Paracladopelma camptolabis (Kieffer, 1913)
Mainland: 49b, 21-Aug-2008. 51; 07-May-2004. 53; 08-May-2003. 56; 21-Apr, 23-Jul-2008. 73; 09-May-2006. 73; 22-Aug-2002. 74; 05-Apr-2006

Comment: Adult male voucher specimen #745 from Site 53 in HCCIC, NMI.

Paracladopelma laminatum (Kieffer, 1921)
Mainland: 37; 07-Apr, 17-Jun-2008. 38; 29-Sep-2004. 73; 22-Aug-2002.

Paracladopelma nigritulum (Goetghebuer, 1942)
Mainland: 73; 09-May-2006.

Paralauterborniella nigrohalteralis (Malloch, 1915)
Mainland: 49b; 21-Aug-2008.

Phaenopsectra flavipes (Meigen, 1818)
[Reported by Grimshaw (1912) as *Tanytarsus flavipes* Mg.]

Clare Island: 20; 21-Aug-2002
Mainland: 46; 24-Jul-2008. 51; 07-May-2004. 55; 08-Aug-2008. 56; 24-Jun, 23-Jul-2008. 72; 25-Aug-2008. 73; 22-Aug-2002. G6; Jul-1911.

Phaenopsectra punctipes (Wiedemann, 1817)
[*Tanytarsus punctipes* Wied. in
Grimshaw (1912)]

Clare Island: 20F; 11-Jun-1998 to 03-Nov-1998 and 10-Jun-1999 to 02-Nov-1999. 23F; 07-Jul-1999, 20-Aug-1999

Mainland: G9, Jul-1911.

Comment: Voucher specimen in FCIC, NMI.

Polypedilum (*Pentapedilum*) *sordens*
(van der Wulp, 1874)
[*Metriocnemus modestus* Mg. in Grimshaw, 1912, misidentification]

Clare Island: 19; 05-Jun-2002. 20; 21-Aug-2002

Mainland: 47; 25-Jun, 18-Sep-2008. 49b; 30-Jun-2008. 55; 24-Jun, 08-Aug-2008. 56; 23-Jul-2008. 72; 26-Jun, 25-Aug-2008. G7 and G9, Jul-1911.

Comment: Grimshaw (1912) drew attention to his record of '*Metriocnemus modestus*' from two mainland sites as a species 'new to the British list' and provided a short description with a figure of the hypopygium from which it is clear that the specimen does not belong to the genus *Metriocnemus* (subfamily Orthocladiinae). Edwards (1929, p.310) examined Grimshaw's specimen and identified it as *Pentapedilum sordens*.

Polypedilum (*Pentapedilum*) *uncinatum*
(Goetghebuer, 1921)

Clare Island: 1, 20-Aug-2002. 7; 20-Aug-2002. 9; 05-Jun-2002. 20; 21-Aug-2002. 29; 21-Aug-2002. 32; 21-Aug-2002

Mainland: 40, 05-Aug-2004.

Comment: This species was reported from Site 9 as *Polypedilum* (*Polypedilum*) sp Pe8 sensu Langton (1991) in Murray and Murray (2003). Voucher specimen from Site 20F in FCIC, NMI.

Polypedilum (*Polypedilum*) *acutum* Kieffer, 1915

Clare Island: 20; 21-Aug-2002. 20F; 11-Jun-1998 to 01-Sep-1998 and 10-Jun-1999 to 07-Sep-1999. 23F; 06-Jul-1998.

Comment: This species was reported from Creggan Lough, Site 20, as *Polypedilum* (*Polypedilum*) sp Pe7 sensu Langton, (1991) in Murray and Murray (2003). Voucher specimen from Site 20F in FCIC, NMI.

Polypedilum (*Polypedilum*) *arundineti*
(Goetghebuer, 1921)

Clare Island: 20F; 11-Jun-1998 to 5-Oct-1998 and 10-Jun-1999 to 07-Sep-1999. 23F; 13-Jul-1998, 10-Jun, 07-Jul, 13-Jul-1999

Mainland: 40; 05-Aug-2004.

Comment: Voucher specimen in FCIC, NMI.

Polypedilum (*Polypedilum*) *nubeculosum*
(Meigen, 1804)
[as *Tendipes nubeculosus* Mg. in Grimshaw (1912)]

Mainland: 56; 21-Apr-2008. 56; 23-Jul, 18-Sep-2008. 72; 25-Aug-2008. 76; 28-Aug-2017. G6; Jul-1911).

Polypedilum (*Polypedilum*) *pedestre* (Meigen, 1830)

Clare Island: 20F; 06-Jul-1998 to 01-Sep-1998 and 02-Jun-1999 to 01-Sep-1999. 23F; 07-Jul-1999, 13-Jul-1999, 13-Aug-1999.

Comment: Voucher specimen in FCIC, NMI.

Polypedilum (*Tripodura*) *pullum* (Zetterstedt, 1838)

Clare Island: 20F; 13-Jul-1998

Mainland: 46; 24-Jul-2008. 73; 12-Jul-2006

Comment: Voucher specimen in FCIC, NMI.

Polypedilum (*Uresipedilum*) *convictum*
(Walker, 1856)

Mainland: 61; 05-Aug-2004. 66; 05-Aug-2004. 73; 09-May-2006.

Polypedilum (*Uresipedilum*) *cultellatum*
Goetghebuer, 1931

Mainland: 49b, 30-Jun-2008. 81; 29-Aug-2017.

Sergentia coracina (Zetterstedt, 1850)

Mainland: 46; 11-Apr-2008

Stictochironomus pictulus (Meigen, 1830)
[*Tendipes pictulus* Mg. in Grimshaw (1912)]

Mainland: 37; 17-Jun-2008. 53; 08-May-2003. 61; 05-Aug-2004. 73; 09-May-2006. 74; 09-May-2006. G7; 1910-1912.

Comment: Adult male voucher specimen #807 from site 53, in HCCIC, NMI.

Stictochironomus sticticus (Fabricius, 1781)
Mainland: 36; 29-Sep-2004, 17-Sep-2008. 47; 22-Apr-2008. 73; 05-Apr, 09-May-2006. 76; 28-Aug-2017.

Comment: Voucher specimen (Pe) #812 from Site 36 in HCCIC, NMI.

Synendotendipes dispar (Meigen, 1830)
[*Tendipes dispar* Mg. in Grimshaw (1912)]
Mainland: G6; G7; Jul-1911.

Comment: Although not recorded in west Mayo in the present survey there are records of *S. dispar* from ten other locations in Ireland (Murray *et al.* 2015)

Xenochironomus xenolabis (Kieffer, 1916)
Mainland: 36; 17-Jun-2008. 36; 05-Aug, 29-Sep-2004, 17-Sep-2008. 46; 24-Jul-2008

Comment: Voucher specimen (Pe) #820 from Site 36 in HCCIC, NMI.

Subfamily Chironominae
Tribe Pseudochironomini

Pseudochironomus prasinatus (Staeger, 1839)
[*Tendipes prasinatus* Staeg. in Grimshaw (1912)]
Clare Island 19; 05-Jun-2002. 20F; 02-Jun-1998, 01-Sep-1998. 23F; 04-Aug-1998

Mainland: 37; 17-Jun-2008. 46; 25-Jun, 24-Jul-2008. 47; 25-Jun-2008. 55; 24-Jun, 08-Aug-2008. 56; 24-Jun, 23-Jul-2008. 57; 24-Jun-2008. G2, G9, Jul-1911.

Comment: Grimshaw (1912) provided records from two mainland sites only, Achill (G2) and 'Lough near Westport (G9)'. Edwards (1929) cited records, attributed to Grimshaw, from 'Clare Island, Castlebar, Ballin and Coolbarren Lough'. Coolbarren Lough lies less than 2km southwest of Ballin Lough (Site 47, above). Voucher specimen in FCIC, NMI.

Subfamily Chironominae
Tribe Tanytarsini

Cladotanytarsus atridorsum Kieffer, 1924
Clare Island: 20F; 11-Jun, 01-Sep, 05-Oct-1998, 10-Jun, 06-Oct-1999. 23F; 02-Jun-1999 to 07-Sep-1999

Mainland: 37; 16-Jul-2008. 46; 22-Apr, 25-Jun, 18-Sep-2008. 49b; 21-Aug-2008. 55; 22-Apr-2008. 55; 24-Jun-2008. 72; 26-Jun-2008.

Comment: Voucher specimen in FCIC, NMI.

Cladotanytarsus iucundus Hirvenoja, 1962
Mainland: 57; 24-Jun-2008.

Comment: This record from L. Moher, Liscarney, in material collected by EPA field staff was the first record of the species from Ireland (Murray 2010).

Cladotanytarsus lepidocalcar Krüger, 1938
Mainland: 37; 16-Jul-2008.

Cladotanytarsus mancus (Walker, 1856)
Mainland: 46; 25-Jun-2008. 49b, 21-Aug, 26-Sep-2008. 56; 23-Jul-2008. 73; 22-Aug-2002.

Comment: Adult male voucher specimen #825 from Site 73 in HCCIC, NMI.

Cladotanytarsus nigrovittatus (Goetghebuer, 1922)
Clare Island: 20F; 02-Jun-1998
Mainland: 72; 26-Jun-2008.

Comment: Voucher specimen from Site 20F in FCIC, NMI.

Cladotanytarsus pallidus Kieffer, 1922
Mainland: 55; 08-Aug-2008. 57; 18-Sep-2008. 73; 22-Aug-2002.

Cladotanytarsus vanderwulpi (Edwards, 1929)
Mainland: 61; 05-Aug-2004.

Micropsectra apposita (Walker, 1856)
Mainland: 57; 18-Sep-2008.

Micropsectra atrofasciata (Kieffer, 1911)
Clare Island: 6; 28-Apr-2000. 12; 20-Aug-2002; 22-Aug-2002. 24; 21-Aug-2002. 30; 21-Aug-2002
Mainland: 34; 05-Aug-2004. 35; 05-Aug-2004.

Micropsectra attenuata Reiss, 1969
Mainland: 54; 08-May-2003.

Comment: Adult male voucher specimen #841 in HCCIC, NMI. This is the only record of the species from Ireland (Murray and Murray 2006). Members of the *attenuata* group are amongst the smallest *Micropsectra* species known, having a wing length of less than 2.0mm. Reiss (1969) considers larvae of *M attenuata* to be denizens of cold stenothermic seepages and lentic regions in the upper courses of small streams. The site for this record at Knappagh is adjacent to a seepage area leading to a small first order stream.

Micropsectra junci (Meigen, 1818)
[? Reported by Grimshaw (1912) as 'T. sp.']
Clare Island: 4; 20-Aug-2002. 11; 04-Jun, 20-Aug-2002. G; ? Jul-1910.

Comments: Grimshaw 1912 cited a record of 'T. sp' (=*Tanytarsus* sp) from Clare Island with the additional comment 'this may be *T. gmundensis* Egger but the condition of the specimens renders this identification doubtful' *T. gmundensis* is a synonym of *Micropsectra junci* and, in spite of Grimshaw's reservations it is likely that his 1912 record is valid. Although it was not found in the mainland collections in this study the species has a widespread distribution in Ireland (Murray *et al*. 2015).

Micropsectra lindebergi Säwedal, 1976
Mainland: 73; 05-Apr-2006.

Micropsectra lindrothi Goetghebuer, 1931
Clare Island: 29; 21-Aug-2002. 32; 21-Aug-2002
Mainland: 36; 29-Sep-2004.

Micropsectra notescens (Walker, 1856)
Mainland: 56; 21-Apr-2008.

Micropsectra pallidula (Meigen, 1830)
Clare Island: 14; 06-Jun-2002

Mainland: 35; 05-Aug-2004. 38; 29-Sep-2004. 41, 29-Sep-2004. 53; 05-Aug-2004. 74; 05-Apr-2006. 77; 28-Aug-2017.

Comment: The record from Clare Island was cited as *M. bidentata* (Goetghebuer, 1921) in Murray and Murray (2003) based on the determination key by Pinder (1978). In a comprehensive review of west Palaearctic *Micropsectra* species Stur and Ekrem (2006) noted that many taxonomic works treating *M. bidentata*, including the keys to the adult males by Pinder (1978) and pupal exuviae by Langton and Visser (2003), were based on an erroneous species concept and that these works actually describe features of *Micropsectra pallidula* (Meigen 1830). This mistaken concept is not corrected in the recent revised keys to the adult males of Chironomidae of Britain and Ireland (Langton and Pinder 2007).

Micropsectra roseiventris (Kieffer, 1909)
Clare Island: 13; 28-Apr-2000, 04-Jun-2002. 14; 28-Apr-2000. 27; 21-Aug-2002. 20F; 01-Sep-1998, 02-Jun-1999. 23F; 10-Aug, 01-Sep-1998, 02-Jun, 13-Aug-1999

Mainland: 41, 29-Sep-2004. 56; 23-Jul-2008.

Comment: Records as *Micropsectra fusca* from Clare Island in Fahy (2002) and Murray and Murray (2003) are reported here as *M. roseiventris* after Stur and Ekrem (2006). Voucher specimen in FCIC, NMI.

Neozavrelia cuneipennis (Edwards 1929)
Mainland: 46; 17-Sep-2008. 55; 18-Sep-2008.

Comment: Reported by Murray (2010) from samples provided by EPA field research staff as the first records of the species from Ireland.

Neozavrelia luteola Goetghebuer, 1941
Mainland: 46; 24-Jul, 17-Sep-2008. 61; 07-May-2004. 73; 12-Jul-2006.

Paratanytarsus bituberculatus (Edwards, 1929)
Mainland: 66; 05-Aug-2004.

Paratanytarsus brevicalcar (Kieffer, 1909)
Clare Island: 7; 20-Aug-2002.

Comment: Recorded as *P. intricatus* (Goetghebuer, 1921) in Murray and Murray (2003) which is now regarded as a synonym of *P. brevicalcar*.

Paratanytarsus dissimilis (Johannsen, 1905)
Mainland: 51; 30-Sep-2004. 56; 21-Apr-2008, 23-Jul-2008.

Paratanytarsus inopertus (Walker, 1856)
Mainland: 47; 22-Apr-2008. 49b, 21-Aug, 26-Sep-2008. 55; 08-Aug-2008. 56; 21-Apr-2008. 57; 21-Apr-2008. 67; 30-Sep-2004.

Comment: Voucher specimen (Pe) #880 from Site 67 in HCCIC, NMI.

Paratanytarsus laccophilus (Edwards, 1929)
Mainland: 40; 05-Aug-2004. 49b; 23-Apr-2008, 30-Jun-2008, 21-Aug-2008. 55; 08-Aug-2008. 80; 29-Aug-2017.

Comment: Voucher specimen (Pe) #884 from Site 40 in HCCIC, NMI.

Paratanytarsus laetipes (Zetterstedt, 1850)
Mainland: 72; 25-Aug-2008.

Paratanytarsus lauterborni (Kieffer, 1909)
Mainland: 81; 29-Aug-2017.

Paratanytarsus penicillatus (Goetghebuer, 1928)
Mainland: 40, 05-Aug-2004. 57; 21-Apr, 18-Sep-2008. 72; 25-Aug-2008. 73; 06-Jun, 22-Aug-2002.
Comment: Voucher specimen (Pe) #890 from Site 40 in HCCIC, NMI.

Paratanytarsus tenuis (Meigen, 1830)
Mainland: 73; 06-Jun, 22-Aug-2002, 09-May, 16-Aug-2006. 74; 09-May-2006.

Rheotanytarsus curtistylus (Goetghebuer, 1921)
Clare Island: 11; 28-Apr-2000. 26; 28-Apr-2002. 30, 21-Aug-2002.

Rheotanytarsus pellucidus (Walker, 1848)
Mainland: 45; 04-Aug-2004. 61; 07-May, 05-Aug-2004. 64; 07-May-2004. 65; 05-Aug-2004.
66; 05-Aug-2004. 68; 06-Jun-2004.

Rheotanytarsus pentapoda (Kieffer, 1909)
Mainland: 56; 23-Jul-2008. 61; 07-May, 05-Aug-2004. 74; 19-May-2005.

Stempellina bausei (Kieffer, 1911)
Mainland: 46; 24-Jul, 17-Sep-2008. 49b; 21-Aug-2008. 56; 23-Jul, 8-Sep-2008. 73; 12-Jul, 16-Aug-2006, 22-Aug-2002. 74; 09-May, 28-Jun, 11-Sep-2006.

Stempellinella brevis (Edwards, 1929)
Clare Island: 21, 05-Jun-2002. 23; 04-Jun-2002
Mainland: 36; 29-Sep-2004, 17-Jun, 16-Jul, 17-Sep-2008. 46; 24-Jul, 17-Sep-2008. 53; 08-May-2003, 05-Aug-2004. 55; 22-Apr, 18-Sep-2008. 56; 21-Apr, 24-Jun-2008. 57; 21-Apr-2008. 61; 30-Sep-2004. 63; 30-Sep-2004. 72; 03-Apr, 26-Jun, 25-Aug, 22-Sep-2008. 73; 09-May,16-Aug-2006. 74; 12-Jul, 11-Sep-2006.
Comment: Voucher specimen (Pe) #911 from Site 53 in HCCIC, NMI.

Stempellinella edwardsi Spies and Sæther, 2004
Mainland: 49b, 26-Sep-2008. 55; 22-Apr, 24-Jun, 08-Aug, 18-Sep-2008. 56; 21-Apr, 24-Jun-2008. 56; 23-Jul, 18-Sep-2008. 57; 24-Jun, 21-Jul-2008. 72; 25-Aug-2008. 73; 12-Jul-2006. 74; 28-Jun-2006.

Tanytarsus brundini Lindeberg, 1963
Mainland: 53; 08-May-2003. 73; 06-Jun, 22-Aug-2002.

Comment: Adult male voucher specimen #921 from Site 53 in HCCIC, NMI.

Tanytarsus buchonius Reiss and Fittkau, 1971
Clare Island: 11; 20-Aug-2002. 13; 28-Apr-2000. 29; 21-Aug-2002
Mainland: 41, 29-Sep-2004. 81; 29-Aug-2017.
Comment: Voucher specimen (Pe) #924 from site 41 in HCCIC, NMI.

Tanytarsus ejuncidus (Walker, 1856)
Mainland: 51; 07-May-2004.

Tanytarsus eminulus (Walker, 1856)
Mainland: 61; 07-May-2004. 73; 22-Aug-2002.

Tanytarsus gracilentus (Holmgren, 1883)
Clare Island: 33; 21-Aug-2002.
Comment: Larvae of this species occur in shallow, brackish-water, rock pools and also in small inland lakes (Lindeberg 1968). The adult male collected on Clare Island was captured in flight adjacent to a small pool, vulnerable to sea spray, above the upper exposed sea shore at Leckacanny.

Tanytarsus gregarius Kieffer, 1909
Clare Island 19; 05-Jun-2002. 20; 21-Aug-2002. 20F; 1999-2002
Mainland: 46; 17-Sep-2008. 73; 06-Jun-2002. 74; 12-Jul-2006, 19-May-2005.
Comment: A voucher slide in FCIC-NMI, from Site 20F, labelled '*Tanytarsus* sp. a' was examined and determined as *T. gregarius*. It has been appropriately re-labelled.

Tanytarsus lactescens Edwards, 1929`
Mainland: 73; 09-May-2006.
Comment: This record from material collected by EPA field staff was the first report of the species from Ireland (Murray 2010).

Tanytarsus lestagei Goetghebuer, 1922
Clare Island: 7; 28-Apr-2000
Mainland: 73; 22-Aug-2002.

Tanytarsus medius Reiss and Fittkau, 1971
Mainland: 51; 30-Sep-2004. 56; 18-Sep-2008. 61; 30-Sep-2004.

Tanytarsus pallidicornis (Walker, 1856)
Clare Island: 20F; 13-Jul-1998 to 05-Oct-1998 and 10-Jun-1999 to 14-Sep-1999.

Comment: Voucher specimen in FCIC, NMI.

[*Tanytarsus pusio* Mg.]
Mainland: G2; G9, Jul-1911.

Comment: This taxon has been regarded as a nomen dubium for over 47 years (Reiss and Fittkau, 1971). Unfortunately Grimshaw's brief note on the specimen as 'a light green, neat little species (2.5–3mm) with three shining black longitudinal stripes on the thorax' is insufficient to determine its identity and the species is omitted from tabulations.

Tanytarsus signatus (van der Wulp, 1859)
Clare Island: 9; 05-Jun-2002. 10; 28-Apr-2000. 11; 20-Aug-2002. 12; 20-Aug-2002. 13; 04-Jun-2002. 20F; 01-Sep-1998. 23F; 01-Sep-1998, 10-Jun-1999, 13-Aug-1999, 20-Aug-1999

Comment: Adult male voucher specimen #971 from site 10 in HCCIC, NMI and an additional voucher specimen in FCIC, NMI.

Tanytarsus striatulus Lindeberg, 1976
Mainland: 40; 05-Aug-2004. 41; 29-Sep-2004

Comment: Voucher specimen (Pe) #974 from site 41 in HCCIC, NMI.

Tanytarsus sylvaticus van der Wulp, 1859
Mainland: 51; 7-Mar-2018

Comment: From a recent collection in March 2018 at Site 51 this is only the seventh record of the species from Ireland and is an unpublished first record for County Mayo. *T. sylvaticus* is a spring emerging species. Previous records in Ireland are from the months of March and April only (Murray *et al.* 2015).

Tanytarsus telmaticus Lindeberg 1959
Mainland: 38; 29-Sep-2004. 74; 19-May-2005.

Tanytarsus usmaensis Pagast, 1931
Mainland: 56; 18-Sep-2008.

Virgatanytarsus triangularis (Goetghebuer, 1928)
Mainland: 45; 04-Aug-2004. 61; 05-Aug-2004.

Zavrelia pentatoma Kieffer, 1913
Clare Island: 20; 21-Aug-2002

Mainland: 62; 07-May-2004. 67; 30-Sep-2004.
Comment: Voucher specimens of adult male (#987) from site 20 and of pupal exuviae (#988) from site 62 in HCCIC, NMI.

CHAPTER 12

SNAIL-KILLING FLIES (SCIOMYZIDAE) OF CLARE ISLAND

Rory Mc Donnell, Mike Gormally and Stephen McCormack

ABSTRACT
The snail-killing marsh flies (Diptera: Sciomyzidae) of Clare Island were surveyed in 2002. In the original survey six species had been recorded from the island. However, in the 2002 survey an additional nine species were collected.

Introduction
The Sciomyzidae, commonly known as marsh flies or snail-killing flies, are small- to medium-sized dipterans (2–14mm), which are predominantly yellowish-brown in colour. They are characterised by porrect antennae and well-developed setae on the head and thorax (Knutson and Lyneborg 1965). There is a strong preapical seta on each tibia and the costal vein is unbroken. In addition, adults usually have a forward-pointing bristle at the centre of the middle femur (Rozkošný 1999).

Sixty species of Sciomyzidae are known from Ireland (Chandler *et al.* 2008; Staunton *et al.* 2008; Gittings and Speight 2008, 2010; Speight and Knutson 2012) and the larvae of the majority of species feed on snails. However, slugs, pea-mussels/fingernail clams and gastropod eggs are attacked by some species (Murphy *et al.* 2012; Knutson and Vala 2011). Considerable research interest has arisen in the biology of the group over the past 50 years as marsh flies may have an important role to play in the biological control of snail hosts of trematode diseases and of slug pests of agricultural and horticultural systems.

In the original Clare Island survey the Sciomyzidae were among 43 dipteran families collected (Grimshaw 1912). Although eighteen sciomyzid species were recorded in total, this included specimens collected from the nearby mainland and only six species were found on Clare Island itself (Grimshaw 1912). These were *Pherbellia (Sciomyza) cinerella* (Fallén 1820), *Tetanocera hyalipennis (laevifrons)* Roser 1840, *Tetanocera fuscinervis (unicolor)* (Zetterstedt 1838), *Dictya umbrarum* (Linné 1758), *Pherbina coryleti* (Scopoli 1763) and *Hydromya dorsalis* (Fabricius 1775). The aim of this work was to resurvey the island, almost a century later, and highlight any changes to the sciomyzid fauna.

Methods
A total of three collections were made (4–5 April 2002, 23–24 July 2002 and 4 September 2002) and during each trip adult sciomyzids were sampled qualitatively by driving/walking around the island and sweep netting vegetation in suitable habitats (e.g. wet flushes, drainage ditches). Flies were killed using ethyl acetate and then stored in 70% ethanol in labelled specimen jars for identification later using Knutson and Lyneborg (1965) and Rozkošný (1987). Adults of some species were found in pitfall traps that were used by Steven McCormack to survey ground dwelling invertebrates, and these flies were also identified and added to the species list.

Results and discussion
A total of 66 individuals comprising thirteen species (21.7% of the Irish fauna) were recorded from Clare Island in this new survey (Table 1). In terms of species richness, this is comparable with other insect groups recorded from the island e.g., aquatic coleoptera (30% of the Irish fauna; see Chapter 8). Nine of these sciomyzid species are new to

Ilione lineata. Photo: Janet Graham

Pherbina coryleti. Photo: Dick Belgers

Table 1
Sciomyzid species collected on Clare Island, April – September 2002

Sciomyzid Species	Total	% Total Catch	Additional Information
Dictya umbrarum (Linné, 1758)	2	3.03%	Recorded by Grimshaw (1912)
Hydromya dorsalis (Fabricius, 1775)	5	7.58%	Recorded by Grimshaw (1912)
Ilione lineata (Fallén, 1820)	15	22.73%	**New to Clare Island**
Limnia unguicornis (Scopoli, 1763)	3	4.55%	**New to Island;** <u>Pitfall trap only</u>
Pherbellia ventralis (Fallén, 1820)	1	1.52%	**New to Clare Island**
Pherbina coryleti (Scopoli, 1763)	13	19.70%	Recorded by Grimshaw (1912)
Pteromicra angustipennis (Staeger, 1845)	3	4.55%	**New to Island;** <u>Pitfall trap only</u>
Sepedon sphegea (Fabricius, 1775)	2	3.03%	**New to Clare Island**
Tetanocera arrogans Meigen, 1830	6	9.09%	**New to Clare Island**
Tetanocera elata (Fabricius, 1781)	1	1.52%	**New to Clare Island**
Tetanocera ferruginea Fallén, 1820	11	16.67%	**New to Clare Island**
Tetanocera hyalipennis Roser, 1840	2	3.03%	Recorded by Grimshaw (1912)
Tetanocera robusta Loew, 1847	2	3.03%	**New to Clare Island**
TOTAL	**66**		

Clare Island (Table 1). However, six were recorded by Grimshaw (1912) from the mainland (7km away) as part of the original survey. In addition, two species (*P. cinerella* and *T. fuscinervis*) recorded in the original survey were not recorded by the authors.

Three species comprised almost 60% of the total catch. They were *Ilione lineata* (Fallén 1820), *P. coryleti* and *Tetanocera ferruginea* Fallén 1820. Larvae of *I. lineata* are relatively unique within the family Sciomyzidae for two reasons. First, they feed on pea mussels within the genera *Sphaerium* and *Pisidium* and second, larvae can remain submerged throughout the course of the first and second larval instars (Knutson and Vala 2011). With the majority of sciomyzid larvae, gaseous exchange occurs at the water surface through the posterior spiracular disc and consequently larvae must maintain regular contact with the surface film. Subsurface gaseous exchange in *I. lineata* is achieved by means of cuticular respiration (Berg and Knutson 1978). This species exhibits a univoltine mode of life and overwinters as a larva. A similar phenology is exhibited by *P. coryleti* (Berg *et al.* 1982), the second most abundant species recorded in this study. Larvae of this species are semi-aquatic and feed on hygrophilous and exposed aquatic snails (Rozkošný 1999). *Tetanocera ferruginea* exhibits a multivoltine lifecycle overwintering in the pupal stage (Berg *et al.* 1982) and larvae are aquatic predators of snails in the families Lymnaeidae, Physidae and Planorbidae (Rozkošný 1999). *Tetanocera elata* (Fabricius 1781) is also of interest. This species is terrestrial and the larvae feed on slugs. The third instar larvae of this species use a neurotoxin produced in the salivary glands to immobilise their slug prey (Rozkošný 1999).

Although sciomyzids are rarely collected in pitfall traps (Knutson and Vala 2011), adults of two species were caught solely in pitfall traps in this study—*Limnia unguicornis* (Scopoli 1763) and *Pteromicra angustipennis* (Staeger 1845). Such records seem to suggest that adults of these species have a microhabitat preference for the lower vegetation strata or ground level and hence they may have been missed by the sweep net. These species were not found by Grimshaw (1912) and may have been missed for this reason. *Limnia unguicornis* is said to be found in moist situations and the larvae are thought to have a preference for snails of the genus *Succinea*. *Pteromicra angustipennis* is a small fly (2.5–4mm), which is easily overlooked and its larvae feed on exposed aquatic and hygrophilous snails (Rozkošný 1999).

Acknowledgements

Sincerest thanks to T.K. McCarthy for his advice and assistance throughout the course of this survey. Many thanks also to the residents of Clare Island for their kind hospitality.

REFERENCES

Berg, C.O. and Knutson, L. 1978 Biology and systematics of the Sciomyzidae. *Annual Review of Entomology* **23**, 239–58.

Berg, C.O., Foote, B.A., Knutson, L., Barnes, J.K., Arnold, S.L. and Valley, K. 1982 *Adaptive differences in phenology in sciomyzid flies*. In W.N. Mathis and F.C. Thompson (eds), *Recent advances in dipteran systematics: Commemorative volume in honor of Curtis W. Sabrosky. Memoirs of the Entomological Society of Washington* **10**, 15–36.

Chandler, P., O'Connor, J. and Nash, R. 2008 *An annotated checklist of the Irish two-winged flies (Diptera)*, 1st ed., 261 pp. Dublin. The Irish Biogeographical Society.

Gittings, T. and Speight, M.C.D. 2010 *Sciomyza simplex* Fallén, 1820 and *Sciomyza testacea* Macquart, 1835, snail-killing flies (Diptera, Sciomyzidae) new to Ireland. *Irish Naturalists' Journal* **31**, 91–3.

Gittings, T. and Speight, M.C.D. 2008 *Colobaea pectoralis* (Zetterstedt, 1847) and *Pherbellia dorsata* (Zetterstedt, 1846) snail-killing flies (Diptera, Sciomyzidae) new to Ireland. *Irish Naturalists' Journal* **29**, 116–8.

Grimshaw, P.H 1912 Clare Island Survey. Diptera. *Proceedings of the Royal Irish Academy* **31**(25) 1–34.

Knutson, L.V. and Lyneborg, L. 1965 Danish acalypterate flies. 3. Sciomyzidae (Diptera). *Entomologiske Meddelelser* **34**(1), 61–101.

Knutson L.V. and Vala, J-C. 2011 Biology of snail-killing Sciomyzidae Flies. Cambridge, UK. Cambridge University Press.

Murphy, W.L., Knutson, L.V., Chapman, E.G., Mc Donnell, R.J., Williams, C.D., Foote, B.A and Vala, J.C. 2012 Key aspects of the biology of snail-killing Sciomyzidae flies. *Annual Review of Entomology* **57**, 425–47.

Rozkošný, R. 1987 A review of the Palaearctic Sciomyzidae (Diptera). *Folia Facultatis Scientiarium Naturalium Universitatis Purkynianae Brunensis, Biologia* **86**, 1–156.

Rozkošný, R. 1999 *Contributions to a manual of Palaearctic Diptera (Higher Brachycera)*. In L. Papp and B. Darvas (eds), 356–82. Budapest. Science Herald.

Speight, M.C.D. and Knutson, L.V. 2012 Species accounts for Sciomyzidae and Phaeomyiidae (Diptera) known from the Atlantic zone of Europe. *Dipterists Digest* **19**, 1–38.

Staunton, J., Williams, C.D., Mc Donnell, R.J., Maher, C., Knutson, L. and Gormally, M.J. 2008 *Pherbellia (Oxytaenia) stackelbergi* Elberg: a sciomyzid (Dip.: Sciomyzidae) new to the British Isles, with comments on generic and subgeneric placement. *Entomologist's Record and Journal of Variation* **120**, 173–7.

CHAPTER 13

THE SOCIAL HYMENOPTERA OF CLARE ISLAND

John Breen, Robert Paxton and Audrey O'Grady

ABSTRACT

The current situation regarding the occurrence of *Bombus* species on Clare Island is largely unchanged from the original 1911 survey by Morley. The following species are reported in both surveys: *Bombus muscorum, B. pascuorum, B. ruderarius, B. jonellus, B. hortorum. Bombus distinguendus* was reported in the 1911 survey but not during the present survey. Notably, species related to *Bombus lucorum* were not mentioned in the 1911 survey. These are now recognised as a group of cryptic species of which *B. cryptarum, B. lucorum* and *B. magnus* were confirmed from Clare Island based on DNA barcoding. Honeybees, *Apis mellifera*, were introduced recently to the island. The social wasps, *Dolichovespula sylvestris, Vespula rufa* and *V. vulgaris* are also additions to the 1911 survey. The ant species are the same as the 1911 survey, namely, *Formica lemani, Lasius flavus, Lasius platythorax, Myrmica ruginodis* and *Myrmica scabrinodis*. Parasitism of two species of *Lasius* by mermithid nematodes is described and the effects of ant activity by *L. flavus* on soil chemistry is also reported.

Introduction

The original Clare Island survey of the Hymenoptera (Morley 1911) was based on a single visit to Clare Island and the Louisburgh district by the author over a two-week period in July 1910 and was supplemented by recording carried out by Rev. W.F. Johnson and J.N. Halbert. Morley's (1911) paper reported on 324 species in the Order Hymenoptera. The Aculeata is one group of this order and includes those insects referred to as bees, wasps and ants. Referring to the Aculeata, Stelfox (1927) commented on the Morley (1911) Clare Island survey, as follows:

A small list was compiled by Halbert, Grimshaw, Johnson and Morley during the Clare Island Survey, 1909–1911. All these collectors were, however, engaged in working out other groups, and consequently the Aculeates were rather neglected. Moreover, some of the records are erroneous, and others somewhat doubtful.

The present study is confined to the Social Hymenoptera, referring to those species of the Aculeata that live in social colonies, and comprise the ants (Formicidae), social wasps (Vespidae) and social bees (family Apidae: bumblebees, genus *Bombus* and honeybee, *Apis mellifera*) (Table 1). For current nomenclature and the current status of *Bombus* species in Ireland, we follow Fitzpatrick *et al.* (2006). For the wasps and ants, we follow the nomenclature of the Bees, Wasps and Ants Recording Society (BWARS 2018).

Morley (1911) did not include any reference to either *Bombus lucorum* or *Bombus terrestris* occurring on Clare Island. Stelfox (1927; see quotation above) noted this omission as these species were—and still are—generally widespread throughout Ireland. One might infer that Stelfox considered the omission to be an error. However, the situation is now more complicated due to recent taxonomic revision. These species are now recognised as four cryptic species in Ireland. Of these, *Bombus terrestris* remains as a single species, with black and yellow bands and a buff tail in the queen caste. Three species of bumble bee are now recognised within *B. lucorum sensu lato*, all with black and yellow bands and a white tail in the queen caste: *Bombus cryptarum, Bombus lucorum sensu stricto* and *Bombus magnus* (Murray *et al.* 2008; Carolan *et al.* 2012; McKendrick *et al.* 2017). All three of these species exist in Ireland and are very difficult to distinguish from each other using morphological characters (Carolan *et al.* 2012). These 'cryptic' species can nevertheless be easily differentiated from each other—and from workers of the fourth species, *B. terrestris*—using DNA barcodes (Murray *et al.* 2008). This practice involves using partial (*circa* 650 base pair) sequences of the mitochondrial cytochrome oxidase subunit I gene (COI) that are used to identify species across the animal kingdom (Hebert *et al.* 2003). DNA barcoding was carried out as part of the present study (by R.P.).

In addition to surveying the species of ants present, the presence of a mermithid nematode in the sexuals of *Lasius* species on Clare Island and the effects of ant activity on soil chemistry were investigated for the present study (A.O'G. and J.B.).

Nematodes of the family Mermithidae are known to parasitise ants of the genus *Lasius* ever since Gould (1747) noted that 'an abundance of Ant-Flies (= winged males and females) are demolished by a white and long Kind of Worm, which is often met with in their bodies. You may frequently take three from the Insides of the large (i.e. queens), but seldom more than one from a small Ant-fly (i.e. males).' Baylis (1921) described the species *Mermis myrmecophila*, which, following taxonomic revision, is now in the genus *Pheromermis* (Poinar *et al.* 1976). The numbers and size of the nematode parasites, and the morphological effects of parasitism on the hosts has been investigated (O'Grady and Breen 2011).

Materials and methods
General records
Collections were made on visits to Clare Island by J.B. on one- to three-day visits, almost annually, between 2002 and 2018. Visits were made in different months between early April and late September in order to avoid any seasonal variation. Voucher specimens have been retained. Some records are supported by photographic records.

DNA barcoding of cryptic species of **Bombus**
For the DNA barcoding, six specimens of the cryptic complex of *Bombus* were collected on 26 August 2014 (five specimens at Fawnglass 53.8065°N, 9.9577°W) and 27 August 2014 (one specimen at Kill 53.7948°N, 9.9942°W) and immediately stored in 99% ethanol to preserve their DNA. The sampling days were both cold and wet and these were the only specimens of this species group seen flying. DNA was subsequently extracted from a rear leg of each bee using a 'high salt' protocol (Paxton *et al.* 1996) and then a *c*. 650bp fragment of the COI gene—the DNA barcode—was amplified by PCR using the standard animal barcoding primers HCO and LCO (see Folmer *et al.* 1994). PCR products were cleaned of primers using a purification kit (Qiagen PCR purification kit) following the manufacturer's recommendations and then sequenced by a commercial company in the forward direction only on a capillary DNA autosequencer (ABI 3730xl).

Trace files of DNA sequences were inspected by eye using 4Peaks (Griekspoor and Groothuis 2006); all six sequences were of excellent quality. Eight additional reference sequences were downloaded from GenBank (http://www.ncbi.nlm.nih.gov), representing two each of *B. cryptarum, B. lucorum, B. magnus* and *B. terrestris*. All sequences were aligned in MEGA (Tamura *et al.* 2011) using the ClustalW algorithm. A Neighbor Joining tree was subsequently generated from the DNA sequence data using the Kimura 2-parameter model, with 2,000 bootstrap replicates to give support to the tree topology. This is a standard approach in DNA barcoding for species identification (Ratnasingham and Hebert 2007).

Soil modification by **Lasius flavus**
Amongst the ants (Formicidae), *Lasius flavus* is abundant on the island, with colonies occurring under stones along road margins and frequent mound nests in the fields. Large numbers of the mound nests of this species occur at north of the wall at Fawnglass (53.8073°N; 9.9583°W) (Plate I) and are frequent within the farmed fields on the island. Soil samples were collected on 1 August 2006 from *Lasius flavus* mound nests in a roadside field at Corr Arru (Capnagower) (53.8033°N, 9.9505°W).

This site is a dry grassland with many mound nests on glacial till derived from carboniferous limestone (Capnagower formation; detailed map in Graham 2001). Samples were taken from ten paired samples of the nest and from 0.5m away from the nest. The following parameters were determined in the laboratory, using standard analytical methods: organic matter (loss on ignition at 500ºC), bulk density (following Jeffrey 1970), soil electrical conductivity and pH. Further analyses were carried out at the National Soils Laboratory, Teagasc, Johnstown Castle, Co. Wexford. Levels of copper, zinc and easily reducible manganese were determined using the EDTA-Quinol method. Extraction of phosphorus and potassium were carried out using Morgan's extraction solution. Cobalt and manganese were analysed by reflux in aqua regia, taken up in 2 M HNO_3, and determined by atomic absorption spectrometry. Further details of the protocols used for the analysis of the soil samples are in O'Grady (2008).

Mermithid parasitism in two species of Lasius

During summer 2002, and annually until 2010, it was observed that many sexuals in most nests of two species of *Lasius* were infected with a mermithid nematode. In the field, infected individuals are differentiated from uninfected individuals by their distended gasters and reduced wings. On 20 August 2005, we collected up to twenty individuals infected with *Pheromermis myrmecophila*, of males and females of *Lasius flavus* and of females of *L. platythorax* (males were not found with mermithids in this species). These were preserved in 95% ethanol. Additional infected individuals were also collected to investigate mermithid emergence from the hosts. In the laboratory, the numbers of parasites per ant host, and the size (length and width using an opisometer) of each parasite was recorded. An Olympus BX60 microscope and ANALYSIS software were used to make the following measurements on infected and normal queen and male ants: head width immediately behind the eyes, femur length, femur width, radial cell length and radial cell width.

Results and discussion
General records

The species recorded by Morley (1911), their current nomenclature and occurrence in the present survey is described in Table 1.

Pl. I Mound nests of yellow garden ants, *Lasius flavus* at Fawnglass, Clare Island. Photo: J. Breen

A Red-tailed bumblebee queen, possibly *Bombus ruderarius*, near the Abbey, Clare Island. 28 April 2006. Photo: J. Breen

Morley (1911) recorded six species of *Bombus*, plus one species in error. Of these species, five were recorded in the present survey, but *Bombus distinguendus* was not found on this survey. As noted previously, Morley (1911) did not mention any records of *Bombus lucorum/terrestris*. Specimens of this group were one of the more common bumblebees encountered during the present survey. However, these species can only be separated by DNA barcoding, which on this occasion unequivocally identified the six Clare Island specimens as *B. cryptarum* (two specimens), *B. lucorum* (two specimens) and *B. magnus* (two specimens) (Fig. 1). This result of a total sample of six yielding two specimens each of the three species was purely coincidental as there was no way of knowing the identity of the specimens until they were DNA barcoded.

Morley (1911) reported that the honey bee, *Apis mellifica* L. (now *Apis mellifera* L.), did not occur on Clare Island during the original survey. This was

Table 1
Species of Social Hymenoptera recorded by Morley (1911), their current nomenclature, and status in the present study. Items highlighted in bold are changes from the original Clare Island survey of 1911.

Morley (1911) records	Current nomenclature* of species recorded in 1911	Present study
Apidae		
Bombus smithianus White and *Bombus venustus* Smith	*Bombus muscorum* (Linnaeus, 1758)	*Bombus muscorum* (Linnaeus, 1758)
Bombus agrorum Fab.	*Bombus pascuorum* (Scopoli, 1763)	*Bombus pascuorum* (Scopoli, 1763)
Bombus derhamellus Kirb.	*Bombus ruderarius* (Müller, 1776)	*Bombus ruderarius* (Müller, 1776)
Bombus jonellus Kirb.	*Bombus jonellus* (Kirby, 1802)	*Bombus jonellus* (Kirby, 1802)
Bombus hortorum L.	*Bombus hortorum* (Linnaeus, 1761)	*Bombus hortorum* (Linnaeus, 1761)
Bombus latreillellus Kirb. var. *distinguendus* Mor.	*Bombus distinguendus* Morawitz 1869	**Not found in present survey**
Bombus latreillellus Kirb. 'type form'	*Bombus subterraneus* (Linnaeus, 1758) This record was dismissed as 'some error' by Stelfox (1927; p. 282). No other Irish records exist of this species.	
		***Bombus cryptarum* (Fabricius, 1775)**
		***Bombus lucorum* (Linnaeus, 1761)**
		***Bombus magnus* Vogt, 1911**
		***Apis mellifera* Linnaeus, 1758**
Vespidae		
None		***Dolichovespula sylvestris* (Scopoli, 1763)**
		***Vespula rufa* (Linnaeus, 1758)**
		***Vespula vulgaris* (Linnaeus, 1758)**
Formicidae		
Formica fusca Latr.	*Formica lemani* Bondroit, 1917	*Formica lemani* Bondroit, 1917
Lasius flavus DeG.	*Lasius flavus* (Fabricius, 1781)	*Lasius flavus* (Fabricius, 1781)
Lasius niger Linn.	*Lasius platythorax* Seifert, 1991	*Lasius platythorax* Seifert, 1991
Myrmica ruginodis Nyl.	*Myrmica ruginodis* Nylander, 1846	*Myrmica ruginodis* Nylander, 1846
Myrmica scabrinodis Nyl.	*Myrmica scabrinodis* Nylander, 1846	*Myrmica scabrinodis* Nylander, 1846

*Nomenclataure follows: Fitzpatrick *et al.* (2006) for *Bombus*, and BWARS (2021) for wasps and ants

the case throughout most of the years of observation of the present survey. During 2016, however, worker honey bees were seen near the harbour area on Clare Island and enquiry suggested that hives had been introduced to the island two years previously. Worker honey bees have been seen on each subsequent visit to the island in the quay to Fawnglass area.

Morley (1911) also stated, 'It is worthy of note that wasps are unknown on Clare Island. The islanders know them only through visits to the mainland; to the island children they are quite unfamiliar.' Two species, the Tree Wasp, *Dolichovespula sylvestris*, and the Red Wasp, *Vespula rufa*, were encountered regularly during the present survey. These species have relatively small colonies and are increasingly scarce on the mainland. Interestingly, these two species also occur on Inis Meáin, in the Aran Islands (J.B., personal observation). A number of workers of the Common Wasp, *Vespula vulgaris*, were seen during 2016 at 53.7999°N, 9.9568°W. This was the

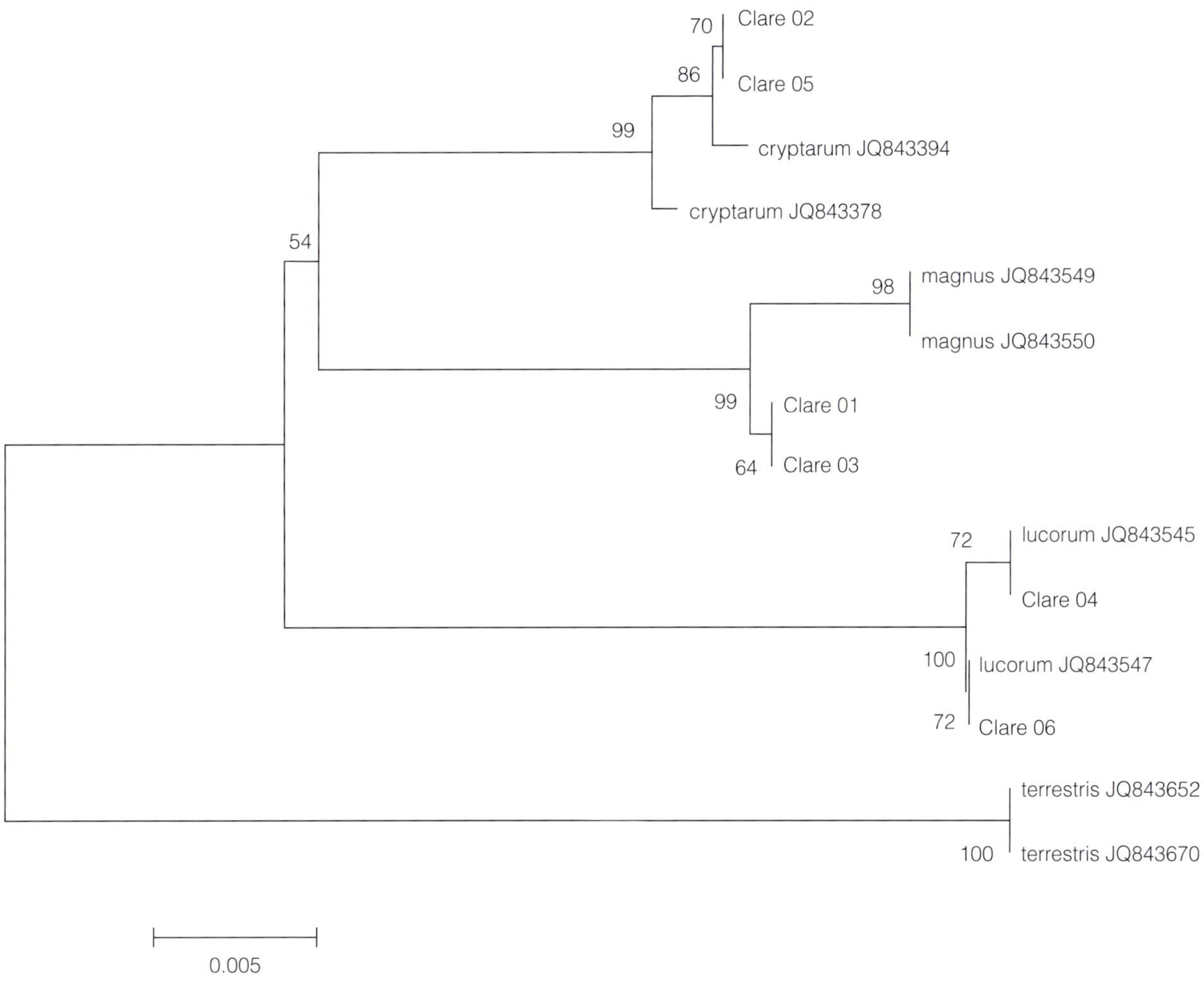

Fig. 1 The optimal unrooted phylogenetic tree using the Neighbor-Joining algorithm, with bootstrap support (values to the immediate left of nodes, 2,000 replicates), and based on 620bp of the mitochondrial COI gene. Evolutionary distances were computed using the Kimura 2-parameter method and are in units of base substitutions per site (scale bar). Branch lengths are drawn to scale. Samples of *Bombus* from Clare Island are labelled Clare 01 – Clare 06. Reference sequences downloaded from GenBank are labelled with their *Bombus* species name and GenBank Accession Number.

Bombus muscorum queen, near the Abbey, Clare Island.
Photo: J. Breen

only time they were seen and may have been from a single nest founded by a single mated queen wasp that managed to make her way to the island.

This species nests underground and the nest was searched for unsuccessfully.

The species of ants found on Clare Island are the same five species reported by Morley (1911), with changes in nomenclature. All five are relatively common in suitable habitats on the mainland. Searches were carried out for two other species. *Tetramorium caespitum* (Linnaeus, 1758) is very locally distributed in Ireland. It is relatively common at one site only, Ballyteigue Burrows, County Wexford, but it occurs occasionally very close to sea cliffs, including on Inis Mór and Inis Meáin in the Aran Islands (J.B., personal observation). The workers of *Leptothorax acervorum* Fabricius, 1793 are very small, and the colonies have very few workers – less than a hundred, often about twenty to thirty. Hence, they are easily overlooked. However, neither species was found during the present survey.

Soil modification by *Lasius flavus*

The results of the soil analyses are in Table 2. Soil pH was significantly higher in the nest samples, compared to the off-nest samples. However, the electrical conductivity values were reduced in the nest samples. The levels of potassium were much higher, and the levels of phosphate were lower in the nest samples when compared to the off-nest samples.

Mermithid parasitism in two species of *Lasius*

Infection by *Pheromermis myrmecophila* was found in both queens and males of *Lasius flavus*, but in females only of *Lasius platythorax* (referred to as *Lasius niger* in O'Grady and Breen (2011); we now believe that true *Lasius niger* does not occur on Clare Island). *Lasius flavus* queens were infected with from one to five mermithids, whereas most infected individuals of the smaller males of that species had only one mermithid (Fig. 2). Infected *Lasius platythorax* queens had mostly just one mermithid, and just two had three parasites (Fig. 2). We did not find mermithids in males of *Lasius platythorax*. It is interesting to note that Gould (1747) mentioned the 'occasional' infection of male *Lasius* with mermithids. However, Crawley (1921) insisted that Gould must be mistaken and Donisthorpe (1927) also questioned the accuracy of Gould's observation. On Clare Island, infected males were relatively

Queen wasp, *Vespula rufa* on *Salix*. Photographed on road to Abbey, Clare Island. Photo: J. Breen

Lasius flavus workers tending pupae in their nest at Kill, Clare Island. Photo: J. Breen

Table 2

Mean levels, with standard errors (SE), of soil variables of soil from *Lasius flavus* nests and off-nests from Clare Island (EC = electrical conductivity; OM = organic matter; BD = bulk density; LR = lime requirement; ER = easily reducible), together with Student's *t* values for paired *t*-tests (n = 10 pairs), and significance values (p)

	Nest		Off-nest			
	Mean	SE	Mean	SE	t	p
pH	5.36	0.03	5.16	0.05	3.88	0.004
EC	167.80	7.59	199.90	12.43	3.20	0.011
OM	84.70	0.82	82.00	0.92	2.20	0.056
BD	0.17	0.00	0.18	0.00	1.77	0.111
Co	6.02	0.58	4.93	0.72	1.75	0.115
Cu	0.65	0.06	0.88	0.08	2.17	0.058
K	217.92	24.99	139.80	6.95	2.91	0.017
LR	7.56	0.50	7.86	0.71	0.33	0.747
Mg	304.90	23.77	294.22	17.23	0.47	0.649
Mn-ER	192.21	21.88	220.99	26.49	1.04	0.326
Mn	835.50	81.03	813.00	86.61	0.41	0.691
P	2.49	0.21	3.04	0.19	2.59	0.029
Zn	2.55	0.21	2.76	0.31	0.59	0.567

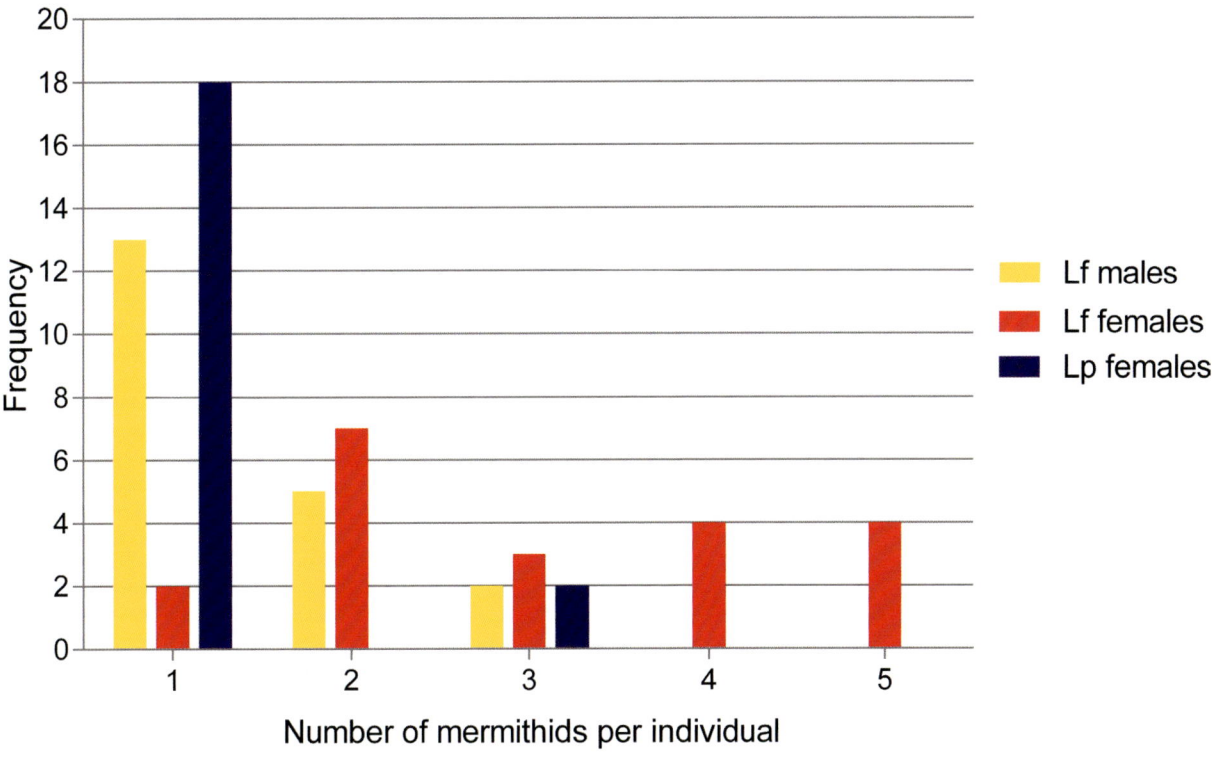

Fig. 2 The numbers of mermithids per individual in *Lasius flavus* (Lf) males and females and in *Lasius platythorax* (Lp) females

Two queens of *Lasius flavus* (lower and upper centre) infected with mermithid nematodes. Note reduced wings compared to two uninfected queens (lower left and lower right). A worker (top right) is carrying a pupa. Photographed at Fawnglass, Clare Island (specimens were isolated in a Petri dish for photograph). Photo: J. Breen

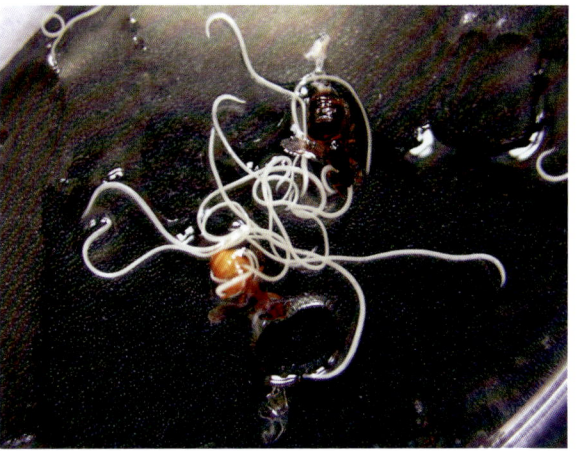

Larval mermithid nematodes emerged from two *Lasius flavus* queens, collected at Fawnglass, Clare Island. The queens were retained in a Petri dish of water under ambient conditions. Photo: J. Breen

common each year of our observations, and this confirms Gould's record.

The mean length of the individual mermithids decreased as their number increased in multiple infections (see O'Grady and Breen 2011). O'Grady and Breen (2011) also showed that mermithid infection resulted in some changes in the relative body sizes of hosts. In both *L. platythorax* and *L. flavus*, the head widths of infected queens were smaller than those of non-infected individuals. Comparisons of the body measurements of uninfected and infected individuals showed one further unexpected finding: in *L. platythorax*, only, the femur lengths of infected queens were considerably and significantly longer. More information is needed on the nature of the mermithid infection and the possible hormonal interruptions that cause these changes.

There are relatively few records of mermithid infection of ants in Ireland. O'Rourke (1946) found queens of *L. flavus* and *L. niger* infected by this mermithid. O'Rourke (1979) described *L. flavus* as being 'massively infected' at Kilmore Quay, Co. Wexford in 1969 but there was no infection at that site a year later. However, J.B. found infected *L. flavus* queens on the nearby Great Saltee Island in 1972 (O'Rourke 1979). Interestingly, we believe that the level of mermithid infection of *Lasius* species observed on Clare Island is one of the richest reported in the ant literature.

Acknowledgements

Dr Robin Niechoj (confirmed by Dr Bernd Seifert) identified *Lasius platythorax*. R.P. thanks Kersin Gößel and Matthias Seidel for help in the laboratory with DNA barcoding.

REFERENCES

Baylis, H.A. 1921 *Mermis* parasitic on ants of the genus *Lasius*. II. Description of *Mermis myrmecophila* Sp. Nov. *Journal of the Royal Microscopical Society of London* **41**, 365–72.

BWARS 2018 Bees, Wasps and Ants Recording Society. Available at: http, //bwars.com/ (Accessed 24 January 2021).

Carolan, J.C., Murray, T.E., Fitzpatrick, Ú., Crossley, J., Schmidt, H., Cederberg, B., McNally, L., Paxton, R.J., Williams, P.H. and Brown, M.J.F. 2012 Colour patterns do not diagnose species: quantitative evaluation of a DNA barcoded cryptic bumblebee complex. *PLoS ONE* **7**, e29251.

Collingwood, C.A. 1958 A survey of Irish Formicidae. *Proceedings of the Royal Irish Academy* **59B**: 213–19.

Crawley, W.C. 1921 *Mermis* parasitic on ants of the genus *Lasius*. I. Account of the occurrence of the worm and its effect on its hosts. *Journal of the Royal Microscopical Society of London* **41**, 353–64.

Donisthorpe, H.S.J.K. 1927 *The guests of British ants*. London. Routledge.

Fitzpatrick, Ú., Murray, T.E., Byrne, A., Paxton, R.J. and Brown, M.J.F. 2006 Available at https://www.npws.ie/sites/default/files/publications/pdf/Fitzpatrick_et_al_2006_Bee_Red_List.pdf (accessed 24 January 2021).

Fitzpatrick, Ú., Murray, T.E., Paxton, R.J., Breen, J., Cotton, D., Santorum, V. and Brown, M.J.F. 2007 Rarity and decline in bumblebees – A test of causes and correlates in the Irish fauna. *Biological Conservation* **136**, 185–94.

Folmer, O., Black, M., Hoeh, W., Lutz, R., Vrigenhoek, R. 1994 DNA primers for amplification of mitochondrial cytochrome c oxidase subunit I from diverse metazoan invertebrates. *Molecular Marine Biology and Biotechnology* **3**, 294–99.

Gould, W. 1747 *An account of English ants*. London. A. Millar.

Graham, J.R. 2001 *New Survey of Clare Island. Volume 2: Geology*. Dublin. Royal Irish Academy.

Griekspoor, A., Groothuis, T. 2006 4Peaks (version 1.7.2). www.mekentosj.com.

Hebert, P.D.N., Cywinska, A., Ball, S.L., deWaard, J.R. 2003 Biological identifications through DNA barcodes. *Proceedings of the Royal Society of London* **270B**, 313–21.

Jeffrey, D.W. 1970 Note on the use of ignition loss as a means for the appropriate estimation of soil bulk density. *Journal of Ecology* **58**, 297–9.

McKendrick, L., Provan, J., Fitzpatrick, U., Brown, M.J.F., Murray, T.E., Stolle, E., Paxton, R.J. 2017 Microsatellite analysis supports the existence of three cryptic species within the bumble bee *Bombus lucorum sensu lato*. *Conservation Genetics* **18**, 573–84.

Morley, C. 1911 Clare Island Survey. Hymenoptera. *Proceedings of the Royal Irish Academy* **31**(24), 1–18.

Murray, T.E., Fitzpatrick, Ú., Brown, M.J.F. and Paxton, R.J. 2008 Cryptic species diversity in a widespread bumble bee complex revealed using mitochondrial DNA RFLPs. *Conservation Genetics* **9**, 653–66.

O'Grady, A. 2008 *The community ecology of ants in Irish limestone grasslands (Hymenoptera; Formicidae)*. PhD thesis. University of Limerick.

O'Grady, A. and Breen, J. 2011 Observations on mermithid parasitism (Nematoda: Mermithidae) in two species of *Lasius* ants (Hymenoptera: Formicidae), *Journal of Natural History* **45**, 2339–45.

O'Rourke, F.J. 1946 The occurrence of three Mermithogynes in Roundstone Connemara, with notes on the ants of the district. *Entomologist's Record and Journal of Variation* **58**, 65–70.

O'Rourke, F.J. 1979 The social Hymenoptera of County Wexford. *Proceedings of the Royal Irish Academy* **79B**, 1–14.

Paxton, R.J., Thorén, P.A., Tengö, J., Estoup, A. and Pamilo, P. 1996 Mating structure and nestmate relatedness in a communal bee, *Andrena jacobi* (Hymenoptera: Andrenidae), using microsatellites. *Molecular Ecology* **5**, 511–19.

Poinar, G.O. Jr, Lane, R.S. and Thomas, G.M. 1976 Biology and description of *Pheromermis pachisoma* (V. Linstow) n. gen., n. com. (Nematoda; Mermithidae) a parasite of yellow jackets (Hymenoptera: Vespidae). *Nematology* **22**, 360–70.

Ratnasingham, S. and Hebert, P.D.N. 2007 BOLD: the barcode of life data system (www.barcodinglife.org). *Molecular Ecology Notes* **7**, 355–64.

Stelfox, A.W. 1927 A list of the Hymenoptera Aculeata (*sensu lato*) of Ireland. *Proceedings of the Royal Irish Academy* **37B**, 201–355.

Tamura, K., Peterson, D., Peterson, N., Stecher, G., Nei, M. and Kumar, S. 2011 MEGA5: molecular evolutionary genetics analysis using maximum likelihood, evolutionary distance, and maximum parsimony methods. *Molecular Biology and Evolution* **28**, 2731–39.

CHAPTER 14

NON-MARINE OSTRACODS OF CLARE ISLAND

Gillian McCall

ABSTRACT

Ostracods were collected from Clare Island, by hand net and scoop, on summer visits over four years in order to report on the present fauna and to compare the findings with those of the Clare Island survey of 1909–11. Seventeen species were identified, seven more than in the previous survey. Many of the species recorded in the previous survey were confirmed present, but some species reported in the previous survey are considered to have been misidentified at that time. Two species collected for this report had been recorded from Co. Mayo in the 1970s but may possibly have been introduced to Clare Island since the early twentieth century, as may the species recorded in Ireland for the first time. One unidentified species was found only as juveniles.

Introduction

The small size of Irish freshwater ostracods—0.45mm to 2.7mm carapace length—and the enclosure of the limbs and mouthparts of preserved specimens within the carapace means that the shell-like cover has to be prised open and the appendages dissected under a microscope to identify the ostracod to species. Water samples almost always need to be hand sorted under a microscope in very small volumes: whether the ostracods are white, yellowish, brown, green, bluish or violet, spotted, striped or plain, opaque or transparent, smooth, ornamented or setose, round, beanlike or elongate, they can be hard to find in muddy or peaty water. The necessity for these techniques may partly explain why there have been so few publications referring to Irish ostracods in the last century, less than twenty since Scourfield (1912) wrote his account of the Freshwater Entomostraca of Clare Island for the 1909–11 survey. Perhaps ten of those authors were reporting on their own fieldwork in Ireland.

Scourfield made his own collections on Clare Island and also received specimens from W.F. deV. Kane, James Murray and J.S. Dunkerly. Scourfield is best known for his expertise in the study of Cladocera. Kane was a lepidopterist who later developed an interest in copepods (Collins 1999). Murray joined the survey to report on tardigrades and rotifers and Dunkerly contributed the reports on Infusoria and Flagellata; these two did not publish any work relating to Ostracoda. When looking at ostracods at that time, Scourfield and Kane did not have the use of the keys of Klie (1938) and Meisch (2000) and other twentieth century papers. Indeed in 1912 some of the ostracods of Britain and Ireland remained to be discovered, let alone described. Scourfield recorded ten species from Clare Island. Greater sampling effort for the present study combined with developments in taxonomy resulted in the collection and identification of seventeen species, with the expectation of more to be found or identified.

Clare Island has a variety of freshwater habitats, with water both still and flowing, where ostracods occur: temporary puddles, cattle troughs, large and small pools on grassland, permanent pools on blanket bog, roadside and field ditches, reedbeds, marshes, streams, flooded patches in once-cutover bog. Most Irish ostracods reproduce asexually and produce drought-resistant eggs; these factors allow continued existence in waterbodies with a tendency to dry out and successful colonisation of an isolated pool by a single individual. Some waterbodies on Clare Island are distinctly acidic (pH 4), others not so (pH 7+). Windborne sea spray can produce conditions suitable for brackish water species. Some of the ostracods were collected from among *Sphagnum*, *Potamogeton* or other macrophytes; others were collected from muddy substrates. Occasionally the water depth exceeded 25cm, but most specimens were collected from depths of 10cm or less, even from wet mud with no overlying water. Glacial till, some of it limestone-rich sediments, covering much of the island below 100m, supported conditions favourable for ostracods. In contrast, no ostracods were found on the blanket bog covered slopes of Knockmore, above 200m.

Scourfield compared his records with those of ostracods from the rest of Mayo and Galway, especially the area around Clew Bay. Collections were not made from Mayo and Galway for this study. However, of the sixteen further species he lists for those counties, three were collected from Clare Island for this report.

The aim of this study was to collect ostracods from a large number of sites in a variety of habitats across Clare Island and to compare the list of those identified with the list recorded by Scourfield in 1912. Reasons for the discrepancies between the two lists are offered. Three possible twentieth century introductions are suggested.

Methods
Samples were taken in May 2003, June 2005, July 2004, August 2004 and September 2002 and 2004. Specimens collected in October 2002 were donated by Stephen McCormack. One hundred and sixty-seven samples were taken from 92 sites in 21 1km squares. Most samples were taken using a plankton net, but small water bodies such as shallow puddles and mud were sampled by scooping with a small container. Vegetation was taken in bags for hand sorting or heat extraction. Where recorded, pH was estimated using three brands of pH paper, Whatman, Johnson and Merck, usually using all three at a site. The southeast of the island was sampled most thoroughly as the geology suggested that the pH of the water would be higher, with the expectation of a more varied ostracod fauna. One of the possible habitats not sampled was water lying out of reach under cattle grids. The sample sites are shown in the map (Fig. 1). The map used for this survey was created by University of Glasgow students in 1992 based on aerial photography of the Ordnance Survey, Dublin (1990) and kindly made available for reference by the Island Development office.

All taxonomic names have been checked and updated with reference to Martens and Savatenalinton (2011). The ostracod records in this report have been submitted to David J. Horne for inclusion in the NODE (Nonmarine Ostracod Distribution in Europe) database and GIS (Horne, Baltanás and Paris 1998). Horne has also seen and confirmed the identity of at least one adult female of every species identified here (not including the *Pseudocandona* specimens).

Results
Ostracods were found at 47 sites. At least seventeen species were present.

Darwinulidae
Penthesilenula malayica (Menzel, 1923)

Candonidae
Candona candida (O.F. Müller, 1776)
Pseudocandona cf. albicans (Brady, 1864)
Cryptocandona vavrai Kaufmann, 1900
Cryptocandona reducta (Alm, 1914)
Cypria ophtalmica (Jurine, 1820)
Cyclocypris globosa (Sars, 1863)
Cyclocypris ovum (Jurine, 1820)

Cyprididae
Eucypris virens (Jurine, 1820)
Bradleycypris obliqua (Brady, 1868)
Bradleystrandesia fuscata (Jurine, 1820)
Herpetocypris chevreuxi (Sars, 1896)
Psychrodromus robertsoni (Brady and Norman, 1889)
Heterocypris incongruens (Ramdohr, 1808)
Heterocypris salina (Brady, 1868)
Cypridopsis lusatica Schäfer, 1943
Potamocypris villosa (Jurine, 1820)

Juvenile ostracods with a spine on each valve may indicate the presence of an eighteenth species.

For Table 1 subdivisions have been chosen to reflect the distribution of limestone-rich

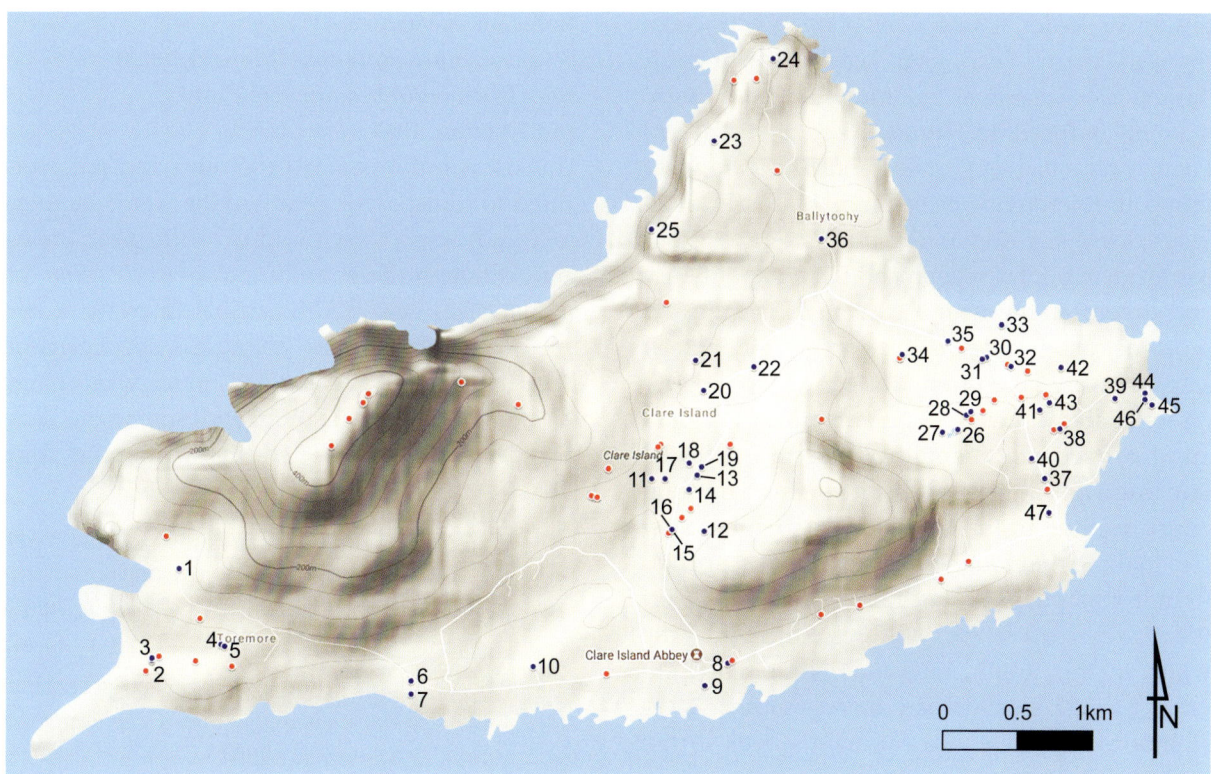

Fig. 1 Map of Clare Island showing the sites sampled. Blue spots with numbers represent sites from which ostracods were collected. Red spots represent sites where ostracods were not found.

Table 1
Distribution of ostracods related to geography, geology and sediments on Clare Island. The site numbers are shown on the map in Fig. 1.

Geographic/soil subdivisions	no. of species.	no. of sites with ostracods	pH
South West	4	5 of 11 sampled (sites 1-5)	4.5 – 8
South	7	5 out of 11 (sites 6-10)	6.1 – 7.4
Centre - pools	4	4 out of 13 (sites 11-14)	4 – 5
Centre - loughs, streams	8 or 9	8 out of 10 (sites 15-22)	5 – 7
Knockmore altitude >200 m	0	0 out of 6	4 – 5.3
North	5	3 out of 6 (sites 23-25)	4.5 – 6.5
East Fawnglas outside wall	6	4 out of 7 (sites 26-29)	6 – 7
East Maum, part Capnagower	5	7 out of 13 (sites 30-36)	4 – 6.8
East Capnagower inside walls	7	8 out of 11 (sites 37-44)	5.8 – 6.1
Saltmarsh, Kinnacorra	5	2 out of 2 (sites 45-46)	6 – 6.4
Harbour	1	1 out of 2 (site 47)	7+

diamictons (Coxon 2001) and the Carboniferous bedrock (Graham 2001) as described in *New Survey of Clare Island. Volume 2: Geology*. The bedrock was not found to affect ostracod distribution, but the south and east of the island, where limestone-rich sediments are found, were relatively rich in ostracods. (The northeast of the island, east of the road to the lighthouse, where limestone-rich diamictons also occur, was not visited for this survey.)

In other parts of the island sampling intensity bore some relation to the number of ostracods found, but more significant was the number of months over which a site was visited. Only seventeen sites were sampled more than once, so in some cases failure to collect certain ostracods may have been caused by absence of adults at that stage of their life cycles.

Knockmore, from 200m to 455m altitude, with shallow pools of low pH, appeared to have no ostracods. Sphagnum pools with low pH on its lower flanks were poor in ostracods, though Lough Merrignagh was interesting for having *C. globosa* at its outlet. (The 'lough' may be reduced in summer to 2m diameter.) *Potamocypris villosa*, *C. ovum* and *B. fuscata* were found to tolerate water below pH5, but in most cases *Sphagnum*-dominated pools were without ostracods. Lough Creggan and L. Leinapollbauty and streams in the central area are less acidic and more ostracods were collected there. Nineteen samples were taken from Lough Leinapollbauty on five dates in unsuccessful attempts to find an adult ostracod that would correspond with the immature specimens with spines; this sampling effort resulted in the collection of six other species making that lough the site with most species found. It is possible that rainwater running off the nearby road above could raise the pH of the water in this lough; limestone chippings were used to surface the road there. Another possibility is that there are calcareous beds in the underlying bedrock of Silurian siltstone (see Graham 2001): this is not an area of limestone-rich diamictons (Coxon 2001).

The saltmarsh samples were mostly mud and extremely time-consuming to examine and it is possible that further salt-tolerant species could have been missed, if their numbers are low. Surprisingly, no ostracods were collected from Toormore, nor from a large pool beyond Loughnaphuca; (both locations are in the south-west of the island). *Heterocypris incongruens* was collected from one puddle; otherwise puddles and ruts on bare earth were more readily colonised by cladocerans and copepods. Three pools near the sea were interesting for the sole occurrence of an ostracod species: the pool near the lighthouse compound for *E. virens* and a pool near the cliff top on Ballytoohey Mor for *B. fuscata*; both species are known to prefer seasonal pools. A small pool in marsh near the harbour was notable for *Heterocypris salina*.

Ostracods occur in flowing water as well as in still water or seepages, but in swiftly running water they are restricted to calm areas or the interstitial habitat (Meisch 2000). In the course of this survey although they were collected from small streams on slight inclines in the centre of the island, they were not found in seven streams sampled along the south road and harbour area roads where the fall is generally steeper.

Notes on species
Penthesilenula malayica (Menzel, 1923)
Syn.: *Darwinula brasiliensis* Pinto and Kotzian, 1961

This darwinulid is small (0.5mm) and pale coloured to transparent. Many of the water samples collected were very muddy with peat staining. Although the strongly coloured *C. ovum*, of similar size, was readily visible in peaty water in a watch glass, many colourless specimens of *P. malayica* may have escaped notice when hand sorting and it could be more widespread on Clare Island than so far detected. It was found at four sites:

(i) Craigmore pool near sea (site 7, L671842) pH 6.1, depth 0.25m, sedge, 4 July 2004
(ii) Roadside stream in centre of island, north of cattle grid, (site 15, L688852), pH 5.1, depth 0.05m, *Apium, Callitriche,* algae, co-occurring with *Pseudocandona sp.* and *Cryptocandona vavrai,* 28 September 2004
(iii) Lough Creggan (site 18, L689856), (pH at other collections 5–6.5), depth 0.1–0.2m (specimen provided by Elvira de Eyto) 27 September 2002
(iv) L. Leinapollbauty (site 19, L690856), pH 6–7, depth 0.3–0.6m, *Sphagnum, Potamogeton polygonifolius, Menyanthes trifoliata, Nitella, Carex, Apium inundatum, Juncus articulatus, Hypericum elodes,* moss, 28 September 2004. *Cyclocypris ovum* and *B. obliqua* were collected in the same sample; three other ostracods have been collected from this lough on other dates.

Eleven adult females and a juvenile were collected. The largest female was 0.49mm in length, with embryo (0.15mm) in the brood pouch and was collected in September as was the juvenile. *Penthesilenula malayica* is reported from Loughs Inagh and Agraffard in Co. Galway (Rossetti and Martens 1998), as *P. brasiliensis*; *P. brasiliensis* (or *Darwinula brasiliensis*) has more recently been found to be a synonym of *P. malayica* Martens and Savatenalinton (2011). This species has previously been misidentified as *Darwinula stevensoni* (Meisch 2000); possibly some pre-1990 records of *D. stevensoni* from Counties Laois, Carlow, Mayo, North Tipperary and Lough Neagh may more properly be referred to *P. malayica*. This species has also been reported from Scotland, southern

France, Brazil, South Africa (Meisch 2000) and Java (Rossetti and Martens 1998).

Candona candida (O.F. Müller, 1776)

Candona candida is widespread on Clare Island; 63 specimens were collected from twelve sites (8, 16, 19, 24, 28, 30, 38, 41, 42, 43, 45 and 46); these included a stream (pH 5), the saltmarsh, turf cuttings (pH5), the lighthouse pool (pH 5.8), a lough (pH 6–7) and a ditch (pH7). No males were found. Females with eggs were noted in September 2002 and 2004; other adult females and juveniles were collected in May, June and September and one juvenile was collected in August. Scourfield reported this species from two sites on Clare Island (Scourfield 1912). It has been recorded from ten counties in Ireland and from Lough Neagh and is common and widespread in Ireland, Britain and Europe.

Pseudocandona cf. *albicans* (Brady, 1864)

Two juvenile specimens (0.52 mm and 0.55 mm in length) were found in the roadside stream described for *P. malayica* (site 15), in September. Adults of *P. albicans* are from 0.68 mm to 0.9 mm in length. The penultimate segment of the cleaning leg was divided, with no medial seta, so these are neither *C. candida* nor *Cryptocandona* sp., but until larger specimens are collected their identity is uncertain. None were found at the site the following June; however, Meisch (2000) states that mature adults of *P. albicans* are most abundant in March and April.

One female specimen (0.79 mm in length), poorly preserved but possibly of this species, was collected from the lighthouse pool, (site 24, L 695881), pH 5.8, depth 0.1 m, altitude 93 m, sedge and *Potamogeton polygonifolius*, in September. The genital lobes protruded moderately, apparently more than those of *P. albicans*; however, the valves did not appear as beaked from dorsal view as those of *P. pratensis* (compared with illustrations in Meisch 2000). The cleaning leg was five segmented, with the terminal seta more than three times the length of the distal segment. There were five setae on the second segment of the mandibular palp. *Candona candida*, *C. ophtalmica* and *E. virens* were collected in the same sample.

P. albicans has been recorded from Co. Wicklow (Henderson 1990) and lives in permanent or temporary, stagnant or slow flowing waters and is mesohalophilic (Meisch 2000).

Cryptocandona vavrai Kaufmann, 1900

Nineteen specimens, all female, of this species were collected from nine sites (15, 19, 21, 32, 36, 39, 40, 41 and 45) in the centre and east of Clare Island in May, August and September. *Cryptocandona vavrai* was found in five streams, L. Leinapollbauty, wet pasture, saltmarsh and a reedbed. It was tolerant of a range of acidity, pH5–7. Previous Irish records were from bogs in Counties Mayo, Louth and Carlow (Douglas and McCall 1992). It is widespread in Britain and generally distributed in Europe (Meisch 2000).

Cryptocandona reducta (Alm, 1914)

This species was found in the saltmarsh (pH 6.4), in the stream flowing through it (sites 45 and 46), in a stream in a field in Capnagower with rushes and flags (site 41) and in the mud of the reedbed in Maum (pH <6.8) (L. Avullin, site 34). Seven specimens, all female, were collected in May, July and September. It is said to co-occur often with *C.vavrai* and this was the case in the saltmarsh and the Capnagower stream. The sole previous Irish record is from a stream on Glenamoy bog, Co. Mayo (McCall 1975). There are records from Britain, Germany, Poland, Norway, Sweden, Czech Republic and France (Meisch 2000).

Cypria ophtalmica (Jurine, 1820)

Given that this species is generally so common in Ireland and in Europe, it might have been expected to occur more frequently on Clare Island. However, numbers were very high at the lighthouse pool (site 24) in September, with over a hundred in the sample; not all were examined and only adult females and juveniles were noted. It was also collected from bog (site 42), streams (38, 41) and marsh with reeds (39) in the walled pastures of Capnagower in May; seven females, one with eggs, three males and a juvenile were found. Scourfield reported it from two localities on Clare Island (Scourfield 1912). It has been recorded from twelve Irish counties. The species is almost cosmopolitan.

Cyclocypris globosa (Sars, 1863)

Two females were collected in May, one from the stream flowing from Lough Merrignagh (site 14) and one from a stream in a field in Capnagower (site 41). Although widely distributed in Europe, Irish records are few, from Counties Mayo (McCall 1976), Galway and Monaghan (Norman 1905).

Cyclocypris ovum (Jurine, 1820)

This was found to be the commonest and most widespread ostracod on Clare Island. Over four hundred specimens were collected at 24 sites (2, 3, 9, 12, 13, 14, 17, 18, 19, 20, 21, 22, 26, 28, 29, 30, 31, 32, 33, 35, 42, 43, 44 and 45), though not on Knockmore, nor from the north of the island, nor from the harbour area. It occurred in still and flowing water at altitudes from 0m to 120m in acidities of pH 4–7. Females were found from May to September. One male was found in July and twelve in September when a female with eggs was also observed. The proportion of males to females was one to thirteen in July and one to eight in September. Irish records are all post-1960 as previously it was confused with other species of *Cyclocypris*, so it is likely to be much more widespread in Ireland than records limited to Counties Antrim, Louth, Mayo, Laois, Carlow and North Tipperary would suggest. Meisch (2000) describes its general distribution as Holarctic.

Eucypris virens (Jurine, 1820)

This species was found at only one site (24) on Clare Island, at the pool below the lighthouse compound wall (pH5.8). Three mature females and a juvenile were collected in September. *Eucypris virens* prefers seasonal pools that dry up in summer and Meisch (2000) describes it as oligohalophilic. The altitude of the lighthouse pool is 93m but it is close (150m) to the sea on an exposed coast. The pool shrank from 60m^2 in September 2004 to 20m^2 in June 2005, but the vegetation suggested it does not dry out entirely; eggs requiring desiccation for development may be deposited at the pool margin before the flooding subsides. As *E. virens* is a common species in Europe more Irish records might be expected; it has been recorded from Counties Antrim, Down, Carlow and Laois. Its general distribution is holarctic (Meisch 2000).

Bradleycypris obliqua (Brady, 1868)

This species occurred in three of the larger waterbodies (pH 5–7+), Lough Creggan (site 18), L. Leinapollbauty (site 19) and Poirtín Fuinch (site 26); this was unexpected, as Scourfield (1912) recorded his specimens from 'two ponds in the south-western portion of Clare Island'. Forty-three females were collected in June, July, August and September; a female with eggs was noted in July. Scourfield believed that his was a new Irish record, so Clare Island is still the only known Irish locality for *B. obliqua*. It is widespread in Britain and reported from Europe, North Africa and North America (Meisch 2000).

Bradleystrandesia fuscata (Jurine, 1820)

This species was found in one pool (site 25) on Clare Island where adult females and juveniles were present in great numbers. The pool was the most remote collecting site visited, being near the cliff edge (altitude 127m) in the northwest of the island. The water was very shallow in September, its surface 14m x 10m, pH 4.5–5, with moss, sedge and *Nitella*; the presence of grass and clover suggests that this was a seasonal flood. *Bradleystrandesia fuscata* is characteristic of seasonal pools on grassland, usually producing large populations (Meisch 2000). Meisch describes it as a pure freshwater form; however, blown spray from the Atlantic must reach the clifftop on occasion. (Salinity was not tested at the time of sampling.) Chydorids and copepods were present, but no other ostracod species. Although it is widespread in Britain, there seems to be no Irish record since 1912 when Scourfield recorded it from Clare Island (one site) and from the Louisburgh/Croaghpatrick area. It is reported from Europe, the Middle East, Turkey and North America (Meisch 2000).

Herpetocypris chevreuxi (Sars, 1896)

This species is widespread on Clare Island; it was collected from eleven sites (5, 10, 14, 17, 19, 26, 27,

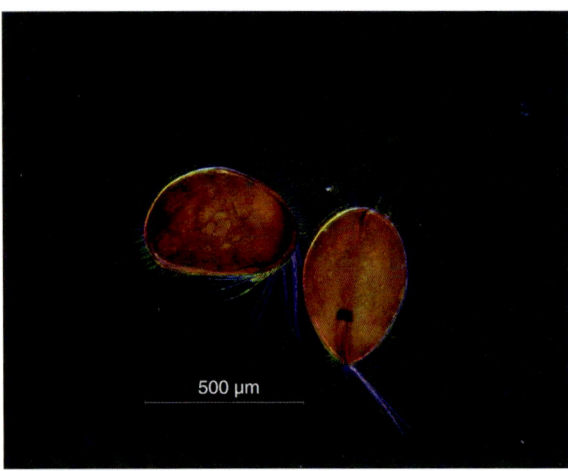

Cyclocypris ovum. Photo: Markus Lindholm, Anders Hobæk/ Norsk institutt for vassforsking. This item is distributed under the terms of the CC BY 3.0 Unported licence

29, 30, 33 and 42). It was numerous only in Billy Gallagher's water trough (site 10) and in marshes on Fawnglass (sites 27 and 29). Most collections were made from still waters but it also occurred in one stream. Over 100 adult females and juveniles were collected in May, July, August, September and October, the juveniles appearing from August to October. Females with eggs were seen in August and in early October.

The first Irish record for this species was from a thermal spring in Co. Meath (Fahy 1974) and it has also been found in Co. Carlow (Douglas and McCall 1992). It is possible that earlier records of *Herpetocypris reptans* should have been assigned to this species and it may be more widespread in Ireland than three records indicate. It is reported from Europe, Africa, South America, Iran and central Asia and is said to prefer slightly salty waters (Meisch 2000).

Psychrodromus robertsoni (Brady and Norman, 1889)

Ten specimens were collected in September, seven adult females from a roadside ditch on Kill (near the abbey, site 8). Three juveniles (1.3mm), which appear to be this species were collected from the salt marsh at Kinnacorra (site 45); adults are from 1.6–1.7mm in length. This species has been recorded from Ireland before, but probably only from Donegal in 1910 (Douglas and McCall 1992). It is normally found in springs and in seepages (D.J. Horne, pers. comm.), but also in pools, ditches and rivers (Meisch 2000). It is endemic to the British Isles and has been recorded from England, Wales and Scotland (Henderson 1990) where, Meisch (2000) states, it is generally considered an endemic or a relict species, but he adds that there is no fossil record of it.

Heterocypris incongruens (Ramdohr, 1808)

This species was found at two sites in July, both typical of its distribution elsewhere. An individual was seen swimming in a puddle (depth 0.02m, pH 8) in the quarry in the west of the island (site 1); this was an egg bearing female. The ubiquitous chydorid *Chydorus sphaericus* was also present in the puddle but no algae or plants were apparent and the substrate was fine mud. There were no ostracods in a second puddle nearby. Rubble dumped in the quarry may have contributed to raising the pH of the water.

Cattle troughs were sought as another likely habitat for *H. incongruens*; the only one found (site10), that of Billy Gallagher on Lecarow West, which is fed by a pipe from a spring, contained many *H. incongruens* females and juveniles and also *Herpetocypris chevreuxi*. Silt had collected on the floor of the trough, algae were present, water depth was 0.15m and pH 7+. This is a common species which has been recorded from eight Irish counties. It is a cosmopolitan species.

Heterocypris salina (Brady, 1868)

Two adult females were collected from marsh close to a tidal drain (site 47) in the harbour area (pH 7+, altitude 3m) in September. Scourfield's record of *Cyprinotus prasinus* from one site on Clare Island presumably referred to this species. It can be common in coastal areas in slightly brackish waters and has been recorded from Counties Clare, Mayo and Dublin but there is also a record (of *Cyprinotus prasinus* (Fischer)) from Co. Monaghan (Norman 1905). Meisch (2000) summarises its distribution as holarctic with introductions in the southern hemisphere.

Cypridopsis lusatica Schäfer, 1943

Six females of this species were found at four sites. They were from 0.57mm to 0.61mm in length and light green-brown in colour with none of the dark bands characteristic of *Cypridopsis vidua*. *Cypridopsis lusatica* has not been reported from Ireland before this survey. It was not possible to identify these ostracods to species using Henderson's key (Henderson 1990) but, as David J. Horne noted (pers. comm.) after dissecting a specimen, they match Meisch's illustration well (see Meisch 2000). In Meisch's opinion *C. bamberi* is a synonym of *C. lusatica* (see also Martens and Savatenalinton 2011); however Meisch gives a size range of 0.60–0.65mm for *C. lusatica* whereas Henderson (1990) who had described *C. bamberi* in 1986, gives the latter's size as up to 0.74mm. Horne, who like Meisch has examined the type material of *C. bamberi*, does not entirely agree that *C. bamberi* is merely a synonym of *C. lusatica*; however it was noted (by this author) that the furca, maxillary palp and maxilliped of these specimens differ from Henderson's illustrations of *C. bamberi*. Horne also observed that the carapaces of these specimens matched *C. lusatica* and not *C. bamberi*.

One specimen was collected in the southwest of Clare Island (site 5, Irish grid reference L659845) from a marsh with *Juncus* and *Myosotis* fed by a small stream, pH 7.1, on 29.9.04, co-occurring with *Herpetocypris chevreuxi*.

One specimen was collected from a pool 20 m x 8 m, depth <15 cm with *Sphagnum*, sedge and *Baldellia ranunculoides*, pH 7.4, near the road at Craigmore (site 6, L672843) 25.9.02.

One specimen was collected on 4.5.03 from a pool leading to a stream on Fawnglass north of the wall, (site 29, L708859) with sedge, algae and *Hypericum elodes*, pH 6–7, depth 10 cm. It co-occurred with *Potamocypris villosa*; *Cyclocypris ovum* and *Herpetocypris chevreuxi* have also been collected at this site on other dates.

Three specimens were collected from mud at the margin of the above pool, (site 28, L707859), 26.9.02, water depth 0.5 cm, pH 6.5–7. *Cyclocypris ovum* and *Candona candida* were found in the same sample.

Cypridopsis lusatica is rare at this latitude, recorded from Britain (as *C. bamberi*), northern France, Germany and Poland; Meisch (2000) suggests that this may be explained by passive introduction, probably by migrating birds, from the south. It is also recorded from Portugal, Spain, Italy, FYR Macedonia, the Azores, Madeira and the Canary Islands.

Potamocypris villosa (Jurine, 1820)

This species was collected at four sites, from two pools (site 11, pH 4.5 and site 17 downstream of Lough Creggan) and from marshes (sites 27 and 29, pH 6.7–7) near Poirtín Fuinch. Over sixty specimens were collected in May, July and September; numbers were most plentiful in July and all specimens were adult females.

Potamocypris villosa occurs in various habitats in Co. Carlow and Co. Laois. It is also reported from Mayo, Galway and Belfast (Norman 1905). It is common in Britain; Meisch (2000) describes it as common and widespread, to be found worldwide except Australia.

Spiny juvenile

Juvenile ostracods were collected from L. Leinapollbauty in May and in September (two years), but not on sampling dates in June, July or August. The valve lengths of the four specimens ranged from 0.33 mm to 0.43 mm and the valves were light blue (after preservation in methanol). There were four pairs of appendages apparent, suggesting fourth larval stage. The specimens had a distinct spine on the lower posterior margin of each valve. Six species have already been identified from this site; the juveniles do not belong to *P. malayica* or *C. ovum* because these species are too small, nor are they likely to be candonids as these are white, nor are they *H. chevreuxi* because immatures of this species are elongate, without spines on the valves. This author can find no reference to spiny juveniles of *B. obliqua* and spines are absent from adults of this species, so these specimens suggest that another species is present.

Adults of *Cypris pubera* O.F. Müller, 1776 have a spine on the postero-ventral margin of the right valve and sometimes also of the left valve. Adults and juveniles from instar V of *Cypris bispinosa* Lucas, 1849 have a spine on each valve (Pons 1981). Pons found the spines very difficult to see in instar V but visible in instar VI; however, he illustrates the spine of instar VI in the postero-ventral quarter of the valve, but not on the margin of the valve. Pons gives the following size ranges for the lengths of juvenile *C. bispinosa*: instar III 0.36–0.47 mm, instar IV 0.48–0.55 mm, instar V 0.63–0.84 mm, instar VI 0.96–1.10 mm. Juveniles of *C. pubera* may be smaller as adults are reported to be 2.0–2.6 mm in length, whereas lengths of *C. bispinosa* generally range from 2.6–3.2 mm (Meisch 2000), and in northern Spain, from where Pons' data came, from 3.03 mm up to 4.04 mm.

Although the spine on each valve initially suggested a possible identification of the specimens as *C. bispinosa*, they are clearly considerably smaller than Pons' collection of juveniles and with spines in a different place. The form of the spines of these specimens is closer to those of *C. pubera*, that is, narrow at the base. L. Leinapollbauty was not sampled between 3 May and 10 June, which is the peak period for adults of *C. pubera* so further sampling in these five weeks may be indicated. However, it would be exceptional to find a juvenile *C. pubera* in September. Both *Cypris* spp. are recorded from Ireland.

Comparison with Scourfield's survey of the Fresh-water Entomostraca of Clare Island

Scourfield recorded Entomostraca from a total of 21 stations (Scourfield 1912); he may have sampled unsuccessfully from additional stations.

Ostracods were found at a maximum of seventeen of his stations (assuming no co-occurrences, so probably less than seventeen); for each species he lists the number of stations at which it was found (never more than three), but does not record if more than one species was found at any station. Of 92 stations visited for this survey, Entomostraca were collected from 78, with ostracods found at 47 of these.

The following five species were collected in the course of both surveys: *Candona candida*, *Cypria ophtalmica*, *Bradleycypris obliqua*, *Bradleystrandesia fuscata* and *Heterocypris salina*. No collection of specimens deposited by Scourfield relating to his survey of Clare Island can be found in either the Natural History Museum in Dublin or the British Museum of Natural History; without being able to check Scourfield's identifications, it is assumed here that Scourfield's record of *Cyprinotus prasinus* (Fischer) refers to *Heterocypris salina*.

Scourfield also reported the presence of *Cyclocypris serena*, *Cyclocypris laevis*, *Herpetocypris reptans*, *Cypridopsis vidua* and *Sarscypridopsis aculeata*.

Cyclocypris ovum, although cosmopolitan and one of the commonest ostracods (Meisch 2000), was often mistaken for *C. serena* or *C. laevis* in the past. In view of the frequency of occurrence of *C. ovum* on Clare Island found in this survey, this author suspects that Scourfield's identifications were erroneous. Scourfield himself described his specimens of *C. serena* from Loughanaphuca as 'somewhat peculiar' in shape and colour, implying they were atypical, and he had his identification checked by Dr Vávra. For this survey, 148 specimens of *C. ovum* were examined and measured. Their lengths ranged from 0.41 mm to 0.52 mm (mean, mode and median 0.47 mm), too small for adult *C. serena*, and in all of them the right valve overlapped the left, ruling out *C. laevis*. Furthermore, the characteristic S-shaped seta on the cleaning leg was observed in each specimen.

Less expected was the discovery of *Herpetocypris chevreuxi* on Clare Island, as it is considered to be less common than *H. reptans*. Those few ostracods that appeared to be immature *Herpetocypris* lacking natatory setae, as does *H. reptans*, were identified by Dr David J. Horne as *Psychrodromus robertsoni*, also a member of the subfamily Herpetocypridinae. (The genus *Psychrodromus* was created in 1977 and includes two species formerly assigned to *Herpetocypris*, or *Erpetocypris* (Martens and Savatenalinton 2011).) In all other specimens the long natatory setae were evident, confirming *H. chevreuxi*. *Herpetocypris chevreuxi* was found to be less common and widespread on Clare Island than *C. ovum*; so although this author feels that *C. serena* or *C. laevis* probably do not occur on the island, it is possible that *H. reptans* could actually be there.

Cypridopsis vidua is another of the commonest ostracods; it could be expected that Scourfield would record it from Clare Island. However, during the course of this survey, few specimens of *Cypridopsis* were collected and they did not fit descriptions of *C. vidua*; Dr David J. Horne has identified them as *C. lusatica*. *Cypridopsis lusatica* was not described until 1943, so Scourfield would not have been wary of confusing the two species. He observed that his specimens 'were not so tumid nor so evidently marked with colour-bands as is usually the case in this species', features that suggest that his specimens may also have been *C. lusatica*. However, he adds 'On the other hand, they showed much more pronounced pitting of the shell', a feature characteristic of *C. obesa* (which Meisch (2000) considers to be a form of *C. vidua*). Scourfield was aware of *C. obesa*, having recorded it from Achill Island, but this species (or form) is more tumid than *C. vidua* or *C. lusatica*. Without being able to examine Scourfield's specimens, one cannot be sure which species of *Cypridopsis* he collected.

Sarscypridopsis aculeata, which Scourfield recorded from two (unnamed) stations on Clare Island, is said to prefer slightly brackish, small water bodies and to be common in coastal rockpools and ponds influenced by marine water, often co-occurring with *Heterocypris salina* (Meisch 2000). Although brackish water was sampled during this survey, and *H. salina* was collected from one site, rock pools have not been visited. Scourfield, or his assistants, did sample this habitat, as he refers to 'little pools on the rocks not far above high-water mark on the north-east coast of Clare Island' where the harpacticoid *Nitocra spinipes* was collected (Scourfield 1912). The valves of *S. aculeata* are usually covered in short spines, so it is unlikely to have been confused with any other species found in this survey. Therefore it is considered that its absence from the list of species recorded in this report is probably due to insufficient sampling of its preferred habitat.

A further eight species are recorded here for the first time from Clare Island, *Penthesilenula malayica, Pseudocandona* cf. *albicans, Cryptocandona vavrai, C. reducta, Cyclocypris globosa, Eucypris virens, Heterocypris incongruens* and *Potamocypris villosa*. It is possible that specimens of the last three could have been mistaken for other species by Scourfield, although he notes that *H. incongruens* and *P. villosa* had previously been reported from Galway and/or Mayo. *Penthesilenula malayica* and *C. globosa* are unlikely to have been mistaken for other species recorded from Clare Island by Scourfield because of their sizes and Scourfield also noted that a darwinulid and *C. globosa* occurred elsewhere in Galway/Mayo, suggesting that he was familiar with these taxa. None of these 5 species was found at more than 4 sites on Clare Island out of the 92 visited for this survey; it is most probable that Scourfield failed to collect them because he only had material from at most 21 sites. Most of Scourfield's material was collected in June 1909, whereas on this survey *C. globosa* was collected only in May and *P. malayica* was most numerous in September and not found in June; it is also possible that he missed species whose life cycles did not produce adults that June.

He is unlikely to have been familiar with the candonids now assigned to the genus *Cryptocandona*. *Cryptocandona vavrai* was described in 1900 but not recorded from Britain until 1931. *Cryptocandona reducta* was not described until 1914 and not recorded from Britain until 1967. Neither species was recorded from Ireland until 1975 (McCall 1975). *Pseudocandona sp.* or *Cryptocandona spp.* were collected from thirteen sites during this survey and any of them may have been mistaken for *Candona candida* in the past. Scourfield wrote 'A certain amount of bottom material was also passed through fine sieves; but it was not found possible to employ this method as frequently as could have been wished'. Insufficient searching through mud could lead to a shortage of Candoninae specimens as these genera crawl or burrow and do not swim. Both *Cryptocandona* species occur at Kinnacorra saltmarsh, which Scourfield did visit, collecting a diaptomid and other copepods.

Apart from the rock pools mentioned above, two stations specified by Scourfield but not visited for this survey are a pond near the old signal tower at the extreme west of the island and a little stream flowing from the lighthouse marsh to the coast. He found these habitats notable for their copepods.

Locations

It should be noted that identifying locations on Clare Island by name can be confusing. Despite townland boundaries shown on the Ordnance Survey's map of 1838 having been reorganised on the Survey's 1913 map (Ó Muraíle 1999), the old usage has persisted in the community.

The location of Loughnaphuca seems uncertain since the original lough either filled with vegetation or was drained. Maps of Scourfield's era would point to its having been at L654845, the location implied in this report, an interpretation supported by the presence of *Littorella uniflora*, a plant associated with lakes (Martyn Rix, pers. comm.). Others are of the opinion that it was at L659845. Both areas are now waterlogged and traversed by small streams, but with no lake to be seen.

This author has not found Poirtín Fuinch located on a map, but the name is in current use for a small lough (or its locality) on Fawnglas at L707858, north of the walled area of that townland. Ó Muraíle (1999) refers to the name being used by residents of Fawnglass and Capnagower but does not place it on a map.

A further problem may arise if using Ó Muraíle's map (Ó Muraíle 1999) to find Lough Leinapollbauty and Lough Merrignagh. Ó Muraíle (pers. comm.) has not actually walked over all the areas named by him. There does not appear to be a lough where his map indicates L. Merrignagh (at any rate between May and September); this author has followed Bald (in Ó Muraíle, 1999) and others in using the name Merrignagh for a small waterbody (L690855) located where Ó Muraíle has marked 'Loch Léana an Phoill Bháite (223A)'. Ó Muraíle (1999) translates the name as 'the rusty lake', and indeed iron stain on the substrate was noted at Bald's L. Merrignagh in September 2002.

Conclusion

The freshwater ostracod fauna of Clare Island seems to be very similar to what it was in the early twentieth century when Scourfield studied it. Any differences are most easily attributed to sampling influences, such as habitats chosen for collection, technique and effort in obtaining ostracods and seasonal variation in faunal assemblages, or to the effects of taxonomic revisions. There is little to suggest that any species has disappeared permanently since the earlier survey. However, the occurrence of *Cryptocandona* in the modern samples is interesting, since species of this genus, although relatively common today (D.J. Horne, pers. comm.), were not recorded in Britain before 1931 or in Ireland before 1975. Further study of the modern and historical distribution of *C. vavrai* and *C. reducta* is needed; there is a possibility that this genus was introduced to Clare Island in the twentieth century. As *Cypridopsis lusatica* was not recorded from Britain before 1986 and occurs only occasionally north of the Mediterranean, it is possible that this species was also introduced since Scourfield's survey.

Acknowledgements

I am greatly indebted to David J. Horne for invaluable information and advice and for examining specimens. I would like to thank Kieran McCarthy, Eoin MacLoughlin, Dariusz Nowak, Margaret Flaherty, Martyn Rix, Phillip Edmonds, Emilia Solano and Donal O'Shea for technical assistance and information and Elvira de Eyto and Stephen McCormack for specimens.

REFERENCES

Collins, T. 1999 The Clare Island Survey of 1909–11: Participants, papers and progress. In C. Mac Cárthaigh and K. Whelan (eds), *New Survey of Clare Island. Volume 1: History and cultural landscape*, 1–40. Dublin. Royal Irish Academy.

Coxon, P. 2001 The quaternary history of Clare Island. In J.R. Graham (ed.), *New Survey of Clare Island. Volume 2: Geology*, 87–112. Dublin. Royal Irish Academy.

Douglas, D.J. and McCall, G. 1992 The freshwater Ostracoda of Ireland. *Bulletin of the Irish Biogeographical Society* **15**, 91–109.

Fahy, E. 1974 Fauna and flora of a thermal spring at Innfield (Enfield), Co. Meath. *Irish Naturalists' Journal* **18**(1), 9–12.

Graham, J.R. 2001 The carboniferous rocks of Clare Island. In J.R. Graham (ed.), *New Survey of Clare Island. Volume 2: Geology*, 75–86. Dublin. Royal Irish Academy.

Henderson, P.A. 1990 *Freshwater ostracods. Synopses of the British Fauna* (New Series) **42** (D.M. Kermack and R.S.K. Barnes, eds), 228 pp. Oegstgeest. Universal Book Services.

Horne, D.J., Baltanás, A. and Paris, G. 1998 Geographical distribution of reproductive modes in living non-marine ostracods. In K. Martens (ed.), *Sex and parthenogenesis: evolutionary ecology of reproductive modes in non-marine ostracods*, 77–99. Leiden. Backhuys.

Klie, W. 1938 Krebstiere oder Crustacea III: Ostracoda, Muschelkrebse. *Die Tierwelt Deutschlands und der angrenzenden Meeresteile nach ihren Merkmalen und nach ihrer Lebensweise* **34**. Teil, 1–230. Jena. Gustav Fischer Verlag.

Martens, K. and Savatenalinton, S. 2011 A subjective checklist of the Recent, free-living non-marine Ostracoda (Crustacea). *Zootaxa* **2855**, 1–79.

McCall, G. 1975 The ostracods *Candona vavrai* and *Candona reducta* in Ireland. *Irish Naturalists' Journal* **18**(7), 223–4.

McCall, G. 1976 The ecology of the freshwater Crustacea of the peatlands of Glenamoy, Co. Mayo. Unpublished PhD thesis. University of Dublin, Trinity College.

Meisch, C. 2000 Freshwater Ostracoda of Western and Central Europe. *Süßwasserfauna von Mitteleuropa* **8**(3), (J. Schwoerbel and P. Zwick, eds), 522 pp. Heidelberg. Spektrum Akademischer Verlag.

Norman, A.M. 1905 Irish Crustacea Ostracoda. *The Irish Naturalist* **14**, 137–55.

Ó Muraíle, N. 1999 The place-names of Clare Island. In C. Mac Cárthaigh and K. Whelan (eds), *New Survey of Clare Island. Volume I: History and cultural landscape*, 99–141. Dublin. Royal Irish Academy.

Pons, J.A. 1981 *Cypris bispinosa* (Crustacea: Ostracoda) en Asturias. Notas sobre su biología. *Revista de la Facultad de Ciencias, Universidad de Oviedo (Seria Biología)* **22**, 55–66.

Rossetti, G. and Martens, K. 1998 Taxonomic revision of the Recent and Holocene representatives of the Family Darwinulidae (Crustacea, Ostracoda), with a description of three new genera. *Bulletin de l'Institut Royal des Sciences Naturelles de Belgique, Biologie* **68**, 55–110.

Scourfield, D.J. 1912 Clare Island Survey Part 46 Freshwater Entomostraca. *Proceedings of the Royal Irish Academy* **31**, 19 pp.

CHAPTER 15

CLADOCERA AND COPEPODA OF THE INLAND WATERS ON CLARE ISLAND

Elvira de Eyto, Adriana Trojanowska-Olichwer, Jens Petter Nilssen and T.K. McCarthy[†]

ABSTRACT

Microcrustaceans are an important component of freshwater ecosystems, and their local diversity is known to be driven by many factors. For island populations, immigration rates, dispersal vectors and distance from the nearest mainland source are crucial factors controlling biodiversity. We sampled freshwater microcrustacea (Cladocera and Copepoda) on Clare Island, Co. Mayo, off the west coast of Ireland. We quantified the faunal richness and compared it to D.J. Scourfield's original Clare Island survey published in 1912. Samples were collected from 21 water bodies on the island in 2001, 2002 and 2008, and we recorded 37 species of Cladocera and 8 species of Copepoda. The Copepoda were only sampled from a small number of water bodies. The cladoceran fauna of the current survey was richer than that recorded in 1912 and included several species that were surprisingly absent in 1912.

Introduction

The Entomostraca of Clare Island were originally surveyed by D.J. Scourfield and his colleagues Mr Kane, Mr Murray and Mr Dunkerly between 1909 and 1911 (Scourfield 1912). They focused on Cladocera, Copepoda and Ostracoda on Clare Island and found a total of 53 species (24 Cladocera, 19 Copepoda and 10 Ostracoda) compared to a 'mainland' species richness of 84. In that case, the 'mainland' was a collection of waterbodies sampled within a 16km distance from the shore of Clew Bay, although significant large water bodies on the mainland (Loughs Feeagh and Beltra) were unfortunately not included. The term Entomostraca is a historical subclass of crustaceans that is no longer in use, since it contained groups that are now considered to be too diverse for inclusion in a single subclass. Instead, freshwater Cladocera, Copepoda and Ostracoda are often grouped under the convenient 'microcrustacean' term with less taxonomic implications.

Cladocera are commonly known as water fleas. They are almost exclusively found in freshwater, and are common inhabitants of lakes, rivers, ponds and ditches. Some families, such as the Daphniidae, are mainly planktonic in nature and are crucial elements of the open water food web, providing an important food source for fish. Other families, such as the Chydoridae, have a more benthic nature, and are more common in the littoral region of lakes, and among vegetation in ponds and ditches. These benthic species often display distinct distributional patterns with regard to substrata, with some preferring vegetation and others sand, mud or rock (Duigan 1992). The composition and richness of the cladoceran assemblages and the relative proportion of different species can therefore provide a lot of information about the ecology and functioning of waterbodies (de Eyto *et al.* 2003). For this reason, they are often used in palaeolimnological studies, tracing the history of lakes and the effects of environmental impacts

(Jeppensen *et al.* 2001). Although recent work on Cladocera in Ireland is limited, Duigan (1992) conducted a detailed survey of the family Chydoridae across Ireland, which documented and reviewed the historic records of 41 species. Hannigan *et al.* (2014) documented the cladoceran biodiversity of small bog pools and found that bog pools on the Atlantic coast of the country were surprisingly species rich. Cladocera have been used to examine ecological quality of Irish water bodies (de Eyto *et al.* 2002; Caroni and Irvine 2010; Chen *et al.* 2010; Drinan *et al.* 2013), while McCarthy *et al.* (2006, 2010) examined the stoichiometric relationships of Irish zooplankton. Discoveries of previously unrecorded species continue (McGavigan 2008), and given the continued evolution of cladoceran taxonomy, further investigations across habitats may prove enlightening.

Like Cladocera, Copepoda are ubiquitous in freshwaters, but unlike Cladocera, their distribution also extends to brackish and marine waters. Copepods constitute a crucial part of the food webs of freshwater ecosystems and are frequently the most abundant microcrustaceans sampled in the open water (Caroni and Irvine 2010). Unfortunately, identification to species is rare, although O'Riordan (1971) documents the distribution of species identified by the late G.P. Farran in the mid-1990s, and also provided a useful reference of previous records. However, recent detailed taxonomic literature related to Irish copepods is practically non-existent.

In theory, island populations of freshwater microcrustaceans are ideal candidates for peripatric speciation resulting from the type of long-distance dispersal envisaged by Darwin (1872), which 'apparently depends in main part on the wide dispersal of their seeds and eggs by animals, more especially by freshwater birds'. Cladocera and calanoid Copepoda both produce resistant resting stages (diapausing eggs) that can survive dessication. In addition, cladoceran ephippial eggs are often sticky, with spines or hooks perfectly suited to attachment onto an obliging transporter (Fryer 1996). In addition to birds, dispersal vectors of cladocerans may include flying insects such as the aquatic Heteroptera in the genera *Notonecta* and *Sigara* (Lansbury 1955; van de Meutter 2008). Once these eggs have been deposited in their new island habitats, colonisation can be quick and efficient, as Cladocera mainly reproduce by parthenogenesis. Detailed species phylogenies are necessary, however, to assess the importance of dispersal in determining speciation, and these have often been lacking for microcrustacean taxa (Taylor *et al.* 1998). The systematics of many microcrustacean taxa are in constant flux, with advances in microscopic and molecular techniques bringing new insights each decade. Species that were previously considered to be cosmopolitan have been shown, with the advance of molecular techniques, to be regionally endemic (Taylor *et al.* 1998; Petrusek *et al.* 2008; Belyaeva and Taylor 2009). Identification of microcrustaceans to species is hampered by phenotypic plasticity on small spatiotemporal scales, outdated or regionally inappropriate taxonomic keys and the lack of detailed species descriptions. These issues all have implications for the resurveying of the microcrustaceans of Clare Island, a century after the publication of Scourfield's species list (Scourfield 1912). It is serendipitous that D.J. Scourfield not only wrote the original chapter on Entomostraca for Clare Island, but also co-authored the taxonomic key that is still widely used in Ireland to identify Cladocera (Scourfield and Harding 1966). The aim of the current survey was to provide an update species list of Cladocera on Clare Island. Copepoda were also recorded at some sites and this information is included here for completeness.

Materials and methods

We sampled microcrustaceans in December 2001, January and September 2002 and September 2008 from 21 waterbodies distributed across Clare Island (Figs 1 and 2; Table 1). Several of the larger waterbodies were sampled on more than one occasion. One site (Lough Leinapollbruty, site 5) was sampled in December 2001 using a light trap. In January 2002, samples (seventeen waterbodies) were taken with a sweep net, and some water bodies were additionally sampled using light traps. In September 2002, samples were taken with a 100-μm sweep net from seven waterbodies. In September 2008 five waterbodies were sampled, including two previously sampled lakes (Lough Leinapollbruty, site 5 and Lough Creggan, site 6) as well as three small ponds in the vicinity of these larger lakes. This study sampled as many vegetated and non-vegetated substrata as possible in each waterbody. Thus, the

CHAPTER 15: CLADOCERA AND COPEPODA OF THE INLAND WATERS ON CLARE ISLAND

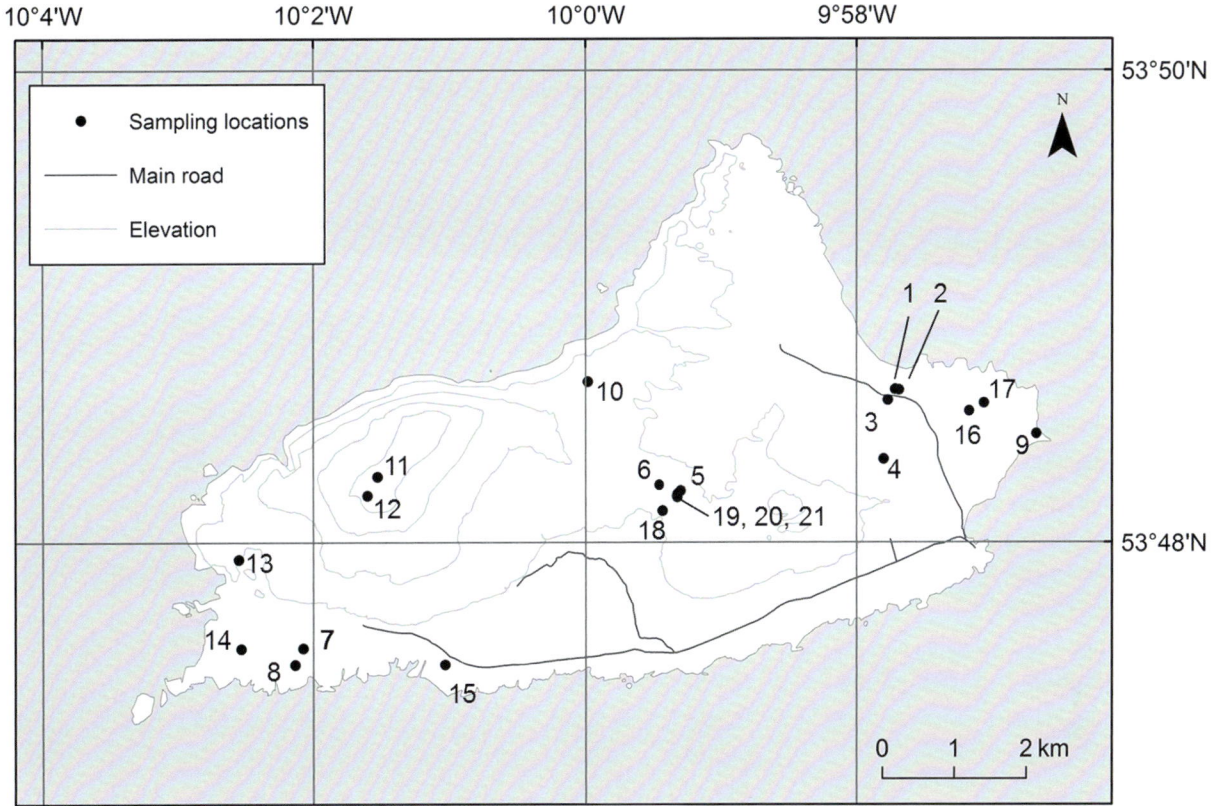

Fig. 1 Map showing microcrustacean sampling locations on Clare Island. Sites were visited between December 2001 and September 2008.

Fig. 2 Some of the microcrustacean sampling locations on Clare Island. A: Site 4, Lough Avullin. B: Sites 7, 8 and 14, Kinatevdilla ponds. C: Site 6, Lough Creggan. D: Site 15, Craigmore.

Table 1
Sampling locations on Clare Island, where microcrustaceans were collected, and the month they were visited

Site	Name	X (°N)	Y (°W)	Dec-01	Jan-02	Sep-02	Sep-08
1	Maum medium pond	53.8108	−9.9619		x		
2	Maum big pond	53.8108	−9.9615		x		
3	Maum small pond	53.8100	−9.9628		x		
4	Lough Avullin	53.8059	−9.9633		x		
5	Lough Leinapollbruty	53.8036	−9.9883	x	x	x	x
6	Lough Creggan	53.8041	−9.9900		x	x	x
7	Kinatevdilla big pond	53.7925	−10.0346		x		
8	Kinatevdilla small pond	53.7913	−10.0356		x		
9	Kinnincorra	53.8077	−9.9443		x	x	
10	Ballytoohy pond	53.8114	−9.9998		x		
11	Knockmore top	53.8046	−10.0255		x		
12	Knockmore top west	53.8033	−10.0267		x		
13	Signal tower	53.7988	−10.0425		x		
14	Kinatevdilla medium pond	53.7925	−10.0422		x		
15	Craigmore	53.7914	−10.0173			x	
16	Poirtin Fuinch	53.8093	−9.9527			x	
17	Marsh near site 16	53.8098	−9.9508			x	
18	Merrignagh	53.8022	−9.9905			x	
19	Pond 1A	53.8034	−9.9887				x
20	Pond 1B	53.8032	−9.9887				x
21	Pond 1C	53.8033	−9.9888				x

samples were qualitative rather than quantitative. Although the main focus of these surveys was Cladocera, Copepoda were also identified from five waterbodies in September 2008. Cladocera were identified according to Scourfield and Harding (1966), Flössner (1972), Amoros (1984), Duigan (1992) and Rybak (1993). Copepods were identified according to Sars (1913), Gurney (1931), Rylov (1963) and Dussart (1967). For copepods, the nomenclature of Kiefer (1973) was followed and for the genus *Daphnia*, Petrusek et al. (2008).

The results of the current survey were compared with Scourfield (1912). For Cladocera, this comparison is valid, as both surveys (1912 and current) sampled a similar range of cladoceran habitats. However, during the earlier survey, Copepoda were collected from a wide range of habitats, including mosses and brackish (= infrahaline) environments. During the current survey, copepods were only identified from five waterbodies in 2008, and so the current species list for these groups is only intended as a first approximation of the current species list.

Results and discussion
Cladocera

Clare Island was found to support diverse assemblages of Cladocera species, with a total of 37 species recorded from the various sampling locations (Table 2). These included 23 Chydoridae species, 6 Macrothricidae species, 5 Daphniidae and 1 member of each of the families Bosminidae and Sididae. Highest species richness was found in Lough Leinapollbruty (site 5), where 21 cladoceran species were found. Lough Creggan (site no. 6), Lough Avullin (site 4) and Poirtin Fuinch (site 16) were also found to have high species richness, with seventeen, eleven and fifteen species respectively. No Cladocera were found in two of the waterbodies: site 12 (west of the summit of Knockmore) and site 8 (a small pond at Kinatevdilla). In general, the larger, more inland water bodies had higher species richness. This is probably a reflection of the higher diversity of habitats, and the presence of diverse macrophyte beds. In 2002, *Chydorus sphaericus* was found in fifteen sites. This is the most common cladoceran

Table 2

The 37 Species of Cladocera found at 21 sites on Clare Island, sampled between 2001 and 2008, the number of sites where each species occurred and whether the species was present (✓) during the survey. The species list from 1912 is also included for comparison. See Table 1 and Figure 1 for site details.

Species	1	2	3	4	5	6	7	8	9	10	11	12	13	14	15	16	17	18	19	20	21	no. of sites	Present	recorded in 1912
Bosminidae																								
Bosmina longirostris (O.F. Müller)				*	*	*										*						4	✓	
Chydoridae																								
Acroperus angustatus G. O. Sars																						0		*
Acroperus harpae (Baird)				*	*	*										*						4	✓	*
Alona affinis (Leydig)					*	*			*							*		*				7	✓	*
Alona costata G. O. Sars			*															*				1	✓	*
Alona guttata G. O. Sars	*			*	*	*										*						5	✓	*
Alona intermedia G. O. Sars					*	*		*								*						3	✓	
Alona quadrangularis (O.F. Müller)				*	*		*															2	✓	*
Alona rectangula G. O. Sars					*	*										*						3	✓	*
Alona rustica T. Scott	*				*		*															3	✓	
Alonella excisa (Fischer)		*		*	*	*							*									7	✓	*
Alonella exigua (Lilljeborg)				*	*	*								*	*			*				7	✓	
Alonella nana (Baird)	*	*	*		*			*														6	✓	*
Alonopsis elongata (G. O. Sars)										*					*							2	✓	
Chydorus barbatus (Brady)																						0		
Chydorus latus G. O. Sars	*									*			*		*							4	✓	?
Chydorus ovalis Kurz	*	*		*									*		*							5	✓	*
Chydorus piger G. O. Sars					*	*			*		*				*	*						5	✓	
Chydorus sphaericus (O.F. Müller)	*	*	*	*	*	*	*				*		*	*	*	*	*			*	*	15	✓	*
Eurycercus lamellatus (O.F. Müller)				*	*											*		*				4	✓	*
Graptoleberis testudinaria (Fischer)					*	*												*				3	✓	*
Leydigia leydigi (Schödler)															*							1	✓	
Pleuroxus aduncus (Jurine)				*															*			2	✓	

(Continued)

Table 2. Cont.

Species	1	2	3	4	5	6	7	8	9	10	11	12	13	14	15	16	17	18	19	20	21	no. of sites	Present	recorded in 1912
Pleuroxus denticulatus Birge				*			*							*								3	✓	
Pleuroxus laevis G. O. Sars		*			*																	2	✓	
Pleuroxus truncata (O.F. Müller)					*	*										*				*		4	✓	*
Daphniidae																								
Ceriodaphnia quadrangula (O.F. Müller)					*	*										*		*	*	*	*	7	✓	*
Daphnia sp.															*							1	✓	
Daphnia hyalina Leydig				*		*																1	✓	
Daphnia hyalina lacustris G. O. Sars																						0		*
Daphnia longispina (O.F. Müller)					*																	1	✓	*
Simocephalus vetulus (O.F. Müller)					*	*													*	*	*	5	✓	*
Macrothricidae																								
Drepanothrix dentata (Eurén)							*									*				*		3	✓	*
Ilyocryptus sordidus (Lievén)																		*				1	✓	
Acantholeberis curvirostris (O.F. Müller)																*					*	2	✓	*
Macrothrix laticornis (Jurine)						*																1	✓	
Macrothrix rosea (Jurine)			*																			1	✓	
Streblocerus serricaudatus (Fischer)	*																					1	✓	
Sididae																								
Sida crystallina (O.F. Müller)					*																	1	✓	*
Diaphanosoma brachyurum (Lievén)					*													*		*	*	4	✓	
Total species	7	5	4	11	21	17	6	0	2	3	1	0	4	3	8	15	1	8	3	6	6	37		24

in Ireland, reflecting its extremely tolerant nature, which enables it to thrive in a wide range of conditions (Fryer 1993). It is therefore not surprising that *C. sphaericus* was the only species found in some of the more inhospitable waterbodies on Clare Island, such as site 11 at the summit of Knockmore, and in a small marsh to the northwest of Poirtin Fuinch (site 17), which had only about one inch of clear water. *Acroperus harpae, Alona affinis, Alonella excisa* and *Alonella nana* were also common in 2002. The more planktonic species, *Bosmina longirostris, Daphnia longispina, Daphnia hyalina,* and *Diaphanosoma brachyurum* were primarily found in the larger water bodies that had a high volume of open water. *Ceriodaphnia quadrangula* (Fig. 3) was frequently observed, even in ponds with small water volumes (sites 19, 20 and 21). In the myriad of coastal ponds along the coast of Fennoscandia another species of *Ceriodaphnia* dominates, *C. reticulata*, but *C. quadrangula* is common in more acidic ponds (Nilssen unpubl. data).

The high diversity of chydorid species is to be expected, given that there are 41 species recorded from the Irish mainland (Duigan 1992). Nearly all of the more common chydorid species were found during this survey, with the exception of *Disporalona rostrata* (Koch) *Rhynchotalona falcata* (Sars) and *Monospilus dispar* Sars. The remaining 15 chydorid species recorded from the mainland, but not during this survey, are rather rare and do not feature often in microcrustacean samples. The lack of the larger, predatory Cladocera *Leptodora kindtii* (Focke) and *Bryothrepes longimanus* Leydig is not surprising, as they are more often found in zooplankton hauls taken from large lakes. However, we were surprised not to find the other predatory Cladocera *Polyphemus pediculus*, as it is often encountered in the littoral samples of Irish lakes and ponds, and usually in very high numbers. The lack of the taxon *Bosmina longispina* (*B. obtusirostris* in Scourfield and Harding (1966)) is also conspicuous, as is the absence of typical pond species of *Daphnia*, such as *D. pulex* and *D. obtusa* (c.f. Fryer 1993).

It seems that there have been several cladoceran species colonisations of the island in the intervening century since Scourfield's 1912 species list was published, as there are fourteen species of Cladocera found in the current survey that were not recorded in 1912 (Table 2). The 1912 survey seems to have been very comprehensive, covering 21 sampling locations, so it is unlikely that the lower species richness was a result of site omissions. We found several species that were notable by their absence in 1912. These were *Diaphanosoma brachyurum, Bosmina longirostris* and *Alonopsis elongata*. These are common taxa on the mainland, so it is not surprising that they eventually made their way to Clare Island. However, no information is available to suggest if any arrived to the island by human agency. Many of the other species recorded in this century, but not in the previous one, are the rarer Chydoridae and Macrothricidae, which may be missed in any survey, as they are generally not locally abundant. As in the 2002 survey, *Chydorus sphaericus* was the most widely distributed Cladocera on Clare Island in 1912, when it was found in two thirds of the 21 sites sampled. Three species (or morphotypes) were found in 1912 but not in 2002; *Acroperus angustatus, Chydorus barbatus* and *Daphnia hyalina lacustris*. *A. angustatus* has a very variable morphology, and some prominent taxonomists (Herbst 1962; Flössner 1972) even consider *A. harpae* conspecific with *A. angustatus*. *Chydorus barbatus* has not been recorded on the Irish mainland by any other researcher, and the record of *C. barbatus* by Scourfield on Clare Island remains a mystery. It was not mentioned in detail in the 1912 paper (only referred to in the species list) and so

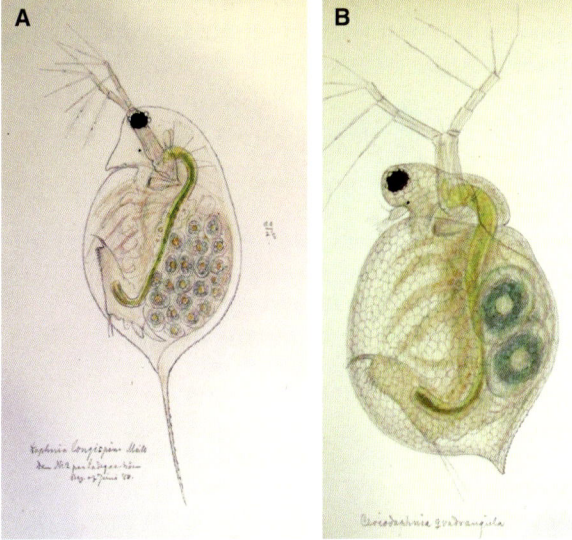

Fig. 3 Unpublished microdrawings by G.O. Sars of *Daphnia longispina* s. str. (the Lough Creggan morphotype has a slightly shorter dorsal spine and longer swimming antennae) and *Ceriodaphnia quadrangula*, one of the most common daphniids on Clare Island

we have no information on where it was found or how abundant this taxon was.

Scourfield (1912) reported both *Daphnia longispina* and *Daphnia hyalina lacustris* from Clare Island. These *Daphnia* taxa have a long taxonomic history from the British Isles and Ireland, and are usually reported from free water regions, especially *Daphnia hyalina lacustris* (Smyly 1958; Fryer 1985). *D. longispina* s.str. can also inhabit the weed region of ponds and lakes (Smyly 1958; Fryer 1985). The taxonomy and nomenclature of these taxa have been updated during the 2010s. *D. hyalina* sensu Leydig is conspecific with *D. longispina* sensu O.F. Müller (Petrusek *et al.* 2008) although Scourfield and Harding's key (1966), which is still widely used, separates these two species. The nomenclature *Daphnia hyalina lacustris* is especially unfortunate, since *D. lacustris* is a separate species of its own (Nilssen *et al.* 2007), known from the early 1860s (Sars 1862). In fact, *Daphnia hyalina lacustris* specimens analysed recently were found to be conspecific with *D. galeata* (Nilssen *et al.* 2007; Petrusek *et al.* 2008; Adamowicz *et al.* 2009). *D. galeata* is very common in the whole of northwest Ireland (Nilssen and Rønning unpublished data), so its occurrence in larger and deeper ponds on the island is not impossible. It is possible, therefore, that Scourfield's 1912 study identified two species, *D. longispina* and *D. galeata* (current taxonomy). *Daphnia* were recorded by three authors (de Eyto, Trojanowska-Olichwer and Nilssen) in the current survey, but some differences in taxonomic resolution have resulted in three *Daphnia* taxa in Table 2. Lough Creggan comprises a typical habitat of *D. longispina* s.str., and this taxon was accordingly recorded from that site (by Nilssen) (Fig. 3). Individuals from the *Daphnia hyalina-longispina* species complex were also recorded by de Eyto in Lough Creggan, and it is likely that these individuals may have been *D. longispina* s.str. *Daphnia* was found at one other site (Craigmore, site 15) by Trojanowska-Olichwer but could only be assigned to genus with confidence.

The current species count for Cladocera in Ireland is 61. Illies (1978) recorded 58 species and McGavigan (2008) added a further 3. The cladoceran species richness (37) recorded from Clare Island therefore represents 61% of the mainland Irish fauna. The slightly depauperate fauna is probably a reflection of the lack of large waterbodies on the island, rather than the isolated nature of the island. In addition, samples were only taken in two months of the year, which could have meant that we missed some of the rarer species. The lower species richness may also be as result of taxonomic anomalies whereby species were recorded earlier in the century but have since been reclassified.

Copepoda

Eight species of Copepoda were identified in the 2002–08 survey of Clare Island (Table 3), albeit from a small number of water bodies. Six of these were from the cyclopoid family, and one each from the calanoid and harpacticoid families. The calanoid *Eudiaptomus gracilis* G.O. Sars was the most abundant microcrustacean in L. Creggan. It is a characteristic species in thousands of small rock-pools and all sorts of lakes along the southern Norwegian coast (Nilssen unpublished data) and is widespread across Ireland. Caroni and Irvine (2010) recorded *E. gracilis* from 31 of 32 lakes sampled across the country between 1996 and 1997 (Caroni and Irvine 2010). This species was

Table 3
Species of Copepoda recorded in 2008 on Clare Island, from five inland waterbodies. Site numbers are indicated in the header. See Table 1 and Figure 1 for site details.

Family	Species	6	5	21	20	19
Calanoida	*Eudiaptomus gracilis* (G. O. Sars)	*	*			
Cyclopoida	*Macrocyclops albidus* (Jurine)		*		*	*
	Eucyclops lilljeborgi (G. O. Sars)		*			
	Megacyclops viridis (Jurine)	*	*		*	
	Acathocyclops vernalis (Fischer)		*			
	Diacyclops languidus (G. O. Sars)				*	
	Diacyclops nanus (G. O. Sars)	*	*	*		
Harpacticoida	*Attheyella crassa* (G. O. Sars)		*			

widespread in 1912, having been recorded in five places by Scourfield.

The cyclopoid genus *Eucyclops* is characterised by a large number of benthic taxa, where some of them have been in nomenclature flux. *Eucyclops lilljeborgi* (= *E. denticulatus*) is a species very commonly recorded from ponds (Fryer 1993) and was found during this study. It was not noted explicitly by Scourfield (1912), but at that time it was included in the taxon *E. macruroides*. Both *Eucyclops serrulatus* and *E. macruroides* ranged among the most common Clare Island copepods by Scourfield (1912). Among harpacticoid copepods, the only species recorded in 2008 was *Attheyella crassa* (=*Canthocamptus crassus* in Scourfield 1912). Only five sites were surveyed for Copepoda in the current survey, and no moors and saline wetlands were included. In contrast, Scourfield (1912) sampled a variety of freshwater ecosystems, including brackish water wetlands, brooks and moors, habitats known to be inhabited by a large numbers of harpacticoid copepod taxa (Lang 1931; Dussart 1967; Kiefer 1973). It is not surprising, therefore that none of the characteristic taxa of these ecosystems, which were listed by Scourfield in 1912, were recorded. These characteristic taxa include those inhabiting mosses (*Diacyclops languidus* G.O. Sars, *Bryocamptus pygmaeus* G.O. Sars, *Bryocamptus zschokkei* Schmeil, *Bryocamptus praegeri* Scourfield), small streams (*Paracyclops prasinus*, *P. affinis*, *P. fimbriatus*), and infrahaline environments (*Canthocamptus hirticornis* Scott, = *Mesocra rapiens* Scott, *Nitocra spinipes* Boeck). However, concerning the copepods species list in the 2008 survey of Clare Island, we were surprised not to find the widespread copepod *Cyclops abyssorum* G.O. Sars, as it is often encountered in Irish and British lakes and ponds, and over the whole northern coastal region of Europe (Nilssen and Elgmork 1977; Nilssen 1979), usually as a dominating species and in very high numbers (Gurney 1931; Smyly 1958).

In conclusion, it seems that the most obvious change to the Clare Island Cladocera in ninety years is the increased species richness. This is almost certainly a result of introductions onto the island over the last century. There has been no apparent shift in dominant chydorid taxa, indicating that there has been little change in the ecological quality of Clare Island waterbodies. This is reassuring, given the increasing impact of anthropogenic influences on mainland lakes and rivers.

Acknowledgements

The authors would like to thank Stephen McCormack, Sharon Fahy and Ann-Helén Rønning (Natural History Museum, University of Oslo, Norway) for assistance in the field, and the Norwegian National Library, Oslo for access to the posthumous scientific collections of the Norwegian carcinologist Georg Ossian Sars (1837–1927).

REFERENCES

Adamowicz, S.J., Petrusek, A., Colbourne, J.K., Hebert, P.D. and Witt, J.D. 2009 The scale of divergence: a phylogenetic appraisal of intercontinental allopatric speciation in a passively dispersed freshwater zooplankton genus. *Molecular Phylogenetics and Evolution* **50**(3), 423–36.

Amoros, C. 1984 Crustacés Cladocères. *Bulletin de la Societe Linnéenne de Lyon* **53,** 72–143.

Belyaeva, M. and Taylor, D.J. 2009 Cryptic species within the Chydorus sphaericus species complex (Crustacea: Cladocera) revealed by molecular markers and sexual stage morphology. *Molecular Phylogenetics and Evolution* **50**, 534–46.

Caroni, R. and Irvine, K. 2010 The potential of zooplankton communities for ecological assessment of lakes: redundant concept or political oversight? *Biology and Environment: Proceedings of the Royal Irish Academy* **110**B, 35–53.

Chen, G., Dalton, C. and Taylor, D. 2010 Cladocera as indicators of trophic state in Irish lakes. *Journal of Paleolimnology* **44**, 465–81.

Darwin, C. 1872 *The Origin of Species*. London. John Murray.

Drinan, T.J., Graham, C.T., O'Halloran, J. and Harrison, S.S.C. 2013 The impact of conifer plantation forestry on the Chydoridae (Cladocera) communities of peatland lakes. *Hydrobiologia* **700**, 203–19.

Duigan, C. 1992 The ecology and distribution of the littoral freshwater Chydoridae (Branchiopoda, Anomopoda) of Ireland, with taxonomic comments on some species. *Hydrobiologia* **241**, 1–70.

Dussart, B. 1967 *Les Copépodes des Eaux Continentales d'Europe Occidentale I: Calanoïdes et Harpacticoïdes*. N. Boubée and Cie, Paris.

de Eyto, E., Irvine, K. and Free, G. 2002 The use of members of the family Chydoridae (Anomopoda, Branchiopoda) as an indicator of lake ecological quality in Ireland. *Biology and Environment, Proceedings of the Royal Irish Academy* **102B**, 81–91.

de Eyto, E., Irvine, K., García-Criado, F., Gyllström, M., Jeppensen, E., Kornijow, R., Miracle, M.R., Nykänen, M., Bareiss, C., Cerbin, S., Salujõe, J.,

Franken, R., Stephens, D. and Moss, B. 2003 The distribution of chydorids (Branchiopoda, Anomopoda) in European shallow lakes and its application to ecological quality monitoring. *Archiv für Hydrobiologie* **156**, 181–202.

Flössner, D. 1972 *Kiemen- und Blattfuesser, Branchiopoda,Fischlause, Branchiura*. Fischer, Jena.

Fryer, G. 1985 The ecology and distribution of the genus Daphnia (Crustacea: Cladocera) in restricted areas: the pattern in Yorkshire. *Journal of Natural History* **19**, 97–128.

Fryer, G. 1993 *The freshwater Crustacea of Yorkshire; a faunistic and ecological survey*. Yorkshire Naturalists' Union and Leeds Philosophical and Literary Society.

Fryer, G. 1996 Diapause, a potent force in the evolution of freshwater crustaceans. *Hydrobiologia* **320**, 1–14.

Gurney, R. 1931 *British fresh-water Copepoda, vol. 3*. London. The Ray Society.

Hannigan, E. and Kelly-Quinn, M. 2014 Aquatic invertebrate communities of ombrotrophic bogs in Ireland with special reference to microcrustaceans. *Biology and Environment: Proceedings of the Royal Irish Academy* **114B**, 249–63.

Herbst, H.V. 1962 *Blattfusskrebse (Phyllopoden: Echte Blattfüsser und Wasserflöhe)*. Kosmos Verlag, Franckh-Stuttgart.

Illies, J. 1978 Limnofauna Europaea. *A checklist of the animals inhabiting European inland waters, with accounts of their distribution and ecology (except Protozoa)*. Stuttgart. Gustav Fischer Verlag.

Jeppensen, E., Leavitt, P., De Meester, L. and Jensen, J.P. 2001 Functional ecology and palaeolimnology: using cladoceran remains to reconstruct anthropogenic impact. *Trends in Ecology and Evolution* **16**, 191–98.

Kiefer, F. 1973 *Ruderfusskrebse (Copepoden)*. Stuttgart. Kosmos-Verlag, Franckh.

Lang, K. 1931 *Schwedische Süsswasser-und Moosharpacticiden*. Stockholm. Almqvist and Wiksell.

Lansbury, I. 1955 Some notes on invertebrates other than Insecta found attached to water bugs (Hempt.-Heteroptera). *Entomologist* **88**, 139–40.

McCarthy, V., Donohue, I. and Irvine, K. 2006 Field evidence for stoichiometric relationships between zooplankton and N and P availability in a shallow calcareous lake. *Freshwater Biology* **51**, 1589–604.

McCarthy, V. and Irvine, K. 2010 A test of stoichiometry across six Irish lakes of low-moderate nutrient status and contrasting hardness. *Journal of Plankton Research* **32**, 15–29.

McGavigan, C. 2008 Three water fleas (Branchiopoda: Cladocera) new to the Irish list. *The Irish Naturalists' Journal* **29**, 139–41.

Nilssen, J.P. 1979 Problems of subspecies recognition in freshwater cyclopoid copepods. *Z. Zool. Syst. Evolut.-Forsch* **17**, 285–95.

Nilssen, J.P. and Elgmork, K. 1977 *Cyclops abyssorum. Life-cycle dynamics and habitat selection*. Memorie dell'Istituto Italiano di Idrobiologia, Dott. Marco de Marchi Verbania Pallanza.

Nilssen, J.P., Hobæk, A., Petrusek, A. and Skage, M. 2007 Restoring Daphnia lacustris G.O. Sars, 1862 (Crustacea, Anomopoda): a cryptic species in the Daphnia longispina group. *Hydrobiologia* **594**, 5–17.

O'Riordan, C.E. 1971 The Freshwater Copepod Work of G.P. Farran Together with Some Other Notes. *Proceedings of the Royal Irish Academy* **71B**, 85–96.

Petrusek, A., Hobæk, A., Nilssen, J.P., Skage, M., ČErný, M., Brede, N., and Schwenk, K. 2008 A taxonomic reappraisal of the European Daphnia longispina complex (Crustacea, Cladocera, Anomopoda). *Zoologica Scripta* **37**, 507–19.

Rybak, J.I. 1993 *Przegląd słodkowodnych zwierząt bezkręgowych: Cladocera*. Biblioteka Monitoringu Środowiska, Warsaw.

Rylov, W.M. 1963 *Freshwater Cyclopoida. Fauna of the USSR. Crustacea III (3)*. 1–314. (Israel Program for Scientific Translations.)

Sars, G.O. 1862 Hr. studios. medic. G.O. Sars fortsatte sit foredrag over de af ham i omegnen af Christiania iagttagne Crustacea Cladocera. *Forhandlinger i Videnskapsselskapet i Kristiania* **1861**, 250–302.

Sars, G.O. 1913 *An account of the Crustacea of Norway with short descriptions and figures of all the species*. Bergen. Bergen Museum.

Scourfield, D.J. 1912 Freshwater Entomostraca. Clare Island Survey. *Proceedings of the Royal Irish Academy* **31B**, 1–18.

Scourfield, D.J. and Harding, J.P. 1966 *A key to the British freshwater Cladocera with notes of their ecology. Second edition*. Windermere. Freshwater Biological Association.

Smyly, W.J.P. 1958 Distribution and seasonal abundance of Entomostraca in moorland ponds near Winderemere. *Hydrobiologia* **11**, 59–72.

Taylor, D.J., Finston, T.L. and Hebert, P.D.N. 1998 Biogeography of a widespread freshwater crustacean: pseudocongruence and cryptic endemism in the North American Daphnia laevis complex. *Evolution* **52**, 1648–70.

van de Meutter, F., Stoks, R. and de Meester, L. 2008 Size selective dispersal of Daphnia eggs by backswimmers (Notonecta maculate). *Biology Letters* **4**, 494–96.

CHAPTER 16

FRESHWATER FISH AND THEIR PARASITES IN CLARE ISLAND

T.K. McCarthy[†], Karen Creed and Éamonn S. Lenihan

ABSTRACT

Three species of fish have been recorded in freshwater habitats on Clare Island. These are European eel *Anguilla anguilla* (L.), brown trout *Salmo trutta* (L.) and flounder *Platichthys flesus* (L.). Only *Anguilla anguilla* occurs in more than one hydrosystem, being found widely in the island's small rivers and streams. *Salmo trutta* was introduced to a mill impoundment in the nineteenth century and still occurs in lotic habitats upstream. *Platichthys flesus* occurs in the extreme lower section of another stream and an associated saltmarsh area. The parasite assemblages of these three host species were investigated. They were found to have very low levels of parasitism and this is probably attributable to a combination of ecological and biogeographical factors. Thirteen metazoan species were recorded, belonging to the following taxonomic groups: Monogenea (two); Digenea (four); Nematoda (five) and Acanthocephala (two).

Introduction

Clare Island has a variety of freshwater habitats, including streams and small loughs, which might be expected to contain freshwater fish species and their associated parasite assemblages. The biodiversity of the island's freshwater fish fauna, however, is extremely low; this conclusion was noted in the original Clare Island survey (Farran 1912) and has since been confirmed in more recent surveys. Only three species have been recorded in freshwater habitats on the island: brown trout *Salmo trutta* (L.), European eel *Anguilla anguilla* (L.) and flounder *Platichthys flesus* (L.). The three-spined stickleback *Gasterosteus aculeatus* (L.), which occurs widely in freshwater and mixohaline habitats in Ireland, has only been observed in intertidal rock pools on Clare Island (R. Cussen, pers. comm.). The general ecology of all of these species in Britain and Ireland has been outlined in various texts (e.g. Maitland and Campbell 1992) and in species reviews on brown trout (e.g. Kennedy and Fitzmaurice 1971; O'Grady *et al.* 2008) and eels (Watson *et al.* 1999; McCarthy 2014).

The composition of Ireland's fauna—a topic discussed intensively in the decades prior to the original Clare Island survey—continues to interest professional Irish zoologists and amateur naturalists. Improved taxonomy, biological record databases, concern about the adverse effects of bioinvasive non-indigenous species and the importance of biomonitoring protocols for protection of natural ecosystems are among the many and diverse factors considered in discussions of the composition of the Irish fauna. The distribution and ecology of the fishes found in Irish freshwater and mixohaline aquatic habitats is generally well known. This is especially true of the more widespread and exploited species. Wilson (1986) and others have

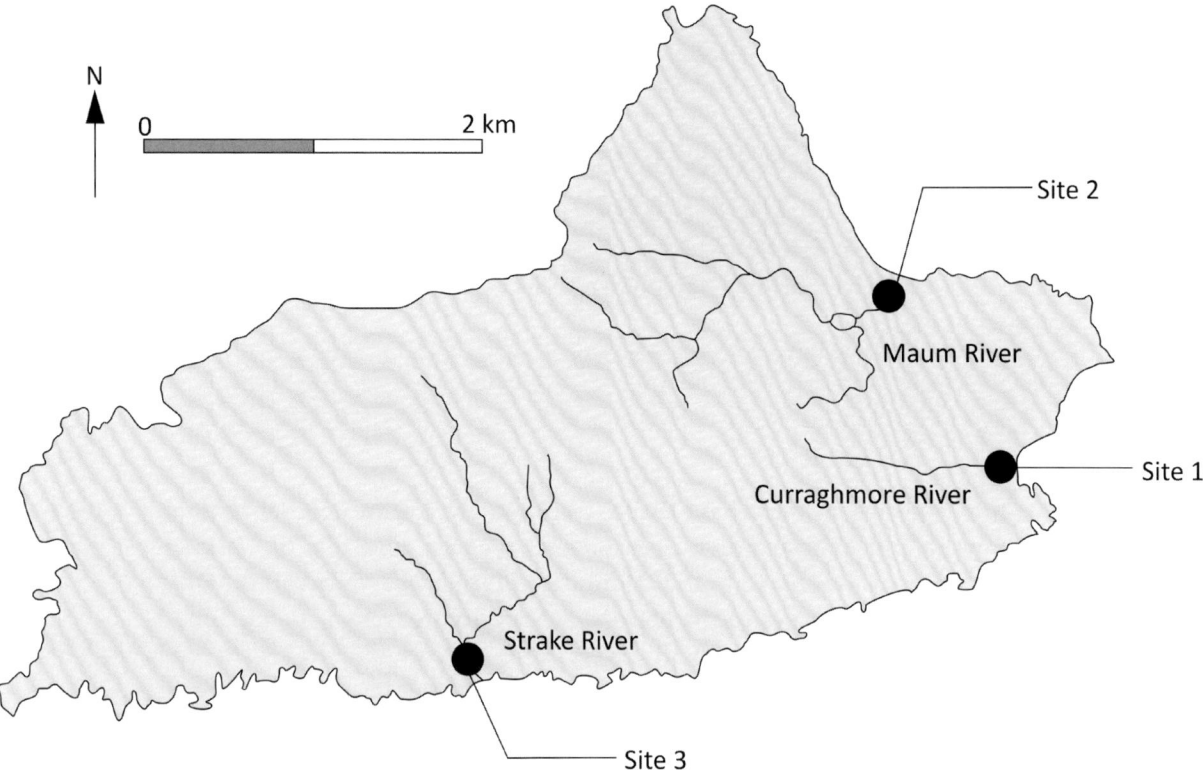

Fig. 1 Map of Clare Island showing the locations of sites samples

noted that, in comparison with mainland Britain and adjacent continental Europe, the Irish freshwater fish fauna is composed of a relatively small number of species. Though there have been some recent introductions of non-indigenous species, the nature of the fish fauna still reflects the conclusions reached by Wilson (1986). In his review, he highlighted the fact that most stenohaline species, such as Cyprinidae, were known to have been deliberately introduced. In contrast, species recognised as being indigenous were euryhaline forms and several of these are either diadromous migratory species or are derived from populations that were once migratory. Some of the species generally thought to have been introduced, such as Northern Pike (*Esox lucius* L.), are now thought to include indigenous fish, based on use of DNA and other molecular taxonomic tools (Pedreschi *et al.* 2014). However, the general ichthyofaunal overview accepted previously is still valid and studies on the fish fauna of offshore islands can contribute to our understanding of the dispersal mechanisms and routes used by freshwater fish.

Advances in ecological parasitology, including seminal contributions by Dogiel *et al.* (1961), Kennedy (1975), Price (1980), Esch *et al.* (1990) and Poulin (1998) have helped fish parasite researchers broaden their horizons in recent decades. Earlier fish parasitologists had focused on species descriptions, provision of checklists for particular hosts or regions, and elucidation of life cycle details. The ecology of fish parasite communities now includes consideration of the species assemblages at the infracommunity (individual host), component community (host population) and entire ecosystem levels (Esch *et al.* 1990) as well as the adoption of a terminology appropriate to the integration of parasitology into a modern ecological framework (Margolis *et al.* 1982; Bush *et al.* 1997).

The present study involved sampling freshwater habitats on Clare Island to determine what fish species were present and to subsequently examine fish samples for metazoan parasites. In this paper we summarise these analyses and discuss the factors that may have determined the composition of the island's fish and their parasite species assemblages.

Methods

Fish samples were obtained from three sites (site 1: unnamed brackish stream, Curraghmore River; site 2: Maum River; site 3: Strake River) on Clare Island (Fig. 1), mostly by electrofishing and use of gill nets on the stream crossing the beach at site 1.

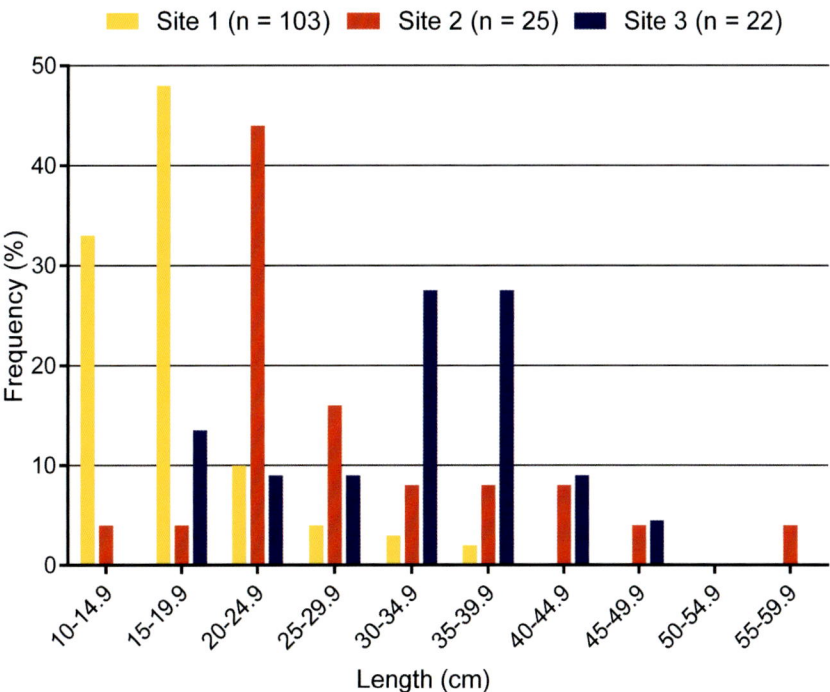

Fig. 2 Length frequency distributions of European eel samples from three sites on Clare Island

In addition, Lough Creggan was sampled using variable mesh gill nets set overnight, but no fish were captured at that site. A total of 210 fish were sampled (30 brown trout at site 2; 150 European eel at sites 1, 2 and 3; 30 flounder at site 1) and examined for metazoan parasites.

The methodology employed in external and internal examination of fishes for parasitology was adopted in earlier studies (e.g. Conneely and McCarthy 1986) and used in a series of subsequent studies on the parasites of Irish freshwater fish populations. This involved dissection of fish and examination of different fish body parts under a binocular microscope. Full counts were undertaken for all species (or operational taxonomic units). Infection levels for each parasite per host are presented as:

Prevalence = (number of infected hosts x 100) / Total number of host fish examined

Mean intensity = Total number of individuals of the parasite / Number of infected hosts

Results
Fish
The brown trout ($n = 30$) examined for parasites were all obtained at site 2. They comprised 21 male and 9 female fish. The trout ranged in size (head to tail fork length) from 9.6cm to 19.8cm (mean ± S.E. = 13.99 ± 0.4cm). Body weights ranged from 9g to 94g (mean ± S.E. = 34.27 ± 3.54g). There were no significant differences between the males and females in respect of either size or weight. Most (93.3%) were assigned to the 0+ year class, with a single fish being 1+ in age. Fulton condition factor indices (K) and the general appearance of the fish prior to and during dissection indicated they all appeared to be healthy fish.

European eel samples were obtained from all three fish sampling sites (site 1, $n = 103$; site 2, $n = 25$; site 3, $n = 22$). The eel population structures varied between sites as is illustrated by length frequencies in Figure 2. Eels sampled at site 1 ranged in length from 7.5cm to 32.7cm (mean ± S.E. = 12.8 ± 0.5cm) and ranged in weight from 0.4g to 64g (mean ± S.E. = 4.8 ± 1.0g). Eels sampled at site 2 ranged in length from 10cm to 52.9cm (mean ± S.E. = 23.6 ± 2.0 cm) and ranged in weight from 1.1g to 290g (mean ± S.E. = 36.7 ± 12.2g). The eels sampled at site 3 ranged in length from 12.5cm to 43.3cm (mean ± S.E. = 27.7 ± 1.7cm) and ranged in weight from 2g to 162g (mean ± S.E. = 43.4 ± 8.6g). All the 150 eels examined appeared to be healthy fish and were in the yellow-phase foraging stage of their life cycle. A subsample of 38 was used for age determination and these were found to vary from five to fifteen years of age, with a mean age of 8.45 years.

Flounder ($n = 30$) were obtained at site 1. They ranged in length from 5.1cm to 12.4cm (mean ± S.E. = 7.8 ± 0.3cm) and in weight from 2g to 22g (mean ± S.E. = 6.9 ± 0.9g). All the flounder captured were 1+ in age and were in good condition.

Parasites

Two species of Monogenetic flukes (Table 1) were found attached to the gill lamellae of eels. *Pseudodacylogyrus anguillae* (Yin and Sproston 1948) was found on a single eel at site 3, and *Pseudodactlyogyrus bini* (Kikuchi 1929) was found on eels at sites 1, 2 and 3. Details of parasitic prevalence, maximum burden and mean intensity of infection for each species of parasite are outlined in Table 1.

Four species of Digenetic flukes (Table 1) were identified from the fish examined: *Crepidostomum metoecus* (Müller, 1784) was recorded in four trout at site 2 and the total number of specimens was twenty. The species was mostly (95%) located in the pyloric caecae and foregut area. Although *Crepidostomum metoecus* (Müller, 1784) is the second most abundant trout parasite, it comprised only 5% of the combined parasites recorded in the host sample. *Crepidostomum farionis* (Braun, 1900) was recorded from two trout at site 2 and the total number of specimens was three. It was found in both the foregut and midgut. *Diplostomum spathaceum* (Rudolphi, 1819) was recorded in the eye lens of a single eel at site 2. *Podocotyle atomon* (Rudolphi, 1802) was recorded in nine flounder at site 1; this intestinal parasite was mostly (63.3%) found in the foregut and it was the dominant species (76.9%) in the combined sample of flounder.

Five species of parasitic Nematoda (Table 1) were identified in the Clare Island fishes examined. *Cystidicola farionis* (Fisher, 1778) was found in the swim bladder of a trout sampled at site 2. *Raphidascaris acus* (Bloch, 1779) was recorded in eel ($n = 1$) at site 1 and trout ($n = 14$) at site 2. In the trout sample, 69% of the infection occurred in the midgut region. *Raphidascaris acus* (Bloch, 1779) was the dominant species in the trout component population, comprising 93.3% of the trout parasites recorded. *Paraquimperia tenerrima* (Linstow, 1878) was recorded in eels at two sites: at site 1 a single eel was infected with three worms; at site 3 a single eel was infected with eight worms. *Anisakis simplex* (Rudolphi, 1809) was recorded in a single founder at site 1. The specimens were larval worms, encysted in the body cavity lining of the intestine. *Cucullanus heterochrous* (Rudolphi, 1802) was found in flounder ($n = 4$) from site 1. This species was found mostly (83.3%) in the foregut.

Two species of Acanthocephala were recorded from Clare Island (Table 1). *Acanthocephalus clavula* (Dujardin, 1845) was recorded in eels ($n = 5$) at site 2. It was the most frequently recorded species in eel on the island. The species was also recorded in the midgut of trout ($n = 2$) from site 2. *Pomphorhynchus laevis* (Müller, 1776) was only recorded in flounder at site 1, and only a single specimen was found.

Species richness or parasite abundances are often correlated with host size and therefore this trend was considered in respect of the three fish species examined. As detailed above, the three host samples differed in length and weight, with eels being larger than trout and these in turn being larger than the flounder. Therefore, at a species or component community level the species richness (number of species) did not reflect the differences in species sizes (Table 1). Likewise, in respect of the eels from the three sites (Fig. 2), no site vs eel size related pattern in species richness was noted. Significant correlations, however, were noted for some more abundant parasites in either trout or flounder. The infection level of *Raphidascaris acus* was significantly correlated (Spearman's Rank Correlation, r_S) with trout weight ($r_S = 0.3832$, $p < 0.05$). Similarly, *Podocotyle atomon* levels were significantly correlated with host individual's length ($r_S = 0.4365$, $p < 0.05$) and weight ($r_S = 0.4408$, $p < 0.05$). *Cuculanus heterochrous* infections in flounder were also correlated with fish length ($r_S = 0.4410$, $p < 0.05$) and weight ($r_S = 0.4189$, $p < 0.05$). The low levels of species richness in brown trout (five species) and flounder (four species) were confirmed by the species accumulation curves constructed (based on five randomisations of the data) for the parasites in the samples of these hosts (Fig. 3).

Discussion

It is generally recognised that the Irish ichthyofauna is characterised by low species richness and that indigenous species are either diadromous migratory forms or have become landlocked

Table 1

Metazoan parasites recorded from Clare Island and their prevalence, maximum burden and mean intensity of infection in eel, trout and flounder

Parasite species	Authority name	Fish species	Site(s) present	Prevalence %			Maximum burden			Mean intensity of infection		
				Eel	Trout	Flounder	Eel	Trout	Flounder	Eel	Trout	Flounder
Monogenea												
Pseudodactylogyrus anguillae	Yin & Sporston, 1948	Eel	3	0.67	-	-	6	-	-	6	-	-
Pseudodactylogyrus bini	Kikuchi, 1929	Eel	1,2,3	2	-	-	17	-	-	3	-	-
Digenea												
Crepidostomum metoecus	Müller, 1784	Trout	2	-	13.3	-	-	1-10	-	-	5	-
Crepidostomum farionis	Braun, 1900	Trout	2	-	6.7	-	-	1-2	-	-	1.5	-
Diplostomum spathaceum	Rudolphi, 1819	Eel	2	0.67	-	-	5	-	-	5	-	-
Podocotyle atomon	Rudolphi, 1802	Flounder	1	-	-	30	-	-	1-15	-	-	3.3
Nematoda												
Cystidicola farionis	Fisher, 1778	Trout	2	-	3.3	-	-	1	-	-	1	-
Raphidascaris acus	Bloch, 1779	Eel, trout	1,2	0.67	46.7	-	1	1-109	-	1	26.2	-
Paraquimperia tenerrima	Linstow, 1878	Eel	1,3	1.3	-	-	3-8	-	-	5.5	-	-
Anisakis simplex	Rudolphi, 1809	Flounder	1	-	-	3.3	-	-	2	-	-	2
Cucullanus heterochrous	Rudolphi, 1802	Flounder	1	-	-	6.7	-	-	1-2	-	-	1.5
Acanthocephala												
Acanthocephalus clavula	Dujardin, 1845	Eel, trout	2	3.3	-	-	1-54	-	-	20	-	-
Pomphorhynchus laevis	Müller, 1776	Flounder	1	-	-	3.3	-	-	1	-	-	1

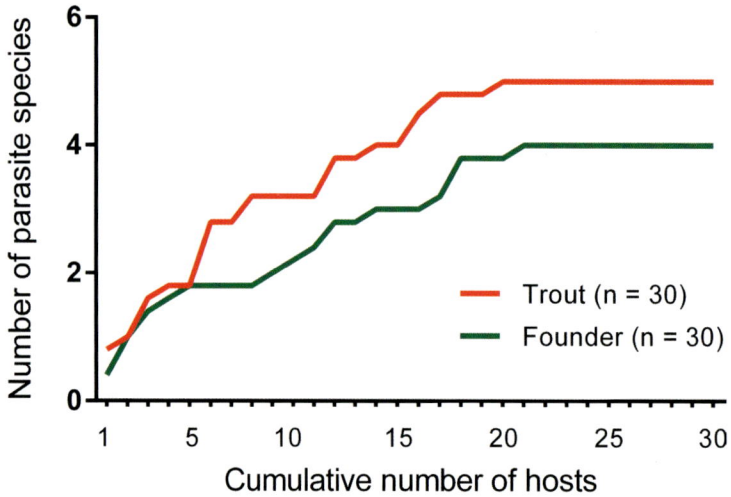

Fig. 3 Cumulative species richness curves for the metazoan parasites recorded from trout and flounder in Clare Island

in freshwater during postglacial time. Wilson (1986) listed 25 freshwater species for Ireland, whereas Maitland and Campbell (1992) published a checklist of 55 species for the British Isles. The freshwater habitats of Clare Island are inhabited by only three species of fish and some sites contain only one species (European eel). The colonisation of Clare Island's freshwater habitats by the catadromous, marine-spawning European eel and flounder has not been influenced in anyway by human activity. In contrast, the freshwater resident brown trout is believed to have been introduced to the island during the nineteenth century. This was apparently done by a landlord who hoped to develop an angling facility in Lough Avullin, a small, impounded waterbody upstream from a small water mill on the lower section of the River Doree. Although the lough, which is now a wetland area, is no longer inhabitable by trout, the population persists in an adjacent river and further upstream. There is no record of the stocking or of the source population on the mainland. In view of the progress made in recent years on the phylogeography of brown trout in Ireland and Britain (e.g. McKeown *et al.* 2010), a molecular genetics study of the Clare Island brown trout might provide an interesting new perspective on the aquatic ecology of the island's freshwater fishes. Similarly, further studies of the three-spined stickleback could reveal useful new information, especially with regard to its ecology and parasites (e.g. Conneely and McCarthy 1986).

European eel, the most widespread fish species in Clare Island, was recorded at all three sites sampled in the present study (Fig. 1; Table 1). It is a panmictic, long-distance migratory species, which annually recruits to its marine, mixohaline and freshwater habitats in Europe and North Africa. In Ireland, it occurs in a wide range of habitats as reflected in otolith microchemical structure and stable isotope ratios (Arai *et al.* 2006; Harrod *et al.* 2005). Some Irish eels move between salinity zones during the foraging (yellow eel) phase of their life cycles. European eel is well researched in Ireland and currently much effort is being devoted to establishing the contribution made by eel populations to the global spawning stock in different river basins such as the River Bann (Rosell *et al.* 2005), River Shannon, (MacNamara and McCarthy 2014), River Erne (McCarthy *et al.* 2014), River

European eel. Photo: © David Pérez (DPC). This item is distributed under the terms of the CC BY 3.0 Unported licence

Foyle (Barry *et al.* 2016) and the Burrishoole river system (Poole *et al.* 2018). Due to the limited eel habitat, the Clare Island populations are not major contributors to eel spawner escapement biomass from Ireland. However, they are unusual in that they occur in fish assemblages characterised by extremely low species richness and in habitats that in other ways reflect the island's biogeography. This study focused mainly on the parasites of Clare Island eels. However, though differences in population structure (Fig. 2) and other aspects of the population ecology were noted, many questions remain unanswered. For example, little is known about the sex ratios, the rate of silver eel production and the habitats of larger pre-silver stage eels. The extent to which the trophic ecology of European eel varies in the absence of the competitors with which it co-occurs on the mainland is still unknown. Likewise, the extent to which the island eel populations benefit from exceedingly low parasite burdens is unclear.

McCarthy *et al.* (2019) presented the most recent taxonomic checklist of the known parasites of European eel in Ireland, which included 44 operational taxonomic units that were mostly species (38). The species richness of Irish metazoan parasite assemblages of European eel is comparable to those investigated in Denmark (Køie 1988a, b), Britain (Kennedy 1990) and other parts of the Europe eel range. In addition, Irish eels are also known to be infected by protozoan, bacterial and viral pathogens (Zintl *et al.* 1997; McCarthy *et al.* 2009, McConville *et al.* 2018). Freshwater fish leeches, such as *Piscicola geometra* L, are vectors of the protozoan *Trypanosoma* in eels in Co. Mayo (Zintl *et al.* 1997). However, as there are no freshwater fish leeches in Clare Island, it can be assumed that the island's European eel populations are not affected by this pathogen. The eel-specific parasite *Anguillicola crassus* (Kuwahar, Nimi and Hagaki, 1974) has not been recorded in Clare Island, though it is now known to occur nearby in Mayo and Galway. In view of its relentless range extension in Ireland, including to river systems not exploited commercially, it can be anticipated that it could eventually colonise Clare Island eel populations and that this may involve natural dispersal mechanisms (Morrissey and McCarthy 2008).

The catadromous flounder is found in mixohaline and coastal waters from the White Sea to the Black Sea. It is the only flatfish found in European freshwater habitats. A general account of the biology of flounder in British and Irish waters is given by Maitland and Campbell (1992). It is common in Irish estuaries, lagoons and coastal areas. Although it always returns to sea to spawn, flounder has also been recorded inland in major river systems, though rarely in great numbers. It has a restricted distribution on Clare Island, and was only sampled at site 1 (Fig. 1), where the fish were all relatively small. Flounder parasites have not been extensively studied in Ireland. Studies on flounder from the River Corrib ($n = 30$) and River Moy ($n = 30$) recorded only six species (Creed and McCarthy, unpublished data). The flounder examined on Clare Island ($n = 30$) were infected by four metazoan parasite species (Table 1) and these were forms recorded in flounder in Ireland and elsewhere except for the nematode *Raphidascaris acus*, which was recorded in Clare Island trout at site 2 (Table 1) as well as in several Irish river systems, and frequently in salmonid fish. Conneely and McCarthy (1984, 1986) recorded *Raphidascaris acus* in brown trout and Arctic char *Salvelinus alpinus* (L.) from Lough Corrib and its tributaries. Doherty and McCarthy (2001) recorded it in Arctic char in Lough Mask, Co. Mayo and in Lough Eske in Co. Donegal. Other Irish records (unpublished data) are from stoneloach, as larval forms, and piscivorous pike (*Esox lucius*), as adults. *Podocotyle atomon* is a widely distributed marine digenetic fluke species that has been found in at least 25 fish hosts in British coastal waters and rockpools (Dawes 1947). In Ireland it has been reported in flounder in the River Corrib and Moy (unpublished data) and in Rusheen Bay near Galway (O'Farrell 2004). *Anisakis simplex,* which is an adult parasite of cetaceans and pinnipeds, occurs as encysted larvae in numerous marine fish intermediate hosts. Its presence in Clare Island flounder is therefore not unexpected. The life cycle and ecology of this parasite is described by Grabda (1991). The record of a single specimen of *Pomphorhynchus laevis* in Clare Island flounder may have been of marine origin, as this acanthocephalan, which is widespread in the lakes and rivers of Ireland, does regularly occur in European marine and estuarine flounder. It was found in flounder in the River Corrib and the River Moy (unpublished data). The low parasite species richness and parasite burdens of Clare Island flounder, where most (57%) of the fish had no parasites, may be attributable to the insularity of the population sampled. In contrast, 73% and 77% of River Corrib and River Moy flounder were found to be infected by at least one species of parasite (unpublished data). The absence in Clare Island of *Acanthocephalus lucii* (Müller, 1776), which uses the freshwater isopod *Asellus aquaticus* (L.) as intermediate host and which infects River Moy flounder, is due to the fact that the only freshwater isopod on the island is *Proasellus meridianus*. This isopod is used by *Acanthocephalus clavula*, which was recorded as the dominant species in brown trout and was also found in the European eel (Table 1).

Several important studies of flounder parasites that have been undertaken in coastal waters in continental Europe (Køie 1999; Chibani *et al.* 2005) and in British coastal areas (Gibson 1972; Kennedy 1984; Munro *et al.* 1989) illustrate the diversity of marine parasites associated with this host species. The potential use of flounder in biomonitoring has also been recognised. However, there have been few studies of the parasitology of this species in Ireland; O'Farrell (2004) surveyed flounder in the shallow Rusheen Bay near Galway city and recorded nine species. This suggests that mixohaline lagoon habitats may be biodiversity hot spots for flounder parasites. No such habitat occurs on Clare Island, and this was reflected in the low parasite species richness. Further research in Irish coastal waters on the ecology of the parasites of this species is required for better interpretation of the Clare Island parasitofauna.

In Clare Island, brown trout ($n = 30$) were found to be infected with only five species of parasites (Table 1, Fig. 3). In contrast, Holland and Kennedy (1997) listed 34 species for this host fish in Ireland, and that list does not include some overlooked freshwater records and marine parasites in anadromous (sea trout) populations. Conneely and McCarthy (1988), in the first major parasitological study of this host species in Ireland, examined brown trout from three sites in the River Corrib catchment area. They recorded sixteen parasite species in the River Drimneen trout ($n = 62$), fifteen species in the River Abbert trout ($n = 139$) and 24 species in Lough Corrib trout ($n = 133$). In total 25 species were recorded in that ecologically diverse river system. The parasite species richness was partly attributed to the fact that brown trout was an indigenous fish species. Many of the parasites recorded from these fish were not host specialist forms and the association of trout in this system with relatively biodiverse fish assemblages also contributed to the parasite richness of brown trout parasite communities. Molloy *et al.* (1993) researched the metazoan parasites of 171 *Salmo trutta*, including resident (brown trout) and anadromous (sea trout) forms from three locations in Galway, Mayo and Sligo. They recorded only twelve parasite species; however, some were only identified to generic level, and the parasite biodiversity may have been underestimated. Surveys in 1995–97 of the parasites of brown trout ($n = 270$) in four western Irish lakes (Loughs Corrib, Mask, Carra and Conn) resulted in records of 23 species (unpublished data). The sample size of the Clare Island brown trout examined for parasites appeared to be sufficient to assess the species richness of the parasite assemblages, as indicated by the species accumulation curve (Fig. 3). Thus, relative to the parasite species richness reported for this host from other western Irish sites, the depauperate nature of the Clare Island trout parasite fauna was clearly established in the present study. The absence of major taxonomic groups such as Monogenea, Cestoda and Crustacea from the limited Clare Island brown trout parasite checklist is significant. Likewise, the fact that *Raphidascaris acus* was the dominant parasite species is unusual. It was also unusual that only one species (*Pomphorhynchus laevis*) of Acanthocephala occurred in the brown trout examined, compared with mainland western Irish populations of this host where three or four Acanthocephala species are regularly found. A review by Thorstad *et al.* (2014) concluded that there was evidence of 'a general and pervasive negative effect of salmon lice on sea trout populations' in areas of Ireland, Norway and Scotland where Atlantic salmon are intensively farmed. The fact that Clare Island brown trout are resident, non-migratory forms has meant that they were not affected by the adverse impact that sea lice infestations of sea trout in the vicinity of western Irish salmon farms (Costello 2006).

In the past century the conceptual framework that underpinned our understanding of the nature of parasitism has changed considerably. The focus was initially on morphology and adaptations typical of parasites. Subsequently, parasitologists considered questions relating to the behavioural

Brown trout, *Salmo trutta*. Photo: © Eric Engbretson. This item is distributed under the terms of the CC BY 4.0 licence

attributes of parasites, such as host-seeking mechanisms, and to analyses of complex parasite life cycles. In the latter half of the twentieth century new definitions of parasitism took account of ecological phenomena, such as the overdispersed pattern of abundance in the individuals of a host population, the non-equilibrium nature of parasite communities, and the significance of parasites as determinants of host mortality rates (Whittaker and Fernandez-Pelacios 2007). Some authors (e.g. Kuris *et al.* 1980) have drawn attention to the parallels between the problems faced by parasites in infecting individual hosts and those faced by animals colonising isolated islands. The effects of parasites on host dynamics, frequently demonstrated in laboratory experiments, can often be seen after deliberate or accidental introductions of parasites (Lafferty *et al.* 2005) and island populations are increasingly threatened by invasive organisms of all sorts. Kennedy (1993) listed examples of introduced fish parasites to Britain. Examples of severe impacts of parasite introductions have been reported by many authors, including Elton (1958) who discussed the adverse effects of an introduced monogetic fluke *Nitzscia sturionis* on sturgeon in the Sea of Aral. In more recent times, the severely adverse effects of the introduced *Gyrodactylus salar* Malmberg on Western Norwegian Atlantic salmon (Bristow and Berland 1991) populations or the introduced nematode *Anguillicola crassus* on European eel (McCarthy *et al.* 2009; Morrissey and McCarthy 2007). However, at least in the case of freshwater fish parasites, few island biogeographical studies have been undertaken. In a scenario redolent of the pattern seen on Clare Island, Kennedy *et al.* (1986) found that the diversity of parasites present within fish species on Jersey Island was always low, often with one species dominating the assemblage. The three Clare Island host populations were also characterised by low species richness and the dominance of a single parasite species in each host species.

Clare Island fish parasitology is of special interest, though ultimately the significance of some results will only be confirmed if comparable studies are undertaken on other Irish offshore islands. Likewise, the parasitology of other host groups, such as small mammals or invertebrates, on Clare Island could provide opportunities for a more comprehensive understanding of the island's biogeographical history. Kennedy *et al.* (1986) drew attention to the differences between the parasite assemblages of fish and bird populations. The importance of allogenic parasites, which use birds or mammals as definitive hosts, has sometimes been linked to the trophic status of lakes. Similarly, it might be thought that autogenic parasites, whose life cycles are completed within a waterbody, might be less likely to colonise isolated island aquatic habitats. However, as can be seen from the checklist in Table 1, most (11) of the 13 parasite species recorded were autogenic forms. The allogenic species were *Diplostomum spathaceum* larvae from eel, with piscivorous birds as definitive hosts, and *Anisakis simplex* larvae from flounder, which uses marine mammal as definitive hosts.

Historical, biogeographical, environmental and ecological factors have been found to be the major determinants of parasite species richness (Poulin 1998). This can be seen in the results of the present study, though many new questions can now be posed in respect of Clare Island's biota. The conservation of Clare Island's aquatic resources can benefit from increased knowledge of the faunal composition, ecosystem structures, natural species turnover and the extent to which its biota may be regarded as an equilibrium species assemblage. Island biogeography is widely recognised as an important tool in the development of appropriate conservation strategies for unique insular communities (Whittaker and Fernandez-Pelacios 2007) and hopefully the present study may partly inform future environmental management on Clare Island.

REFERENCES

Arai, T., Kotake, A. and McCarthy, T.K. 2006 Habitat use by the European eel *Anguilla anguilla* in Irish waters. *Estuarine, Coastal and Shelf Science* **67**(4), 569–78.

Barry, J. Newton, M., Dodd, J.A., Lucas, M.C., Boylan, P. and Adams, C. 2016 Freshwater and coastal migration patterns in the silver-stage eel *Anguilla Anguilla*. *Journal of Fish Biology* **88**, 676–89.

Bristow, G.A. and Berland, B. 1991 A report on some metazoan parasites of wild marine salmon (*Salmo salar* L.) from the west coast of Norway with comments on their interactions with farmed salmon. *Aquaculture* **98**(1–3), 311–18.

Bush, A.O., Lafferty, K.D., Lotz, J.M. and Shostak, A.W. 1997 Parasitology meets ecology on its own terms: Margolis *et al*. revisited. *The Journal of Parasitology* **83**(4), 575–83.

Chibani M.O., Kijewska, A.G. and Rokicki, J.E. 2005 Sex and age of flounder *Platichthys flesus* (L.) and parasitic infection in the Gulf of Gdańsk. *Oceanological and Hydrobiological Studies* **34**(3), 85–96.

Conneely, J.J. and McCarthy, T.K. 1984 The metazoan parasites of freshwater fishes in the Corrib catchment area, Ireland. *Journal of Fish Biology* **24**(4), 363–75.

Conneely, J.J. and McCarthy, T.K. 1986 Ecological factors influencing the composition of the parasite fauna of the European eel, *Anguilla anguilla* (L.) in Ireland. *Journal of Fish Biology* **28**, 207–19.

Conneely, J.J. and McCarthy, T.K. 1988 The metazoan parasites of trout (*Salmo trutta* L.) in western Ireland. *Polskie Archiwum Hydrobiologii* **35**, 443–60.

Costello, M.J. 2006 Ecology of sea lice parasitic on farmed and wild fish. Trends in Parasitology **22**, 475–83.

Dawes, B. 1947. *The Trematoda of British fishes*. London. B. Quaritch.

Dogiel, V.A., Petrushevski, G.K. and Polyanski, Y.I. 1961 *Parasitology of fishes*. 384 pp. Edinburgh. Oliver and Boyd.

Doherty, D. and McCarthy, T.K. 2001 The metazoan parasites, diets and general biology of arctic char, *Salvelinus alpinus* L., in two Irish Lakes. *Internationale Vereinigung fur Theoretische und Angewandte Limnologie Verhandlungen* **27**(2), 1056–61.

Elton, C.S. 1958 *The ecology of invasions by plants and animals*. London. Methuen.

Esch, G.W., Shostak, A.W., Marcogliese, D.J. and Goater, T.M. 1990 Patterns and processes in helminth parasite communities: an overview. In G.W. Esch, A.O. Bush and J.M. Aho (eds), *Parasite communities: patterns and processes*, 1–19. London. Chapman Hall.

Farran, G.P. 1912 Clare Island Survey: Pisces. *Proceedings of the Royal Irish Academy* **31**(19), 3.

Gibson, D.I. 1972 Flounder parasites as biological tags. *Journal of Fish Biology* **4**, 1–9.

Grabda, J. 1991 *Marine fish parasitology: an outline*. Warszawa. PWN, Polish Scientific Publishers.

Harrod, C., Grey, J., McCarthy, T.K. and Morrissey, M. 2005 Stable isotope analyses provide new insights into ecological plasticity in a mixohaline population of European eel. *Oecologia* **144**, 673–83.

Holland, C.V. and Kennedy, C.R. 1997 A checklist of parasitic helminth and crustacean species recorded in freshwater fish from Ireland. *Biology and Environment: Proceedings of the Royal Irish Academy*, **97B**(3), 225–43.

Kennedy, C.R. 1975 *Ecological animal parasitology*. Oxford. Blackwell.

Kennedy, C.R. 1984 The status of flounders, *Platichthys flesus* L., as hosts of the acanthocephalan *Pomphorhynchus laevis* (Müller) and its survival in marine conditions. *Journal of Fish Biology* **24**(2), 135–49.

Kennedy, C.R. 1990 Helminth communities in freshwater fish: structured communities or stochastic assemblages. In G.W. Esch, A.O. Bush and J.M. Aho (eds), *Parasite communities, patterns and processes*, 131–56. London. Chapman Hall.

Kennedy, C.R. 1993 Introductions, spread and colonization of new localities by fish helminth and crustacean parasites in the British Isles: a perspective and appraisal. *Journal of Fish Biology* **43**(2), 287–301.

Kennedy, M and Fitzmaurice, P. 1971 Growth and food of trout *Salmo trutta* (L.) in Irish waters. *Proceedings of the Royal Irish Academy* **71B**, 269–352.

Kennedy, C.R., Bush, A.O. and Aho, J.M. 1986 Patterns in helminth communities: why are birds and fish different? *Parasitology* **93**(1), 205–15.

Kennedy, C.R., Laffoley, D.D.A., Bishop, G., Jones, P. and Taylor, M. 1986 Communities of parasites of freshwater fish of Jersey, Channel Islands. *Journal of Fish Biology* **29**(2), 215–26.

Køie, M. 1988a Parasites in European eel *Anguilla anguilla* (L.) from Danish freshwater, brackish and marine localities. *Ophelia* **29**, 93–118.

Køie, M. 1988b Parasites in eels *Anguilla anguilla* (L.) from eutrophic Lake Esrum, Denmark. *Acta Parasitologica Polonica* **33**, 89–100.

Køie, M. 1999 Metazoan parasites of flounder *Platichthys flesus* (L.) along a transect from the southwestern to the northeastern Baltic Sea. *ICES Journal of Marine Science: Journal du Conseil* **56**(2), 157–63.

Kuris, A.M., Blaustein, A.R. and Alio, J.J. 1980 Hosts as islands. *The American Naturalist* **116**(4), 570–86.

Lafferty, K.D., Smith, K.F., Torchin, M.E., Dobson, A.P. and Kuris, A.M. 2005 *The role of infectious disease in natural communities: what introduced species tell us*. In Dov F. Sax, John J. Stachowicz and Steven D. Gaines (eds), *Species invasions: insights into ecology, evolution and biogeography,* 111–34. Sunderland, Mass. Sinauer Associates.

MacNamara, R. and McCarthy, T.K. 2014 Silver eel (*Anguilla anguilla*) population dynamics and production in the River Shannon, Ireland. *Ecology of Freshwater Fish* **23**(2), 181–92.

Maitland, P.S. and Campbell, R.N. 1992 *Freshwater fishes*. London. Harper Collins.

Margolis, L., Esch, G.W., Holmes, J.C., Kuris, A.M. and Schad, G. 1982 The use of ecological terms in parasitology (report of an ad hoc committee of the American Society of Parasitologists). *The Journal of Parasitology* **68**(1), 131–33.

McCarthy, T.K. 2014 Eels and people in Ireland: from mythology to international stock conservation. In K. Tsukomoto and M. Karoki (eds) *Eels and humans, humanity and the seas*, 13–40. Tokyo. Springer.

McCarthy, T.K, Creed, K., Naughton, O., Cullen, P. and Copley, L. 2009 The metazoan parasites of eels in Ireland: zoogeographical, ecological and fishery management perspectives. *American Fisheries Society Symposium* **58**, 175–87.

McCarthy, T.K., Nowak, D., Grennan, J., Bateman, A., Conneely, B. and MacNamara, R. 2014 Spawner escapement of European eel (*Anguilla anguilla*) from the River Erne, Ireland. *Ecology of Freshwater Fish* **23**(1), 21–32.

McCarthy, T.K., Morrissey, M., Creed, K., Conneely, J. and Lenihan, E.S. 2019 Parasite assemblages of European eel in Irish freshwater, mixohaline and marine habitats. In A. Don and P. Coulson (eds) *Eels biology, monitoring, management, culture and exploitation: proceedings of the first international eel science symposium*. 293–304. Sheffield, UK. 5M Publishing.

McConville, J., Fringuelli, E., Evans, D. and Savage, P. 2018 First examination of the Lough Neagh European eel (*Anguilla anguilla*) population for eel virus European, eel virus European X and Anguillid Herpesvirus-1 infection by employing novel molecular techniques. *Journal of Fish Diseases* **41**(12), 1783–91.

McKeown, N.J., Hynes, R.A., Duguid, R.A., Ferguson, A. and Prodöhl, P.A. 2010 Phylogeographic structure of brown trout *Salmo trutta* in Britain and Ireland: glacial refugia, postglacial colonization and origins of sympatric populations. *Journal of Fish Biology* **76**(2), 319–47.

Molloy, S., Holland, C. and Poole, R. 1993 Helminth parasites of brown and sea trout Salmo trutta L. from the west coast of Ireland. *Biology and Environment: Proceedings of the Royal Irish Academy* **93B**(3), 137–42

Morrissey, M. and McCarthy, T.K. 2007 The occurrence of *Anguillicola crassus* (Kuwahar, Nimi, and Hagaki, 1974), an introduced nematode, in an unexploited western Irish eel population. *Biology and Environment: Proceedings of the Royal Irish Academy*, **107B**(1), 13–18.

Morrissey, M. and McCarthy, T.K. 2008 A first record of the parasitic nematode *Daniconema anguillae* Moravec et Køie, 1987 (Spirurida, Dracunculoidea: Daniconematidae) from European Eels (*Anguilla anguilla*) in Ireland. *The Irish Naturalists' Journal*, **29**, 99–101.

Munro, M.A., Whitfield, P.J. and Diffley, R. 1989 Pomphorhynchus laevis (Müller) in the flounder, *Platichthys flesus* L., in the tidal River Thames: population structure, microhabitat utilization and reproductive status in the field and under conditions of controlled salinity. *Journal of Fish Biology* **35**(5), 719–35.

O'Farrell, L. 2004 The biology and parasitology of the flounder *Platichthys flesus* (L.) from three sites in Ireland. Unpublished final year B.Sc (Marine Science) thesis. National University of Ireland, Galway.

O'Grady, M.F., Kelly, M. and O'Reilly S. 2008 *Brown trout in Ireland*. Irish Freshwater Fisheries Ecology and Management Series No.6. Dublin. Central Fisheries Board.

Pedreschi, D., Kelly-Quinn, M., Caffrey, J., O'Grady, M. and Mariani, S. 2014 Genetic structure of pike (*Esox lucius*) reveals a complex and previously unrecognized colonization history of Ireland. *Journal of Biogeography* **41**(3), 548–60.

Poole, W.R., Diserud, O.H., Thorstad, E.B., Durif, C.M., Dolan, C., Sandlund, O.T., Bergesen, K., Rogan, G., Kelly, S. and Vøllestad, L.A. 2018 Long-term variation in numbers and biomass of silver eels being produced in two European river systems. *ICES Journal of Marine Science* **75**(5), 1627–37.

Poulin, R. 1998 Comparison of three estimators of species richness in parasite component communities. *The Journal of Parasitology* **84**(3), 485–90.

Price, P.W. 1980 *Evolutionary biology of parasites. Monographs in Population Biology*. New Jersey. Princeton University Press.

Rosell, R., Evans, D. and Allen, M. 2005 The eel fishery in Lough Neagh, Northern Ireland – An example of sustainable management. *Fisheries Management and Ecology* **12**, 377–85.

Thorstad, E.B., Todd, C.D., Bjorn, P.A., Gargan, P.G., Vollset, K.W., Halttunen, E., Kalas, S., Uglem, I., Berg, M., and Finstad, B. 2014 Effects of salmon lice on sea trout–A literature review. *NINA Report* **1044**, 1–168.

Watson, L., Moriarty, C. and Gargan, P. 1999 *Development of the Irish eel fishery: proceedings of a national workshop-Dun Laoghaire, 7 July 1998*. Dublin. Marine Institute.

Whittaker, R.J. and Fernández-Palacios, J.M. 2007 *Island biogeography: ecology, evolution, and conservation*. Oxford. Oxford University Press.

Wilson, J.P.F. 1986 The postglacial colonization of Ireland by fish, amphibians and reptiles. *Occasional Publication of the Irish Biogeographical Society* **1**, 53–58.

Zintl, A., Poole, R.R., Voorheis, H.P. and Holland, C. 1997 Naturally occurring *Trypanosoma granulosum* infections in the European eel *Anguilla anguilla* from the west coast of Ireland. *Journal of Fish Diseases* **20**, 333–41.

CHAPTER 17

THE BIRDS OF CLARE ISLAND

Thomas C. Kelly

ABSTRACT
The data relating to the birds of Clare Island, collected since the publication of the first survey (Ussher 1912), is reviewed. Trends in the numerical abundance of breeding seabirds and in the species richness of land birds are summarised. Six species of seabird have joined the Clare Island colony since 1912, and one has departed. There has been considerable 'turnover' in the species composition of land birds over the same interval. The equilibrium number of land bird species is estimated to be 24. Approximately 130 species have now been recorded on Clare Island.

Introduction
The idea of surveying the flora and fauna of Clare Island in the early twentieth century was both pioneering and prescient in the sense that the general topics of insular ecology, evolution and biogeography were to become—and continue to be—a major focus of investigation for researchers from a wide range of disciplines. A more recent surge in interest followed the publication of R.H. MacArthur and E.O. Wilson's highly original and quantitative monograph *The theory of island biogeography* (1967), which has been cited more than 13,500 times in scientific books and papers. There is now a voluminous literature (laced, it must be said, with some heated debate) on the general topic of 'island biogeography' (see especially Losos and Ricklefs 2010), which has dealt *inter alia* with the main tenets of MacArthur and Wilson's theory, including the species–area relationship, dispersal, isolation, equilibrium faunas and the phenomenon of 'turnover'.

MacArthur and Wilson's (1967) theory was based to a considerable extent on data collected from remote 'oceanic' islands, i.e. those of volcanic origin that emerged at plate margins—the so-called 'hot-spots' such as the Hawaiian chain. In this chapter, however, we are dealing with what is termed a 'continental shelf island', i.e. a land mass that was isolated by the post glacial rise in sea levels (Whittaker *et al.* 2010). In this context, Clare Island is a classical continental shelf island (also referred to as a 'fragment island' (Warren *et al.* 2015)), as are all of the islands lying off the coast of Ireland. Although these islands vary considerably in size and shape (e.g. Berry 2009), the distances between them and the mainland and, therefore, from the potential source of colonising propagules, are relatively short, and so in this sense their biogeography is not strictly comparable with those of the true 'oceanic' isles, especially in terms of endemicity and local speciation.

While the distance to a source of propagules is a variable of major importance to non-volant taxa it is obviously much less of a barrier to flying animals, including many insects, bats and birds—as noted by Ussher (1912).

Birds, the subject of this account, have been well—though haphazardly—studied on Clare Island (e.g. Kelly 2020). In the original Clare Island survey, Ussher (1912) detailed the timing and duration of visits made by various expeditions and also incorporated information sent to R.M. Barrington by lightkeepers and from J.J. McCabe, 'owner of the Granuaile Hotel', relating to observations made in the nineteenth century (see also Barrington 1900). Some 33 years elapsed before the next ornithological expedition (Barlee and Ruttledge 1945; Ruttledge 1950), and a further 37 years before the detailed survey of Dr C.S. Lloyd (1982); the latter account included unpublished data collected by the late Dr Tony Whilde in May 1982. D'Arcy (1992) presents data collected by groups who visited the island in July 1989 and June 1990. The island was surveyed as part of the research conducted from 1968 to 1972 for the first British Trust for Ornithology Breeding Atlas (Sharrock 1976), the subsequent Atlas spanning the 1988 to 1990 interval (Gibbons *et al.* 1993) and also surveyed between 1981 and 1984 for what came to be known as the 'Winter Atlas' (Lack 1986). The island's seabirds were surveyed between 1985 and 1987 as part of a major country-wide census of colonies that took place in Britain and Ireland (Lloyd *et al.*1991; see also the detailed summary of Forsyth and Allen 1997).

The most recent Clare Island survey of the Aves began with expeditions by Ian Forsyth and Pamela Allen (Forsyth and Allen 1997) and studies of the island's seabirds by R. Cussen and others, mainly in 1999; subsequent expeditions by D. Coombes, P. Winters, E. McGreal and T.C. Kelly between 1999 and 2019 provided additional information on terrestrial species (Kelly 2020). Further information on breeding and wintering birds was gathered by surveyors associated with the most recent Bird Atlas 2007–11 (Balmer *et al.* 2013).

The results of these expeditions are analysed in detail in the *New Survey of Clare Island. Volume 9: Birds* (which includes a review of the birds of Ireland's islands), but it is important to enter one caveat: it is difficult to estimate 'turnover' when surveys are not repeated year-on-year, or when the seasonal timing between successive visits leads to the generation of non-comparable data. These often-unavoidable difficulties may result in the mistaken conclusion that a species has become extinct or has recently colonised an area (e.g. Diamond and May 1977), or that undetected extinctions and colonisations may have occurred in the interval between successive surveys (e.g. the 'crypto-turnover' of Simberloff (1974) (see also Schoener 2010 for an authoritative review)). Nevertheless, an accurate estimate of the equilibrium number of species and its relationship with the area of an island is crucially dependent on reliable knowledge of the extent and detail of the turnover, i.e. the immigration and extinction of taxa.

The aim of this contribution is to present a brief review of the recent findings relating to the birds of Clare Island and to discuss these in the context of the historical data as well as from the general perspective of island ecology.

Avian habitats on Clare Island
The coastline

With an area of 1,550ha, Clare Island is larger than most of those lying off the coast of Ireland (e.g. see data in Berry 2009). Although there is an extensive coastline with the cliffs beneath Cnoc Mór rising to 463m, seabird colonies beneath several sections are too dangerous to census from the clifftop. Robert Cussen surveyed the entire accessible coastline in 1999 and a further survey of the inaccessible cliff faces was conducted from a boat with the late Commandant Michael Hartnett and T.C. Kelly in June of that year. The results showed that Clare Island has a large seabird colony and that thirteen species were confirmed to be breeding, with an estimated 8,000 pairs based on apparently occupied nests (AON). Most notable is the fulmar *Fulmarus glacialis* (Linnaeus) colony, which Ussher (1912) stated was 'not found on Clare Island cliffs but should be looked for there'. How right he was, because the island was colonised by the fulmar relatively soon after—in 1935—according to Lloyd (1984) and its numbers had risen to 2,555 in 1983, based on apparently occupied sites (AOS) data collected by the late Oscar Merne (Hutchinson 1989). However, the total number of AOS in 1999

Fulmar *Fulmarus glacialis* (Linnaeus). Photo: R.T. Mills

was estimated to be 4,029, making it one of the largest, if not the largest, fulmar colony in Ireland (see Lloyd 1984).

Subsequent surveys, up to and including May 2017, completed by Eoin McGreal and his colleagues, have shown that major declines have occurred since 1999 in the numerical abundance of the shag, fulmar, kittiwake and common gull (Cussen *et al.* 2020), with an overall loss of some 4,000 individuals or 30% of the total number of seabirds in the colony. The fulmar population decreased by approximately 38% over this interval while that of kittiwake declined by 54%, shag by 16% and common gull by 73%. On the other hand, the gannet population has grown from 3 to 352 AONs since 1999, and there have also been substantial increases in the numbers of the larger white-headed gull species.

Since the original Clare Island survey, one seabird species—the arctic tern *Sterna paradisaea* Pontoppidan—has become extinct as a breeder, as the 'small colony on the detached rocky western end of Clare Island' (Ussher 1912) no longer exists and has not been found since (Lloyd 1984). However, six species of seabird have joined the colony since 1912: the fulmar (sometime between 1911 and 1945); common gull *Larus canus* Linnaeus and gannet *Morus bassanus* Linnaeus (in 1976); great cormorant *Phalacrocorax carbo carbo* (Linnaeus) (in 1982); lesser black-backed gull *Larus fuscus* Linnaeus, which was first found to be breeding in 1969 (Lloyd 1984); and, most recently, the great skua *Stercocorarius skua* (Brünnich) (in 2015) (Eoin McGreal, pers. comm; Martin Deithrick, pers. comm).

The arrival of the great skua—a notorious predator of seabirds—may have an adverse influence on the marine bird community of Clare Island (e.g. Votier *et al.* 2004). However, notably, this species did not breed on Clare Island in 2019 (Eoin McGreal, pers. comm).

Finally, the status of two seabird species, namely the European storm petrel *Hydrobates pelagicus* (Linnaeus) and manx shearwater *Puffinus puffinus* (Brünnich) remains unresolved (Ussher 1912; Barlee and Ruttledge 1945; Ruttledge 1950; Lloyd 1984; Forsythe and Allen 1997; this study). Given that both species are nocturnal and live in cavities and/or burrows, it is conceivable that both the manx shearwater and storm petrel could be breeding in inaccessible places on Clare Island; nocturnal calling by *Puffinus puffinus* was heard by the 'lighthouse keepers' in the early 1900s and a single bird was heard in 1969 (Lloyd 1984).

It is debatable whether true seabirds should be included in species area curves or in the estimates of turnover of different islands, as most will only use these isolated habitats for breeding and not for provisioning their young. The larger white-headed gulls, especially *Larus fuscus*, *L. argentatus* and the common gull (*L. canus*), are all well known to forage in terrestrial habitats including, in some cases, the islands on which they breed, to provide food for their chicks (e.g. Crème 1995; Kelly *et al.* 2012). However, even in these cases the overwhelming bulk of their prey is likely to be taken at considerable distances from the nesting colony (e.g. Camphuysen 2013); therefore, even these species should probably be omitted from a formal island biogeographical analysis.

The coastline on Clare Island is mostly composed of rocky shores, and consequently soft estuarine mud suitable for wading birds (Charadriidae and Scolapcidae) is almost completely absent. There are some boulder beaches that can be exploited by Eurasian whimbrel *Numenius pheopus* (Linnaeus) and the short section of sandy beach receives occasional visits of ringed plover *Charadrius hiaticula* Linnaeus, redshank *Tringa tetanus* (Linnaeus) and Eurasian curlew *Numenius arquata* (Linnaeus). Most sightings of wading birds, especially during the breeding season, are from the margins of small wetlands and areas of abandoned cut away blanket bog.

The terrestrial (including wetlands) on Clare Island

The vital importance of habitat for the survival and reproductive success of birds is well documented (Nairn and O'Halloran 2012; Fuller 2012). Habitat structure (O'Halloran and Kelly *et al.* 2012; Fuller 2012) is one of several key variables explaining the distribution and abundance of different bird species and communities. In island ecology studies, 'area' is generally seen as a good predictor of the number of different habitats; 'increasing area' is 'a general surrogate for island carrying capacity K' (Whittaker *et al.* 2010) though others have questioned the relationship (e.g. Simberloff 1976). However, as recent trends have shown, human impacts such as the intensification in agriculture (e.g. Wilson *et al.* 2009) and the increasing urbanisation of large areas of previously undeveloped natural habitat—including those associated with recreation and tourism—are also causing major changes in avian communities (e.g. Kelcey and Rheinwald 2005; Ní Lamhna *et al.* 2012). In addition, climate change is not only impacting phenology but also the numerical abundance and the geographical distribution of species, though in the latter case the effects, while profound, are emerging rather slowly (Moller *et al.* 2006; Balmer *et al.* 2013). Finally, changes in the type of crops being sown and harvested—as occurred in Ireland during the nineteenth century—may also have had a major impact on bird communities including those of Clare Island (Hutchinson 1989).

It is immediately obvious to the visiting ornithologist that despite its relatively large size Clare Island has a limited variety of habitats, especially wetlands; areas of open water are few, and all of these are very small (<2ha) (see especially Vullings 2013 and Ryle 2013). The transformation of Lough Avullin from an open body of water in the time of the first survey (Praeger 1911; Ussher 1912) into a large reed bed at Maum—known as a 'lough' by Barlee and Ruttledge (1945)—is perhaps the most notable change in the availability of wetland habitats (Vullings 2013; Ryle 2013; Coxon 2001; Coxon *et al.* 2013; see also the intriguing history of this habitat and Lassau Wood in Feehan 2019). Nevertheless, Ussher's (1912) paper does not indicate a rich water bird diversity during the first survey other than mallard *Anas platyrhynchos* Linnaeus, moorhen *Gallinula chloropus* (Linnaeus), common snipe *Gallinago gallinago* (Linnaeus), water rail *Rallus aquaticus* Linnaeus, lapwing *Vanellus vanellus* (Linnaeus) and the common sandpiper *Actitis hypoleucos* (Linnaeus). Mallard and common snipe continue to breed on Clare Island and there have been occasional sightings of moorhen; the water rail has not been detected in recent times, but this species is notoriously difficult to see and a targeted survey of the Lough Avullin reed bed will be necessary to exclude the possibility of its still being absent (see also Lloyd 1984). These data may be compared with the water bird diversity on the Inishkea Islands (Cabot 1963) where mallard, teal, red breasted merganser *Mergus serrator* Linnaeus, moorhen and coot *Fulica atra* Linnaeus as well as dunlin *Calidris alpina* Linnaeus, ringed plover *Charadrius hiaticula* Linnaeus, lapwing *Vanellus vanellus* (Linnaeus) and common snipe were all found to be breeding in 1961, and moorhen, teal and eider duck

Lapwing *Vanellus vanellus* (Linnaeus). Photo: R.T. Mills

Somateria mollissima (Linnaeus) were found on Inishmurray Co. Sligo, also in 1961 (Cabot 1962). While drainage of small wetlands is still underway on Clare Island, the reed bunting *Emberiza schoeniculus* (Linnaeus) continues to breed and, remarkably, the sedge warbler *Acrocephalus schoenobaenus* (Linnaeus) is still to be found, as it was in June1910 and July 1911, 'in marshy ground near Clare Island harbour' (Ussher 1912).

Structured vegetation (i.e. with a distinct tree, shrub and ground layer) on Clare Island is scarce and localised, and confined to gardens, mainly around the harbour area, Ballytoohy More, where a planted wood has attracted a small colony of breeding lesser redpolls. This species was first discovered on the island by D'Arcy (1992), and to be breeding (in 1999) by Robert Cussen, and still present in 2013 (Kelly *et al.* 2020), in Ballytoohey Beg. Although the remarkable Lassau Wood has been on the island since perhaps 1850 (see Feehan 2019 for an authoritative account of its probable history), it remains to be thoroughly surveyed for breeding and migratory birds. For the most part however the modern-day island is heavily grazed and so a compact grass sward alternating with the occasional patch of sedge (*Juncus* sp.), bracken (*Pteridium equilinum*) and heather (*Calluna vulgaris*) are the only vegetation layers. However, this habitat supports large numbers of breeding meadow pipits *Anthus pratensis* (Linnaeus); and—where there are stone walls—breeding wheatears *Oenanthe oenanthe* Linnaeus; and, in much lower abundance, skylarks *Alauda arvensis* Linnaeus. Hooded crows *Corvus cornix* Linnaeus, choughs *Pyrrhocorax pyrrhocorax* (Linnaeus) and ravens *Corvus corax* Linnaeus also frequent this habitat.

The gardens of the island support blackbirds *Turdus merula,* Linnaeus robin *Erithacus rubecula*

Meadow pipit *Anthus pratensis* (Linnaeus). Photo: R.T. Mills

Willow warbler *Phylloscopus trochilus* (Linnaeus). Photo: R.T. Mills

(Linnaeus), song thrush *T. philomelos* C.L Brehm and other species typical of this habitat on the mainland.

'Turnover' of species on Clare Island 1900 to 2013
Approximately 59 bird species have been recorded as breeding on Clare Island at one time or another (e.g. Kelly *et al.* 2020; Winters 2006). However, among the land birds there have been some definite losses including the dipper *Cinclus cinclus hibernicus* Hartert, corn bunting *Emberiza calandra* (Linnaeus), whitethroat *Sylvia communis* Latham (Barlee and Ruttledge 1945), yellowhammer *Emberiza citronella* Linnaeus (Lloyd 1984), twite *Carduelis flavirostris* Linnaeus and sparrow hawk *Accipiter nisus* Linnaeus based on more recent surveys (Eoin McGreal and others unpublished data; see Kelly *et al.* 2020). New arrivals include tree sparrow *Passer montanus* Linnaeus, chaffinch *Fringilla coelebs* (Linnaeus), willow warbler *Phylloscopus trochilus* (Linnaeus) and lesser redpoll *Carduelis flammea* (P.L.S. Müller). The equilibrium number of land bird species is estimated to be at 24 (see Kelly *et al.* 2020).

Migrant species
During spring and autumn, field ornithologists regularly visit offshore islands such as Inishmore (Aran Islands), Inishboffin (Co. Mayo), the Saltee Islands (Co. Wexford), Tory Island (Co. Donegal) as well as the bird observatories on Cape Clear (Co. Cork) and the Copeland Islands (Co. Down). Clare Island, however, has not been a focus of such surveys despite the pioneering investigation of migration at the lighthouse conducted by Barrington in 1900. Consequently, the systematic list of species recorded on the island is rather limited. Some scarce and relatively rare species have been recorded there in recent years including the lapland longspur *Calcarius lapponicus* (Linnaeus), European pied flycatcher *Ficedula hypoleuca* (Pallas) and the yellow browed warbler *Phylloscopus inornatus* (Blyth) (see www.irishbirding.com). The low number of non-breeding bird species documented on Clare Island to date may prove an incentive to intrepid ornithologists in the future who will find it relatively easy to add new taxa to the existing list.

Introduced species
The pheasant *Phasanius colchicus* Linnaeus and Canada goose *Branta canadensis* (Linnaeus) have been introduced relatively recently to Clare Island.

REFERENCES

Balmer, D., Gillings, S., Caffrey, B., Swann, B., Downie, I. and Fuller, R. 2013 *Bird atlas 2007–2011. The breeding and wintering birds of Britain and Ireland*. Thetford. BTO Books.

Barlee, J., and Rutledge, R.F. 1945 Notes on the present status of birds on Clare Island. *The Irish Naturalists' Journal*, 311–13.

Barrington, R.M. 1900 *Migration of birds as observed at Irish lighthouses and lightships*. London. R.H. Porter; Dublin. Edward Ponsonby.

Berry, R.J. 2009 *The natural history of islands*. New Naturalist Library. London. Collins.

Cabot, D.B. 1962 An ornithological expedition to Inishmurray, Co. Sligo. *The Irish Naturalists' Journal* **14**(3), 59–61.

Cabot, D. 1963 The breeding birds of the Inishkea Islands, Co. Mayo. *The Irish Naturalists' Journal* **14**(6), 113–15.

Camphuysen, C.J. 2013 *A historical ecology of two closely related gull species (Laridae): Multiple adaptations to a man-made environment*. Ph.D. thesis, University of Groningen, Groningen.

Coxon, P. 2001 The quaternary history of Clare Island. In J.R. Graham (ed.), *New Survey of Clare Island. Volume 2: Geology*, 87–112. Dublin. Royal Irish Academy.

Coxon, P., Corcoran, R., Gibson, P. and McCarron, S. 2013 A Holocene pollen diagram from Lough Avullin, Clare Island, Western Ireland. In D. Synnott (ed.), *New Survey of Clare Island. Volume 7: Plants and fungi*, 1–26. Dublin. Royal Irish Academy.

Crème, G. 1995 *Aspects of the biology of an expanding population of lesser black-backed gulls Larus fuscus in Ireland*. Unpublished PhD thesis, University College Cork.

Cussen, R.E., McGreal, E., Warner S. and Kelly T.C. 2020 The seabirds of Clare Island. In T.C. Kelly (ed.), *New Survey of Clare Island. Volume 9: Birds*, 43–56. Dublin. Royal Irish Academy.

D'Arcy, G. 1992 Fauna observed on Clare Island, Co. Mayo, June 14–18 1990. *Irish Naturalists' Journal* **24**, 33–4.

Diamond, J.M and May, R. 1977 Species turnover on islands: Dependence on census interval. *Science* **197**: 266–70.

Feehan, J. 2019 *Clare Island*. Dublin. Royal Irish Academy.

Forsyth, I. and Allen, P. 1997 *Report to the Royal Irish Academy on the breeding birds of Clare Island, Co. Mayo, 1996 and 1997*. Unpublished. 9 pages.

Fuller, R.J. (ed.) 2012 *Birds and habitat. Relationships in changing landscapes*. Cambridge. Cambridge University Press.

Gibbons, D.W., Reid, J.B. and Chapman, R.A. 1993 *The new atlas of breeding birds in Britain and Ireland: 1988–1991*. London. T. & A.D. Poyser.

Hutchinson, C.D. 1989 *Birds in Ireland*. Calton. T. & A.D. Poyser.

Kelcey, J.G. and Rheinweld, G. (eds) 2005 *Birds in European cities*. St Katharinen. Ginster Verglag.

Kelly, M., Oxbrough, A. and Kelly, T.C. 2012 Some observations on the diet of adult common gulls (*Larus canus*) at a coastal breeding colony in County Kerry. *Irish Birds* **9**, 357–60.

Kelly, T.C., Smiddy, P. and Keating, U. 2020 Breeding land birds on the offshore islands of Ireland – some patterns and trends. In T.C. Kelly (ed.), *New Survey of Clare Island. Volume 9: Birds*. 85–105. Dublin. Royal Irish Academy.

Lack, P.C. 1986 *The atlas of wintering birds in Britain and Ireland*. Calton. T. & A.D. Poyser.

Lloyd, C.S. 1984 The birds of Clare Island, Co. Mayo in June 1982. *The Irish Naturalists' Journal* **21**(5), 212–16.

Lloyd, C.S., Tasker, M.L. and Partridge, K. 1991 *The status of seabirds in Britain and Ireland*. Calton. T. & A.D. Poyser.

Losos, J.B. and Ricklefs, R.E. (eds) 2010 *The theory of island biogeography revisited*. Princeton N.J. Princeton University Press.

MacArthur, R.H. and Wilson, E.O. 1967 *The theory of island biogeography*. Princeton N.J. Princeton University Press.

Moller, A.P., Fiedler, W. and Berthold, P. (eds) 2006 *Birds and climate change*. London and New York. Academic Press.

Nairn, R., and O'Halloran, J. 2012 *Bird habitats in Ireland*. Cork. The Collins Press.

Ni Lamhna, E., Nairn, R., Benson, L. and Kelly, T.C. 2012 Urban habitats. In R. Nairn and J. O'Halloran (eds) *Bird habitats in Ireland*, 196–212. Cork. The Collins Press.

O'Halloran, J. and Kelly T.C. 2012 Bird habitats: structure and complexity. In R. Nairn and J. O'Halloran (eds) *Bird habitats in Ireland*, 14–23. Cork. The Collins Press.

Praeger, R.L. 1911 Clare Island survey part 10 Phanerogamia and Pteridophyta. *Proceedings of the Royal Irish Academy* **31**, 1–112.

Ruttledge, R.F. 1950 A list of the birds of the counties Galway and Mayo showing their status and distribution. *Proceedings of the Royal Irish Academy* **52B**, 315–81.

Ryle, T. 2013 Vegetation-environment interactions on Clare Island. In D. Synnott (ed.). *New Survey of Clare Island. Volume 7: Plants and Fungi*, 27–178. Dublin. Royal Irish Academy.

Schoener, T.W. 2010 The MacArthur-Wilson equilibrium model. In J.B. Losos and R.E. Ricklefs (eds) *The theory of island biogeography revisited*, 52–87. Princeton and Oxford. Princeton University Press.

Simberloff, D.S. 1974 Equilibrium theory of island biogeography and ecology. *Annual Review of Ecology and Systematics* **5**, 161–82.

Simberloff, D. 1976 Experimental zoogeography of islands: effects of island size. *Ecology* **57**, 629–48.

Sharrock, J.T.R. 1976 *The atlas of breeding birds in Britain and Ireland*. A and C Black. British Trust for Ornithology.

Ussher, R.J. 1912 A biological survey of Clare Island in the County of Mayo, Ireland and of the adjoining district. *Proceedings of the Royal Irish Academy* **31**, Section 11, Part 20, 1–54.

Votier, S.C., Furness, R.W., Bearhop, S., Crane, J.E., Caldow, R.W.G., Catry, P., Ensor, K., Hamer, K.C., Hudson, A.V., Kalmbach, E., Klomp, N.I., Pfeiffer, S., Phillips, R.A., Prieto, I. and Thompson, D.R. 2004 Changes in fisheries discard rates and seabird communities. *Nature* **427**, 727–30.

Vullings, W., Collins, J.F. and Smillie, G. 2013 *New Survey of Clare Island. Volume 8: Soils and soil associations*. Dublin. Royal Irish Academy.

Warren, B.H., Simberloff, D., Ricklefs, R.E., Aguile, R., Condamine, F.L., Gravel, D., Morlon, H., Mouquet, N., Rosindell, J., Casquet, J., Cornault, J., Fernandez-Palacios, J.M., Hengl, T., Norder, S.J., Rijsdijk, K.F., Sanmartin, D., Triantis, K.A., Valente, L.M., Whittaker, R.J., Gillespie, R.G., Emerson, B.C. and Thebaud, C. 2015 Islands as model systems in ecology and evolution: prospects fifty years after MacArthur-Wilson. *Ecology Letters* **18**, 200–17.

Whittaker, R.J., Triantis, K.A. and Ladle, R.J. 2010 A general dynamic theory of oceanic island biogeography: extending the MacArthur and Wilson theory to accommodate the rise and fall of volcanic islands. In J.B. Losos and R.E. Ricklefs (eds) *The theory of island biogeography revisited*, 88–115. Princeton and Oxford. Princeton University Press.

Wilson, J.D., Evans, A.D. and Grice, P.V. 2009 *Bird conservation and agriculture* (Vol. 6). Cambridge. Cambridge University Press.

Winters, P. 2006 *A summer bird survey of Clare Island*. Unpublished M.Sc. thesis. Trinity College, University of Dublin.

CHAPTER 18

THE MAMMALS OF CLARE ISLAND

Colin Lawton and Kate McAney

ABSTRACT

The mammals of Clare Island were surveyed through a variety of techniques including live capture of small rodents and bats, surveys for tracks and signs, and interviews with local residents. Twelve terrestrial or semi-aquatic mammal species were detected, and a further seven cetacean species were reported to the Irish Whale and Dolphin Group. The survey was in broad agreement with the 1911–15 survey by Barrett-Hamilton, although the house mouse may no longer be present. None of the recently introduced mammals on the mainland of Ireland (bank vole, American mink, great white toothed shrew) were detected.

Introduction

In general Ireland has a relatively low number of mammal species (34 species) when compared with the rest of Europe, and even its closest neighbour Britain (57 species) (Harris and Yalden 2008). It is now believed that any land bridge that existed between Ireland and Britain following the Last Glacial Maximum was very short-lived, and probably occurred while ice still dominated (Edwards and Brooks 2008). This means that those animals that arrived in Ireland without the help of human settlers were either species that arrived very early, before the ice had fully retreated (Irish hare *Lepus timidus hibernicus* and stoat *Mustela erminea*) or those that arrived on floating debris or by swimming (e.g. otter *Lutra lutra*) (Searle 2008). The remaining Irish mammal species are introduced animals, with some brought in at a very early stage, including those that seem to have been brought directly from continental Europe rather than via Britain (the 'Lusitanian' element), and other much more recent introductions.

The number of mammal species drops even further on examination of the offshore islands surrounding mainland Ireland as access to the islands requires overcoming further barriers. Some mammals do have means of reaching islands themselves, through swimming or flying; others may have made the move to an outlying island by a one-off, chance event. Some species have been introduced by humans, either accidentally as is the case with many rodent species arriving by ships, or deliberately, for hunting or aesthetic reasons. Upon reaching the islands, many species, such as the larger carnivores and ungulates, are unable to find sufficient resources and as a result do not establish themselves as a viable population. There may be a lack of suitable habitat, in particular a lack of cover, and there may be inclement exposed conditions that prevent other species, in particular small mammals, from establishing themselves.

The mammalian fauna of Clare Island, excluding the cetaceans, was reviewed for the 1911–15 survey by Barrett-Hamilton. He listed ten species as being present at that time: three rodent species (wood mouse *Apodemus sylvaticus*, brown rat *Rattus norvegicus* and house mouse *Mus domesticus*); two lagomorphs (rabbit *Oryctolagus cuniculus* and Irish hare); three semi-aquatic species (common seal *Phoca vitulina*, grey seal

Table 1
Terrestrial and amphibious Irish mammal species (based on Irish mammal species listed by Hayden and Harrington (2000)) recorded on Clare Island

Mammal species	Present	Absent
House mouse *Mus domesticus*	X	
Wood mouse *Apodemus sylvaticus*	X	
Bank vole *Myodes glareolus*		X
Brown rat *Rattus norvegicus*	X	
Black rat *Rattus rattus*		X
Red squirrel *Sciurus vulgaris*		X
Grey squirrel *Sciurus carolinensis*		X
Red fox *Vulpes vulpes*		X
Stoat *Mustela erminea*		X
Badger *Meles meles*		X
Pine Marten *Martes martes*		X
Mink *Neovison vison*		X
Otter *Lutra lutra*	X	
Grey seal *Halichoerus grypus*	X	
Harbour seal *Phoca vitulina*	X	
Irish hare *Lepus timidus hibernicus*	X	
Brown hare *Lepus europaeus*		X
Rabbit *Oryctolagus cuniculus*	X	
Red deer *Cervus elaphus*		X
Sika deer *Cervus nippon*		X
Fallow deer *Dama dama*		X
Feral goat *Capra hircus*		X
Pygmy shrew *Sorex minutus*	X	
Hedgehog *Erinaceous europaeus*		X
Daubenton's bat *Myotis daubentonii*	X	
Brown long-eared bat *Plecotus auritus*	X	
Natterer's bat *Myotis nattereri*	X	

Halichoerus grypus and otter); the pygmy shrew *Sorex minutus* and an unidentified bat species. That review is repeated here, identifying the animals present on Clare Island, and those known not to be present (Table 1). Where relevant, recent research on the species in the area is examined.

Cetacea

Table 2 details the records of cetacean sightings from Clare Island and the surrounding area (using records found between Achill Island and Inishturk, and within Clew Bay) received by the Irish Whales and Dolphin Group (IWDG) between 1991 and 2013. The species listed are all relatively common in Ireland, with the harbour porpoise *Phocaena phocaena*, common dolphin *Delphinus delphis* and bottlenose dolphin *Tursiops truncatus* being the most recorded and abundant species off the coast of Ireland. The bottlenose dolphin in particular is regularly recorded near Clare Island. There is evidence to suggest the collection of bottlenose dolphins in the Connemara-Mayo region is a genetically distinct aggregation with a degree of site fidelity (Mirimin *et al.* 2011), and Clare Island is within the West Connacht Coast Special Area of Conservation for the species (S. Berrow, pers. comm.). Risso's dolphins *Grampus griseus* are another relatively common Irish species that have also been known to be quite susceptible to stranding. The killer whale *Orcinus orca* is a large, charismatic and easily recognisable odontocete, and as such is readily identified and recorded. They are likely to prey on other mammal species using the marine environment, including otters moving from island to

Table 2
Cetacean records from waters surrounding Clare Island (including waters between the south of Achill Island, the north of Inishturk and into Clew Bay) as reported up to November 2013. All records are validated and available on www.iwdg.ie

Species	Most recent record	No. of records since 1991	Median animals per record	Nearest location to Clare Island
Bottlenose dolphin	2013	85	8	Clare Island
Harbour porpoise	2013	9	2	Clare Island
Common dolphin	2013	6	12.5	Clare Island
Fin whale	2013	4	3	Clare Island
White-beaked dolphin	2010	1	35	Clare Island
Killer Whale	2010	3	3	Clare Island
Risso's dolphin	2006	1	10	Achill Island

island or feeding offshore (de Jong 2012). Only one stranding is recorded by IWDG, a common dolphin in 1992.

Fin whales *Balaenoptera physalis* are ordinarily deep-water species, but they do breach frequently, making them more commonly recorded than most other baleen whale species. Conspicuous in its absence is the minke whale *Balaenoptera autorostrata*, the most common of the Irish baleen whales.

Two further records for a single beluga whale *Delphinapterus leucas* are listed from 1948 for Clare Island on the IWDG website (www.iwdg.ie). Beluga whales are rarely seen in Ireland, and it is likely that these sightings were of an individual who had strayed from its usual range.

Rodentia

Reports were received of brown rats being quite common on the island, but house mice have not been seen recently by homeowners on the island (M. Moran, pers. comm.). This is despite earlier records of house mice on Clare Island by Fairley (2001), Montgomery and Ferron (1994) and Barrett-Hamilton (1911–15). Fairley noted that house mice are more common away from human habitation on offshore islands and are caught at ratios of 1:2 in comparison with wood mice in available cover on such landscapes, when live trapping (Fairley and Smal 1987). In general, the house mouse is more competitive with wood mice on Irish islands than on the mainland, where they generally require human presence for long-term survival. The perceived absence of house mice by residents may reflect this tendency to adopt home ranges away from houses rather than an actual absence of the species on the island. In saying that an attempt at live trapping small mammals in covered areas on the island by C. Lawton did not yield any captures of house mice.

Brown rats have broad, opportunistic diets and will feed in the intertidal zone on molluscs and crustaceans (Navarrette and Castilla 1993) as well as raiding birds' nests for eggs and chicks. Cats, dogs, rats and mink are all considered important predators of sea birds (Moore 2002); rats, for example, eradicated puffin populations on islands off Scotland. Brown rats pose a threat to hole-nesting and colonial seabirds and attempts to poison rats can lead to inadvertent poisoning of scavenger bird species such as corvids (Moore 2002). Black rats *Rattus rattus* (Linnaeus) have disappeared from Ireland and its offshore islands, with the exception of Lambay Island off the east coast, where they have caused problems for nesting birds (Newton 2002).

One hundred Longworth traps were set and baited on Clare Island over two nights, with fifty set in the only available wooded area (Pl. I) and fifty in a gorse area during the first night, and fifty again in the wooded area and fifty along the wall of a field with high levels of ground cover during the second night. Just five wood mice were caught over the two hundred trap nights, reflecting the suboptimal nature of the habitat, but at least confirming the presence of the species on the island. Their presence has also been noted on Clare Island in previous studies (Fairley 1984; Kelly *et al.* 1982; Kelly and Fairley 1981; Barrett-Hamilton (1911–15)). A similar trapping survey by Montgomery and Ferron (1994) also only caught wood mice, and again in very low numbers (twelve captured across 1,000 trap nights). Wood mice found on other islands (e.g. Rathlin and Great Blasket Islands (Fairley *et al.* 1978)) have been noted as being very large when compared to mainland individuals. However, there is generally a greater variation in size between wood mice populations from various islands than between different mainland populations, meaning extreme examples are more likely (Kelly *et al.* 1982). The weight and various body measurements of the five mice captured here are listed in Table 3, together with average morphometrics listed in Harris and Yalden (2008). Fairley (1984) considered 25g to be a large size for a wood mouse

Pl. I Wooded scrub, in which fifty Longworth traps were set to catch wood mice. Photo: C. Lawton

Table 3
Morphometrics of wood mice captured on Clare Island in April 2006 (Mean ± 95% Confidence Limit) with expected body measurements (Mean and range), as listed in Harris and Yalden (2008)

	Clare Island (n=5)	Harris and Yalden (2008)
Weight (g)	29.44 ± 2.91	18.6 (13-27) (n=33)
Head Body length (mm)	96.72 ± 1.97	92.6 (81-103) (n=33)
Tail length (mm)	90.90 ± 4.04	82.7 (71-95) (n=33)
Hindfoot length (mm)	22.54 ± 0.93	21.8 (20.2-23) (n=33)
Condylobasal length (mm)	29.70 ± 0.39	23.3 (21.5-24.9) (n=31)

caught on the mainland, with an individual over 30g to be exceptionally large. In this study the smallest mouse captured was 24.3g, with three individuals weighing over 30g. The other body measurements taken (head/body length, tail length, hindfoot length and condylobasal length), also are much larger than the standard values expected. Montgomery and Ferron (1994) similarly reported large hindfeet in the sample of wood mice they captured. Although the sample size in both studies is too small to draw definite conclusions it appears that Clare Island wood mice follow the pattern of island wood mice being larger than mainland wood mice. Fairley (1984) hypothesises that the lack of ground predators may mean that evasion is no longer an important factor for the wood mice, allowing larger individuals to prevail. The timing of the current study may also have been a factor, as the traps were set in April, meaning the mice were probably older, and therefore larger, overwintered mice, rather than individuals born in that calendar year. Barrett-Hamilton (1911–15) experienced similar low capture rates in his attempt to investigate the wood mice of Clare Island.

Two flea species were recorded on the wood mice captured: *Typhloceras poppei poppei* and *Ctenophthalamus nobilis*. Both species are common on wood mice in Ireland, with *T. poppei poppei* previously recorded in Mayo (O' Mahoney 1939; Fairley and West 1975; Sleeman 1981; Sleeman *et al.* 1996; Whitaker 2007) and *C. nobilis* in both Galway (Langley and Fairley 1982; Whitaker 2007) and Clare (Sleeman 1986; Whitaker 2007) in the past.

Carnivora

As expected, few carnivore species are found on Clare Island, with foxes *Vulpes vulpes*, badgers *Meles meles*, stoats, pine martens *Martes martes* and the invasive American mink *Neovison vison* all being confirmed as absent following a survey of island residents and a search for tracks and signs of these animals. The surface area of the island would be expected to be too small to sustain viable populations, with the available habitat further reducing the number of individuals that could live there. However, badgers have been recorded previously on Irish islands, such as Little Island, Co. Waterford (Sleeman 2010); Coney Island, Co. Sligo (Sleeman *et al.* 2002); and Rutland Island, Co. Donegal (Sleeman *et al.* 2009). Although Little Island is smaller than Clare Island, it is not isolated from the mainland in the same way. The two western islands are similarly closer to the mainland than Clare Island.

The three carnivore species represented on the island are aquatic animals, with both Irish seal species and the semi-aquatic otter recorded. In a recent survey of harbour seal distributions in Ireland (Cronin *et al.* 2004), grey seals are listed as hauling out on the island, whereas harbour seals are in the waters around the island, hauling out at the harbour on Roonagh Quay. In a 2005 survey of grey seal pups (O'Cadhla *et al.* 2007) grey seals were noted as breeding on the island, with a total population size of between 151 and 500 individuals.

Otters are classified as vulnerable by the International Union for Conservation of Nature, and Ireland holds one of the most important otter populations in Europe (Sleeman and Moore 2005). Otters are reported as being present in 93.6% of potential sites in Ireland, after correction for sources of bias and error, in the latest otter survey of Ireland (Reid *et al.* 2013). Otters can be found on a number of the islands around the coast of Ireland, using the coastal waters as their feeding grounds (e.g. Kingston *et al.* 1999). Although otters supposedly require a source of fresh water in order to wash salt from their fur, they are found on Inish Mor, where there are no natural fresh water sources, implying they can survive with very low amounts of fresh water (C. Lawton, pers. obs.).

The diet of otters feeding off Clare Island was examined by Curran Cournane (2006), with 202 spraints analysed from an area of coast along the southeast of the island (Pl. II), where grassland and exposed rocky cliffs predominated, broken by minor freshwater streams. Rockling (Gadidae), blenny (*Lipophrys pholis*) and wrasse (Labridae) were all consumed heavily (Table 4), with butterfish

Pl. II Coastal area providing the bulk of otter spraints. Photo: C. Lawton

Table 4
Composition of otter spraints collected on Clare Island (n=202) from April to September 2006, expressed as relative frequency of occurrence (%Fr) (number of occurrences of a food category expressed as a percentage of the total number of occurrences of all food categories)

Food category	%Fr
Rockling	20.56
Blenny	14.95
Wrasse	5.6
Butterfish	3.5
Goby	2.57
Eel	0.93
Sand smelt	0.23
Sea scorpion	0.46
Unidentified fish	2.8
Mussel	14.95
Limpet	4.91
Barnacle	4.21
Isopod	9.35
Vegetation	1.4
Other objects	5.84
Other Crustaceans	7.71

(*Pholis gunnellus*), goby (*Pomatoschistus microps*), eel (*Anguilla anguilla*), sand smelt (*Atherina presbyter*) and sea scorpion (*Taurulus bubalis*) also featuring. Mussels (*Mytilis* spp.) and limpets (*Patella* spp.) were important to otter diet, with barnacles (*Chthalamus montaguii*), isopods (*Idotea pelagica*) and 'other' crustaceans (common shrimp (*Palaemon serratus*), shore crab (*Carcinus maenus*), porcelain crab (*Porcellina platycheles*) and velvet swimmer (*Necora puber*)) also found. Some insects also featured in very low numbers. There was some evidence detected of a seasonal pattern to the otter diet, particularly in the more commonly eaten fish, blenny and rockling. Blenny was more frequently eaten in the summer months, with rockling increasing in importance as autumn approached. The crustacean group when examined as a whole also appeared to display a seasonal pattern, with very few eaten during the summer months. This project only ran for six months (April to September) so a longer study would be required to examine this further.

A number of reports of feral cats *Felis catus* were received, which are apparently thriving on the island, in the absence of other terrestrial predators. They are considered to be a pest by many of the island's inhabitants, and like other predators may be influencing abundance and distribution of prey items, particularly birds.

Chiroptera
There was no prior information available on bat species present on Clare Island. Barrett-Hamilton (1911–15) referred to bats being present during the original survey, but no species were listed. Bats are found on several islands around Ireland, both offshore and closer to the mainland. Sleeman (1988) recorded a female Leisler's bat (*Nyctalus leisleri*) on Great Saltee Island, off the Wexford coast. The soprano pipistrelle (*Pipistrellus pygmaeus*) is a resident bat species on Cape Clear Island and brown long-eared bats (*Plecotus auritus*) have been mist-netted during bird monitoring by the Bird Observatory, so may also be resident, while a survey of Sherkin Island conducted in August 2006 by the Cork County Bat Group revealed the presence of the soprano pipistrelle, common pipistrelle (*P. pipistrellus*), brown long-eared and Leisler's bat (C. Kelleher, pers. comm.). On Garinish Island brown long-eared, soprano and pipistrelle bats are resident; Leisler's bat is known from droppings in a night roost (C. Kelleher, pers. comm.) while

Brown Long-Eared Bat. Photo: Conor Kelleher

Daubenton's Bat. Photo: Martin McKenna

Table 5
Bat species heard (using bat detectors) or caught (using mist nets) during 2002 survey on Clare Island. Bat detectors were deployed in seven locations, and mist nets used in three locations.

Area	Detector	Mist net
Craigmore/Abbey	0	-
Lassau	0	0
Portlea	*Plecotus* sp.	6 *Plecotus auritus*
Ooghcorragaun	Myotid sp.	-
Ballytoohy Beg	Myotid sp. *Plecotus* sp.	
Maum	Myotid sp.	1 *Myotis daubentonii* 1 *Plecotus auritus*
Harbour/Castle	0	-

lesser horseshoe bat droppings have also been found in a night roost (P. Scott, pers. comm.).

Bat surveys were conducted on Clare Island in July 1993 and July 2002. Although little information was gained on the possible bat community on the island during the 1993 survey, there were indications of the presence of a Myotid species on the island, based on a short echolocation call and the presence of droppings (McAney 1993). In the two months following this survey, two bat species were confirmed on the island from dead specimens: a brown long-eared bat in August in Portlea (P. Gill, pers. comm.) and a Natterer's bat (*Myotis nattereri*) in September in the Kill area (J.S. Fairley, pers. comm.).

In 2002 a team of researchers deployed bat detectors and mist nets on 5–6 July in the two areas of the island where the dead specimens had been recorded in 1993, and also in the only area of woodland on the island, known as Lassau. The group divided into three groups for the first night of fieldwork: one team with bat detectors patrolled the roadway between Craigmore and the Abbey while the other two teams erected mist nets at Lassau and Portlea, as well as walking the areas with detectors. The three teams were in position just prior to sunset and remained until 01.00 hrs. During daylight hours on 5 July suitable bridges and buildings were searched for bats or signs of bats (droppings, staining on stone/timber work), and study areas for the second night's work were selected. Four groups monitored bat activity between dusk and 01.00 hrs on 6 July with detectors and one mist net in the following areas: Ooghcorragaun, Ballytoohy Beg, Maum and around the harbour/castle.

In total seven brown long-eared bats and a Daubenton's bat were caught in mist nets. The bat detectors picked up echolocation signals of both *Myotis* and *Plecotus* bats. Details of the bats recorded in the 2002 survey are given in Table 5.

Daubenton's bat is strongly associated with water—which it uses for foraging purposes—and in using bridges, other stone structures and caves close to water as roosting sites. It is a specialist at feeding over water, skimming insects off the surface with its large feet or taking insects out of the air (Neuweiler 2000). It generally emerges quite late in the evening once it is fully dark, and will spread out to local waterways, ponds and lakes to feed. It is possible only to speculate that there is a population of Daubenton's bats roosting in sea caves along the eastern coastline of the island that fly inland at dusk to feed; further research is needed to test this theory.

Brown long-eared bats are most often found in older houses with large roof spaces, which allow the bats to fly around inside before emerging late in the evening. They are woodland animals and often hunt by picking insects off foliage, rather than catching them in flight. The area where the six long-eared bats were recorded contained vegetation representing the most suitable foraging habitat on the island. Swift and Racey (1983) observed *P. auritus* marked with reflective rings foraging in deciduous woodland, birch scrub and a garden containing many mature coniferous and deciduous trees, and Barataud (1990) mist-netted individuals in village gardens, scattered woodlands, orchards and parkland among meadows. Radio tracking studies in Germany (Fuhrmann and Seitz 1992) and in Scotland (Entwistle *et al.* 1996) have confirmed the importance of woodland to this species. Natterer's bats are similarly considered to be a woodland species.

Lagomorpha

Rabbits have been recorded since 1900 on various Irish offshore islands, including Clare Island, as well as eight other Co. Mayo islands, eleven Co. Galway islands and eleven Co. Donegal islands (Fairley 2001). Rabbits were originally introduced to many islands as they required little tending and were unable to escape. The lack of ground predators and the availability of sandy soil (ideal for burrowing) ensure their survival in the long term (Flux and Fullager 1992). Conditions can be so favourable that rabbits have been known to reach plague proportions on islands, with as many as 100 per hectare on Lambay Island, Co. Dublin. This can be followed by a subsequent crash in numbers (Baring 1907). Rabbits were observed by C. Lawton on Clare Island, and their burrows are easily observed in the sandy coastal soil (Pl. III).

Pl. III   Rabbit burrows found in the sandy soil near the coast

Irish Hare. Photo: Ruth Hanniffy

Irish hares are also frequently observed. Barrett-Hamilton (1911–15) gives very specific details of the introduction of both Irish hares and rabbits to the island by J.J. McCabe, with six of each introduced respectively in 1906 and 1907. By the time of his survey, Barrett-Hamilton observed large numbers of both species, and it seems likely that the current populations of both derive from these introductions. The availability of rabbits, and indeed rats, provides a large biomass of prey for invasive carnivore species. If the American mink were to arrive at Clare Island for example, they would have a sufficient food source to allow them to establish on the island in large numbers, which could have implications for nesting birds and domestic animals on the island. As non-burrowers the substrate is of less importance to the Irish hare.

They were traditionally used for coursing as well as being a food source in Ireland, and so their distribution in the whole of Ireland has been influenced by man. They are, however, an endemic Irish species.

Eulipotyphla

The old classification of order Insectivora has recently been abandoned, with most of the previous animals of the order now placed in the order Eulipotyphla. Three Irish species belong to this group: the European hedgehog *Erinaceous europaeus*, the pygmy shrew and the newly introduced greater white-toothed shrew *Crocidura russula*. Only the pygmy shrew is found on Clare Island, with the greater white-toothed shrew only found on a regional basis in Ireland. No signs of hedgehogs were found in a search for tracks and signs by C. Lawton, and no records received from local inhabitants. They were also absent in the original survey one hundred years ago (Barrett-Hamilton 1911–15). A dead pygmy shrew, a male just reaching sexual maturity, was found during the search for tracks and signs. Pygmy shrews have scent glands along the length of their body that render them unpalatable to predators, and so they can be found having been abandoned, most commonly by domestic or feral cats. Pygmy shrews have been recorded in other Irish island studies such as Tory Island (Cabot 1962; O'Gorman 1965). The pygmy shrew is quickly disappearing in areas of Ireland where the greater white-toothed shrew is found (Montgomery *et al.* 2012), and so these island populations may prove important in its long-term survival in Ireland. The pygmy shrew in particular is considered a very early introduction to Ireland, with closest ties to pygmy shrew populations in northern Spain (the Lusitanian element). It has been very successful in further spreading to many offshore islands.

Conclusions

In total twelve terrestrial mammal species (including the seals) were recorded on Clare Island, along with seven cetacean species. These include definite native species of Ireland (the Irish hare, two seal species and three bat species), and three other species that, if not early natives, were very early Mesolithic arrivals (otters, wood mice and pygmy shrews). Three of the species introduced more recently to Ireland are also present; the brown rat and house mouse, which have been introduced to islands worldwide by human settlers, and the rabbit, brought in as a source of food. Many of the species not recorded can be considered definitely absent, as they are conspicuous or leave obvious tracks or signs. It is possible that other bat or cetacean species exist on or near the island but have not yet been recorded.

The list is very similar to that of Barrett-Hamilton in the 1911–15 survey, with the extra detail on the bat and cetacean species found on the island. None of the recent introductions to Ireland (such as the bank vole *Myodes glareolus* or American mink) have yet managed to invade this offshore island. Similarly, other species absent during the original survey remain so, and are unlikely to reach the island unless directly brought in by humans.

Acknowledgements

Many thanks to the residents of Clare Island, in particular Mary Moran. Thanks to Prof. Ian Montgomery and Dr Simon Berrow for providing information on mammals of Clare Island, and the Irish Whales and Dolphins Group for permission to use records submitted to their website. Special thanks also to the staff and students at National University of Ireland, Galway, especially Eoin MacLoughlin, Prof. James Fairley (retired), Fiona Curran Cournane and Dr T.K. McCarthy. Thanks are also due to an anonymous reviewer whose comments on an early draft helped in the production of the final document. The 1993 bat survey team members were J. Dawes, S. Hassett, A. Hopkirk, B. Keeley, E. Lalloway, K. McAney, F. McGowan, G. O'Donnell, P. Smiddy and D. Tangney. The 2002 bat survey team members were S. Biggane, D. Buckley, S. Gueguen, A. Hopkirk, C. Kelleher, D. Keogh, K. McAney, E. Mullen, R. Nash and A. Taylor.

REFERENCES

Barataud, M. 1990 Eléments sur le comportement alimentire des Oreillards brun et gris, *Plecotus auritus* (Linnaeus, 1758) et *Plecotus austriacus* (Fisher, 1829). *Le Rhinolophe* **7**, 3–10.

Baring, C. 1907 Contributions to the natural history of Lambay Island. Mammals. *Irish Naturalist* **16**, 19–23.

Barrett-Hamilton, G.E.H. 1912 A Biological Survey of Clare Island in the County of Mayo, Ireland

and of the Adjoining District. Part 17. Mammalia. *Proceedings of the Royal Irish Academy* **31**, 1–14

Cabot, D. 1962 The pygmy or lesser shrew (*Sorex minutus* L.) on Tory Island, Co. Donegal. *Irish Naturalists' Journal* **14**, 82.

Cronin, M., Duck, C., Ó Cadhla, O., Nairn, R., Strong, D. and O'Keeffe, C. 2004 Harbour seal population assessment in the Republic of Ireland: August 2003. *Irish Wildlife Manuals*, No. 11. National Parks and Wildlife Service, Department of Environment, Heritage and Local Government, Dublin, Ireland.

Curran Cournane, F. 2007 Diet of marine-feeding European otters, *Lutra lutra*, in Clare Island, County Mayo. B.Sc. (Hons) Thesis, National University of Ireland, Galway.

De Jong, A. 2012 GPS tracking of mustelids. Abstract at 30th European Mustelid Colloquium, National Museum of Ireland.

Edwards, R. and Brooks, A. 2008 The island of Ireland; drowning the myth of an Irish land-bridge. In J.L. Davenport, D.P. Sleeman and P.C. Woodman (eds), *Mind the gap: postglacial colonization of Ireland*, 19–34. *Irish Naturalists' Journal*.

Entwistle, A.C., Racey, P.A. and Speakman, J.R. 1996 Habitat exploitation by a gleaning bat, *Plecotus auritus*. *Philosophical Transactions of the Royal Society of London* **351B**, 921–31.

Fairley, J.S. 1984 *An Irish beast book, second edition*. Belfast. Blackstaff press.

Fairley, J.S. 2001 *A basket of weasels*. Belfast. Private publication.

Fairley, J.S., McCarthy, T.K. and Andrews, J.F. 1978 Notes on the fieldmice of Inishkea North and a large race of fieldmouse from Great Blasket Island. *Irish Naturalists' Journal* **19**, 270–1.

Fairley, J.S. and Smal, C.M. 1987 Feral house mice in Ireland. *Irish Naturalists' Journal* **22**, 284–90.

Fairley, J.S. and West A.B. 1975 Fieldmice on Inishkea South, Co. Mayo. *Irish Naturalists' Journal* **18**, 196–8.

Flux, J.E.C. and Fullager, P.J. 1992 World distribution of the rabbit, *Oryctolagus cuniculus*, on islands. *Mammal Review* **22**, 151–205.

Fuhrmann, M. and Seitz, A. 1992 Nocturnal activity of the brown long-eared bat (*Plecotus auritus* L. 1758): data from radiotracking in the Lenneburg forest near Mainz (Germany). In I.G. Priede and S.M. Swift (eds), *Wildlife telemetry. Remote monitoring and tracking of animals*. Chichester. Ellis Horwood, 358–548.

Harris, S. and Yalden, D.W. (eds) 2008 *Mammals of the British Isles: handbook, fourth edition*. Southampton, UK. The Mammal Society.

Hayden, T. and Harrington, R. 2000 *Exploring Irish mammals*. Dublin. Town House.

Kelly, P.A. and Fairley, J.S. 1981 Field mice and other mammals on islands off Galway and Mayo. *Irish Naturalist's Journal* **8**, 352–3.

Kelly, P.A., Mahon, G.A.T. and Fairley, J.S. 1982 An analysis of morphological variation in the fieldmouse *Apodemus sylvaticus* (L.) on some Irish islands. *Proceedings of the Royal Irish Academy* **82B**, 39–51.

Kingston, S., O'Connell, M. and Fairley, J.S. 1999 Diet of otters, *Lutra lutra*, on Inishmore, Aran Islands, west coast of Ireland. *Biology and Environment*: *Proceedings of the Royal Irish Academy* **99B**, 173–82.

Langley, R. and Fairley, J.S. 1982 Seasonal variation in infestations of parasites in a woodmice *Apodemus sylvaticus* populations in the woodlands of Ireland. *Journal of Zoology* **198**, 237–48.

McAney, K. 1993 New Clare Island Survey: *Report on bat survey 3–4 July 1993*. Unpublished report.

Mirimin, L., Miller, R., Dillane, E., Berrow, S.D., Ingram, S., Cross, T.F. and Rogan, E. 2011 Fine-scale population genetic structuring of bottlenose dolphins in Irish coastal waters. *Animal Conservation* **14**, 342–53.

Montgomery, W.I. and Ferron, L. 1994 *Clare Island mammal survey preliminary report*. Unpublished report, School of Biology and Biochemistry, Queen's University, Belfast.

Montgomery, W.I., Lundy, M.G. and Reid, N. 2012 'Invasional meltdown': evidence for unexpected consequences and cumulative impacts of multispecies invasions. *Biological Invasions* **14**, 1111–25.

Moore, P.G. 2002 Mammals in intertidal and marine ecosystems: interactions, impacts and implications. In R.N. Gibson, M. Barnes and R.J.A. Atkinson (eds) *Oceanography and Marine Biology: an annual review* **40**, 491–608.

Navarrette, S.A. and Castilla, J.C. 1993 Predation by Norway rats in the intertidal zone in central Chile. *Marine Ecology Progress* **92**, 187–99.

Neuweiler, G. 2000 *The biology of bats*. Oxford. Oxford University Press.

Newton, S.F. 2002 Manx Shearwaters *Puffinus puffinus* proved breeding on Lambay, County Dublin. *Irish Birds* **7**, 140–1.

Ó Cadhla, O., Strong, D., O'Keeffe, C., Coleman, M., Cronin, M., Duck, C., Murray T., Dower, P., Nairn, R., Murphy, P., Smiddy, P., Saich, C., Lyons, D. and Hiby, A.R. 2007 An assessment of the breeding population of grey seals in the Republic of Ireland. *Irish Wildlife Manuals* No. 34. Dublin. National Parks and Wildlife Service, Department of Environment, Heritage and Local Government.

O'Gorman, F. 1965 The mammals of Tory Island, Donegal. *Proceedings of the Zoological Society of London* **145**, 155–8.

O'Mahoney, E. 1939 A preliminary list of Irish fleas. *Entomologist's Monthly Magazine* **75**, 124–6.

Reid, N., Hayden, B., Lundy, M.G., Piotravalle, S., McDonald, R.A. and Montgomery, W.I. 2013 National otter survey of Ireland 2004/05. *Irish Wildlife Manuals*, No. 76. Dublin. National Parks and Wildlife Service, Department of Environment, Heritage and Local Government.

Searle, J.B. 2008 The colonization of Ireland by mammals. In J.L. Davenport, D.P. Sleeman and P.C. Woodman (eds), *Mind the gap: postglacial colonization of Ireland*, 109–15. *Irish Naturalists' Journal*.

Sleeman, D.P. 1981 Fleas from Irish mammals. *Irish Naturalists' Journal* **20**, 244–7.

Sleeman, D.P. 1986 Recent records of Irish fleas (Siphonaptera). *Irish Naturalists' Journal* **22**, 80–1.

Sleeman, D.P. 1988 *Recent records of Leisler's bat from the south coast. Irish Naturalists' Journal* **22**, 416.

Sleeman, D.P. 2010 The badgers (*Meles meles* (L)) of Little Island, Co Waterford. *Irish Naturalists' Journal* **31**, 94–9.

Sleeman, D.P., Cussen, R.E., Southey, A.K. and O'Leary, D. 2002 The badgers *Meles meles* (L.) of Coney Island, Co. Sligo. *The Irish Naturalists' Journal* **27**, 10–18.

Sleeman, D.P., Davenport, J., Cussen, R.E. and Hammond, R.F. 2009 The small-bodied badgers (*Meles meles* (L.)) of Rutland Island, Co. Donegal. *The Irish Naturalists' Journal* **30**, 1–6.

Sleeman, D.P. and Moore, G.P. 2005 Otters (*Lutra lutra*) in Cork City. *Irish Naturalists' Journal* **28**, 73–8.

Sleeman, D.P., Smiddy, P. and Moore, P. 1996 The fleas of Irish terrestrial mammals: a review. *Irish Naturalists' Journal* **25**, 237–48.

Swift, S.M., and Racey, P.A. 1983 Resource partitioning in two species of vespertilionid bats (Chiroptera) occupying the same roost. *Journal of Zoology* **200**, 249–59.

Whitaker, A.P. 2007 *Handbooks for the Identification of British Insects: Fleas (Siphonaptera)*, London. Royal Entomological Society.

TAXONOMIC INDEX

Note: (t) after a page reference denotes a table; (app) denotes appendix to a chapter.

Page references in italics denote illustrations

A

Abax parallelepipedus, 103(t)
Ablabesmyia
    A. longistyla, 140 (t), 154 (t) (app), 161 (app)
    A. monilis, 140 (t), 154 (t) (app), 161 (app)
    A. nebulosa, 139 (t), 141, 162 (app)
    A. phatta, 139 (t), 140 (t), 154 (t) (app), 161 (app)
    A. pygmaea, 139 (t), 164 (app)
Abraxas grossulariata, 124 (app)
Abrostola
    A. tripartita, 127 (app)
    A. triplasia, 127 (app)
Acamptocladius
    A. reissi, 140 (t), 142, 155 (t) (app), 165 (app)
    A. submontanus, 155 (t) (app), 165 (app)
Acanthinula aculeata, 8 (t), 10 (t)
Acanthocephalus
    A. clavula, 222 (t), 223, 225
    A. lucii, 225
Acantholeberis curvirostris, 214 (t)
Acari, 74
Acasis viretata, 126 (app)
Acathocyclops vernalis, 216 (t)
Accipiter nisus, 235
Achorutes, 71 (t)
    A. armatus, 71 (t)
    A. longispina, 71 (t)
    A. viaticus, 71 (t)
Acicula fusca, 10 (t), 12, 14
Acilius sulcatus, 85 (t), 89, 96, 98
Acleris
    A. aspersana, 119 (app)
    A. emargana, 119 (app)
    A. hastiana, 119 (app)
    A. hyemana, 119 (app)
    A. variegana, 119 (app)
Acricotopus lucens, 140 (t), 155 (t) (app), 165–166 (app)
Acrocephalus schoenobaenus, 234
Acrogalumna longipluma, 76 (t)
Acronicta rumicis, 114, 126–127 (app)
Acroperus
    A. angustatus, 213 (t), 215
    A. harpae, 213 (t), 215
Actitis hypoleucos, 234
Adoristes ovatus, 76 (t)
Adropion scoticum scoticum, 25, 36, 36 (t)

Aegopinella
    A. pura, 8 (t), 10 (t), 14
    A.nitidula, 8 (t), 10 (t), 14
Aepus robinii, 102, 103(t)
Aethes
    A. cnicana, 119 (app)
    A. rubigana, 111 (t)
Agabus
    A. affinis, 85 (t), 89, 97
    A. bipustulatus, 85 (t), 90
    A. nebulosus, 85 (t), 89, 90
    A. paludosus, 85 (t), 89
    A. sturmii, 85 (t)
Agapeta hamana, 119 (app)
Aglais
    A. io, 123 (app)
    A. urticae, 123 (app)
Agonopterix
    A. heracliana, 117 (app)
    A. nervosa, 117 (app)
    A. subpropinquella, 117 (app)
Agonum
    A. fuliginosum, 103(t), 105
    A. gracile, 103(t), 105
    A. marginatum, 103(t)
    A. muelleri, 103(t)
    A. nigrum, 103(t), 104
    A. thoreyi, 103(t), 105
Agriolimax caruanae, 13
Agriphila
    A. geniculea, 122 (app)
    A. inquinatella, 122 (app)
    A. straminella, 122 (app)
    A. tristella, 121 (app)
Agrochola lunosa, 128 (app)
Agrotis
    A. exclamationis, 132 (app)
    A. ipsilon, 132 (app)
    A. trux, 113, 132 (app)
    A. vestigialis, 132 (app)
Agyneta
    A. cauta, 45, 53 (app)
    A. conigera, 53 (app)
    A. decora, 54 (app)
    A. mossica, 45, 49, 54 (app)
    A. saxatilis, 45, 54 (app)
    A. subtilis, 54 (app)
Alauda arvensis, 234
Alcis repandata, 124 (app)
Allomengea vidua, 45, 49, 54 (app)
Alona
    A. affinis, 213 (t), 215
    A. costata, 213 (t)
    A. guttata, 213 (t)
    A. intermedia, 213 (t)
    A. quadrangularis, 213 (t)
    A. rectangula, 213 (t)
    A. rustica, 213 (t)

Alonella
    A. excisa, 213 (t), 215
    A. exigua, 213 (t)
    A. nana, 213 (t), 215
Alonopsis elongata, 213 (t), 215
Alopecosa pulverulenta, 62 (app)
Amara
    A. communis, 103(t)
    A. lunicollis, 103(t), 105
    A. plebeia, 103(t)
Amaurobius
    A. ferox, 46, 52 (app)
    A. similis, 46, 52 (app)
Amphipoea
    A. crinanensis, 129 (app)
    A. lucens, 129 (app)
Ampullaceana balthica, 8 (t), 10 (t)
Anacaena
    A. globulus, 86 (t), 89, 91
    A. limbata, 86 (t), 89, 90
    A. lutescens, 86 (t), 89
Anania
    A. fuscalis, 122 (app)
    A. hortulata, 122 (app)
Anas platyrhynchos, 234
Anchomenus dorsalis, 103(t)
Ancylis
    A. badiana, 120 (app)
    A. unguicella, 120 (app)
Ancylus fluviatilis, 8 (t), 10 (t), 168 (app)
Androniscus dentiger, 81, *81*, 82
Anguilla anguilla, 219, 241
Anguillicola crassus, 224, 227
Anisakis simplex, 222 (t), 223, 225, 227
Anisus leucostoma, 8 (t), 10 (t)
Anthophila fabriciana, 120 (app)
Anthus pratensis, 234, 235
Antistea elegans, 45, 53 (app)
Anurida
    A. granaria, 71 (t)
    A. maritima, 70, 71 (t)
Anuridella marina, 71 (t)
Apamea
    A. crenata, 128 (app)
    A. lithoxylaea, 128 (app)
    A. monoglypha, 128 (app)
    A. remissa, 128 (app)
    A. sordens, 111 (t)
    A. sublustris, 111 (t), 113
Aphantopus hyperantus, 124 (app)
Aphelia viburnana, 111 (t), 113
Aphileta misera, 45, 49, 54 (app)
Aphomia sociella, 121 (app)
Apis mellifera, 189, 191, 192 (t)
Apodemus sylvaticus, 237, 238 (t)
Aporophyla lueneburgensis, 113, 128 (app)
Apotomis semifasciana, 119 (app)

Apsectrotanypus trifascipennis, 154 (t) (app), 162 (app)
Araeoncus humilis, 48
Araneus diadematus, 46, 52 (app)
Araniella opistographa, 46, 52 (app)
Archips rosana, 111 (t), 113
Arctia caja, 132 (app)
Arctopelopia
  A. barbitarsis, 140 (t), 154 (t) (app), 162 (app)
  A. griseipennis, 140 (t), 154 (t) (app), 162 (app)
Arctosa
  A. leopardus, 62 (app)
  A. perita, 48
Arctoseius
  A. cetratus, 72 (t)
  A. minutus, 72 (t)
Argenna subnigra, 47, 53 (app)
Argyresthia
  A. bonnetella, 117 (app)
  A. brockeella, 116 (app)
  A. goedartella, 116 (app)
Argyroneta aquatica, 47, 47, 53 (app)
Argyroploce
  A. olivana, 111 (t), 114
  A. palustrana, 111 (t), 114
Arion
  A. ater, 8 (t), 10 (t)
  A. circumscriptus silvaticus, 10 (t), 14
  A. distinctus, 8 (t), 10 (t), 12
  A. hortensis, 8 (t), 10 (t), 12–13, 14
  A. intermedius, 8 (t), 10 (t)
  A. owenii, 8 (t), 10 (t), 13, 14
  A. rufus, 8 (t), 10 (t)
  A. subfuscus, 8 (t), 10 (t), 14
Asellus aquaticus, 225
Aspilapteryx tringipennella, 116 (app)
Atherina presbyter, 241
Atropacarus wandae, 75 (t)
Attheyella crassa, 216 (t), 217
Autographa
  A. gamma, 127 (app)
  A. jota, 127 (app)
  A. pulchrina, 127 (app)
Axylia putris, 130 (app)

B
Bactra
  B. furfurana, 119 (app)
  B. lancealana, 119 (app)
Badister bullatus, 103(t)
Balaenoptera
  B. autorostrata, 239
  B. physalis, 239
Balea
  B. heydeni, 8, 8 (t), 10 (t)
  B. perversa, 8
Ballistrua schoetti, 71 (t)
Banksinoma lanceolata, 76 (t)
Baryphyma trifrons, 54 (app)
Bathyphantes
  B. approximatus, 45, 54 (app)
  B. gracilis, 54 (app)
  B. nigrinus, 44, 54 (app)
  B. parvulus, 54 (app)
Bembidion
  B. aeneum, 102, 103(t)
  B. bipunctatum, 102, 103(t), 104
  B. bruxellense, 102, 103(t)
  B. doris, 102, 103(t)
  B. lampros, 103(t)
  B. mannerheimi, 103(t)
  B. pallidipenne, 103(t)
  B. saxatile, 102, 103(t), 104
  B. stephensii, 102, 103(t)
  B. tetracolum, 103(t)
Benthalia carbonaria, 158 (t) (app), 174 (app)
Blaniulus guttulatus, 79, 80
Blastobasis adustella, 118 (app)
Bolyphantes luteolus, 46, 54 (app)
Bombus
  B. cryptarum, 190, 191, 192 (t)
  B. distinguendus, 191, 192 (t)
  B. hortorum, 192 (t)
  B. jonellus, 192 (t)
  B. lucorum, 189, 190, 191, 192 (t)
  B. magnus, 190, 191, 192 (t)
  B. muscorum, 192 (t), *193*
  B. pascuorum, 192 (t)
  B. ruderarius, 192 (t)
  B. terrestris, 189, 190, 191
Boreoiulus tenuis, 79, 80
Bosmina
  B. longirostris, 213 (t), 215
  B. longispina, 215
  B. obtusirostris, 215
Brachychthonius berlesei, 75 (t)
Brachydesmus superus, 79
Bradleycypris obliqua, 198, 200, 202, 204, 205
Bradleystrandesia fuscata, 198, 200, 202, 205
Bradycellus, 105
  B. caucasicus, 103(t)
  B. collaris, 105
  B. ruficollis, 103(t), 105
  B. verbasci, 103(t), 105
Branta canadensis, 235
Brillia
  B. bifida, 140 (t), 155 (t) (app), 166 (app)
  B. longifurca, 155 (t) (app), 166 (app)
Broscus cephalotes, 103(t)
Bryocamptus
  B. praegeri, 217
  B. pygmaeus, 217
  B. zschokkei, 217
Bryodelphax parvulus, 35 (t)
Bryophaenocladius subvernalis, 155 (t) (app), 166 (app)
Bryothrepes longimanus, 215
Bryotropha
  B. politella, 110
  B. terrella, 110, 118 (app)

C
Cabera
  C. exanthemata, 124 (app)
  C. pusaria, 124 (app)
Calathus
  C. fuscipes, 103(t)
  C. melanocephalus, 103(t), 105
  C. mollis, 103(t)
Calcarius lapponicus, 235
Calidris alpina, 234
Callophrys rubi, 112, 123 (app)
Calohypsibius
  C. ornatus, 36 (t)
  C. verrucosus, 35, 36 (t)
Caloptilia elongella, 116 (app)
Camisia
  C. segnis, 75 (t)
  C. spinifer, 75 (t)
Campaea margaritata, 124 (app)
Camptocladius stercorarius, 140 (t), 155 (t) (app), 166 (app)
Camptogramma bilineata, 125 (app)
Candidula intersecta, 10 (t)
Candona candida, 198, 201, 204, 205, 206
Canthocamptus
  C. crassus, 217
  C. hirticornis, 217
Capra hircus, 238 (t)
Carabodes
  C. elongatus, 75 (t)
  C. marginatus, 75 (t)
  C. willmanni, 75 (t)
Carabus, 106
  C. clatratus, 103(t), 106, *106*
  C. granulatus, 103(t), 106
  C. problematicus, 103(t), 104, 106, *106*
Caradrina clavipalpis, 127 (app)
Carcinus maenus, 241
Cardiocladius, 142
  C. fuscus, 140 (t), 142, 155 (t) (app), 166 (app)
Carduelis
  C. flammea, 235
  C. flavirostris, 235
Carychium
  C. minimum, 10 (t)
  C. tridentatum, 8 (t), 10 (t)
Caryocolum marmoreum, 111 (t)
Celypha
  C. cespitana, 120 (app)
  C. lacunana, 120 (app)
Centromerita
  C. bicolor, 54 (app)
  C. concinna, 55 (app)
Centromerus
  C. dilutus, 55 (app)
  C. prudens, 46, 55 (app)
Cepaea nemoralis, 8 (t), 10 (t)
Cepheus tegeocranus, 76 (t)
Ceramica pisi, 129 (app)
Cerapteryx graminis, 130 (app)
Cerastis rubricosa, 131 (app)
Ceratinella
  C. brevipes, 55 (app)
  C. brevis, 45, 55 (app)
Ceratophysella
  C. armata, 71 (t)
  C. denticulata, 71 (t)
  C. longispina, 71 (t)
Ceratoppia
  C. bipilis, 76 (t)
  C. quadridentata, 76 (t)
Ceratozetes
  C. gracilis, 76 (t)
  C. medio, 76 (t)
Cercyon
  C. depressus, 86 (t), 90
  C. littoralis, 86 (t)
  C. ustulatus, 86 (t)

Ceriodaphnia
  C. quadrangula, 214 (t), 215, *215*
  C. reticulata, 215
Cernuella virgata, 8 (t), 10 (t), 13
Cerura vinula, 126 (app)
Cervus
  C. elaphus, 238 (t)
  C. nippon, 238 (t)
Ceryon littoralis, 90
Chaetarthria seminulum/simillima agg., 86 (t)
Chaetocladius
  C. dentiforceps, 140 (t), 155 (t) (app), 166 (app)
  C. melaleucas, 140 (t), 155 (t) (app)
  C. melaleucus, 166 (app)
  C. perennis, 140 (t), 155 (t) (app), 166 (app)
Chamobates cuspidatus, 76 (t)
Charadrius hiaticula, 233, 234
Cheiracanthium erraticum, 46, 52 (app)
Cheiroseius (Posttrematus) serratus, 72 (t)
Chironomus
  C. alpestris, 139 (t), 141 (t), 142, 143, 144, 158 (t) (app), 174 (app), 175 (app)
  C. annularius, 139 (t), 144, 158 (t) (app), 174 (app)
  C. anthracinus, 158 (t) (app), 174 (app)
  C. aprilinus, 139 (t), 141 (t), 158 (t) (app), 175 (app)
  C. commutatus, 141 (t), 143, 158 (t) (app), 174 (app), 175 (app)
  C. dorsalis, 139 (t), 143, 174 (app), 175 (app)
  C. nuditarsis, 141 (t), 143, 158 (t) (app), 175 (app)
  C. piger, 158 (t) (app), 175 (app)
  C. plumosus, 139 (t), 143, 158 (t) (app), 175 (app)
  C. pseudothummi, 141 (t), 158 (t) (app), 175 (app)
  C. riparius, 139 (t), 141 (t), 158 (t) (app), 175 (app)
  C. salinarius, 144, 158 (t) (app), 175 (app)
  C. tentans, 139 (t), 141 (t), 142, 143, 158 (t) (app), 175 (app)
Chrysoteuchia culmella, 121 (app)
Chthalamus montaguii, 241
Chthonius
  C. ischnocheles, 81
  C. tetrachelatus, 81
Chydorus
  C. barbatus, 213 (t), 215–216
  C. latus, 213 (t)
  C. ovalis, 213 (t)
  C. piger, 213 (t)
  C. sphaericus, 203, 212, 213 (t), 215
Cicindela campestris, 103(t)
Cinclus cinclus hibernicus, 235
Cladopelma
  C. krusemani, 141 (t), 158 (t) (app), 175 (app)
  C. virescens, 158 (t) (app), 175 (app)
  C. viridulum, 158 (t) (app), 175 (app)
Cladotanytarsus
  C. atridorsum, 141 (t), 159 (t) (app), 180 (app)

  C. iucundus, 144, 159 (t) (app), 180 (app)
  C. lepidocalcar, 159 (t) (app), 180 (app)
  C. mancus, 159 (t) (app), 180 (app)
  C. nigrovittatus, 141 (t), 159 (t) (app), 180 (app)
  C. pallidus, 159 (t) (app), 180 (app)
  C. vanderwulpi, 159 (t) (app), 180 (app)
Clausilia bidentata, 8 (t), 10 (t)
Clinotanypus
  C. nervosus, 139 (t), 154 (t) (app), 162 (app)
Clivina fossor, 103(t)
Clubiona
  C. comta, 44, 52 (app)
  C. neglecta, 52 (app)
  C. phragmitis, 45, 53 (app)
  C. reclusa, 53 (app)
  C. stagnatilis, 45, 53 (app)
  C. trivialis, 46, 53 (app)
Clunio marinus, 140 (t), 142, 144, 155 (t) (app), 166 (app)
Cnephalocotes obscurus, 55 (app)
Cnephasia
  C. conspersana, 119 (app)
  C. incertana, 119 (app)
Cochlicella, 13
  C. acuta, 8 (t), 10 (t), 13, *14*
  C. barbara, 13
Cochlicopa
  C. lubrica, 8 (t), 10 (t)
  C. lubricella, 8 (t), 10 (t)
Coelostoma
  C. orbicualre, 97
  C. orbiculare, 86 (t)
Coenonympha pamphilus, 112, 124 (app)
Coleophora
  C. albidella, 118 (app)
  C. alticolella, 118 (app)
  C. deauratella, 118 (app)
  C. discordella, 118 (app)
  C. lusciniaepennella, 118 (app)
  C. saxicolella, 112, 118 (app)
  C. serratella, 118 (app)
  C. sternipennella, 112
  C. taeniipennella, 118 (app)
  C. tamesis, 118 (app)
  C. virgaureae, 112, 118 (app)
Colias croceus, 123 (app)
Columella
  C. aspera, 10 (t), 12, 14
  C. edentula, 12
Conchapelopia
  C. melanops, 154 (t) (app), 162 (app)
  C. pallidula, 154 (t) (app), 162 (app)
  C. viator, 154 (t) (app), 162 (app)
Cornu aspersum, 8 (t), 10 (t)
Corvus
  C. corax, 234
  C. cornix, 234
Corynoneura
  C. arctica, 155 (t) (app), 166 (app)
  C. carriana, 155 (t) (app), 166 (app)
  C. celeripes, 166 (app)
  C. celtica, 155 (t) (app), 166 (app)
  C. edwardsi, 140 (t), 155 (t) (app), 166 (app)
  C. gratias, 155 (t) (app), 166 (app)
  C. lobata, 140 (t), 155 (t) (app), 166 (app)

  C. paludosa, 140 (t), 155 (t) (app), 167 (app)
  C. Pe2a, 155 (t) (app), 166 (app)
Cosmolaelaps claviger, 72 (t)
Crambus
  C. lathoniellus, 121 (app)
  C. pascuella, 121 (app)
  C. perlella, 121 (app)
  C. uliginosellus, 111 (t), 114
Crepidostomum
  C. farionis, 221, 222 (t)
  C. metoecus, 221, 222 (t)
Cricotopus, 139 (t), 144
  C. albiforceps, 140 (t), 155 (t) (app), 167 (app)
  C. annulator, 139 (t), 140 (t), 155 (t) (app), 167 (app)
  C. bicinctus, 139 (t), 155 (t) (app), 167 (app)
  C. brevipalpis, 155 (t) (app), 167 (app)
  C. ephippium, 140 (t), 155 (t) (app), 167 (app)
  C. festivellus, 140 (t), 155 (t) (app), 167 (app)
  C. flavocinctus, 155 (t) (app), 167 (app)
  C. fuscus, 155 (t) (app), 167 (app)
  C. (Isocladius) Pe 2 sensu, 156 (t) (app), 168 (app)
  C. (Isocladius) Pe 5 sensu, 156 (t) (app), 168 (app)
  C. (Isocladius) sylvestris, 139 (t), 140 (t), 156 (t) (app), 168 (app)
  C. (Isocladius) tricinctus, 139 (t), 144, 156 (t) (app), 168 (app)
  C. (Isocladius) trifasciatus, 139 (t), 156 (t) (app), 168 (app)
  C. intersectus, 155 (t) (app), 167 (app)
  C. laricomalis, 155 (t) (app), 167 (app)
  C. motitator, 139 (t), 167 (app)
  C. ornatus, 140 (t), 155 (t) (app), 167 (app)
  C. pallidipes, 155 (t) (app), 167 (app)
  C. pilosellus, 140 (t), 155 (t) (app), 167 (app)
  C. polaris, 155 (t) (app), 167 (app)
  C. (Paratrichocladius) rufiventris, 140 (t), 156 (t) (app), 168 (app)
  C. (Paratrichocladius) skirwithensis, 156 (t) (app), 168 (app)
  C. pulchripes, 140 (t), 155 (t) (app), 167 (app)
  C. silvestris, 139 (t), 168 (app)
  C. similis, 155 (t) (app), 167 (app)
  C. tibialis, 139 (t), 140 (t), 142, 155 (t) (app), 167 (app)
  C. tremulus, 155 (t) (app), 167 (app)
  C. trifascia, 155 (t) (app), 167 (app)
  C. trifasciatus, 139 (t), 168 (app)
  C. tristis, 155 (t) (app), 167 (app)
Crocallis elinguaria, 124 (app)
Crocidura russula, 244
Cryphoeca silvicola, 53 (app)
Cryptocandona, 201, 206
  C. reducta, 198, 201, 206, 207
  C. vavrai, 198, 201, 206, 207
Cryptochironomus
  C. albofasciatus, 158 (t) (app), 175 (app)
  C. obreptans, 158 (t) (app), 175 (app)

C. *psittacinus*, 139 (t), 158 (t) (app), 175 (app)
C. *supplicans*, 158 (t) (app), 176 (app)
Cryptops hortensis, 80
Ctenophthalamus nobilis, 240
Cucullanus heterochrous, 222 (t), 223
Cucullia umbratica, 127 (app)
Cupido minimus, 111 (t), 113
Curtonotus aulicus, 103(t)
Cychrus, 105, 106
　C. *caraboides*, 103(t), 104, 105, 106
Cyclocypris
　C. *globosa*, 198, 200, 201, 206
　C. *laevis*, 205
　C. *ovum*, 198, 200, 202, *202*, 204, 205
　C. *serena*, 205
Cyclops abyssorum, 217
Cylindroiulus
　C. *latestriatus*, 79, 80, *80*
　C. *londinensis*, 80
　C. *punctatus*, 80
Cylisticus convexus, 81, 82
Cymindis vaporariorum, 107
Cyphon
　C. *coarctatus*, 86 (t)
　C. *hilaris*, 86 (t)
　C. *laevipennis*, 86 (t), 89
　C. *ochraceus*, 86 (t), 90
　C. *padi*, 86 (t)
　C. *palustris*, 86 (t)
　C. *variabilis*, 86 (t), 89
Cypria ophtalmica, 198, 201, 205
Cypridopsis
　C. *bamberi*, 203, 204
　C. *lusatica*, 198, 203–204, 205, 207
　C. *obesa*, 205
　C. *vidua*, 203, 205
Cyprinotus prasinus, 203, 205
Cypris
　C. *bispinosa*, 204
　C. *pubera*, 204
Cyrtolaelaps
　C. *kocki*, 73 (t)
　C. *nemorensis*, 73 (t)
　C. *transisalae*, 73 (t)
Cystidicola farionis, 222 (t), 223

D
Dama dama, 238 (t)
Damaeus (Paradamaeus) clavipes, 76 (t)
Daphnia, 214 (t)
　D. *galeata*, 216
　D. *hyalina*, 214 (t), 215
　D. *hyalina lacustris*, 214 (t), 215, 216
　D. *lacustris*, 216
　D. *longispina*, 214 (t), 215, *215*, 216
　D. *obtusa*, 215
　D. *pulex*, 215
Darwinula stevensoni, 200
Deilephila elpenor, 123 (app)
Delphinapterus leucas, 239
Delphinus delphis, 238
Delplanqueia dilutella, 121 (app)
Demeijerea rufipes, 139 (t), 141 (t), 158 (t) (app), 176 (app)
Demicryptochironomus
　D. *(Irmakia) neglectus*, 144, 158 (t) (app), 176 (app)

D. *(Demicryptochironomus) vulneratus*, 158 (t) (app), 176 (app)
D. *(s. str.) vulneratus*, 139 (t)
Denticucullus pygmina, 129 (app)
Depressaria
　D. *badiella*, 117 (app)
　D. *daucella*, 117 (app)
　D. *radiella*, 117 (app)
Deroceras
　D. *caruanae*, 13
　D. *invadens*, 9 (t), 10 (t), 13
　D. *laeve*, 9 (t), 10 (t)
　D. *panormitanum*, 13
　D. *reticulatum*, 9 (t), 10 (t)
Diachrysia chrysitis, 127 (app)
Diacrisia sannio, 111 (t), 113, 114
Diacyclops
　D. *languidus*, 216 (t), 217
　D. *nanus*, 216 (t)
Diamesa
　D. *ammon*, 139 (t), 142, 165 (app)
　D. *insignipes*, 140 (t), 142, 154 (t) (app), 165 (app)
Diaphanosoma brachyurum, 214 (t), 215
Diarsia
　D. *mendica*, 130 (app)
　D. *rubi*, 130 (app)
Dichrorampha plumbana, 111 (t), 114
Dicranopalpus ramosus, 81
Dicrotendipes
　D. *lobiger*, 158 (t) (app), 176 (app)
　D. *nervosus*, 139 (t), 141 (t), 144, 158 (t) (app), 176 (app)
　D. *notatus*, 158 (t) (app), 176 (app)
　D. *pulsus*, 158 (t) (app), 176 (app)
　D. *tritomus*, 158 (t) (app), 176 (app)
Dictya umbrarum, 185, 186 (t)
Dicymbium
　D. *nigrum*, 55 (app)
　D. *nigrum* f. *brevisetosum*, 42
　D. *tibiale*, 43 (t), 44, 55–56 (app)
Dicyrtoma fusca, 72 (t)
Dicyrtomina minuta, 72 (t)
Dinocheirus panzeri, 81
Diphascon, 18, 24
　D. *chilenense*, 24, 35, 36, 36 (t)
　D. *langhovdense*, 24
　D. *pingue*, 24
　D. *puniceum*, 24
　D. *sanae*, 24
　D. *stappersi*, 24
　D. *tenue*, 24
Diplocephalus
　D. *cristatus*, 56 (app)
　D. *permixtus*, 56 (app)
Diplopoda, 79–80
Diplostomum spathaceum, 222 (t), 223, 227
Diplostyla concolor, 56 (app)
Discopoma integra, 73 (t)
Discus rotundatus, 9 (t), 10 (t)
Dismodicus bifrons, 56 (app)
Disporalona rostrata, 215
Dissorhina ornata, 76 (t)
Ditula angustiorana, 111 (t)
Dolichovespula sylvestris, 192, 192 (t)
Donacaula mucronella, 122 (app)
Donacia
　D. *simplex*, 86 (t), 90

D. *thalassina*, 86 (t), 90
D. *versicolorea*, 86 (t)
D. *vulgaris*, 86 (t), 96
Drassodes
　D. *cupreus*, 48, 53 (app)
　D. *lapidosus*, 48
Drepanothrix dentata, 214 (t)
Drepanotylus uncatus, 56 (app)
Dryops luridus, 86 (t)
Dyschirius globosus, 103(t)
Dyscia fagaria, 124–125 (app)
Dysdera crocata, 48
Dysstroma truncata, 125 (app)
Dytiscus
　D. *marginalis*, 85 (t), 90, *90*
　D. *semisulcatus*, 86 (t)

E
Eana penziana, 119 (app)
Echiniscoides, 35 (t)
Echiniscus, 18
　E. *columinis*, 18, 35, 35 (t)
　E. *granulatus*, 35 (t)
　E. *militaris*, 18, 35, 35 (t)
　E. *quadrispinosus quadrispinosus*, 23, 35, 35 (t)
　E. *testudo*, 35 (t)
　E. *trisetosus*, 35 (t)
Ecliptopera silaceata, 125 (app)
Ectoedemia occultella, 116 (app)
Edwardzetes edwardsi, 76 (t)
Elachista
　E. *alpinella*, 117 (app)
　E. *argentella*, 117 (app)
　E. *atricomella*, 117 (app)
　E. *canapennella*, 117 (app)
　E. *consortella*, 117 (app)
Elaphrus cupreus, 103(t)
Elophila nymphaeata, 122 (app)
Ematurga atomaria, 124 (app)
Emberiza
　E. *calandra*, 235
　E. *citronella*, 235
　E. *schoeniculus*, 234
Endochironomus
　E. *albipennis*, 158 (t) (app), 176 (app)
　E. *tendens*, 141 (t), 158 (t) (app), 176 (app)
Endothenia quadrimaculana, 119 (app)
Endrosis sarcitrella, 117 (app)
Enochrus
　E. *affinis*, 86 (t)
　E. *coarctatus*, 86 (t)
　E. *fuscipennis*, 86 (t), 97
Enoplognatha
　E. *ovata*, 46, 66 (app)
　E. *thoracica*, 66 (app)
Entelecara erythropus, 46, 56 (app)
Entomobrya nicoletii, 71 (t)
Epermenia chaerophyllella, 120 (app)
Epiblema
　E. *cirsiana*, 120 (app)
　E. *costipunctana*, 111, 111 (t)
　E. *scutulana*, 120 (app)
Epicrius mollis, 72 (t)
Epilohmannia sp., 75 (t)
Epinotia
　E. *brunnichana*, 120 (app)

E. caprana, 120 (app)
E. cruciana, 111 (t)
E. immundana, 120 (app)
E. sordidana, 109, *109*, 112, 120 (app)
E. subocellana, 120 (app)
E. trigonella, 120 (app)
Epirrhoe alternata, 125 (app)
Eratigena, 47
Erigone
  E. arctica, 48
  E. atra, 49, 56 (app)
  E. dentipalpis, 43 (t), 46, 49, 56 (app)
  E. promiscua, 56 (app)
Erigonella hiemalis, 56–57 (app)
Erinaceous europaeus, 238 (t), 244
Erithacus rubecula, 234
Esolus parallelepipedus, 86 (t), 91
Esox lucius, 220, 225
Euconulus cf. *fulvus*, 10 (t), 14
Eucosma
  E. campoliliana, 120 (app)
  E. cana, 120 (app)
Eucyclops
  E. denticulatus, 217
  E. lilljeborgi, 216 (t), 217
  E. macruroides, 217
  E. serrulatus, 217
Eucypris virens, 198, 200, 201, 202, 206
Eudiaptomus gracilis, 216, 216 (t)
Eudonia
  E. mercurella, 121 (app)
  E. pallida, 121 (app)
  E. truncicolella, 121 (app)
Euglesa, 12
  E. casertana, 10 (t), 12
  E. hibernica, 9 (t), 10 (t), 12, 14
  E. lilljeborgii, 9 (t), 10 (t), 12, 14
  E. milium, 10 (t), 12
  E. nitida, 9 (t), 11 (t)
  E. obtusalis, 9 (t), 11 (t)
  E. personata, 9 (t), 11 (t)
  E. subtruncata, 9 (t), 11 (t)
Eukiefferiella
  E. ancyla, 156 (t) (app), 168 (app)
  E. brevicalcar, 156 (t) (app), 168 (app)
  E. claripennis, 140 (t), 156 (t) (app), 168 (app)
  E. clypaeata, 156 (t) (app)
  E. clypeata, 168 (app)
  E. coerulescens, 156 (t) (app), 168 (app)
  E. cyanea, 140 (t), 156 (t) (app), 168 (app)
  E. devonica, 140 (t), 156 (t) (app), 168 (app)
  E. dittmari, 156 (t) (app), 168 (app)
  E. ilkleyensis, 156 (t) (app), 168 (app)
  E. minor / fittkaui, 140 (t), 169 (app)
  E. tirolensis, 156 (t) (app), 168 (app)
Eulaelaps stabularis, 72 (t)
Eulithis populata, 125 (app)
Euophrys
  E. frontalis, 48
  E. petrensis, 47, 49, 64 (app)
Eupelops
  E. acromios, 77 (t)
  E. plicatus, 77 (t)
Euphyia unangulata, 125 (app)
Eupithecia
  E. absinthiata, 126 (app)
  E. centaureata, 126 (app)
  E. nanata, 126 (app)
  E. pulchellata, 126 (app)
  E. subfuscata, 126 (app)
Euplexia lucipara, 127 (app)
Eupoecilia angustana, 111 (t), 113
Eurycercus lamellatus, 213 (t)
Euthrix potatoria, 123 (app)
Euxoa tritici, 131 (app)
Euzetes globulus, 76 (t)
Exoteleia dodecella, 118 (app)

F
Felis catus, 241
Ficedula hypoleuca, 235
Folsomia
  F. candida, 70, 71 (t)
  F. fimetaria, 71 (t)
  F. quadrioculata, 70, 71 (t)
Formica lemani, 192 (t)
Fractonotus caelatus, 35, 36 (t)
Friesea mirabilis, 70, 71 (t)
Fringilla coelebs, 235
Fulica atra, 234
Fulmarus glacialis, 232, *232*

G
Galba truncatula, 9 (t), 11 (t)
Gallinago gallinago, 234
Gallinula chloropus, 234
Gamasodes
  G. fimbriatus, 73 (t)
  G. spiniger, 73 (t)
  G. spinipes, 73 (t)
Gamasus
  G. (Pergamasus) crassipes, 73 (t)
  G. (Ologamasus) inornatus, 73 (t)
  G. (Pergamasus) lapponicus, 73 (t)
  G. (Ologamasus) pollicipatus, 73 (t)
  G. (Pergamasus) robustus, 73 (t)
  G. runcatellus, 73 (t)
Gasterosteus aculeatus, 219
Geholaspis
  G. aeneus, 72 (t)
  G. longispinosus, 72 (t)
Geophilus
  G. easoni, 80
  G. flavus, 80
  G. insculptus, 80
  G. truncorum, 80
Georthocladius luteicornis, 156 (t) (app), 169 (app)
Glomeris marginata, 79
Glyphipterix thrasonella, 117 (app)
Glyptotendipes
  G. barbipes, 158 (t) (app), 177 (app)
  G. cauliginellus, 141 (t), 158 (t) (app), 177 (app)
  G. (Caulochironomus) foliicola, 158 (t) (app), 176 (app)
  G. gripkoveni, 177 (app)
  G. (Caulochironomus) imbicilis, 176 (app)
  G. pallens, 141 (t), 158 (t) (app), 177 (app)
  G. paripes, 141 (t), 158 (t) (app), 177 (app)
  G. (Caulochironomus) scirpi, 158 (t) (app), 176 (app)
  G. (Caulochironomus) viridis, 139 (t), 144, 158 (t) (app), 176 (app)
Gnathonarium dentatum, 43 (t), 44, 45, 57 (app)
Gnophos obfuscata, 112, *112*, 124 (app)
Gonatium rubens, 57 (app)
Gongylidiellum vivum, 57 (app)
Gongylidium rufipes, 44, 57 (app)
Gracillaria syringella, 116 (app)
Grampus griseus, 238
Graptoleberis testudinaria, 213 (t)
Gymnoscelis rufifasciata, 126 (app)
Gypsonoma dealbana, 120 (app)
Gyraulus crista, 11 (t)
Gyrinus
  G. marinus, 85 (t)
  G. minutus, 85 (t), 98
  G. substriatus, 85 (t), 98
Gyrodactylus salar, 226

H
Hada plebeja, 129 (app)
Hadena perplexa capsophila, 129 (app)
Hahnia montana, 48
Halichoerus grypus, 238, 238 (t)
Haliplus
  H. confinis, 85 (t), *91*
  H. fulvus, 85 (t), 90, 98
  H. lineatocollis, 85 (t), 98
  H. ruficollis, 85 (t), 88, 90
  H. sibircus, 85 (t), 88, 90, 91
Halocladius
  H. fucicola, 140 (t), 142, 144, 156 (t) (app), 169 (app)
  H. variabilis, 140 (t), 142, 144, 156 (t) (app), 169 (app)
  H. varians, 144, 156 (t) (app), 169 (app)
Halorates reprobus, 47, 57 (app)
Haplodrassus signifer, 48
Haplophthalmus mengei, 81
Harnischia curtilamellata, 158 (t) (app), 177 (app)
Harpactea hombergi, 46, 53 (app)
Harpalus
  H. affinis, 103(t)
  H. latus, 103(t)
  H. rufipes, 103(t)
Hedya
  H. nubiferana, 119 (app)
  H. pruniana, 119 (app)
Heleniella ornaticollis, 156 (t) (app), 169 (app)
Helicella itala, 9 (t), 11 (t)
Heliophanus cupreus, 46, 65 (app)
Heliozela hammoniella, 116 (app)
Hellinsia tephradactyla, 112, 121 (app)
Helochares punctatus, 86 (t), 96, 97
Helophorus
  H. aequalis, 86 (t), 89
  H. aquaticus, 89
  H. brevipalpis, 86 (t), 90
  H. flavipes, 86 (t), 96
  H. grandis, 86 (t), 89
  H. granularis, 86 (t), 89, 90
  H. obscurus, 86 (t)
Helotropha leucostigma, 129 (app)
Hepialus humuli, 111 (t)

Hermannia
   H. bistriata, 75 (t)
   H. convexa, 75 (t)
   H. gibba, 75 (t)
   H. nanus, 75 (t)
   H. reticulata, 75 (t)
   H. scabra, 75 (t)
Herpetocypris
   H. chevreuxi, 198, 202–203, 204, 205
   H. reptans, 203, 205
Heterocypris
   H. incongruens, 198, 200, 203, 206
   H. salina, 198, 200, 203, 205
Heterotanytarsus apicalis, 140 (t), 156 (t) (app), 169 (app)
Heterotrissocladius
   H. grimshawi, 156 (t) (app), 169 (app)
   H. marcidus, 140 (t), 156 (t) (app), 169 (app)
Hilaira excisa, 48
Hipparchia semele, 111 (t), 113
Hofmannophila pseudospretella, 118 (app)
Holoparasitus
   H. inornatus, 73 (t)
   H. stramenti, 73 (t)
Holostaspis
   H. longispinosus, 72 (t)
   H. longulus, 72 (t)
Hoploderma magnum, 75 (t)
Hoplodrina
   H. blanda, 127 (app)
   H. octogenaria, 127 (app)
Hydraecia micacea, 129 (app)
Hydriomena
   H. furcata, 125 (app)
   H. ruberata, 125 (app)
Hydrobates pelagicus, 233
Hydrobius fuscipes, 86 (t), 90, 97
Hydromya dorsalis, 185, 186 (t)
Hydroporus
   H. discretus, 85 (t), 90
   H. erythrocephalus, 85 (t), 90
   H. gyllenhalii, 85 (t), 90, 97
   H. incognitus, 85 (t)
   H. longulus, 85 (t), 89
   H. memnonius, 85 (t), 90
   H. nigrita, 85 (t), 91
   H. obscurus, 85 (t), 91, 96, 97, 98
   H. obsoletus, 85 (t)
   H. palustris, 85 (t), 89, 90
   H. planus, 85 (t), 90
   H. pubescens, 85 (t)
   H. tessellatus, 85 (t)
   H. tristis, 85 (t), 90
Hydroschendyla submarina, 80
Hygrotus
   H. impressopunctatus, 85 (t), 96
   H. inaequalis, 85 (t), 90, 98
Hypechiniscus
   H. exarmatus, 35 (t)
   H. gladiator gladiator, 35 (t)
Hypena proboscidalis, 127 (app)
Hypenodes humidalis, 127 (app)
Hypochthonius rufulus, 75 (t)
Hypogastrura viatica, 71 (t)
Hypomma bituberculatum, 42 (t), 43, 43 (t), 44, 45, 46, 57 (app)

Hypsibius, 25
   H. arcticus, 36 (t)
   H. dujardini, 24, 36, 36 (t)
   H. microps, 24
   H. pallidoides, 17, 24–25, 35, 36 (t)
   H. pallidus, 25
Hypsosinga pygmaea, 48

I

Idaea
   I. aversata, 125 (app)
   I. biselata, 125 (app)
   I. dimidiata, 125 (app)
Idotea pelagica, 241
Ilione lineata, 186, *186*, 186 (t)
Ilybius
   I. fuliginosus, 85 (t)
   I. guttiger, 85 (t), 89
   I. montanus, 85 (t), 89, 90, 97
Ilyocryptus sordidus, 214 (t)
Isohypsibius
   I. annulatus annulatus, 36 (t)
   I. papillifer bulbosus, 35, 36 (t)
   I. sattleri, 17, 25, 35, 36 (t)
   I. schaudinni, 36 (t)
   I. tuberculatus, 36 (t)
Isotoma
   I. anglicana, 71 (t), 74, *74*
   I. olivacea, 71 (t)
   I. sensibilis, 71 (t)
   I. viridis, 70, 71 (t), 74, *74*
Isotomiella minor, 71 (t)
Isotomurus palustris, 70, 71 (t)
Iulus luscus, 79

K

Kloosia pusilla, 139 (t), 144, 158 (t) (app), 177 (app)
Krenopelopia nigropunctata, 140 (t), 154 (t) (app), 162 (app)
Krenosmittia camptophleps, 156 (t) (app), 169 (app)

L

Lacanobia oleracea, 129 (app)
Laccobius
   L. atratus, 86 (t), 89
   L. bipunctatus, 86 (t)
   L. minutus, 86 (t), 89
   L. ytenensis, 86 (t), 89
Laccophilus minutus, 85 (t), 90
Lacinius ephippiatus, 81
Laelaps
   L. agilis, 72 (t)
   L. (Pseudoparasitus) meridionalis, 72 (t)
   L. (Eulaelaps) stabularis, 72 (t)
   L. (Ololaelaps) tumidulus, 72 (t)
Lamyctes emarginatus, 80
Laothoe populi, 123 (app)
Larinioides cornutus, 46, 52 (app)
Larsia
   L. atrocincta, 140 (t), 154 (t) (app), 162 (app)
   L. curticalcar, 154 (t) (app), 162 (app)
Larus
   L. anus, 233
   L. argentatus, 233
   L. canus, 233
   L. fuscus, 233

Lasiocampa quercus, 122 (app)
Lasiommata megera, 112, 124 (app)
Lasius
   L. flavus, 190–191, 192 (t), 194, *194*, 195, *195*, 196
   L. niger, 192 (t), 194, 196
   L. platythorax, 191, 192 (t), 194, 195, 196
Lauria cylindracea, 9 (t), 11 (t)
Lauterborniella agrayloides, 141 (t), 159 (t) (app), 177 (app)
Lehmannia marginata, 9 (t), 11 (t)
Leiobunum blackwalli, 81
Leiobunum rotundum, 81
Leiostyla anglica, 9 (t), 11 (t), 12, 15
Leistus fulvibarbis, 103(t), 104, 106
Lepidocyrtus cyaneus, 71 (t)
Lepthyphantes leprosus, 48
Leptodora kindtii, 215
Leptoiulus belgicus, 80
Leptorhoptrum robustum, 57 (app)
Leptothorax acervorum, 193
Lepus
   L. europaeus, 238 (t)
   L. timidus hibernicus, 237, 238 (t)
Leydigia leydigi, 213 (t)
Liacarus
   L. coracinus, 76 (t)
   L. ovatus, 76 (t)
Liebstadia
   L. humerata, 74, 76 (t)
   L. similis, 76 (t)
Ligia oceanica, 81
Limacus
   L. flavus, 13
   L. maculatus, 9 (t), 11 (t), 13
Limax
   L. cinereoniger, 11 (t), 12, 14–15
   L. maximus, 9 (t), 11 (t)
Limnebius truncatellus, 86 (t)
Limnia unguicornis, 186 (t), 187
Limnophyes
   L. angelicae, 140 (t), 142, 156 (t) (app), 169 (app)
   L. asquamatus, 144, 156 (t) (app), 169 (app)
   L. gurgicola, 140 (t), 156 (t) (app), 169 (app)
   L. habilis, 140 (t), 156 (t) (app), 169 (app)
   L. minimus, 140 (t), 156 (t) (app), 169–170 (app)
   L. natalensis, 140 (t), 156 (t) (app), 170 (app)
   L. pentaplastus, 140 (t), 156 (t) (app), 170 (app)
   L. pumilio, 156 (t) (app), 170 (app)
Linyphia triangularis, 46, 57 (app)
Liocranoeca striata, 47, 49, 53 (app)
Liopterus haemorrhoidalis, 85 (t), 89
Lipophrys pholis, 241
Lithobius
   L. borealis, 80
   L. crassipes, 80
   L. forficatus, 80
   L. melanops, 80
   L. microps, 81
   L. variegatus, 80, *80*
Litoligia literosa, 128 (app)
Lobesia littoralis, 120 (app)

Lobochironomus dissidens, 174 (app)
Lomaspilis marginata, 124 (app)
Longicheles mandibularis, 72 (t), 74
Lophomma punctatum, 57 (app)
Loricera pilicornis, 103(t)
Luperina testacea, 128–129 (app)
Lutra lutra, 237, 238 (t)
Lycaena phlaeas, 123 (app)
Lycophotia porphyrea, 131 (app)
Lyonetia clerkella, 117 (app)
Lysigamasus
   L. armatus, 73 (t), 74
   L. celticus, 73 (t), 74
   L. lapponicus, 73 (t), 74
   L. runcatellus, 73 (t)

M
Macrobiotus, 18, 25, 31, 34
   M. almadi, 32
   M. crenulatus, 25, 36, 36 (t)
   M. echinogenitus, 36 (t)
   M. harmsworthi, 34, 36
   M. harmsworthi harmsworthi, 36 (t)
   M. hufelandi, 17, 25, 35, 36, 36 (t), 37–38
      egg types, 32–34, *33*, 37
   M. hufelandi hufelandi, 32, 37
   M. hufelandi sensu stricto, 37
   M. iharosi, 32
   M. joannae, 33–34
   M. martini, 32
   M. occidentalis occidentalis, 36 (t)
   M. sandrae, 32
   M. sapiens, 32
   M. serratus, 33
   M. terminalis, 32
   M. trunovaue, 33
   M. virgatus, 36 (t)
   M. vladimiri, 32
   morphotypes, 26, *26*, 27, *27*, *28*, 28–29, *29*, 29–30, *30*, 30–31, 31–32, 37–38
Macrocheles
   M. glaber, 73 (t)
   M. opacus, 73 (t)
   M. submotus, 73 (t), 74
   M. tridentinus, 73 (t)
Macrocyclops albidus, 216 (t)
Macropelopia, 141–142
   M. adaucta, 140 (t), 141, 154 (t) (app), 162–163 (app), 162 (app)
   M. nebulosa, 139 (t), 140 (t), 141, 154 (t) (app), 162–163 (app)
   M. notata, 140 (t), 141, 154 (t) (app), 162–163 (app), *163* (app)
Macrosternodesmus palicola, 79, 80
Macrothrix
   M. laticornis, 214 (t)
   M. rosea, 214 (t)
Macrothylacia rubi, 111 (t), 113, 114
Malaconothrus monodactylus, 75 (t)
Mamestra brassicae, 129 (app)
Maniola jurtina, 124 (app)
Martes martes, 238 (t), 240
Maso sundevalli, 48
Medoppia obsoleta, 76 (t)
Megabunus diadema, 81
Megacyclops viridis, 216 (t)
Megalothorax minimus, 72 (t)
Megasternum concinnum, 86 (t)

Melanchra persicariae, 129 (app)
Melanozetes
   M. meridianus, 74, 76 (t)
   M. mollicomus, 76 (t)
   M. stagnatilis, 76 (t)
Meles meles, 238 (t), 240
Mergus serrator, 234
Mermis myrmecophila, 190
Mesapamea
   M. secalella, 128 (app)
   M. secalis, 128 (app)
Mesaphorura krausbaueri, 71 (t)
Mesobiotus cf. *harmsworthi*, 34
Mesocra rapiens, 217
Mesostigmata, 69, 72, 74
Mesotype didymata, 126 (app)
Meta menardi, 48
Metellina
   M. mengei, 43, 46, 65 (app)
   M. merianae, 46, 65 (app)
   M. segmentata, 46, 65 (app)
Metriocnemus
   M. ephemerus, 144, 145, 156 (t) (app), 170 (app)
   M. eurynotus, 140 (t), 156 (t) (app), 170 (app)
   M. fuscipes, 139 (t), 140 (t), 144, 156 (t) (app), 170 (app), *170* (app)
   M. hygropetricus, 170 (app)
   M. modestus, 139 (t), 144, 170 (app), 179 (app)
   M. picipes, 140 (t), 156 (t) (app), 170 (app)
   M. sibericus, 144
Micaria pulicaria, 48
Micrargus herbigradus, 57 (app)
Micrargus subaequalis, 58 (app)
Microctenonyx subitaneus, 47, 58 (app)
Micropsectra
   M. apposita, 180 (app)
   M. atrofasciata, 141 (t), 160 (t) (app), 180 (app)
   M. attenuata, 160 (t) (app), 180–181 (app)
   M. bidentata, 181 (app)
   M. fusca, 181 (app)
   M. junci, 139 (t), 141 (t), 143, 160 (t) (app), 181 (app)
   M. lindebergi, 160 (t) (app), 181 (app)
   M. lindrothi, 141 (t), 160 (t) (app), 181 (app)
   M. notescens, 160 (t) (app), 181 (app)
   M. pallidula, 141 (t), 160 (t) (app), 181 (app)
   M. roseiventris, 141 (t), 160 (t) (app), 181 (app)
Micropterix
   M. aruncella, 116 (app)
   M. aureatella, 111 (t), 114
   M. calthella, 116 (app)
Microtendipes
   M. chloris, 141 (t), 159 (t) (app), 177 (app)
   M. diffinis, 141 (t), 159 (t) (app), 177 (app)
   M. pedellus, 139 (t), 141 (t), 159 (t) (app), 177 (app)
   M. rydalensis, 159 (t) (app), 177 (app)

Milax gagates, 9 (t), 11 (t)
Milnesium, 18, 36, 36 (t)
   M. tardigradum, 36
   M. tardigradum tardigradum, 36
Minibiotus
   M. allani, 34
   M. aquatilis, 34
   M. continuus, 34
   M. crassidens, 34
   M. decrescens, 34
   M. floriparus, 34
   M. hispidus, 34
   M. intermedius, 34, 36–37, 36 (t)
   M. maculartus, 34
   M. milleri, 34
   M. stuckenbergi, 34
   M. subintermedius, 34
   M. taiti, 34
   M. weinerorum, 34
Minonthozetes pseudofusiger, 76 (t)
Minunthozetes semirufus, 76 (t)
Mitopus morio, 81
Mniotype adusta, 128 (app)
Monocephalus
   M. castaneipes, 44, 49, 58 (app)
   M. fuscipes, 43 (t), 44, 58 (app)
Monochroa
   M. cytisella, 118 (app)
   M. lucidella, 112, 118 (app)
Monopelopia tenuicalcar, 140 (t), 154 (t) (app), 163 (app)
Monopis laevigella, 116 (app)
Monospilus dispar, 215
Morus bassanus, 233
Murrayon, 21
   M. hastatus, 36 (t)
   M. hibernicus, 18, 35, 36 (t)
Mus domesticus, 237, 238 (t)
Mustela erminea, 237, 238 (t)
Mycobates sp., 76 (t)
Myodes glareolus, 238 (t), 244
Myotis
   M. daubentonii, 238 (t)
   M. nattereri, 238 (t), 242
Myrmica
   M. ruginodis, 192 (t)
   M. scabrinodis, 192 (t)
Mythimna
   M. conigera, 129 (app)
   M. impura, 130 (app)
   M. unipuncta, 113, 130 (app)
Mytilis spp., 241

N
Nanhermannia nana, 75 (t)
Nanocladius
   N. balticus, 156 (t) (app), 170 (app)
   N. dichromus, 156 (t) (app), 170 (app)
   N. rectinervis, 156 (t) (app), 170 (app)
Nanogona polydesmoides, 79, 80
Natarsia punctata, 140 (t), 154 (t) (app), 163 (app)
Neanura muscorum, 71 (t)
Nebria brevicollis, 104(t)
Nebria salina, 104(t), 105
Nebrioporus assimilis, 85 (t), 89, 91
Nebula salicata latentaria, 111 (t)
Necora puber, 241

Nelima
    N. gothica, 81
    N. gothica Lohmander, 81
Nemastoma bimaculatum, 81
Neobisium
    N. carcinoides, 81
    N. maritimum, 81
Neofaculta ericetella, 118 (app)
Neon reticulatus, 48
Neovison vison, 238 (t), 240
Neozavrelia
    N. cuneipennis, 144, 160 (t) (app), 181 (app)
    N. luteola, 160 (t) (app), 181 (app)
Neriene
    N. clathrata, 58 (app)
    N. montana, 43 (t), 44, 49, 58 (app)
    N. peltata, 44, 58 (app)
Nesovitrea hammonis, 9 (t), 11 (t)
Nesticus cellulanus, 64 (app)
Nilotanypus dubius, 140 (t), 154 (t) (app), 163 (app)
Nitocra spinipes, 205, 217
Nitzscia sturionis, 226
Noctua
    N. comes, 131 (app)
    N. interjecta, 131 (app)
    N. janthe, 131 (app)
    N. pronuba, 130–131 (app)
Nomophila noctuella, 122 (app)
Notaspis
    N. bipilis, 76 (t)
    N. exilis, 77 (t)
    N. lucorum, 77 (t)
    N. similis, 76 (t)
Noterus clavicornis, 85 (t), 89, 90, 91, 98
Nothrus
    N. anauniensis, 74, 75 (t)
    N. palustris, 75 (t)
    N. segnis, 75 (t)
    N. silvestris, 75 (t)
    N. spinifer, 75 (t)
Notiophilus
    N. aquaticus, 104(t)
    N. biguttatus, 104(t)
    N. palustris, 104(t)
    N. substriatus, 104(t)
Notodonta ziczac, 126 (app)
Nubensia nubens, 159 (t) (app), 177 (app)
Nudaria mundana, 132 (app)
Numenius
    N. arquata, 233
    N. pheopus, 233
Nyctalus leisleri, 241

O
Obscuriphantes obscurus, 58 (app)
Ochropleura plecta, 130 (app)
Ochsenheimeria taurella, 112, 117 (app)
Ochthebius
    O. lejolisii, 86 (t), 89
    O. minimus, 86 (t), 89
    O. punctatus, 86 (t), 89
Odeles marginata, 86 (t)
Odontocepheus elongatus, 75 (t)
Odontomesa fulva, 165 (app)
Oedothorax
    O. fuscus, 42 (t), 43 (t), 45, 58 (app)
    O. gibbosus, 59 (app)
    O. gibbosus f. tuberosus, 42
    O. retusus, 43, 45, 59 (app)
Oenanthe oenanthe, 234
Oligia fasciuncula, 128 (app)
Oligolophus tridens, 81
Olisthopus rotundatus, 104(t)
Ololaelaps venetus, 72 (t)
Oniscus, 81
    O. asellus, 81
Oonops pulcher, 46, 64 (app)
Ophyiulus pilosus, 80
Opiliones, 79, 81
Opisthograptis luteolata, 124 (app)
Orchesella
    O. cincta, 71 (t)
    O. villosa, 71 (t)
Orcinus orca, 238
Orgyia antiqua, 132 (app)
Oribata
    O. alpina, 76 (t)
    O. cuspidata, 76 (t)
    O. edwardsi, 76 (t)
    O. fusigera, 76 (t)
    O. globula, 76 (t)
    O. gracilis, 76 (t)
    O. mollicoma, 76 (t)
    O. ovalis, 76 (t)
Oribatida, 74, 75–77 (t)
Orthocladius
    O. (Euorthocladius) ashei, 156 (t) (app), 171 (app)
    O. (Pogonocladius) consobrinus, 157 (t) (app), 171 (app)
    O. dentifer, 140 (t), 157 (t) (app), 171 (app)
    O. (Mesorthocladius) frigidus, 157 (t) (app), 171 (app)
    O. (Eudactylocladius) fuscimanus, 140 (t), 156 (t) (app), 170 (app)
    O. glabripennis, 157 (t) (app), 171 (app)
    O. (Symposiocladius) holsatus, 144, 157 (t) (app), 171 (app)
    O. oblidens, 140 (t), 157 (t) (app), 171 (app)
    O. (Eudactylocladius) olivaceus, 144, 156 (t) (app), 170 (app)
    O. (Euorthocladius) rivicola, 156 (t) (app), 171 (app)
    O. rivinus, 157 (t) (app), 171 (app)
    O. rubicundus, 157 (t) (app), 171 (app)
    O. ruffoi, 157 (t) (app), 171 (app)
    O. sordidellus, 139 (t), 172 (app)
    O. wetterensis, 157 (t) (app), 171 (app)
Orthonama vittata, 125 (app)
Orthosia
    O. cerasi, 130 (app)
    O. gothica, 130 (app)
    O. gracilis, 130 (app)
    O. incerta, 130 (app)
Orthotaenia undulana, 119 (app)
Oryctolagus cuniculus, 237, 238 (t)
Oxychilus
    O. alliarius, 9 (t), 11 (t)
    O. cellarius, 9 (t), 11 (t)
Oxyloma elegans, 9 (t), 11 (t)
Ozyptila
    O. atomaria, 67 (app)
    O. brevipes, 67 (app)
    O. trux, 67 (app)

P
Pachygnatha
    P. clercki, 42 (t), 43, 43 (t), 44, 45, 65 (app)
    P. degeeri, 42 (t), 43, 43 (t), 45, 46, 65–66 (app)
Pachylaelaps, 73 (t)
Pachyseius humeralis, 73 (t)
Pagastiella orophila, 141 (t), 159 (t) (app), 177–178 (app)
Palaemon serratus, 241
Palliduphantes ericaeus, 59 (app)
Pandemis
    P. cerasana, 119 (app)
    P. heparana, 119 (app)
Parachipteria
    P. italicus, 76 (t)
    P. ovalis, 76 (t)
Parachironomus, 143
    P. arcuatus, 178 (app)
    P. biannulatus, 143
    P. cinctellus, 141 (t), 143, 159 (t) (app), 178 (app)
    P. frequens, 159 (t) (app), 178 (app)
    P. gracilor, 159 (t) (app), 178 (app)
    P. mauricii, 141 (t), 143, 146, 159 (t) (app), 178 (app)
    P. parilis, 141 (t), 143, 159 (t) (app), 178 (app)
    P. Pe 2a, 178 (app)
    P. tenuicaudatus, 141 (t), 143, 159 (t) (app), 178 (app)
    P. vitiosus, 141 (t), 143, 159 (t) (app), 178 (app)
Paracladius conversus, 157 (t) (app), 171 (app)
Paracladopelma
    P. camptolabis, 159 (t) (app), 178 (app)
    P. laminatum, 159 (t) (app), 178 (app)
    P. nigritulum, 159 (t) (app), 178 (app)
Paracyclops
    P. affinis, 217
    P. fimbriatus, 217
    P. prasinus, 217
Paracymus scutellaris, 86 (t)
Paradromius linearis, 104(t)
Parakiefferiella
    P. bathophila, 157 (t) (app), 171 (app)
    P. coronata, 157 (t) (app), 171 (app)
    P. scandica, 157 (t) (app), 171 (app)
    P. smolandica, 157 (t) (app), 171 (app)
Paralauterborniella nigrohalteralis, 159 (t) (app), 178 (app)
Paramacrobiotus
    P. areolatus, 36 (t)
    P. richtersi, 18, 35, 36 (t)
Parametriocnemus stylatus, 140 (t), 157 (t) (app), 171 (app)
Paranchus albipes, 104(t)
Parapelecopsis nemoralis, 44, 59 (app)
Paraphaenocladius
    P. impensus, 140 (t)
    P. impensus impensus, 157 (t) (app), 172 (app)
    P. irritus, 140 (t)
    P. irritus irritus, 157 (t) (app), 172 (app)
    P. pseudirritus, 140 (t)
    P. pseudirritus subsp. pseudirritus, 172 (app)

Paraquimperia tenerrima, 222 (t), 223
Pararge aegeria tircis, 123 (app)
Paraseius serratus sp. nov., 72 (t)
Paratanytarsus
    P. bituberculatus, 160 (t) (app), 181 (app)
    P. brevicalcar, 141 (t), 160 (t) (app), 181 (app)
    P. dissimilis, 160 (t) (app), 181 (app)
    P. inopertus, 160 (t) (app), 181 (app)
    P. laccophilus, 160 (t) (app), 181 (app)
    P. laetipes, 160 (t) (app), 181 (app)
    P. lauterborni, 160 (t) (app), 182 (app)
    P. Paratanytarus intricatus, 181 (app)
    P. penicillatus, 160 (t) (app), 182 (app)
    P. tenuis, 160 (t) (app), 182 (app)
Pardosa
    P. amentata, 43 (t), 45, 62 (app)
    P. nigriceps, 43 (t), 46, 62–63 (app)
    P. palustris, 49, 63 (app)
    P. pullata, 42 (t), 43, 43 (t), 45, 46, 63 (app)
Parisotoma notabilis, 70, 71 (t)
Paroligolophus agrestis, 81
Passer montanus, 235
Patella spp., 241
Patrobus assimilis, 104(t), 105
Pelopia, 139 (t), 144, 164 (app)
    P. (Tanypus) culiciformis, 139 (t), 144, 163 (app)
Pelops
    P. acromias, 77 (t)
    P. fuliginosus, 77 (t)
Pentapedilum sordens, 144
Penthesilenula
    P. brasiliensis, 200
    P. malayica, 198, 200, 201, 204, 206
Peponocranium ludicrum, 46, 59 (app)
Pergamasus
    P. crassipes, 73 (t), 74
    P. longicornis, 73 (t)
    P. robustus, 73 (t)
    P. septentrionalis, 73 (t)
Perizoma
    P. albulata, 111 (t)
    P. alchemillata, 126 (app)
    P. blandiata, 126 (app)
Petrophora chlorosata, 124 (app)
Phaenopsectra
    P. flavipes, 139 (t), 141 (t), 159 (t) (app), 178 (app)
    P. punctipes, 139 (t), 141 (t), 159 (t) (app), 179 (app)
Phalacrocorax carbo carbo, 233
Phalangium opilio, 81
Phalera bucephala, 114, 126 (app)
Pharmacis fusconebulosa, 116 (app)
Phasanius colchicus, 235
Phauloppoa lucorum, 77 (t)
Pheosia gnoma, 126 (app)
Pherbellia
    P. (Sciomyza) cinerella, 185, 186
    P. ventralis, 186 (t)
Pherbina coryleti, 185, 186, *186*, 186 (t)
Pheromermis myrmecophila, 191, 194
Philodromus cespitum, 46, 64 (app)
Philoscia muscorum, 81
Phlogophora meticulosa, 128 (app)
Phocaena phocaena, 238
Phoca vitulina, 237, 238 (t)

Pholcomma gibbum, 66 (app)
Pholcus phalangioides, 47, 64 (app)
Pholis gunnellus, 241
Photedes minima, 128 (app)
Phragmatobia fuliginosa, 111 (t), 113
Phrurolithus festivus, 46, 64 (app)
Phthiacarus
    P. affinis, 75 (t)
    P. arduus, 75 (t)
Phycitodes saxicola, 121 (app)
Phylloneta sisyphia, 46, 67 (app)
Phyllonorycter
    P. quinqueguttella, 112, 116 (app)
    P. salicicolella, 116 (app)
    P. ulmifoliella, 116 (app)
Phylloscopus
    P. inornatus, 235
    P. trochilus, 235
Pieris
    P. brassicae, 123 (app)
    P. napi, 123 (app)
    P. rapae, 123 (app)
Pipistrellus
    P. pipistrellus, 241
    P. pygmaeus, 241
Pirata piraticus, 42 (t), 43, 43 (t), 45, 63 (app)
Piratula hygrophila, 43 (t), 44, 45, 63 (app)
Piscicola geometra, 224
Pisidium spp., 12
Plateumaris sericea, 86 (t), 97
Platichthys flesus, 219
Platicrista angustata, 36 (t)
Platynothrus peltifer, 75 (t)
Platyptilia
    P. isodactyla, 114
    P. isodactylus, 111 (t)
Plecotus auritus, 238 (t), 241, 243
Pleurota bicostella, 118 (app)
Pleuroxus
    P. aduncus, 213 (t)
    P. denticulatus, 214 (t)
    P. laevis, 214 (t)
    P. truncata, 214 (t)
Plusia festucae, 127 (app)
Plutella xylostella, 117 (app)
Pocadicnemis pumila, 59 (app)
Podocotyle atomon, 222 (t), 223, 225
Poeciloneta variegata, 59 (app)
Poecilus
    P. cupreus, 104(t)
    P. versicolor, 104(t)
Polydesmus
    P. angustus, 79
    P. coriaceus, 80
    P. inconstans, 79
Polyommatus icarus, 123 (app)
Polypedilum
    P. acutum, 141 (t), 143, 159 (t) (app), 179 (app)
    P. arundineti, 141 (t), 159 (t) (app), 179 (app)
    P. (Uresipedilum) convictum, 159 (t) (app), 179 (app)
    P. (Uresipedilum) cultellatum, 159 (t) (app), 179 (app)
    P. nubeculosum, 139 (t), 159 (t) (app), 179 (app)

P. (Pentapedilum) nubens, 177 (app)
P. pedestre, 141 (t), 159 (t) (app), 179 (app)
P. (Tripodura) pullum, 141 (t), 159 (t) (app), 179 (app)
P. (Pentapedilum) sordens, 139 (t), 141 (t), 159 (t) (app), 170 (app), 179 (app)
P. sp Pe7 sensu, 179 (app)
P. sp Pe8 sensu, 179 (app)
P. uncinatum, 143
P. (Pentapedilum) uncinatum, 141 (t), 159 (t) (app), 179 (app)
Polyphemus pediculus, 215
Pomatoschistus microps, 241
Pomphorhynchus laevis, 222 (t), 223, 225, 226
Porcellina platycheles, 241
Porcellio
    P. dilatatus, 81, 82
    P. scaber, 81
    P. spinicornis, 81, 82
Porcellionides cingendus, 82
Porobelba spinosa, 76 (t)
Porrhomma montanum, 47, 60 (app)
Potamocypris villosa, 198, 200, 204, 206
Potamopyrgus antipodarum, 9 (t), 11 (t), 12
Potthastia
    P. gaedii, 139 (t), 140 (t), 142, 155 (t) (app), 165 (app)
    P. longimanus, 155 (t) (app), 165 (app)
Proasellus meridianus, 225
Procladius
    P. (Holotanypus) choreus, 140 (t), 154 (t) (app), 163 (app)
    P. (Holotanypus) crassinervis, 144, 154 (t) (app), 163 (app)
    P. (Holotanypus) culiciformis, 139 (t), 154 (t) (app), 163 (app)
    P. (Psilotanypus) flavifrons, 154 (t) (app), 164 (app)
    P. (Psilotanypus) lugens, 154 (t) (app), 164 (app), 164 (app)
    P. nervosus, 139 (t), 162 (app)
    P. (Psilotanypus) rufovittatus, 139 (t), 144, 154 (t) (app), 164 (app)
    P. (Holotanypus) sagittalis, 140 (t), 154 (t) (app), 163–164 (app)
    P. (Holotanypus) signatus, 154 (t) (app), 164 (app)
    P. (Holotanypus) simplicistilus, 154 (t) (app), 164 (app)
Prodiamesa
    P. obscurimana, 139 (t), 142, 165 (app)
    P. olivacea, 139 (t), 140 (t), 142, 155 (t) (app), 165 (app)
Proisotoma minuta, 71 (t)
Protanypus morio, 155 (t) (app), 165 (app)
Proteroiulus fuscus, 79
Psectrocladius
    P. (Mesopsectrocladius) barbatipes, 157 (t) (app), 172 (app)
    P. fennicus, 140 (t), 157 (t) (app), 172 (app)
    P. limbatellus, 140 (t), 157 (t) (app), 172 (app)
    P. (Allopsectrocladius) obvius, 140 (t), 157 (t) (app), 172 (app)
    P. octomaculatus, 157 (t) (app), 172 (app)

P. oligosetus, 140 (t), 157 (t) (app), 172 (app)
P. oxyura, 157 (t) (app), 172 (app)
P. (Allopsectrocladius) platypus, 140 (t), 157 (t) (app), 172 (app)
P. psilopterus, 157 (t) (app), 172 (app)
P. sordidellus, 134, 139 (t), 140 (t), 142, 157 (t) (app), 172 (app)
P. ventricosus, 141(t), 144, 157 (t) (app), 172–173 (app)
Pseudechiniscus suillus, 23–24, 35 (t), 36
Pseudocandona, 206
   P. albicans, 198, 201, 206
   P. pratensis, 201
Pseudochironomus prasinatus, 139 (t), 141 (t), 143, 144, 159 (t) (app), 180 (app)
Pseudodactylogyrus
   P. anguillae, 221, 222 (t)
   P. bini, 221, 222 (t)
Pseudoisotoma sensibilis, 71 (t)
Pseudoparasitus meridionalis, 72 (t)
Pseudorthocladius
   P. curtistylus, 141(t), 157 (t) (app), 173 (app)
   P. filiformis, 141(t), 157 (t) (app), 173 (app)
   P. macrovirgatus, 141(t), 157 (t) (app), 173 (app)
Pseudoscorpiones, 79, 81
Pseudosmittia
   P. oxoniana, 141(t), 157 (t) (app), 173 (app)
   P. trilobata, 141(t), 157 (t) (app), 173 (app)
Pseudoterpna pruinata, 125 (app)
Psychrodromus robertsoni, 198, 203, 205
Pteromicra angustipennis, 186 (t), 187
Pterostichus
   P. diligens, 104(t), 105
   P. melanarius, 104(t)
   P. niger, 104(t)
   P. nigrita, 104(t)
   P. rhaeticus, 104(t)
   P. strenuus, 104(t)
   P. vernalis, 104(t)
Puffinus puffinus, 233
Punctum pygmaeum, 11 (t)
Pyrrhocorax pyrrhocorax, 234

R
Rallus aquaticus, 234
Ramazzottius oberhauseri, 37
Raphidascaris acus, 222 (t), 223, 225, 226
Rattus
   R. norvegicus, 237, 238 (t)
   R. rattus, 238 (t), 239
Rhantus
   R. exsoletus, 85 (t), 89
   R. suturellus, 85 (t)
Rheocricotopus
   R. (Psilocricotopus) chalybeatus, 141(t), 157 (t) (app), 173 (app)
   R. fuscipes, 157 (t) (app), 173 (app)
Rheopelopia maculipennis, 154 (t) (app), 164 (app)
Rheotanytarsus
   R. curtistylus, 141 (t), 160 (t) (app), 182 (app)

R. pellucidus, 160 (t) (app), 182 (app)
R. pentapoda, 160 (t) (app), 182 (app)
Rhigognostis annulatella, 112, 117 (app)
Rhodacarus roseus, 73 (t)
Rhopobota naevana, 120 (app)
Rhynchotalona falcata, 215
Rhysotrichia ardua, 75 (t)
Rilaena triangularis, 81
Rivula sericealis, 127 (app)
Robertus
   R. arundineti, 46, 66–67 (app)
   R. lividus, 66 (app)
Rodentia, 239–240
Rugathodes instabilis, 44, 46, 67 (app)

S
Saaristoa abnormis, 60 (app)
Salmo trutta, 219, 224, 226
Salvelinus alpinus, 225
Sarscypridopsis aculeata, 205–206
Saturnia pavonia, 123 (app)
Schendyla nemorensis, 80
Sciurus
   S. carolinensis, 238 (t)
   S. vulgaris, 238 (t)
Scoparia ambigualis, 121 (app)
Scopula immutata, 125 (app)
Scotina gracilipes, 53 (app)
Scotopteryx chenopodiata, 125 (app)
Scutovertex sculptus, 77 (t)
Segestria senoculata, 65 (app)
Seiulus minutus, 72 (t)
Selenia dentaria, 124 (app)
Sepedon sphegea, 186 (t)
Sergentia coracina, 159 (t) (app), 179 (app)
Sesia bembeciformis, 112, 119 (app)
Sida crystallina, 214 (t)
Sideridis rivularis, 129 (app)
Silometopus elegans, 45, 49, 60 (app)
Simocephalus vetulus, 214 (t)
Simulium reptans, 142
Sminthurides aquaticus, 70, 72 (t)
Sminthurinus elegans, 72 (t)
Smittia pratorum, 141(t), 157 (t) (app), 173 (app)
Sorex minutus, 238, 238 (t)
Spermodea lamellata, 11 (t), 12, 15
Sphaeridia pumilis, 72 (t)
Spilosoma
   S. lubricipeda, 113, 132 (app)
   S. lutea, 132 (app)
Stagnicola fuscus, 9 (t), 11 (t)
Standfussiana lucernea, 131 (app)
Steatoda grossa, 47, 67 (app)
Steganacarus magnus, 75 (t)
Stempellina bausei, 160 (t) (app), 182 (app)
Stempellinella
   S. brevis, 141 (t), 160 (t) (app), 182 (app)
   S. edwardsi, 160 (t) (app), 182 (app)
Stenaphorura denisi, 71 (t)
Stenoptilia bipunctidactyla, 111 (t), 114
Stercocorarius skua, 233
Sterna paradisaea, 233
Stictochironomus
   S. pictulus, 139 (t), 159 (t) (app), 179 (app)
   S. sticticus, 159 (t) (app), 180 (app)
Stictonectes lepidus, 85 (t)

Stictotarsus 12-pustulatus, 85 (t), 90
Stigmatogaster subterranea, 80
Stigmella
   S. continuella, 116 (app)
   S. lapponica, 116 (app)
   S. salicis, 116 (app)
Stilbia anomala, 127 (app)
Streblocerus serricaudatus, 214 (t)
Strigamia maritima, 80
Subacronicta megacephala, 126 (app)
Supraphorura furcifera, 70, 71 (t)
Swammerdamia pyrella, 116 (app)
Sylvia communis, 235
Synendotendipes dispar, 139 (t), 144, 159 (t) (app), 180 (app)
Synorthocladius semivirens, 141(t), 157 (t) (app), 173 (app)
Synuchus vivalis, 104(t)

T
Tachypodiulus niger, 80
Tandonia budapestensis, 9 (t), 11 (t), 13
Tanypus
   T. kraatzi, 144, 154 (t) (app), 164 (app)
   T. vilipennis, 154 (t) (app), 164 (app)
Tanytarsus, 142
   T. brundini, 160 (t) (app), 182 (app)
   T. buchonius, 141 (t), 160 (t) (app), 182 (app)
   T. ejuncidus, 160 (t) (app), 182 (app)
   T. eminulus, 160 (t) (app), 182 (app)
   T. flavipes, 139 (t), 178 (app)
   T. gmundensis, 139 (t), 142, 143, 181 (app)
   T. gracilentus, 141 (t), 160 (t) (app), 182 (app)
   T. gregarius, 141 (t), 160 (t) (app), 182 (app)
   T. lactescens, 144, 160 (t) (app), 182 (app)
   T. lestagei, 141 (t), 160 (t) (app), 182 (app)
   T. medius, 160 (t) (app), 182 (app)
   T. pallidicornis, 141 (t), 160 (t) (app), 183 (app)
   T. punctipes, 139 (t), 179 (app)
   T. pusio, 139 (t), 183 (app)
   T. signatus, 141 (t), 160 (t) (app), 183 (app)
   T. striatulus, 160 (t) (app), 183 (app)
   T. sylvaticus, 160 (t) (app), 183 (app)
   T. telmaticus, 160 (t) (app), 183 (app)
   T. usmaensis, 160 (t) (app), 183 (app)
Tapinopa longidens, 48
Taranucnus setosus, 44, 60 (app)
Taurulus bubalis, 241
Tegenaria domestica, 47, 52 (app)
Teleiopsis diffinis, 112, 118 (app)
Telmatogeton murrayi, 146
Tendipes, 134
   T. annularis, 139 (t), 144, 174 (app)
   T. aprilinus, 139 (t), 175 (app)
   T. brevitibialis, 139 (t), 144, 176 (app)
   T. dispar, 139 (t), 144, 180 (app)
   T. dorsalis, 139 (t), 141 (t), 142, 143, 144, 174 (app)
   T. errugineovittatus, 139 (t)
   T. ferrugineovittatus, 143, 175 (app)
   T. nigrimanus, 139 (t), 176 (app)

TAXONOMIC INDEX

T. nubeculosus, 139 (t), 179 (app)
T. pedellus, 139 (t), 177 (app)
T. pictulus, 139 (t), 179 (app)
T. plumosus, 139 (t), 175 (app)
T. prasinatus, 139 (t), 143, 180 (app)
T. psittacinus, 139 (t)
T. pusillus, 139 (t), 144, 177 (app)
T. pusio, 143
T. riparius, 139 (t)
T. rufipes, 139 (t), 176 (app)
T. tentans, 139 (t), 142, 175 (app)
T. viridis, 139 (t), 144, 176 (app)
Tenuiphantes
  T. alacris, 42 (t), 43, 43 (t), 44, 49, 60 (app)
  T. flavipes, 60 (app)
  T. mengei, 60 (app)
  T. tenebricola, 44, 60 (app)
  T. tenuis, 60 (app)
  T. zimmermanni, 42 (t), 43, 43 (t), 44, 46, 60–61 (app)
Tetanocera
  T. arrogans, 186 (t)
  T. elata, 186, 186 (t)
  T. ferruginea, 186, 186 (t)
  T. fuscinervis (unicolor), 185, 186
  T. hyalipennis, 186 (t)
  T. hyalipennis (laevifrons), 185
  T. robusta, 186 (t)
Tetracanthella brachyuran, 72 (t)
Tetragnatha
  T. extensa, 46, 66 (app)
  T. montana, 46, 66 (app)
Tetramorium caespitum, 193
Textrix denticulata, 52 (app)
Thalassomya frauenfeldi, 140 (t), 142, 154 (t) (app), 165 (app)
Thalassosmittia thalassophila, 141(t), 142, 144, 157 (t) (app), 173 (app)
Thera obeliscata, 125 (app)
Thienemannia gracilis, 141(t), 157 (t) (app), 173 (app)
Thienemanniella
  T. acuticornis, 157 (t) (app), 173 (app)
  T. clavicornis, 157 (t) (app), 173 (app)
  T. majuscula, 141 (t), 157 (t) (app), 173 (app)
  T. Pe2a, 141 (t), 158 (t) (app), 174 (app)
  T. vittata, 157 (t) (app), 173–174 (app)
Thienemannimyia
  T. laeta, 154 (t) (app), 164 (app)
  T. northumbrica, 154 (t) (app), 164 (app)
Thulinius augusti, 36 (t)
Thyreosthenius parasiticus, 48, 61 (app)
Tinea pallescentella, 116 (app)
Tiso vagans, 42 (t), 43, 43 (t), 45, 46, 49, 61 (app)
Tomocerus
  T. longicornis, 72 (t)
  T. minor, 70, 72 (t)
Trachyuropoda formicaria, 73 (t)
Trechus
  T. obtusus, 104(t)

T. quadristriatus, 104(t)
Trichocellus placidus, 104(t), 105
Trichoniscoides
  T. albidus, 82
  T. saeroeensis, 81
Trichoniscus
  T. pusillus, 81
  T. pygmaeus, 81
Trichopternoides thorelli, 61 (app)
Trimalalconothrus tardus, 75 (t)
Tringa tetanus, 233
Trissopelopia longimana, 154 (t) (app), 164 (app)
Trochosa
  T. ruricola, 48
  T. terricola, 42 (t), 43, *43*, 43 (t), 45, 46, 64 (app)
Trochulus
  T. hispidus, 11 (t), 14
  T. striolatus, 9 (t), 11 (t), 13–14
Troxochrus scabriculus, 61 (app)
Trypanosoma, 225
Turdus
  T. merula, 234
  T. philomelos, 235
Tursiops truncatus, 238
Tvetenia
  T. bavarica, 158 (t) (app), 174 (app)
  T. calvescens, 158 (t) (app), 174 (app)
  T. discoloripes, 158 (t) (app), 174 (app)
  T. verralli, 158 (t) (app), 174 (app)
Typhloceras poppei poppei, 240
Tyria jacobaeae, 132 (app)

U

Udea
  U. ferrugalis, 122 (app)
  U. lutealis, 122 (app)
Uropoda
  U. (Cilliba) cassida, 73 (t)
  U. halberti, 73 (t)
  U. minima, 73 (t)
Urotrachytes formicarius, 73 (t)

V

Vallonia cf. *excentrica*, 9 (t), 11 (t)
Vallonia costata, 11 (t)
Valvata piscinalis, 11 (t), 12
Vanellus vanellus, 234, *234*
Vanessa
  V. atalanta, 123 (app)
  V. cardui, 123 (app)
Veigaia
  V. agilis, 73 (t)
  V. kochi, 73 (t)
  V. nemorensis, 73 (t)
  V. transisalae, 73 (t)
Vertigo
  V. antivertigo, 9 (t), 11 (t)
  V. pygmaea, 9 (t), 11 (t)
  V. substriata, 9 (t), 11 (t)
Vespula
  V. rufa, 192, 192 (t), *194*

V. vulgaris, 192–193, 192 (t)
Virgatanytarsus triangularis, 160 (t) (app), 183 (app)
Vitrea crystallina, 11 (t), 14
Vitrina pellucida, 9 (t), 11 (t)
Vulpes vulpes, 106, 238 (t), 240

W

Walckenaeria
  W. acuminata, 61 (app)
  W. antica, 61 (app)
  W. clavicornis, 62 (app)
  W. cuspidata, 62 (app)
  W. nudipalpis, 62 (app)
  W. unicornis, 45, 49, 62 (app)
  W. vigilax, 43 (t), 45, 62 (app)

X

Xanthia togata, 128 (app)
Xanthorhoe
  X. designata, 125 (app)
  X. ferrugata, 125 (app)
  X. montanata, 111 (t)
Xenillus tegeocranus, 76 (t)
Xenochironomus xenolabis, 159 (t) (app), 180 (app)
Xenylla
  X. humicola, 71 (t)
  X. maritima, 71 (t)
Xestia
  X. baja, 131 (app)
  X. castanea, 131 (app)
  X. xanthographa, 131 (app)
Xysticus
  X. cristatus, 67 (app)
  X. erraticus, 46, 67 (app)

Y

Ypsolopha dentella, 117 (app)

Z

Zavrelia pentatoma, 141 (t), 160 (t) (app), 183 (app)
Zavrelimyia
  Z. barbatipes, 140 (t), 154 (t) (app), 164 (app)
  Z. (Paramerina) cingulata, 139 (t), 140 (t), 154 (t) (app), 164 (app)
  Z. (Paramerina) divisa, 154 (t) (app), 164 (app)
  Z. melanura, 154 (t) (app), 164 (app)
Zenobiellina subrufescens, 11 (t), 12, 14
Zerconopsis remiger, 72 (t)
Zercon triangularis, 73 (t)
Zercoseius spathuliger, 72 (t)
Zonitoides
  Z. excavatus, 9 (t), 11 (t), 15, *15*
  Z. nitidus, 9 (t), 11 (t)
Zygaena filipendulae, 119 (app)
Zygiella
  Z. atrica, 46, 52 (app)
  Z. x-notata, 46, 47, 52 (app)
Zygoribatul exilis, 77 (t)

GENERAL INDEX

Note: This is an index to persons, places and subjects treated in the text. Authors are included only when discussed in the text. Authors cited in parentheses, and persons and places etc. listed in appendices to chapters are not included in this index (*see the separate Taxonomic Index for all species listed*).

Page references: (t) after page reference denotes a table; (app) denotes appendix; page references in *italics* denote illustrations

A
Abbert River, 226
Abbey, 47, 194, 203, 242
Acarina (soil mites), 69–70
   Mesostigmata, 72–73 (t)
   Oribatida, 75–77 (t)
   species recorded, 74–77
Achill Island, 18, 34–35, 101, 133, 136, *138* (map), 142, 144, 145, 161, 205, 238
Aculeata, 189; *see also* Social Hymenoptera
Adams, L.E., 12
Africa, 203
Agnarsson, L., 146
agriculture, 49, 98, 106, 233–234
Alaska, 21
Allen, P., 232
American mink, 238 (t), 240, 243
Amoros, C., 212
Anderson, R., 3, 8
Antarctic expedition (1907-09), 18–19
Anthropocene, 3
Antrim, County, 202
ants, 2, 189, 190–191
   mermithid parasitism (genus *Lasius*), 190, 191, 194–196, *195*
   soil modification by *Lasius flavus*, 190–191, 194
   species recorded, 192 (t), 193–194
aquatic beetles (coleoptera), 2, 83–91, 93–99, 185
   dominant families, 84
   habitats, 83, *84*, 93
   intra-habitat spatial variation study: Poirtín Fhuinch Lough, 93–96, *94*
      results, 96–99
   methodology, 83–84
   results, 84–88
      checklists of species recorded, 84, 85–86 (t), 88–89
      comparisons, 84, 87 (t)
      discussion, 88–91
      dispersal ability island faunas, 87
      flight abilities, 90–91
      relative composition, 84–85, 88 (t)
      trophic categories, 87–88, 91

aquatic carnivores, 240–241
aquatic Heteroptera, 210
aquatic mosses, 34–35
arachnida. *see* spiders
Aran Islands, 192, 193, 235
archipelagos, 1
Arctic expedition (1913-16), 20–21
Arctic Ocean, 21
Arctic tern, 3, 233
Arctiscoida, 17, 18
Armagh, County, 114
Arthropoda, 17
ash-grey slug, 14–15
Ashe, P., 84, 145, 163, 165
Asia, 203
Atlantic salmon, 226
Australasia, 13
Australia, 204
Azores, 204

B
badger, 238 (t), 240
baleen whales, 239
Balfour-Browne, F., 84, 87, 88, 89, 90, 133
Ballyteigue Burrows, 193
Ballytoohy, 8–9 (t), 14, 106, 151 (list), 200, 212 (list), 234, 242
bank vole, 238 (t), 244
Bann River, 12, 144, 224
Barataud, M., 243
Barlee, J., 234
barnacles, 241
Barrett-Hamilton, Gerald Edwin, 237, 239, 240, 241, 243, 244
Barrington, R.M., 232, 235
Bartlett, Captain *(Karluk)*, 21
Bathymetrical Survey of the Freshwater Lochs of Scotland, 18
bats, 238, 238 (t), 241–243
   species recorded, 242 (t)
   surveys, 242
Baxter, W.H., 37
Baylis, H.A., 190
bees, 2, 189–190
   species recorded, 191–192, 192 (t)
Bees, Wasps and Ants Recording Society (BWARS), 189
Beetle Rocks, 12, 14
beetles
   ground. *see* Carabidae
   parasites, 91
   water. *see* aquatic beetles (coleoptera)
Belclare, 18
Belfast, 204
beluga whales, 239
Bertolani, R., 32, 37
Binda, MG, 24, 25, 32

biodiversity, 1, 2–3, 69–70, 77, 146, 210, 219, 225, 226
   assessment, 2, 145, 147
biogeography. *see* island biogeography
biology, 1, 19, 185, 225
Bird Atlas 2007-11, 232
Bird Observatory, 241
birds, 1, 2, 3, 204, 210, 231–235
   'turnover' of species, 3, 235
   climate change and, 234
   habitats
      coastline, 232–233
      terrestrial and wetland, 233–235
   introduced species, 235
   land birds, 233–235
   migrant species, 235
   parasites, 227
   predators of, 233, 239, 241, 243
   scarce and rare species, 235
   seabirds, 233–234
   surveys and expeditions, 231–233, 235
      first Clare Island survey, 231–232
   waterbirds, 234
Biserov, V., 32
black rat, 238 (t), 239
Black Sea, 225
blackbirds, 234
blenny, 241
Bog Myrtle, 113
Bolger, T., 3
Bolivia, 19
Bolivia Boundary Commission, 19, 20
Bolton, S.J., 136
Bombus. *see* bumblebees; Social Hymenoptera
Bond, K.G.M., 2, 110, 118
bottlenose dolphin, 238, 238 (t)
Bradley, J.D., 110
Brazil, 19, 201
Breen, J., 190, 194, 195, 196
Britain, 12, 13–14, 37, 202, 204, 206, 207, 220, 224, 226
   landbridge with Ireland, 237
British Isles, 12, 203, 216, 223
British Museum of Natural History, 144, 205
British Trust for Ornithology Breeding Atlas, 232
brown hare, 238 (t)
brown long-eared bat, 238 (t), 241, 242, *242*, 243
brown rat, 237, 238 (t), 239, 244
brown snail, 14
brown trout, 219, 220, 223–224, *224*, 226
   parasites, 221–223, 225–226
Budapest slug, 13
Bulgaria, 13

bumblebees, 189–190, *191*; *see also* Social
  hymenoptera
  DNA barcoding, 190
  species recorded, 191, 192 (t)
Buncrana, 14
Burren, 112, 113
Burrishoole River, 224
butterfish, 241
butterflies, 3, 112, 113; *see also* Lepidoptera
ButterflyIreland, 113
Byrne, A., 14

C
Cameron, R.A.D., 7
Campbell, R.N., 223, 225
Canadian Arctic Expedition (1913-16),
  20–21
Canary Islands, 204
Cape Clear, 235
Capnagower, 142, 190–191, 201, 206
Carabidae (ground beetles), 3, 101–107, *106*
  methodology, 101–102, 105
  results and discussion, 102–107
  riparian species, 102
  sampling sites, *102* (map)
  seacoast species, 104
  species recorded, 103–104 (t)
    comparisons with mainland, 106
  wetland species, 105–106
  woodland species, 104–105
Carlow, County, 200, 201, 202, 203, 204
carnivores, 237, 240–241, 243
Caroni, R., 216
Carpenter, G.H., 70
Castlebar Lough, Co. Mayo, 18, 34,
  35, 133, 135, 144, 152 (list), 164 (list),
  180 (list)
cats, 239, 241, 244
Cawley, Martin, 42, 81
cellar slugs, 13
centipedes (Chilopoda), 79, *80*, 80–81
Ceratopogonidae (biting midges), 134
Cetacea, 238–239, 238 (t)
chaffinch, 235
Chilopoda (centipedes), 79, 80–81
Chironomidae (non-biting midges), 2, 3,
  133–183
  methodology, *135*
    field collections, 135
    preparations and specimen
      identifications, 135–136
  nomenclature, 133–134, 136, 138,
    143, 145
  pupal exuviae, 134–135, *135, 136*
  results, 138–144
    discussion, 145–147
    extant chironomidae, 138, 141–144
    sampling sites and collections, 136–138,
      151–153
      Achill Island, *138* (map), 144
      Clare Island, 136, 136–137, *137* (map),
        138, 140–141 (t), 141–143, 151 (t)
      Mayo mainland, 136, *137* (map), 138,
        143–144, 152 (t), 153 (t)
    species recorded, 138, 139–140 (tables),
      154–160 (app)
      inventory, 161–183 (app)
Chiroptera. *see* bats

choughs, 234
Cladocera (water fleas), 197, 200,
  209–212
  species recorded, 212–216,
    213–214 (t), 217
Clare Island Survey (1909-11), 1, 2, 3, 12,
  18, 244
  aquatic species, 12
    coleoptera, 84
  bats, 241
  birds, 231, 233
  Carabidae, 101, 102
  Chironomidae, 135, 136, 161
  Cladocera and Copepoda, 209
  Collembola and mites, 70
  fish, 219
  Hymenoptera, 189, 191
  Lepidoptera, 109, 110–111, 113
  Sciomyzidae, 185, 186
  spiders, 41, 48
  Tardigrades, 17–18, 34–36
Clare, county, 114, 203
Clew Bay, 101, 110, 198, 209, 238
climate change, 15, 234
Clogher, 133
Coleoptera (beetles), 101–102, 106;
  see also aquatic beetles; Carabidae
  (ground beetles)
Collembola (springtails), 69–70
  original survey, 70
  results and discussion, 70–77
  species recorded, 71–72 (t)
Collins, T., 1
common dolphin, 238, 238 (t), 239
common gull, 233
common pipistrelle, 241
common seal, 237
common shrimp, 241
common snipe, 234
common wasp, 192
Conchological Society, 12
Coney Island, Co. Sligo, 240
Conneely, J.J., 225
Connemara, 238
continental shelf islands, 231
Convention on Biological Diversity
  (CBD), 2
Coombes, D., 232
coot, 234
Copeland Islands, 235
Copepoda, 200, 209, 210, 212
  species recorded, 216–217, 216 (t)
Cork County Bat Group, 241
Cork, County, 13, 80, 114
corn bunting, 235
Corrib River, 225
corvids. *see* crows
Costello, M.J., 146–147
Courchamp, F., 147
Cournane, Curran, 241
Cox, M.L., 83
crabs, 241
Craigmore, 204, *211*, 242
Crawley, W.C., 194
Creggan Lough, 89, 136, 142, 143, 161,
  200, 202, 204, 210, *211*, 212, 216, 220
Croaghpatrick, 202
crows (corvids), 234, 239

crustaceans, 209, 239, 241; *see also*
  microcrustaceans
curlew, 233
Curragh, 49
Curraghmore River, 220, 221 (map)
Cussen, R., 232, 234
Czech Republic, 201

D
D'Arcy, G., 232, 234
daddy long-legs spider, 47
Darwin, Charles, 1, 210
Daubenton's bat, 238 (t), *242*, 242–243
De Jong, A., *Fauna Europaea*, 110
deer, 238 (t)
DeMilio, E., 17, 24
Denmark, 224
Denton, Jonty, 84
Dettinger-Klem, A., 136
Devil's-bit Scabious, 114
Diamond, J.M., 1
Diplopoda (millipedes), 79–80
dipper, 235
Diptera
  non-biting midges. *see* Chironomidae
  snail-killing/marsh flies.
    see Sciomyzidae
Dogiel, V.A., 220
dogs, 239
dolphins, 238, 238 (t), 239
Donegal, County, 14, 47, 112, 203, 225, 243
Donisthorpe, H.S.J.K., 194
Doogue, D., 80
Doree River, 142, 223
Down, County, 202
Doyle, G.J., 98
Drimneen River, 225
Drost, M.P.B., 83
Dublin, County, 203
Duff, A.G., 102
Dugort, 144
Duigan, C., 210, 212
Dunkerly, J.S., 197, 209
dunlin, 234
Dussart, B., 212

E
earthworms, 2
Eastern Europe, 13
ecological parasitology, 220
ecological research
  local surveys, contribution of, 1–3
Edwards, F.W., 143, 144, 163, 164, 180
Edwards, J., 88
eels, 224; *see also* European eel
eider duck, 234
Ekrem, T., 136
Elton, C.S., 226
Emmet, A.M., 113
England, 203
English chrysalis snail, 15
Entomostraca, 209, 210
Environmental Protection Agency, 136,
  144, 161
Erne River, 224
EU Habitats Directive, 144
Eulipotyphla, 244
Eurasian curlew, 233

Eurasian whimbrel, 233
Europe, 8, 13, 15, 37, 202, 203, 220, 224, 240
European eel, 219, 220–221, 223, 224, *226*
    parasites, 224, 225, 226–227
European hedgehog, 244
European pied flycatcher, 235
European storm petrel, 233
Evans, J.G., 12
evolutionary theory, 1, 2
Eyre, M.D., 99

F
Fahy, S., 84, 135, 136, 138, 142, 143, 161, 162, 166, 169, 170, 177, 178
Fairley, J.S., 239, 240
fallow dear, 238 (t)
false scorpions. *see* Pseudoscorpiones
Faroe Islands, 12
Farran, G.P., 210
Fauna Europaea (de Jong), 110
Fawcett, Percy, 19, 20
Fawnglass, 190, *191*, 192, 203, 204, 206
Feehan, J., 234
feral cats, 241, 244
feral goat, 238 (t)
Fermanagh, county, 114
Ferron, L., 239, 240
fin whale, 238 (t), 239
fingernail clams, 185
fish, 241; *see also* freshwater fish
fish parasites, 2, 219, 221–223, 222 (t), 224, 225–227
fish parasitology, 225, 226, 227
Fittkau, E.J., 136, 141
Fitzpatrick, Ú., 189
Fjellberg, A., 74
fleas, 240; *see also* Cladocera
Fletcher, D.S., 110
flies, snail-killing. *see* Sciomyzidae
Florida Keys, 1
Flössner, D., 212
flounder, 219, 220, 221, 223, 225
    parasites, 222 (t), 223, 225, 227
Fontaine, B., 147
Fontoura, P., 24, 32
Formicidae. *see* ants
Forsyth, Ian, 232
Foss, P.J., 98
Foster, G.N., 83–84, 87
Foster, N.H., 79, 82
Fox Moth, 114
foxes, 238 (t), 240
Foyle River, 224
France, 201, 204
freshwater fish, 2, 219–227
    methodology, 220
    parasites, 220, 221–223, 224–227 passim
    results and discussion, 220–227
freshwater microcrustaceans, 209–212;
    see also Cladocera; Copepoda;
    non-marine ostracods
Friday, L.E., 83
fulmar, 3, *232*, 232–233
FYR Macedonia, 204

G
Galapagos Islands, 1
Gallagher, Billy, 203

Galway, county, 113, 114, 198, 201, 204, 206, 224, 226, 243
gannet, 233
Garinish Island, 241
gastropods. *see* non-marine molluscs
Germany, 13, 135, 201, 204, 243
GIS, 198
Glasgow School of Art, 18
Glen, 142
Glenamoy bog, 201
Glöer, P., 7
goby, 241
Gould, W. 190, 194–195
Grabda, J., 225
Grann, D., 20
Granuaile Hotel, 232
Granuaile's Tower, 12–13, 14
Gravesend, Britain, 12
grayling butterfly, 113
Great Blasket Islands, 239
great cormorant, 233
Great Famine, 15
great skua, 233
greater white-toothed shrew, 244
green cellar slug, 13
Green Hairstreak butterfly, 112
Greer, Thomas, 110
Gregory, S., 82
grey seal, 237–238, 238 (t), 240
grey squirrel, 238 (t)
Grimshaw, P.H., 133, 134, 136–137, 138, 141, 142, 143, 144, 145, 146, 147, 161, 186, 187, 189
ground beetles. *see* Carabidae
Guidetti, R., 35
Gurney, R., 212

H
Halbert, J.N., 2, 70, 74, 84, 89, 90, 102, 104, 133, 134, 144, 189
Hamilton (Scotland), 18
Hammond, A.R., 134
Hannigan, E., 210
Hansen, J.G., 83, 87
harbour porpoise, 238, 238 (t)
harbour seal, 238 (t), 240
Harding, J.P., 212, 216
hares. *see* Irish hare
Harris, S., 239
Hartnett, Cmmdt Michael, 232
harvestmen (Opiliones), 79, 81, *81*
Hay, W.P., 17
hedgehog, 238, 238 (t), 244
Helsdingen, P.J. Van, 42
Hendel, F., 134
Henderson, P.A., 203
Heneghan, L., 1
Heritage Council Collection of Irish Chironomidae (HCCIC), 136, 161
Heteroptera, aquatic, 210
Holland, C.V., 225
hollowed glass snail, 15, *15*
Holmen, M., 83, 87
Holyoak, Geraldine, 12, 15
honeybees, 189, 191–192, 192 (t); *see also* Social Hymenoptera
hooded crow, 234
Horne, David J., 198, 203, 205

house mice, 237, 238 (t), 239
Humphries, Carmel, 134–135
Hymenoptera, 189; *see also* Social Hymenoptera
Hyndman Collection (now Ulster Museum), 12

I
ichthyofauna. *see* freshwater fish
Inis Meáin, 192, 193
Inishbofin, 235
Inishkea Islands, 234
Inishkea North, 106
Inishmore (Inis Mór), 193, 235, 240
Inishmurray, 234
Inishowen, 14
Inishtearaght Island, 142
Inishturk, 18, 238
Insectivora, 244
insects, 231, 241, 243; *see also* Chironomidae; Coleoptera
Intergovernmental Platform for Biodiversity and Ecosystem Services (IPBES), 2, 3
International Commission on Zoological Nomenclature (ICZN), 134
International Union for Conservation of Nature, 240
Iran, 203
Irish hare, 237, 238, *243*, 243–244
Irish Whales and Dolphin Group (IWDG), 238, 239
Irvine, K., 216
island biogeography, 91, 224, 227, 231, 233
Islay, 84, 90
isopods, 225, 241; *see also* woodlice (Land Isopoda)
Italy, 18, 204

J
Jackson, D.J., 87, 90
Java, 201
Jenkin's spire shell, 12
Jersey Island, 227
Johnson, Rev W.F., 79, 79–80, 84, 89, 90, 102, 104, 110, 133, 189

K
Kaczmarek, L., 24
Kane, W.F. deV., 110, 113, 197, 209
Karluk expedition, 20, 21
Keel Lough, 144
Kelly, T.C., 3, 232, 236
Kennedy, C.R., 220, 225, 226, 227
Kevan, D.K., 83
Kew, H.W., 81
Kidney Vetch, 113
Kiefer, F., 212
Kieffer, J.J., 141
Kilkee, 142
Kill, 190, 194, 242
Killala, 113
Killeen, I., 7
killer whale, 238–239, 238 (t)
Kilmore Quay, Co. Wexford, 196
Kinacorra, 18, 23, 35, 89, 203
Kinatevdilla, *211*
kittiwake, 233

Klie, W., 197
Knockmore (Cnoc Mór), 7, 12, 14, 15, 18, 23, 101, 105, *105*, 198, 199, 232
Knotgrass Moth, *114*
Kuntner, M., 146

L
La Paz, 20
Lachnacranny, 7, 12, 101; *see also* Lighthouse
Laghta, 133
Lagomorpha (rabbits and hares), 243–244
Lambay Island, 1, 239, 243
land birds, 233–235
Land Isopoda, 79, 81–82
Langton, P.H., 136, 143, 144, 145, 161
Laois, County, 200, 202, 204
lapland longspur, 235
lapwing, 234, *234*
Lassau Wood, 234, 242
Last Glacial Maximum, 237
Lavery, A., 42, 52
Lawton, C., 239, 243, 244
Lecarrow, 12, 203
Leckacanny, 136
Lee, P., 80
Leisler's bat, 241
Lepidoptera, 2, 109–114
   butterflies, 3, 112, 113
   methodology, 109–110
   moths, 109, 113–114
   nomenclature and arrangement, 110
   results of survey, 110–113
      losses, 111 (t), 113
      species recorded, 110–114, 110 (t)
         important species, 111, 112–113
         list, 116–132 (app)
         number of, by families, 110 (t)
lesser black-headed gull, 233
lesser redpoll, 234, 235
Letterfrack, 143
Lighthouse (Lacnacranny), 7, 12, 101, 105, 199, 200, 233, 235
limpets, 241
Lisi, O., 32
Little Island, 240
Lloyd, C.S., 232
local surveys, value of, 1–3
Londonderry, County, 14
Lough Agraffard, 200
Lough Aille, 144
Lough Avullin, 98, 201, *211*, 223–224, 234
   Phragmites swamp, 41, 44–45
Lough Carra, 226
Lough Conn, 226
Lough Corrib, 225, 226
Lough Creggan. *see* Creggan Lough
Lough Eske, 225
Lough Inagh, 200
Lough Knappabeg, 144
Lough Leinapollbauty, 200, 202, 204, 206, 210, 212
Lough Mask, 225, 226
Lough Merrignagh, 98, 200, 201, 206
Lough Moher, 144
Lough Neagh, 12, 200
Lough Poirtín Fhuinch. *see* Poirtín Fhuinch Lough

Lough Sruhill, 144
Loughnaphuca, *135*, 200, 206
Louisburgh, 18, 133, 202
Louth, County, 201, 202
Luff, M.L., 102
Lunar Hornet Moth, 112, 119
Lunar Underwing, 128

M
McCabe, J.J., 232, 243
McCarthy, T.K., 2, 93, 224, 225
McCarthy, V., 210
McCormack, Stephen, 2, 3, 41, 42, 79, 87, 93, 110, 198
McGrath, D., 136
McGreal, E., 232, 233
MacArthur, R.H. and Wilson, E.O., *The theory of island biogeography*, 1, 90, 231
Mackay, Alistair Forbes, *20*, 21
Madeira, 204
Maitland, P.S., 223, 225
mallard, 234
Malloch, J.R., 134
Malta, 13
mammals, 2, 237–238
   aquatic and semi-aquatic, 237–238, 238–239, 244
   Carnivora, 237, 240–241
   Cetacea, 238–239, 238 (t), 244
   Chiroptera, 238, 241–243, 242 (t)
   Eulipotyphla, 238, 244
   first Clare Island survey, 237–238, 244
   Lagomorpha, 237, 243–244
   native species, 237
   Rodentia, 237, 239–240
   terrestrial and amphibious species, 238 (t), 244
manx shearwater, 233
marsh flies. *see* Sciomyzidae
Marston, George, 19
Martens, K., 198, 200
Maum, 12, 15, 143, 201, 234, 242
Maum River, 220, 221 (map)
Mayo, County, 18, 41, 42, 48, 198, 200, 201, 202, 203, 204, 206, 224, 226, 238, 243; *see also* Achill Island; Clew Bay
   Chironomidae, 134, 135, 136, *137* (map), 138, 142, 143, 143–144, 161
meadow pipit, 234, *235*
Meath, County, 203
Mediterranean, 13, 207
Meigen, J.W., 134
Meisch, C., 197, 202, 203, 204
Menzies, L.S., 83
mermithid parasitism
   ants (genus *Lasius*), 190, 191, 194–196, *195*
Merne, Osar, 232
Mesostigmata (mites). *see* Acarina
Miall, L.C., 134
mice. *see* house mice; wood mice
Michalczyk, L., 24
microcrustaceans, 209–217; *see also* Cladocera; Copepoda; non-marine Ostracods
   sampling locations, 211 (map), 212 (t)
Middle East, 202
midges

biting, 134
non-biting. *see* Chironomidae
Mill Fen, 8–9 (t), 12, 13, 15, 106
Mill, the, 15, 105, *105*
millipedes (Diplopoda), 79–80, *80*
mink, 238 (t), 239, 240, 243
minke whale, 239
mites (Oribatida and Mesostigmata). *see* Acarina
Molloy, S., 226
molluscs, 106, 239; *see also* non-marine molluscs
Monaghan, County, 201, 203
Montgomery, W.I., 239, 240
moorhen, 234
Moraza, M.L., 74
Morgan, C.I., 36, 37
Morley, C., 133, 189, 191, 192, 193
moths, 109, 113–114; *see also* Lepidoptera
Moths and Butterflies of Great Britain and Ireland, 111
MothsIreland, 111
Mourne Park, 15
Moy River, 225
Mull, 84, 90
Mullet, 106
Murray, Sir John, 18
Murray, D.A., 136, 138, 142, 143, 144, 145, 161, 165, 171, 173, 174, 175, 179
Murray, James, 2, 17–18, 19, *19*, 20, *20*, 21, 23, 34–35, 197, 209
   Antarctic days, 19
   life of, 18–21
Murray, W.A., 138, 142, 143, 144, 165, 171, 173, 174, 175, 179
mussels, 241; *see also* pea mussels
Myers, A., 136
myriapods, 79, 80

N
National Museum of Ireland (Natural History), 1, 70, 109, 110, 111, 205
National Museum of Ireland (NMI), 84, 136, 142, 143, 161
National Soils Laboratory, Johnstown Castle, 191
National University of Ireland, Galway, 41
National Vegetation Classification Scheme, 94
Natterer's bat, 238 (t), 242, 243
Natural History Museum of Denmark, Copenhagen, 25
Natural History Society of Glasgow, 18
Naturalist, 1–2
Nelson, Brian, 84
New Survey of Clare Island, 1, 2, 3
New Survey of Clare Island. Volume 2: Geology, 199
New Survey of Clare Island. Volume 9: Birds, 232
New Zealand, 12
Nilsson, A.N., 83, 87
Nimrod expedition (1907-09), 18–19, *19*
Niven, J., 21
Nolan, Myles, 2, 41, 42
Nome (Alaska), 21

non-marine molluscs, 7–15, 185, 186
 aquatic species, 7, 12
 declining species, 12, 14
 endangered species, 14–15
 gaps in survey coverage, 12
 incursive or invasive species, 7, 12–14
 literature, 7
 methodology, 7–8
 nomenclature, 7, 8, 12, 13
 results: species recorded, 7, 8–9 (t), 10–11 (t), 12
 terrestrial, 7
 wood fauna, 12, 14–15
non-marine ostracods, 197–207
 distribution, 199 (t)
 habitats, 197–198
 results, 198–200
 sampling and sample sites, 198, 199, *199* (map), 206
 species recorded, 198–200
 notes on, 200–206
Nonmarine Ostracod Distribution in Europe (NODE), 198
North Africa, 202, 221
North America, 13, 202
Northern Deep-brown Dart, 113, 128
Northern Irish Natural Heritage Areas, 49
Northern Pike, 220
Norway, 201, 216, 226

O
O'Connor, J.P., 110, 118, 146, 163, 165
O'Grady, A., 191, 194, 195
O'Riordan, C.E., 210
O'Rourke, F.J., 196
Ó Muráile, N., 206
oceanic isles, 231
Ooghcorragaun, 242
Opiliones (harvestmen), 79, 81
Ordnance Survey, Dublin, 198
Oribatida (soil mites). *see* Acarina
ornithological expeditions, 231–233, 235
ostracods, 109; *see also* non-marine ostracods
otters, 237, 238–239, 238 (t), 240–241, 244
 spraints, 241 (t)
Owen's slug, *13*, 14

P
Pack-Beresford, Dennis R., 41, 42, 48–49, 79, 81
palaeolimnological studies, 209
parasites. *see* fish parasites; mermithid parasitism
pea mussels, 12, 185, 186
Peru, 19
Petrusek, A., 212
Phalangida (harvest-men), 79
Phragmites swamp, 41, 42, 44
Pilato, G., 24, 25, 32
Pinder, L.C.V., 136
pine marten, 238 (t), 240
pipistrelle bats, 241
plaited snail, 15
plume moths, 112, 114
point snail, 14
pointed snail, 13

Poirtín Fhuinch Lough, 93–94, *94*, 136, 138, 142, 143, 202, 204, 206, 212, 215
 aquatic beetles, study of, 93–99
Poland, 201, 204
Pons, J.A., 204
porcelain crab, 241
porpoises, 238, 238 (t)
Portlea, 12, 15, 41, 44, 242
Portnakilly, 136
Portnakilly Harbour, 142
Portugal, 204
Poulin, R., 220
Praeger Committee (RIA), 136, 144
Praeger, Robert Lloyd, 1, 113
Price, P.W., 220
Pseudoscorpiones (false scorpions), 79, 81
puffins, 239
pygmy shrew, 238, 238 (t), 244

R
rabbits, 237, 238, 238 (t), 243
 burrows, *243*
Racey, P.A., 243
Ramazzotti, G., 18
Rathlin Island, 239
rats, 237, 238 (t), 239, 244
ravens, 234
Rebecchi, L., 32
red deer, 238 (t)
red fox, 238 (t), 244
red squirrel, 238 (t)
red wasp, 192
red-breasted merganser, 234
Reddish Light Arches moth, 113
redshank, 233
reed bunting, 234
Reise, H., 13
ringed plover, 233
Risso's dolphin, 238, 238 (t)
Roberts, M.J., 42
robin, 234–235
rockling, 241
rodents, 237, 238 (t), 239–240, 240 (t), 244
Romania, 13
Roonagh Quay, 133, 240
Ross Island, 18
Rostrevor Forest, 15
Rotifers, 18, 197
Rowson, B., 7, 8
Royal Geographic Society, 20
Royal Irish Academy, 41, 134, 136, 144
 State of Taxonomy in Ireland (seminar, 1987), 146
Royal Society of Edinburgh, 18
Ruse, L.P., 135, 145
Rusheen Bay, 225
Russia, 13
Rutland Island, 240
Rutledge, R.F., 234
Rybak, J.J., 212
Ryle, T., 98
Rylov, W.M., 212

S
Saether, O.A., 136
Sala, O.E., 3
Saltee Islands, 196, 235, 241
sand smelt, 241

Sars, G.O., 212, 216
Savatenalinton, S., 198
Scandinavia, 12
Scharff, R.F., 13
Schikora, H.-B., 42
Schultze, C.A.S., 17, 18
Sciomyzidae (snail-killing/marsh flies), 185
 collection methods, 185, 187
 results and discussion, 185–187
 species recorded, 186 (t)
Scotch Annulet butterfly, 112, *112*
Scotland, 14, 18, 90, 144, 200, 203, 226, 239, 243
Scourfield, D.J., 197, 198, 201, 202, 203, 204, 205, 206, 209, 212, 215, 216, 217
sea lice, 226
Sea of Aral, 226
sea scorpion, 241
sea trout, 226
seabirds, 232–233, 239
seals, 237–238, 238 (t), 240
sedge warbler, 234
Shackleton, Ernest, 18–19, 21
 The heart of the Antarctic, 19
shag, 233
Shannon River, 224
Sherkin Island, 241
shore crab, 241
shrews, 238, 238 (t), 244
Siberia, 21
Sicily, 13
sika deer, 238 (t)
Silva, F.L., 136
Simberloff, D., 1, 232
skylarks, 234
Sleeman, D.P., 241
Slievemore, 34–35
Sligo, County, 114
slugs, 13, 14–15; *see also* non-marine molluscs
Slyne Head, 113
Small Blue butterfly, 113
Small Heath butterfly, 112
snail-killing flies. *see* Sciomyzidae
snails, 13–14, 15; *see also* non-marine molluscs
Social Hymenoptera, 189–196; *see also* ants; bees; wasps
 collection methods, 190
 DNA barcoding, 190, 191
 nomenclature, 189–190, 192 (t)
 results and discussion, 191–195
 species recorded, 192 (t)
soil fauna, 3, 69–70; *see also* Acarina (soil mites); Collembola (springtails)
 methodology, 70
 results and discussion, 70–77
soil modification by ants. *see* ants
song thrush, 235
soprano bats, 241
soprano pipistrelle, 241
South Africa, 201
South America, 19, 20, 24, 203
South Pole, 18–19
Spain, 204, 244
Spallanzani, G., 17
sparrow hawk, 235

Sphagnum moss, 83, 94, 97–98, 99, 198, 199–200
spiders, 2, 41–67; *see also* harvest-spiders
   collection and trapping methods, 41–42, 48, 49, 51 (t)
      hand collections, 41–42, 46–47
      malaise tray traps, 41, 42, 44–45
      pitfall traps, 41, 45–46
   most abundant species, 43, 49
   nomenclature, 42
   pastoral/agricultural species, 49
   rarer spiders, 47, 49
   results of survey, 42–49
      conclusions, 48–49
      differences from original survey, 48–49
   sites
      Lough Avullin Phragmites swamp, 41, 44–45
      Portlea woodland, 41, 44
   species and habitat types, 42–43 (t)
   species catalogue, 42, 52–67 (app)
   synanthropic species, 47–48
   wetland species, 44–45, 49
   woodland species, 43, 44, 49
Spies, M., 136
springtails. *see* Collembola
squirrels, 238 (t)
Stefansson, Vilhjalmur, 20, *20*, 21
Stelfox, A.W., 7, 12, 189
stoat, 237, 238 (t)
storm petrel, 233
Strake River, 220, 221 (map)
strawberry snail, 13–14
Strenzke, K., 136
striped snail, 13
Stur, E., 136
Sublette, J.E., 136
Sweden, 201
Swift, S.M., 243

T
Tardigrades (water bears), 17–38, 197
   classes and genera, 17–18
   methodology, 21–23
   Murray's contribution to study of, 17–18, 19–21, 34–35
   nomenclature, 17–18
   original survey, 17–18, 34–35
   results of survey, 23
      discussion, 34, 36–38
      dominant group species, 37
      notes on species recorded, 23–34
teal, 234
Tendipedidae. *see* Chironomidae
Thienemann, A., 134, 135, 141
Thomson, S.A., 147
Thorstad, E.B., 226
three-spined stickleback, 219
Thulin, G., 18, 24, 25
Tipperary (North), 200, 202
Toormore, 200
Tory Island, 235, 244
tortrix moth, 109
tramp slugs, 13
tree sparrow, 235
tree wasp, 192
tree-trunk mosses, 44
trout. *see* brown trout
Turkey, 13, 202
twite, 235
two-winged insects
   non-biting midges. *see* Chironomidae
Tyrone, County, 14

U
UKMoths, 111
Ulster Museum, 12
ungulates, 237
United States, 18
University College Dublin, 135
University of Glasgow, 198
urbanisation, 233–234
Ussher, R.J., 231–232, 234

V
van Berge Henegouwen, A.L., 89
Vavra, Dr, 205
velvet swimmer, 241
vertebrate species, decline in, 3
Victoria, British Columbia, 21
Visser, H., 136
voles, 238 (t), 244

W
wading birds, 233
Wales, 203
Wall Brown butterfly, 112
wasps, 189, 192–193, 192 (t), *194*
water bears. *see* Tardigrades
water beetles. *see* aquatic beetles (coleoptera)
water birds, 234
water fleas. *see* Cladocera
water rail, 234
water spider, 47, *47*
Welch, R.J., 12
West Connacht Coast Special Area of Conservation, 238
Westport, 18, 133, 136, 144, 161
wetland habitats, 233, 234
wetland species
   ground beetles, 105–106
   spiders, 44–45, 49
whales, 238–239, 238 (t), 239
wheatears, 234
Whilde, Tony, 232
whimbrel, 233
White Ermine moth, 113
White Sea, 225
white-beaked dolphin, 238 (t)
white-headed gull, 233
whitethroat, 235
Wicklow, County, 201
Wiederholm, T., 136
willow warbler, 235, *235*
Wilson, E.O., 1–2, 90, 231
Wilson, J.P.F., 220, 223
Wilson, R.S., 135
Winters, P., 232
Wirth, W.W., 142
wood mice, 238 (t), 239
   flea species on, 240
   morphometrics, 239–240, 240 (t)
   trapping surveys, 239
woodland fauna
   Carabidae (ground beetles), 104–105
   non-marine mollusca, 7, 12, 14–15
   spiders, 41, 43, 44, 49
woodlice (Land Isopoda), 79, *81*, 81–82
World Spider Catalog, 42
worms, 2, 190, 223
wrasse, 241
Wyse, Bonaparte, 111

Y
Yalden, D.W., 239
yellow browed warbler, 235
yellow eel, 224
yellow garden ants, 191
yellowhammer, 235

Z
Zoological Museum, Copenhagen, 25